Contributions to Economics

The series *Contributions to Economics* provides an outlet for innovative research in all areas of economics. Books published in the series are primarily monographs and multiple author works that present new research results on a clearly defined topic, but contributed volumes and conference proceedings are also considered. All books are published in print and ebook and disseminated and promoted globally. The series and the volumes published in it are indexed by Scopus and ISI (selected volumes).

More information about this series at http://www.springer.com/series/1262

Umit Hacioglu
Editor

Blockchain Economics and Financial Market Innovation

Financial Innovations in the Digital Age

Editor
Umit Hacioglu
School of Business
Ibn Haldun University
Basaksehir, Istanbul, Turkey

ISSN 1431-1933 ISSN 2197-7178 (electronic)
Contributions to Economics
ISBN 978-3-030-25274-8 ISBN 978-3-030-25275-5 (eBook)
https://doi.org/10.1007/978-3-030-25275-5

© Springer Nature Switzerland AG 2019
This work is subject to copyright. All rights are reserved by the Publisher, whether the whole or part of the material is concerned, specifically the rights of translation, reprinting, reuse of illustrations, recitation, broadcasting, reproduction on microfilms or in any other physical way, and transmission or information storage and retrieval, electronic adaptation, computer software, or by similar or dissimilar methodology now known or hereafter developed.
The use of general descriptive names, registered names, trademarks, service marks, etc. in this publication does not imply, even in the absence of a specific statement, that such names are exempt from the relevant protective laws and regulations and therefore free for general use.
The publisher, the authors, and the editors are safe to assume that the advice and information in this book are believed to be true and accurate at the date of publication. Neither the publisher nor the authors or the editors give a warranty, expressed or implied, with respect to the material contained herein or for any errors or omissions that may have been made. The publisher remains neutral with regard to jurisdictional claims in published maps and institutional affiliations.

This Springer imprint is published by the registered company Springer Nature Switzerland AG.
The registered company address is: Gewerbestrasse 11, 6330 Cham, Switzerland

Foreword

Blockchain technology is on its fast lane and soon it will be a part of our daily life. In this piece of work edited by Dr. Hacioglu, it is discussed from different aspects of how blockchain technology and its variations will affect our monetary and financial activities.

The first part, which contains five distinguished articles by scholars from different academies, deals with the economy of blockchain and the related innovations in financial markets. In this part, the concept of digital money is explained from different perspectives and a model is presented for redesigning of the banknotes in circulation. Are humans going to shop with an empty physical wallet in their pocket but with a digital unseen wallet full of digital or crypto monies? The answer is there.

The second part deals with the issues related to handling the investment activities in digital or cryptocurrencies and/or markets. Are classic investment strategies valid in digital markets? Is it possible to value a digital currency? What are the differences between present banknote monies and digital currencies? These questions and many more are discussed in this part.

The third part deals with the economic and financial assessment of digital currencies as analyzing the relationship between cryptocurrencies and financial parameters, stock exchanges, and precious metals. Although it is not easy to do such analyses due to the relatively little data, interesting findings are presented.

Are (returns on) cryptocurrencies taxed traditionally or digitally? What is the process of taxation? What is the process of taxation related to these new digital financial activities in emerging economies? The fourth part that contains five articles focuses on presenting answers to these questions.

The last part dealing with political and other issues (cryptocurrency derivatives, Industry 4.0, and speculation) in digital money markets seals the end of this non-digital book that will probably go online though not as a part of the blockchain system. It is your choice how to get it. But do have it.

School of Applied Sciences
Kirklareli University
Kirklareli, Turkey

Mehmet Hasan Eken

Preface

In the last decade, scholars assessed the ramifications of the global financial crisis on economic growth, stability, and prosperity. Major studies analyzing the roots of the financial crisis demonstrated the fact that financial stress in advanced economies had been transferring to emerging economies via an integrated financial system. In these studies, it was also highlighted that the transmission of financial stress caused a meltdown in global economic activity. To which extent the massive collapse of financial institutions, regulated monetary, and fiscal policies were linked to this global economic meltdown? Was it possible to prevent this collapse while integrating a more secured digital system without any intervention of the governmental institutions? Could blockchain economics be part of the more stabilized global financial system? Answering these questions is not so easy without knowing more about the components of blockchain technologies and its ability to transform the traditional financial systems. Many scholars today desire to have a deeper focus on this issue with a distinguished interdisciplinary perspective to understand the role of blockchain technologies in this transformational change in financial markets. Undoubtedly, the newest technology in the blockchain ecosystem has been shaping our understanding of traditional business and financial activities. Blockchain technologies are referred to as the decentralized integration of computers and distributed networks. These computers and networks are linked together safely based on the new growing list of records, so-called blocks, connecting the world to the future of business without regulation of any central authority.

As we know that the largest intermediaries in financial systems are the banking institutions with the largest funds and assets. This intermediary role is based on a fundamental record-keeping function and transactions are made among fundraisers and demanders. Could it be efficiently managed by advanced technology like a distributed ledger technology? Could it be possible to decrease the cost of transactions, improve data security, and improve correctness by a modern transaction than a traditional centralized intermediary function? Through a macro-perspective, similar questions could be directed to the economy itself. If traditional money for an economic system is defined as the medium of exchange, a unit of account, or a store of value, then it is time to redefine money in a digital economic system. Classical functions of money with existing record-keeping or general ledger system cannot increase the efficiency of the monetary system or banking without the integration of new digital money with a modern distributed ledger system based on blockchain technology. Blockchain economies are referred to as a potential future environment by which cryptocurrencies replace the traditional monetary system based on blockchain technology without the intervention of any intermediary institution.

This novel book emphasizes on the blockchain economics and new investment strategies in crypto-markets. Recently, technological developments associated with the financial services industry have been introducing new investment models and techniques. With the integration of blockchain technologies in the financial services industry, a new crypto asset—Bitcoin—sparked the attention of many investors in crypto-markets apart from traditional investment strategies. Blockchain economies in this digital innovation path will continue to design a new financial model for many actors. Crypto-asset investment models, Crypto-Lending, Bitcoin Transfer, Cloud Mining, Lightning Networks, Hard-wallets, ICOs, and crowd sales are just some of the hot topics which we have recently occupied within our daily lives. Is blockchain economy a threat or opportunity for the integrated financial system? What about the economic and political agenda of FIAT money countries for new crypto-markets? How will the financial system be evolved in the next decade? Coping with this new issue of blockchain economies, scholars and researchers from different disciplines are gathered together in this novel book and tried to answer these challenging questions.

In this novel book, distinguished authors of different disciplines from economics to finance gave satisfactory answers to these challenging questions. The authors of the chapters in this publication have contributed to the success of this book by the inclusion of their respective studies. Contributors in this study formulated the new insights for the blockchain economics and questioned its future for a globalized and transforming financial system.

This book is composed of five contributory parts with 27 chapters. The first part outlines the components of *blockchain economics and financial market innovation*. Chapters in this part made assessments on a new blockchain-based financial market innovation from an economic perspective. This book continues with part two outlining *Cryptocurrency Investment Strategies and Crypto-Market Components*. The present and future of cryptocurrency investment strategies have been assessed. The third part develops a deeper understanding of the *Economic and Financial Role of Crypto-Currencies*. In this part, the authors assessed the correlations between financial conditions and cryptocurrency price changes within the international arbitraging mechanism. In the fourth part, *Cryptocurrency Taxation System in Emerging Markets* has been introduced. In this part, contributors analyzed the correlation of the effective taxation system and accounting practices via blockchain technology. In the final part, *Related Subjects and Political Agenda* for this hot topic have been assessed from a multidisciplinary perspective.

Chapter 1 evaluates the transformation of corporate finance in the new business ecosystem in the digital age. Prof. Gurunlu aims to explore the impacts of blockchain technology—a foundationally innovative technology—reshaping and even revolutionizing corporate finance. From her explanation, this underlying technology in the heart of all financial innovations has the great disruptive potential for the financial sector. In this context, her study focuses on how corporate finance is being transformed by blockchain technology and its extensions (distributed ledgers and smart contracts) and underlines what the reflections will be on corporate governance, capital markets, corporate voting, and accounting matters in corporate finance.

Chapter 2 features an economic approach to new digital money as the global financial system's new tool. Dr. Dayi, in this chapter, underlines that the developments in information and Internet technology have led to profound changes in the global financial system. Dr. Dayi advocates that cryptocurrency technology aims to make transactions reliable and provide money control with the encryption technique. Due to the high-security encryption technique of the network structure, it is not possible to infiltrate into the system. In addition to being reliable, the new currency recently has been more effective as an investment tool rather than being a medium of exchange in daily life anywhere in the world.

Chapter 3 proposes a model with redesigning current banknotes with blockchain infrastructure. In the proposed model, Dr. Erdem and Dr. Altun aimed to propose a new hybrid system that can be applied with fewer efforts and requiring little modifications in existing structures by combining the advantages of both digital currency system and paper-based banknote system. They offered a model that uses QR Coded banknotes, digital wallets, and blockchain technology for ensuring and confirming the ownership of the banknotes. This model is considered as applicable at the macro level by governmental policies.

Chapter 4 introduces the tokens as innovative financial assets in crypto-markets and evaluates ICOs. Dr. Adhami and Professor Giudici describe the new phenomenon of initial coin offerings (ICOs), i.e., unregulated offerings of digital tokens, built on the innovative blockchain technology, as to provide a means to collect finance for a project on the Internet, disintermediating any external platform, payment agent, or professional investor. ICO tokens allow access to platform services, may serve as cryptocurrencies, or grant profit rights; they are traded on electronic exchanges and represent a new financial asset. They highlighted the issues raised with respect to information asymmetries and moral hazard, and we review the nascent empirical literature exploring the ICO token market.

Chapter 5 develops a futuristic view on the blockchain question: Can this new technology enhance social, environmental, and economic sustainability? In this chapter, Dr. Semen Son Turan explores the nature of blockchain and discusses how it may contribute to or obstruct sustainability. To this end, first, blockchain technology is introduced. Next, a short discussion on sustainability is presented, including how it is defined, measured, reported, and understood in theoretical frameworks. After that, the 2015 United Nations Sustainable Development Goals are briefly explained. This is followed by a systematic literature review, which highlights the scarcity of literature linking blockchain to sustainability. Finally, the author offers her own reflections on the potential of blockchain to revolutionize the financial services industry and weighs up the pros and cons of vis-a-vis sustainable development.

Chapter 6 develops a practical approach to herding behavior in cryptocurrency market by using CSSD and CSAD analysis. Dr. Gumus and her colleagues focus on the cryptocurrency index and cryptocurrencies, which have existed since the arbitrarily set starting date of the index. In addition to the CCI 30 Index, as a proxy for the market, Bitcoin, Litecoin, Stellar, Monero, Dogecoin, and Dash are used for empirical analysis. To the best of the author's knowledge, the CCI 30 Index is used for the first time as a proxy for market return. Despite the growing literature on cryptocurrencies, there is still a gap in herding behavior in the cryptocurrency market. Results indicate no evidence of herding behavior in the cryptocurrency market in both CSSD and CSAD approaches. The findings of both approaches are in line with the findings of the previous literature regarding the herding behavior in cryptocurrencies.

Chapter 7 analyzes cryptocurrency volatility. Dr. Cankaya and his colleagues, in this chapter, contribute to the field of research by examining the relationship between cryptocurrency's volatile returns and the effects of different types of news on selected cryptocurrencies. This chapter categorizes the news about cryptocurrencies and determines the effect of news from each category on the return structure of each cryptocurrency. By using 1054 news sources, 22 categories are created, and a clustering analysis is used to set these categories into six groups. These groups are modelized in proper ARCH family models, which are created for different cryptocurrencies to analyze the effect on volatility. The results show that different cryptocurrencies react differently to various news categories. News about regulations from national authorities exhibits a significant effect on all selected cryptocurrencies.

Chapter 8 analyzes Bitcoin market price movements and develops an empirical comparison with main *currencies, commodities, securities, and altcoins.* Dr. Haslak and her colleagues analyze the Bitcoin (BTC) market prices and to answer the question of whether there is a relationship between BTC and other asset prices, where other assets include currencies, commodities, securities, and altcoins. In the empirical part, they evaluate the lead-lag relationships among each type of asset. Their result shows that BTC does not have a long-run relationship with any asset type, but that it has a short-run relationship with gold and especially altcoins, which are both significant and bidirectional. While BTC and altcoins are closely interrelated with each other, BTC price variation is mostly borne by its own prices in all cases.

Chapter 9 initially assesses the causal relationship between returns and trading volume in cryptocurrency markets with a recursive evolving approach. Dr. Efe Cagli examines the time-varying causal relationship between trading volume and returns in cryptocurrency markets. The chapter employs a novel Granger causality framework based on a recursive evolving window procedure. The procedures allow detecting changes in causal relationships among time series by considering potential conditional heteroskedasticity and structural shifts through recursive subsampling. The chapter analyzes the return–volume relationship for Bitcoin and seven other altcoins:

Dash, Ethereum, Litecoin, Nem, Stellar, Monero, and Ripple. The results suggest rejecting the null hypothesis of no causality, indicating bidirectional causality between trading volume and returns for Bitcoin and the altcoins except Nem and Stellar. The findings also highlight that the causal relations in cryptocurrency markets are subject to change over time. The chapter may conclude that trading volume has predictive power on returns in cryptocurrency markets, implying the potential benefits of constructing volume-based trading strategies for investors and considering trading volume information in developing pricing models to determine the fundamental value of the cryptocurrencies.

Chapter 10 underlines a piece of empirical evidence from unit root tests with different approximations on the assessment of the crypto-market efficiency. Dr. Iltas and his colleagues examine whether the weak form of the efficient market hypothesis (EMH) is valid for the Bitcoin market. To that end, they consider the recent developments in unit root analysis utilizing daily data from February 2, 2012, to November 23, 2018. More specifically, they employ unit root tests with and without sharp breaks and also a unit root test with gradual breaks in order to obtain the efficient and unbiased output. Major findings show that the EMH appears to be valid for the Bitcoin market. We discuss the theoretical and practical implications of these findings.

Chapter 11 develops a critical approach to forecasting the prices of cryptocurrencies using GM(1,1) Rolling Model. Dr. Kartal and Dr. Bayramoglu explain the functioning of the cryptocurrencies as an investment tool in the market and to share information about the types of investors who have transferred their funds to cryptocurrencies by providing statistical information. Then, it is aimed to share the theoretical knowledge about GM(1,1) Rolling Model which has been proved by the literature in which it produces successful results especially in forecasting problems in an uncertainty environment. The results may be considered that the model was successful in forecasting the prices but unsuccessful in the direction forecasting.

Chapter 12 questions the possibility to understand the dynamics of cryptocurrency markets using econophysics in crypto-econophysics. Dr. Ulusoy and Dr. Celik advocate that following the second law of thermodynamics, the Carnot cycle was written from a new point of view: whether the amount of work given to the system in the cryptocurrency reserve can explain the possible trading (exchange) prices that occur or are likely to occur with the exchange of money.

Chapter 13 initially evaluates the linkages between cryptocurrencies and macro-financial parameters: a data mining approach. Dr. Arzu Bayramoglu and Dr. Basarir address the importance of digital currencies which have increased their effectiveness in recent years and have started to see significant demand in international markets. Bitcoin stands out from the other cryptocurrencies in considering the transaction volume and the rate of return. In this study, Bitcoin is estimated by using a decision tree method which is among the data mining methodologies.

Chapter 14 assesses the impact of digital technology and the use of blockchain technology from the consumer perspective. In this chapter, cryptocurrencies, specifically Bitcoin and the underlying technology blockchain, are discussed by

Dr. Bumin Doruk, from the consumer point of view. Awareness of cryptocurrencies, attitudes toward it, purchase intentions, user profiles, and usage motivation around the world and in Turkey are also assessed. Following this part, the adoption of blockchain technology and Bitcoin is analyzed by different technology acceptance and adoption models.

Chapter 15 demonstrates the empirical evidence of the relationships between Bitcoin and stock exchanges with a case of return and volatility spillover. Dr. Kamisli and his colleagues explain the new investment vehicle which is also used for portfolio diversification. But to provide the desired benefits, the relationships between the Bitcoin and asset or assets will be included in the portfolio. Therefore, the purpose of this study is to analyze the return and volatility relationships between Bitcoin and stock markets from different regions. For this purpose, Diebold and Yilmaz's spillover tests are applied to the return series. The empirical results indicate both return and volatility spillovers between the Bitcoin and the selected stock markets that should be considered in portfolio and risk management processes.

Chapter 16 focuses on the asymmetric relationships between Bitcoin and precious metals. Dr. Kamisli analyzes the causality relationships between the most popular cryptocurrency Bitcoin and gold, silver, platinum, palladium, ruthenium, rhodium, iridium, osmium, and rhenium by asymmetric causality in frequency domain approach.

Chapter 17 develops an economic and institutional approach to effective taxation system by blockchain technology. Dr. Dermirhan discusses the applicability of blockchain technology for use in a tax system. This chapter, therefore, attempts to explain the applicability of blockchain technology in relation to taxation, and it clarifies (1) how blockchain technology represents a new approach to taxation, (2) how blockchain technology reduces tax expenditure, (3) how blockchain technology increases both transparency and accountability, (4) how tax evasion can be reduced using blockchain technology, and (5) how blockchain technology can reduce the administrative tax burden.

Chapter 18 examines the impact of size and taxation of cryptocurrency with an assessment for emerging economies. Dr. Teyyare and Dr. Ayyildirim attempt to discuss the dimensions of cryptocurrencies in developing countries and show the debates on the taxation of these currencies. As countries' tax systems and taxable incomes may differ, under which income category cryptocurrencies and incomes to be gained through them will be treated and how they will be taxed are still being debated. In this regard, it is aimed to determine the current situation in certain developing countries and in Turkey and to put forward some policy recommendations about taxation.

Chapter 19 explains the framework of the accounting and taxation system of cryptocurrencies in emerging markets. Dr. Kablan makes recommendations on accounting and taxation of cryptocurrencies by providing examples of accounting records. Finally, the study recommends that common definitions are made for these new assets throughout the world, and globally accepted international cryptocurrency standards for the accounting and taxation of these currencies are established and implemented.

Chapter 20 evaluates the cryptocurrency and tax regulation by assessing the global challenges for tax administration. Dr. Yalaman and Dr. Yildirim investigate whether the government should tax cryptocurrency or not by using the game theoretical framework. In this game, both government and cryptocurrency investors will determine the strategies to maximize their own benefits. This chapter also investigates various countries taxation policy on cryptocurrency. It is clear that there is no consensus among countries about legal status and taxation process of cryptocurrencies.

Chapter 21 proposes a model for using smart contracts via blockchain technology for effective cost management in health services. In this chapter, the health services where E-government applications, tele-medicine, and artificial intelligence are reviewed and the effects of the sharing of the data about patients and diseases among health sector parties with the blockchain technology through smart contracts have been investigated by Dr. Oflaz. The theoretical framework of blockchain technology has been also investigated within the existing framework, and the applications of countries such as Estonia, Sweden, and the USA, who use blockchain technology in the health sector, have been analyzed and their effects on the costs of health services were evaluated.

Chapter 22 analyzes the role of technological trust and its international effects and evaluates the future of cryptocurrencies in the digital era. Dr. Dilek has assessed the evolution of money in the era of digital transformation and the repositioning of cryptocurrencies with a focus on Bitcoin. The study analyzes the global effects of blockchain technology and cryptocurrencies and the risk, opportunities, and environmental effects of mining.

Chapter 23 takes a contrary view and discusses the existence of speculative bubbles for us at times of two major financial crises in the recent past with an econometric check of Bitcoin prices. In this study, Dr. Mukherjee uses a speculative bubble tracker, based on Wiener stochastic process, at times of two major financial crises, i.e., during the 2008–2009 US subprime mortgage market crisis and the global recession that started from 2010 onward.

Chapter 24 examines the relationship between international interest rates and cryptocurrency prices with a case for Bitcoin and LIBOR. In this study, the change in weekly USD LIBOR Rate and USD Bitcoin Price for 2013–2018 was analyzed by Dr. Erdogan and Dr. Dayan. According to the results of their study, the variables are stationary at the I(0) level. The VAR model was stationary and significant. According to the ARDL model, short-term deviations have stabilized in the long run. The Granger causality test was one-way significant.

Chapter 25 evaluates cryptocurrency derivatives in blockchain economics with a case of Bitcoin. Dr. Soylemez states that "cryptocurrencies are recognized by individuals, institutions, and governments as an economic asset. However, the high price volatility of cryptocurrencies shows that they have significant risks. Cryptocurrency derivatives are used to hedge against and benefit from price movements." This study provides a basic framework for cryptocurrency derivatives. In this study, the most traded cryptocurrency type, Bitcoin derivatives, are used.

Chapter 26 questions that how a machine learning algorithm is *now-casting* stock returns. Dr. Sorhun's study focuses on measuring the performance of algorithmic trading in the now-casting of stock returns using machine learning techniques. The main findings are: (1) the decision tree algorithm performs better than K-nearest neighbors, logistic regression, Bernoulli naïve Bayes alternatives; (2) the now-casting model allowed us to realize an 18% of yield over the test period; and (3) the model's performance metrics (accuracy, precision, recall, f1 scores, and the ROC-AUC curve) that are commonly used for classification models in machine learning takes values just in the acceptance boundary.

Chapter 27 draws a comprehensive framework for accounting 4.0 with the implications of Industry 4.0 in the digital era. Prof. Aslanterlik and Dr. Yardimci underline that accounting systems, which have a very important function for businesses, need to adapt to Industry 4.0 by redefining the whole accounting system, as well as redesigned strategies. Industry 4.0 offers new potential for the transformation of the accounting process through digitalization and application of new tools of Industry 4.0 such as big data analytics, networking, and system integration. The main objective of this chapter is to offer a conceptual framework for a newly designed accounting process in terms of procedures, technology, and accounting professionals.

This book gathers colleagues and professionals across the globe from multicultural communities to design and implement innovative practices for the entire global society of business, economics, and finance. The authors of the chapters in this premier reference book developed a new approach to economic and financial issues in the digital era with an elaborate understanding of financial innovation on the basis of blockchain economics and crypto-markets.

Finally, distinguished authors and professionals with respect to their studies in the field contributed to the success of existing literature with their theoretical and empirical studies from multidisciplinary perspectives in this novel book.

Istanbul, Turkey — Umit Hacioglu

Acknowledgment

In this novel book, I have many colleagues and partners to thank for their impressive contribution to this publication. First of all, I would like to praise the people at Springer International Publishing AG: Editors Mr. Prashanth Mahagaonkar and Mr. Philipp Baun, who have the attitude and substance of a genius—they continually and convincingly conveyed a spirit of adventure in regard to this research at each stage of our book development process; our Project coordinator and all Springer team, without their persistent help this publication would not have been possible; and others who assisted us to make critical decisions about the structure of the book and provided useful feedback on stylistic issues.

I would like to express our appreciation to the Editorial Advisory Board Members. The members who helped with the book included Dursun Delen, Ekrem Tatoglu, Ekrem Tatoglu, Hasan Eken, Idil Kaya, Ihsan Isik, Martie Gillen, Michael S. Gutter, Nicholas Apergis, Ozlem Olgu, Ulas Akkucuk, and Zeynep Copur. The excellent advice from these members helped us to enrich the book.

I would also like to thank all of the authors of the individual chapters for their excellent contributions.

I would particularly like to thank the Bussecon International Academy Members for their highest level of contribution in the editorial process.

The final words of thanks belong to my family and parents separately. I would like to thank my wife Burcu, my son Fatih Efe, my girl Zeynep Ela, as well as my parents, my father Ziya and my mother Fatma. Their pride in this challenging accomplishment makes it even more rewarding to me.

Umit Hacioglu

Editorial Advisory Board

Dursun Delen, Oklahoma State University, USA
Ekrem Tatoglu, Ibn Haldun University, Istanbul, Turkey
Erkan Bayraktar, American University of the Middle East, Kuwait City, Kuwait
Hasan Eken, Kirklareli University, Kirklareli, Turkey
Idil Kaya, Galatasaray University, Turkey
Ihsan Isik, Rowan University, NJ, USA
Martie Gillen, University of Florida, USA
Michael S. Gutter, University of Florida, USA
Nicholas Apergis, University of Piraeus, Greece
Ozlem Olgu, Manchester Metropolitan University, UK
Ulas Akkucuk, Bogaziçi Üniversitesi, İstanbul, Turkey
Zeynep Copur, Hacettepe Üniversitesi, Ankara, Turkey

Contents

About the Editor

Umit Hacioglu is professor of finance at Ibn Haldun University, School of Business, Istanbul, Turkey. Dr. Hacioglu has BA degrees in Business Administration and International Relations (2002). He received PhD in Finance and Banking with his thesis entitled "Effects of Conflict on Equity Performance". Corporate finance, strategic management, and international political economy are the main pillars of his interdisciplinary studies. As a blockchain enthusiast, he is recently working on strategic fit and design in business ecosystem with blockchain-based applications. He is the editor of *International Journal of Research in Business and Social Science* (IJRBS), *International Journal of Business Ecosystem and Strategy*, and *Bussecon Review of Finance and Banking*. Dr. Hacioglu is the founder member of the Society for the Study of Business and Finance (SSBF) and BUSSECON International Academy.

Part I
Blockchain Economics and Financial Market Innovation

Chapter 1
Corporate Finance in the New Business Ecosystem in the Digital Age

Meltem Gürünlü

Abstract In the digital age of continuous technological evolution, business ecosystems are being renovated. This study aims to explore the impacts of blockchain technology—a foundationally innovative technology—which is reshaping and even revolutionizing corporate finance. This underlying technology in the heart of all financial innovations has a great disruptive potential for the financial sector. In this context, it is deeply focused on how corporate finance is being transformed by blockchain technology and its extensions (distributed ledgers and smart contracts) and what the reflections will be on corporate governance, capital markets, corporate voting, and accounting matters in corporate finance.

1.1 Introduction

We are living in a digital age where technology is changing at a greater pace than ever before in the history of mankind, triggering important and radical changes in the business ecosystem. Technological evolution triggers changes in market demands, inviting the incumbents of the financial infrastructure to innovate and enhance time and cost efficiency. In this context, "Financial Technologies (FinTech)" is a trendy topic which has a great disruptive potential for reshaping and even revolutionizing corporate finance. FinTech is a technologically enabled financial innovation that could result in new business models, applications, processes, products, or services with an associated material effect on financial markets and institutions and the provision of financial services. According to this definition, FinTech includes such innovations as online marketplace lending (or peer-to-peer lending), crowdfunding, robo-advice, financial applications of blockchain and distributed ledger technology, financial applications of artificial intelligence, and digital currencies (Schindler, 2017).

M. Gürünlü (✉)
Department of International Trade and Finance, Faculty of Economics and Administrative Sciences, Istanbul Arel University, Büyükçekmece, Istanbul, Turkey
e-mail: meltemgurunlu@arel.edu.tr

© Springer Nature Switzerland AG 2019

U. Hacioglu (ed.), *Blockchain Economics and Financial Market Innovation*, Contributions to Economics, https://doi.org/10.1007/978-3-030-25275-5_1

Online platforms for peer-to-peer lending and crowdfunding enable retail investors to lend money directly to retail customers, easing the way money flows and removing the formal barriers for financing investments. The growing transactions in digital currencies, such as bitcoin, change the method of payment, enabling people to make monetary transactions without a trusted third party (e.g., government-backed currencies). These innovations also enlarge the scope of financial inclusion through their ability to reach people and businesses in remote and marginalized regions (Patwardhan, Singleton, & Schmitz, 2018).

In the heart of these financial innovations lies the blockchain or distributed ledger technology enabling FinTech innovations to go further. According to International Monetary Fund (IMF) Chief Christine Lagarde (2018), "Blockchain innovators are shaking the traditional financial world and having a clear impact on incumbent players. In order to ensure stability and trust in the financial markets, the transformative potential of blockchain-based technologies and assets should be broadly embraced by regulators and central banks, who recognize the positive effect new inventions can offer for the business model of commercial banks".

Starting from the year 2008, blockchain technology has attracted attention all over the world as it has significant power to radically change the financial ecosystem. Blockchain is a new disruptive information technology that enables many users to complete their own financial transactions without any need for further approval by an intermediating party which has a central power to supervise all transactions. Distributed ledgers empowered by blockchain technology provide a secure way of storing data that make it largely unchangeable. With this technology, information about a history of transactions can be stored in a decentralized way. Blockchain technology can be utilized in many areas such as easing the ways the payments are completed in remote regions of the world, obtaining financial sources as debt or equity instruments, collateralized asset management, legal reporting activities to guarantee the adherence to related laws and regulations, and voting in absence in the annual shareholders meetings of the companies (World Economic Forum, 2016).

This chapter aims to investigate the impact of this foundationally innovative technology in an effort to shed light on how this new financial ecosystem will continue to grow and evolve in the future.

1.2 New Financial Landscape in the Digital Era of Financial Innovations

Blockchain technology and distributed ledgers have extensive usage areas in many different industries. However, the financial industry seems to be the forerunner among other industries. This is not only due to the fact that the most well-known application of this technology is the crypto-currency bitcoin, but it is also driven by substantial process inefficiencies and a massive cost base issue specifically in this industry. On top of this is the need for tracing ownership over a longer chain of changing buyers in global financial transaction services. The blockchain technology promises to overcome problems and inefficiencies related to the traditional intermediation process, which include possible human errors, representing "a shift from trusting people to trusting math" since human interventions are no longer necessary (Nofer, Gomber, Hinz, & Schiereck, 2017).

The blockchain eschews a bank or other intermediary and allows parties to transfer funds directly to one another, using a peer-to-peer system. This disruptive technology has done for money transfers what email did for sending mail—by removing the need for a trusted third party just as email removed the need for using the post office to send mail (Lee, 2016).

Blockchain introduced by Satoshi Nakamoto in Nakamoto, 2008, is popular due to the first cryptocurrency, bitcoin; it was created and supported by blockchain technology. Going further, this new technology underlying bitcoin and other cryptocurrencies has also started to shake the world of finance by transforming corporate governance, shareholder activism and dialogue, corporate voting, double bookkeeping accounting, and stock exchange transactions (Nakamoto, 2008).

Blockchain technology not just provides a reliable infrastructure for bitcoin and other cryptocurrencies, it is also a breakthrough in future technologies which will disrupt everything with a great pace in similar way to the world wide web (www) technologies which were introduced in the early 90s. In the beginning of 90s, IP-based technologies led by world wide web (www), http and html protocols became publicly available and since then a great wave of change such as development and commercialization of internet applications have swept away traditional business models. In the same vein as the world wide web during 90s, blockchain technology will also be a revolution changing the financial ecosystem by lowering the cost of transactions, providing transparency of ownership, and accurate financial record keeping.

The 2008 financial crisis was a breaking point for blockchain technology as this crisis apparently became a motivating factor for the creation of bitcoin and other cryptocurrencies. The 2008 financial crisis awakened mistrust in fiat currencies (US dollar or British pound), or currencies created and backed solely by faith in a government. This underscored the fragility of the modern financial system with heavy reliance on banks and other financial institutions showed the importance of high costs of financial intermediation which later have become a burden on national governments and tax payers. In the aftermath of the 2008 financial crisis, many banks and financial institutions failed and had to be bailed out by national governments, being a burden on tax payers.

As a result of this turbulent environment after the 2008 financial crisis, investments in the blockchain technology have risen sharply. Blockchain Technology (also called Distributed Ledger Technology (DLT)) allows for the entire financial services industry to dramatically optimize business processes by sharing data in an efficient, secure, and transparent manner. In many countries, central banks are investing in projects improving blockchain technology which will shape the future of financial sector. Although central banks are among the most cautious and prudent institutions in the world, it is indicated that these institutions, perhaps surprisingly, are among the first to implement blockchain technology. Central banks around the world are actively investigating whether blockchain can help solve long-standing issues in banking, such as payment-system efficiency, payment security and resilience, as well as financial inclusion (Lannquist, 2019).

There are many possible application fields of blockchain in various industries such as e-government practices (property registry, voting and e-citizen applications), media, healthcare, manufacturing, supply chain management, energy and finance. Financial industry including capital markets, banking and insurance is making the hugest investments in blockchain technologies in order to ensure stability and security for their customers.

Global spending on blockchain solutions is forecast to be nearly $2.9 billion in 2019, an increase of 88.7% from the $1.5 billion spent in 2018. It is expected that blockchain spending will grow at a robust pace over the 2018–2022 forecast period with a 5-year compound annual growth rate (CAGR) of 76.0% and total spending of $12.4 billion in 2022 (IDC, 2019).

1.3 Benefits of the Blockchain System in Comparison with Traditional Financial System

A blockchain is a digital, immutable, distributed ledger that chronologically records transactions in near real-time (Bhattacharyya, 2018; Fig. 1.1). The prerequisite for each subsequent transaction to be added to the ledger is the respective consensus of the network participants (called nodes), thereby creating a continuous mechanism of control regarding manipulation, errors, and data quality. It creates a digital ledger of transactions and thereby allows for it to be shared among a distributed network of computers and maintains a continuously-growing list of records called “blocks” which are secured from tampering and revision (Shah & Jani, 2018). The existing traditional financial system which depends on intermediation has two basic problems. First, it is difficult to monitor and evaluate asset ownership and its transfer in a trusted business network. Second, it is inefficient, expensive, and more sensitive to human risks and errors. Blockchain provides decentralization of the powers of the decision-makers and makes the system less prone to human based errors such as bounded rationality and opportunistic behaviour. When there are more than one party involved in making decisions, any possible mistake will be reduced to it’s minimum level. Similarly, in decentralized decision-making, since parties would have less power, they could accommodate the needs of other parties more easily in their decisions, which could ultimately deal with the bounded rationality problem. Hence, the hierarchy of decision-makers in a decentralized system, may serve as a system of checks and controls, which could reduce the negative effects of the opportunistic behaviour (Avdzha, 2017; Fig. 1.2).

When compared with traditional systems, blockchain has the advantage that a record which is maintained in a ledger is available to each party. This distributed

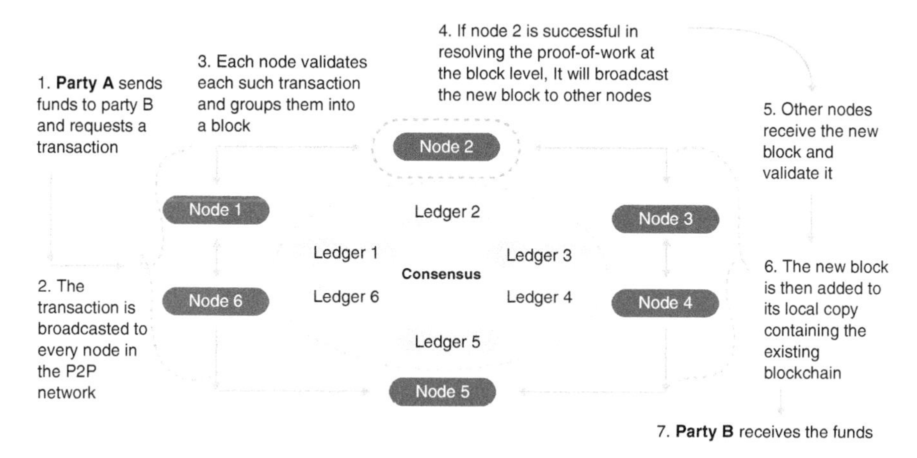

Fig. 1.1 How a transaction takes place in a blockchain. Source: Bhattacharyya (2018)

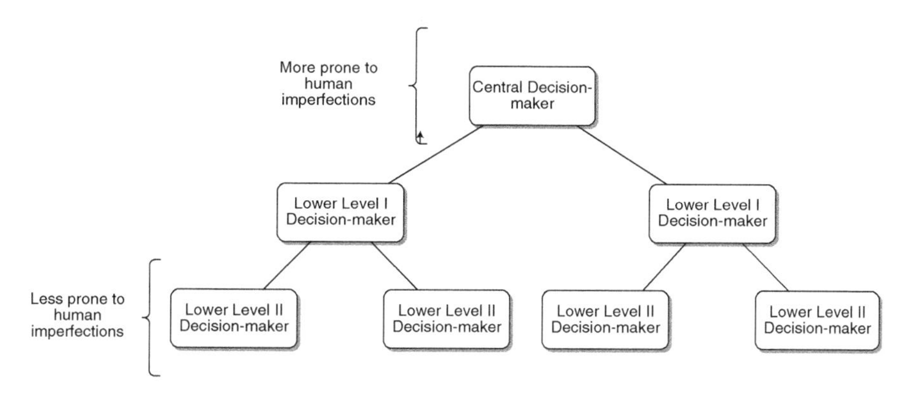

Fig. 1.2 Traditional centralized decision-making versus decentralized decision-making. Source: Avdzha (2017)

ledger can be widely passed between multiple users and creates a shared database for all users who have the access right for it. The distributed nature of blockchain increases transparency in processing by decreasing the need for manual verification and authorization. The main advantages of the blockchain are (Shah & Jani, 2018; Tapscott & Tapscott, 2017) can be mentioned as follows.

Real-time settlement of recorded transactions is ensured, decreasing the level of human risk and errors. There is no intermediation. Blockchain provides cryptographic proof which eliminates the need for a trusted third-party intermediation. Every participant on a blockchain can reach to the database and its history archive completely. There is no central power controlling the database. Each participant in a transaction can approve the records of its transaction partners without any need to an approval by a central intermediating authority. A shared database (distributed ledger) of public history of peer-to-peer transactions between the parties is distributed and made available to all users or parties to transactions. The blockchain contains a certain and verifiable record of every single transaction ever made. The records related to a transaction can not be changed since all records are linked to each other historically on a blockchain. Smart contracts can be used. They are stored procedures that are executed in a blockchain to process pre-defined business steps and execute a commercially and legally enforceable transaction without involvement of an intermediary. As a result, blockchain increases transparency and accountability in decision-making while decreases the occurrence of human related errors in decision-making.

1.4 How Blockchain Is Transforming Corporate Finance

Blockchain technology will affect how trades are taking place in between financial institutions in a number of ways. Firstly, financial transactions will be processed and settled more securely and faster than before due to the peer-to-peer network and cryptographic security properties of blockchain technology. Secondly, for legal compliance matters, it will be much more easier for regulating parties to carry out auditing of transactions since they will be able to reach each step in a transaction in real-time manner. Thirdly, blockchain technology allows for automatically executed contracts or smart contracts in between parties to a transaction. In the future, blockchain will be critically important to a firm's survival in the long-run. Because it is going to transform all mechanisms of the financial industry and will have huge impacts on corporate governance, capital markets, corporate voting and corporate accounting (Bhattacharyya, 2018).

1.4.1 Impact on Corporate Governance

As long as there are human imperfections such as bounded rationality and benevolence, corporate governance mechanisms will be unable to solve agency problems between different stakeholders within a firm. Corporate governance mechanisms such as independent directors, executive renumeration packages, and a debt concentrated financial structure are not adequate as human related errors based on bounded rationality and benevolence will still remain though they may be reduced by making use of these specially designed corporate governance tools. A total solution for all types of agency related problem areas, which would work in different types of companies, seemed like a fantasy until blockchain technology arrived and offered alternative solutions due to its potential to make corporate transactions more accurate, transparent, and efficient than ever. The pioneering article by Yermack (2017) asserts that blockchains will noticeably lower wasteful uses of corporate sources and misbehavior by managers.

Irreversibility and immutability properties of blockchain make records of transactions transparent and open to all available parties. Since transparency and full disclosure are the bottom lines for a good corporate governance environment, all stakeholders will be informed with more knowledge simultaneously, limiting the scope of agency costs related with asymmetric information. Changes in ownership can now be easily tracked as the authorized users only have to reach to the shared ledger, allowing for the accurate and timely transmission of information to stockholders (Lafarre & der Elst, 2018)

A blockchain produces copies of each block transactions simultaneously and forms an archive which are made available to all parties in a transaction. This is the distributed ledger property of the blockchain technology. This property makes the functions such as auditing and verification of the transactions unnecessary since all participants of a financial transaction would be able to see the changes in the ownership structure at any time as they occurred (Yermack, 2017).

Removal of verification and validation by trusted third party intermediaries, such as banking or clearing, will make the financial system work faster, effectively, and efficiently. There will be less corruption and errors. Blockchain transactions will be safer than traditional financial transactions which need time for settlement and clearing before being completed.

Smart contracts empowered by blockchain technology enable the removal of possible errors in principal-agency relations, minimizing the agency costs for the parties. This is due to the property of smart contracts that they run as coded, preventing any opportunistic behavior of the agent. Thus, smart contracts provide automatically executed protocols and eases the verification, monitoring of contracts between the principal and the agent, making information asymmetries between interest holders in a company, breaches of contract terms and fraud impossible. All information is publicly available in a transparent manner. For example, a company's financing position can be seen by anyone on the blockchain. This information is not limited to the use by company's insiders or managers. Smart agency contracts run on a custom built blockchain which permits all interested parties (managers and investors) to view records of company debts (Kaal, 2018).

1.4.2 Impact on Capital Markets

In many countries, stock exchanges are making every effort in order to exploit possible advantages of the usage of blockchain technology in the financial sector. These advantages focus on minimization of transaction costs and on-time settlements of financial transactions. National stock exchanges such as NASDAQ, the Australia Securities Exchange, the Tallinn Stock Exchange, the London Stock Exchange, and the Korea Stock Exchange are the pioneers on this front.

In these stock exchanges, trade settlements will be made by the confirming parties in the peer-to-peer network of buyers and sellers enabled by blockchain technology. There will be no need for a custody for confirmation of trade settlements. This will lead the trade settlements to be made almost simultaneously, instead of the existing t +3 trade settlement scheme and the related costs will be minimized. The settlement process will be carried out on blockchain which will reduce the time and manual work needed for the settlement of financial transactions by intermediaries in the existing centralized system.

Great attention has been paid by stock exchanges to the use of blockchain technology in order to eliminate the costly and inefficient dependency of traditional stock exchanges on centralized systems which make the existence of a stock transfer agent or a trusted third party for verification and completion of the transactions necessary. This dependency also prevents the true pricing of stocks as full information cannot be provided to the parties because of counter-party risks.

There are many benefits for stock exchanges in the areas of reconciliation, trade validation, reference data, faster settlement, collateral management, regulatory reporting, and audit trail (Bhattacharyya, 2018).

Reconciliation There will be no need for third parties or intermediaries such as stock agents. This is one of the primary benefits expected to be derived from blockchain technology in the capital market. The related costs will be cut severely.

Trade Validation Blockchain enables smart contracts which makes the trade validation process simpler and more efficient. The parties in the peer-to-peer network can view and trace history of ownership structure and contract terms in a more effective and efficient way.

Reference Data Share price information, security data, calendar days and client data are essential reference data for completing a trade between buyers and sellers. Storage of this huge data and providing its validation across a number of participants involved in the trade is a very hard and long process. Instead of this time-taking process, the shared reference data can be collected on blockchain, with running it's confirmation codes among the participants in the network and parties in a transaction can trace changes in the reference data records. Blockchain enables validation of data creation by parties to the transaction in a peer-to-peer network instantaneously. Hence, supervision of the real-time data in the ledger would be much easier.

Faster Settlement Blockchain technology could decrease the time spent in the custody and settlement processes. When a transaction is validated and sent to the the distributed ledger, the digital wallet of the owner can be updated instantaneously. Increased speed in settlement will lower associated costs and risks.

Guarantees Management Smart contracts enabled by blockchain technology can run coded rules for automated margin calls for a trade. The seller and buyer take place on the same blockchain network. Hence, the movement of digital recording of assets through tokens makes possible the tracking of asset movement in between the parties and automated smart contracts allow exchange of assets for collateral purposes.

Regulatory Reporting As there will be a single, reliable and transparent archive including records of all transactions made by all parties, the costs related to the reconciliation and verification of the data will decrease sharply, leading to huge savings. Regulators will monitor trade transactions in real-time, increasing the efficiency of supervision.

Audit Trail Distributed ledgers enabled by blockchain technology make every entry transparently seen by participants of the trade activity and there is no way to hide or falsify records. As all transactions are completely digital, auditing of such transactions can also be done digitally. This will reduce much of the manual work, and the time and cost associated with it. Since on the blockchain, each node on the block can see full history of all transactions, audit trail can be traceable transparently from beginning to the end of the trading process.

Blockchains will allow for a fully transparent share registration system so that the market makers can view investors' ownership and shareholding positions in various shares. This will lead to pricing of assets reflecting all available information and efficient capital markets in which capital is allocated more effectively and efficiently with assets fairly priced and decision takers can make better decisions about their investments in the capital markets (Yermack, 2017).

1.4.3 Impact on Corporate Voting

Blockchain is a technology that can offer smart solutions for classical corporate governance inefficiencies, especially in the relationship between shareholders and the company. Blockchain technology can reduce shareholder voting costs and the organization costs for companies substantially by increasing the speed of decision-making, facilitating the fast and efficient involvement of shareholders in company meetings for all corporate governance matters. Moreover, the main obstacles with the existing chains of intermediaries and the remote voting system can be solved with transparency, verification, and identification advantages which are uniquely provided by blockchain technology (Lafarre & der Elst, 2018).

Inefficiencies inherent in the existing system of corporate voting like incomplete voting, errors in the distribution of ballots, and vote listing problems can be completely overcome by blockchain supported distributed ledgers and smart contracts as the safe recording of the votes will be ensured by this new technology. Furthermore, digital voting in the annual shareholders meetings will enhance shareholder participation in company decision-making, modernizing dialog with shareholders by removing the barriers to physical attendance. This will contribute to the democratic environment in the company, protecting shareholders rights by motivating speedy and accurate voting. More practically, voting by using blockchain system would be performed by the help of eligible voters tokens or "vote coins" which will represent the voting power of each shareholder (Piazza, 2017).

In 2016, NASDAQ Tallinn (Estonia) Stock Exchange became the first stock exchange to apply blockchain voting in company meetings for publicly traded corporations on stock exchange. Speedy, transparent and accurate usage of shareholder votes provided by blockchain voting can further motivate shareholders to be more demanding and participative in corporate governance issues (Yermack, 2017).

This specific use of blockchains and smart contracts in voting process provides increased transparency, accountability and decentralization which can directed to eliminate human-related agency problems and conflicts of interest between principals and agents of companies. If this technology could be adopted prevalently by companies, in the future, it may eliminate all human related mistakes from the voting process.

1.4.4 Impact on Corporate Accounting

Since the medieval period, accounting has relied on double-entry principles and has not changed its core principles based on mathematical integrity and accountability. For example, the technology of data representation in documents has changed the generalization of information from simple accounting to double accounting—where the events have are entered into documents along with the grounds for their implementation. Such records were made (and are still made today) in paper registers, later they were transferred into the digital environment with the wide use of software of all kinds. Thus, the method of data accumulation has changed. However, some approaches remained unchanged such as the need for processing by auditors, and checking the accounting data, regardless of their form (paper or digital). Blockchain, providing a distributed ledger of all transactions immediately in an immutable and irreversible manner, eliminates the need for a third-party approval or assurance by auditing (Melnychenko & Hartinger, 2017).

Blockchain technology as an infrastructure to support the transactions of bitcoin, the most widespread cryptocurrency, has exceeded beyond its initial role from when it was first initiated in 2008. Blockchain is a distributed digital ledger with the chronological records of transactions available to all those who have the access to view recorded data. All data in the blockchain can be transparently seen by the parties in a transaction but copying or changing data is not allowed on blockchain system. This technology is predicted to affect many accounting practices such as auditing, cybersecurity, and financial planning and analysis. All transactions can be fully automated by being programmable as smart contracts between the parties. When the transactions are validated by nodes, the nodes will have a confirmation and guarantee that prepared financial statements will be free from any substantial error. Use of blockchain technology can optimally lower the need for the accounting and auditing processes, leading to a revolution in corporate accounting (Milosavljevic, Joksimovic, & Milanovic, 2019). As a result, the blockchain will decrease the costs related to bookkeeping and auditing functions since all information about the company will be available without delay on a real-time basis with accurate and transparent recording.

The use of blockchain technology will be a breakthrough in accounting having far-reaching and important results. Real-time accounting without any delay will signal suspicious assets transfers in the company, severely limiting the scope of abuse by managers. This will prevent tunneling by managers who have insider information and will guarantee the health of the company's financial position to the creditors by enabling a real-time surveillance against fraudulent attempts. On the other hand, window-dressing or earnings management activities that are directed to manipulate financial reports in order to pay less tax or make the financial outlook attractive to outside investors can be eliminated by irreversible recording of time-stamped transactions. Investors can make their own judgements about the company with certainty, free of any bias or the judgement of managers or auditors (Yermack, 2017).

1.5 Conclusion

In the past few years, words like FinTech, blockchain and cryptocurrency went from being used by only a few experts in the field to words that are used daily. The development in the financial technology infrastructure enabled by the technological breakthroughs in the last decades triggered a fast-paced and technology-driven financial environment (Davradakis & Santos, 2019). In the digital age where technology is changing at a great pace, FinTech has the power of shaping the financial ecosystem going forward by reducing imperfections and inefficiencies in the traditional financial intermediation. This can take many forms: faster and safer payments and settlements through the use of distributed ledgers supported by blockchain technology, greater real-time control of personal and business finances by consumers and small businesses, simple person-to-person transfers including cross-border remittances, easier mobilization of savings to fund investments through crowdfunding, cheaper investment management for small investors through robo-advice, and better and faster credit decisions through big data analytics (Demekas, 2018).

Some FinTech innovations have deeper impacts on the financial ecosystem. Blockchain is the technology that supported the infrastructure for bitcoin. Most importantly, it is a secure way of storing data that makes it largely unchangeable so that a history of transactions can be recorded in a decentralized way. It is commonly agreed by researchers that blockchain and its extensions, distributed ledger and smart contracts, are foundational innovations which will change the traditional centralized financial system in a radical and revolutionary way (Schindler, 2017).

References

Avdzha, A. K. (2017). *The coming age of blockchain technology in corporate governance*. Tilburg University, Master Thesis. Retrieved 01/11/2018, from http://arno.uvt.nl/show.cgi?fid=143457

Bhattacharyya, D. B. (2018). *How blockchain is transforming capital market* (Whitepaper by LTI). Retrieved 2/2/2019, from https://www.lntinfotech.com/wp-content/uploads/2018/05/How-Blockchain-is-Transforming-Capital-Market.pdf

Davradakis, E., & Santos, R. (2019). *Blockchain, FinTechs and their relevance for international financial institutions* (EIB Working Papers). doi:https://doi.org/10.2867/11329

Demekas, D. (2018). *Emerging technology-related issues in finance and the IMF—A stocktaking*. IEO Background Paper Independent Evaluation Office of the International Monetary Fund. Retrieved 02.03.2019, from https://ieo.imf.org/~/media/IEO/Files/evaluations/completed/01-15-2019-financial-surveillance/fis-bp-18-02-07-emerging-technology-related-issues-in-finance-and-the-fund-a-stocktaking.ashx?la=en

Hilleman, G. & Rauchs, M. (2017). *Global blockchain benchmarking study*. Cambridge Centre for Alternative Finance, University of Cambridge. Retrieved 01/03/2019, from https://www.jbs.cam.ac.uk/fileadmin/user_upload/research/centres/alternative-finance/downloads/2017-09-27-ccaf-globalbchain.pdf

IDC (International Data Corporation). (2019). *Worldwide semiannual blockchain spending guide*. Retrieved 12.07/2019, from https://www.idc.com/getdoc.jsp?containerId=prUS44898819

Kaal, W. (2018). *Blockchain solutions for agency problems in corporate governance*. Retrieved 10/11/2018, from https://kaal.io/dynamic-regulation/blockchain-solutions-for-agency-problems-in-corporate-governance/

Lafarre, A. & Van der Elst (2018). *Blockchain technology for corporate governance and shareholder activism* (ECGI Paper Series in Law, No. 390). Retrieved 01/12/2018, from https://ssrn.com/abstract=3135209

Lagarde, C. (2018). *Winds of change: The case for new digital currency* (Speech by, IMF Managing Director, Singapore Fintech Festival). Retrieved 23/04/2019, from https://cointelegraph.com/news/imf-chief-lagarde-distributed-ledger-technologies-are-shaking-the-system

Lannquist, A. (2019). *Blockchain and distributed ledger technology*. World Economic Forum. Retrieved July 13, 2019, from https://www.weforum.org/agenda/2019/04/blockchain-distrubuted-ledger-technology-central-banks-10-ways-research/

Lee, L. (2016). New kids on the blockchain: How bitcoin's technology could reinvent the stock market. *Hastings Business Law Journal, 12*(2), 81–132.

Melnychenko, O., & Hartinger, R. (2017). Role of blockchain technology in accounting and auditing. *European Cooperation Journal, 9*(28), 27–34.

Milosavljevic, M., Joksimovic, N. Z., & Milanovic, N. (2019). Blockchain accounting: Trailblazers' response to a changing paradigm. In S. Drezgić, S. Žiković, & M. Tomljanović (Eds.), *Economics of Digital Transformation* (pp. 425–441). Rijeka: University of Rijeka Press.

Nakamoto, S. (2008). *Bitcoin: a peer-to-peer electronic cash system* (unpublished manuscript). Retrieved 12/12/2018, from https://bitcoin.org/bitcoin.pdf

Nofer, M., Gomber, P., Hinz, O., & Schiereck, D. (2017). Blockchain. *Business Information System Engineering Journal, 59*(3), 183–187. https://doi.org/10.1007/s12599-017-0467-3

Patwardhan, A., Singleton, K. & Schmitz, K. (2018). *Financial inclusion in the digital age* (IFC, Stanford Business School and Credit Ease Report). Retrieved 15/02/2019, from https://www.ifc.org/wps/wcm/connect/f5784538-6812-4e06-b4db-699e86a0b2f2/Financial+Inclusion+in+the+Digital+Age.pdf?MOD=AJPERES

Piazza, F. S. (2017). Bitcoin and the blockchain as possible corporate governance tools: Strengths and weaknesses. *Penn State Journal, 5*(2), 262–301.

Schindler, J. (2017). *FinTech and financial innovation: Drivers and depth* (Federal Reserve System Finance and Economics Discussion Series, No.081). https://doi.org/10.17016/FEDS.2017.081

Shah, T., & Jani, S. (2018). *Applications of financial technology in banking and finance* (Technical Report). https://doi.org/10.13140/RG.2.2.35237.96489

Tapscott, A., & Tapscott, D. (2017). How blockchain is changing finance. *Harvard Business Review*, 2–7.
World Economic Forum (2016). *The future of financial infrastructure: An ambitious look at how blockchain can reshape financial services*. Retrieved 01/02/2019, from http://www3.weforum.org/docs/WEF_The_future_of_financial_infrastructure.pdf
Yermack, D. (2017). Corporate governance and blockchains. *Review of Finance, 21*(1), 7–31.

Meltem Gürünlü is an Assistant Professor of Finance at the Department of International Trade & Finance in Istanbul Arel University, Turkey. She has received her Ph.D. degree in Accounting & Finance from Marmara University in Istanbul in 2008. Between 2011 and 2012, she worked at the Uni-East Project funded by Unicredit as a Post-Doc researcher at the University of Torino. Her research interests include corporate finance, behavioral finance, corporate governance and strategic decision making (such as IPOs and FDI), capital structure decisions, emerging markets finance, ownership structure, and entrepreneurial finance.

Chapter 2
The Global Financial System's New Tool: Digital Money

Faruk Dayi

Abstract The developments in information and internet technology have led to profound changes in the global financial system. In the new financial system, modern financial instrument are used more than conventional money and financial instruments. With the emergence of new financial instruments, economic and financial crises were experienced in both money and capital markets and significant structural problems were observed in the economy. Those who believed that the financial system caused the crisis of 2008 were in search of a new financial system. Satoshi Nakamoto, about one and a half months after the crisis with encryption technique invented a highly reliable currency that eliminates intermediary service as a solution to many problems caused by real money, which allows monetary parties to conduct their transactions directly. Cryptocurrency technology aims to make transactions reliable and provide money control with the encryption technique. Due to the high security encryption technique of the network structure it is not possible to infiltrate into the system. In addition to being reliable, the new currency recently has been more effective as an investment tool rather than being a medium of exchange in daily life anywhere in the world. In order to increase the use of cryptocurrencies serious infrastructural preparation is needed. Therefore, its use for investment purposes has become widespread. The use of cryptocurrencies as a financial instrument and their impact on the money and capital markets remain to be seen.

2.1 Introduction

Thinking in terms of the functions of exchange, investment and value gain "money" is one of the most discussed financial instruments. Everyone wants to have more money and to buy more goods and services with the money they have. However, in today's condition with inflation in economies, prices are constantly increasing and the purchase power of money is decreasing. The borrowers cannot pay their loans to

F. Dayi (✉)
Department of Business and Administration, Faculty of Economics and Administrative Sciences, Kastamonu University, Kastamonu, Turkey
e-mail: fdayi@kastamonu.edu.tr

© Springer Nature Switzerland AG 2019
U. Hacioglu (ed.), *Blockchain Economics and Financial Market Innovation*, Contributions to Economics, https://doi.org/10.1007/978-3-030-25275-5_2

the banks which weaken the bank's collection power. When the banks are unable to collect their receivables from their customers, their debts to banks may cause financial crisis. The crisis that arises when loans and other debts are not paid will be reflected on other people in the market just like the domino effect. The crises in the global financial system show that there are some problems in the current financial system. Each crisis creates a new knowledge and introduces a new product development process. The financial system has begun to be rebuilt with the impact of liberal policies of global markets and the collapse of trade barriers between the countries. Because, those who are thought that the 2008 financial crisis was due to the current monetary system stated that with the development of electronic commerce, issuance of a new electronic payment instrument was needed. It is stated that cryptocurrency emerged as a new mean of payment in electronic commerce (Ates, 2016: 351). However, Satoshi Nakamoto a person or a group people introduced the idea of a new currency that would change the world order. That was Bitcoin. Bitcoin, even though ignored in the early days, has become the most valuable currency recently. The newly created currency was named cryptocurrency and extracted as Bitcoin for the first time. In fact, the reason behind the emergence of cryptocurrencies is the global financial crisis of 2008 (IMF, 2018: 14). Since the current monetary order is thought to cause crisis, Bitcoin was extracted as a cornerstone of a new global financial system (Bradbury, 2013: 5). Yet, Bitcoin is not a money even if it is traded as a financial instrument today. As a result of the combination of cryptography and block chain technology, the global economic system is introduced to a new monetary system. In the new order, money is no longer a physical means of circulation rather it has been used in the electronic environment and has become a payment and investment tool transferred thought the internet. In the era when saving are institutionalized, new financial investment instruments which are different from traditional instruments are developed and investors started to assess their saving with these tools (Yalciner, 2012: 9). Virtual currencies have become as one of the new financial instruments of our age. The emergence of cryptocurrencies caused a shock to the real money. Cryptocurrencies become a tool that is used in the virtual environment and out of the control of the states. Cryptocurrencies, are transferred from once place to another without legal regulations, as if they were transformed into a black money transfer system that is not subjected to any tax or monetary and capital transactions. In this case, it is necessary to recognize and investigate the working order of the cryptocurrency system which is likely to be used in the future with the current monetary system and to be preferred as a financial asset. This study examines the use of cryptocurrency in the new global financial system and its use as a financial instrument.

2.2 Transition from Real Money to Cryptocurrency

Money is an indispensable part of life. Money can be used as a measure of value in shopping and as a mean of exchange, and can be used in investments to accumulate wealth and value gain (Vora, 2015: 816). Money is not a medium that just is used by people. Companies, non-governmental organizations, banks and governments also using money. For example, money is an important asset that is used in the exchange of goods or services or in the exchange of goods or services in interbank markets (Alpago, 2018: 418). In today's banking system, funds between banks can be transferred through systems such as EFT and SDR (Special Drawing Rights). The transfer of money through the internet decreases the transfer cost and transaction time. One of the most important functions of money is the use of the bank money in interbank markets. The bank money is actually traded as a financial instrument. Money is not just a means of payment, but it is also regarded as a symbol of independence and freedom for the countries with national currencies. More than 160 countries in the world now have their own national currencies (Turan, 2018: 2).

Developments in technology have led to innovations in many fields as well as leading to a new era in money and financial markets (Inshyn, Mohilevskyi, & Drozod, 2018: 170). In today's economy, real money has been replaced by different tool payment tools. Even if their transactions seem similar to those of look like money, they cannot take the place of real money, because of the fact that money is an asset that has the highest liquidity. For instance, how can a customer who wants to shop with a credit card pay to a vendor without a post device? Customers who want to pay with cryptocurrency anywhere in the world need to find a vendor with cryptocurrency technology. Therefore, the convertibility of cryptocurrency is not as high as real money. In daily life, everyone may not have the technological infrastructure to use cryptocurrency. Some segments, such as the elderly population in the community, do not use internet and smart phones, so it is not possible to exchange with cryptocurrency. In addition, cryptocurrency is not easy to use in small units such as peddlers. Also, when you see someone in need walking down the street and, you want to give him or her money, you may not be able to help because the person in need do not have a technological infrastructure. For these reasons, the use of real money is very practical in social life. Studying the layout of cryptocurrency before and after its emergence is considered to be very useful for understanding its convertibility.

2.3 Status of the Global Financial System After World War II

After World War II, a new global financial system was created. Describing itself as a superpower in the new order, the United States has played a leading role. USA with the meeting in Bretton Woods laid the foundation of a new financial system. In the system, fixed exchange rate system was applied and the US dollar value was pegged to the gold. So, the new financial system was established by linking the value of the US dollar to the gold and other currencies by indexing to the value of the US dollar. In the financial system by applying the fixed exchange rate, all currencies were indirectly pegged to the gold. In the fixed exchange rate system, the price of the US dollar was to change 1% in upper and lower band. If the upper and lower support points were broken, the central bank would intervene in the market. For this, the Central Bank's foreign exchange reserves had to be adequate. If there were not enough foreign exchange reserves, the devaluation of the money would come to the fore (Yalciner, 2012: 23). As most of the gold reserves were in the USA, the term "the gold owner put the rules" came true and the United States indexed the monetary system to the US dollar (Yalciner, 2012: 22). It made it hard to print money in exchange for the small amount of gold reserves. The depreciation of the gold was causing the US dollar to depreciate too. In order to ensure the continuity of the financial system, new financial institutions such as the IMF and the World Bank were established. When there was a problem in the system, states would obtain the needed funds from these institutions, which was aimed to ensure the continuity of the system (Yalciner, 2012: 23). However, The Bretton Woods system was terminated by the president of the United States in 1971 (Yagci, 2018: 18). After the collapse of the Bretton Woods system, Great Oil Crisis happened, damaging the world economy. The 1970s were dominated by the ideas of Maynard Keynes for an effective fiscal policy. Due to the fact that the fiscal policy of the 1970s were not efficient enough Milton Friedman proposed the monetarist approach (Yagci, 2018: 17). Consequently, many states started to seek for a new monetary system. By the end of the 1970s, the basis of a common currency called "euro", which was planned to be used in the European region, had been laid.

After World War II, Eurobond was issued as a new debt instrument. Accordingly in the European region, the debts were being exchanged by selling bonds through foreign currencies. However, before the Eurobond issuance, banks operating in London were trading short-term debts through foreign currencies. As a result, with transactions like these the basis of swap has been laid down. With the development of the Eurobond market, making loans was possible through the US dollar in Europe. For the first time bond was issued by an Italian company (Yagci, 2018: 18). In the following years, the development of securities markets, issuance of financial derivative instruments, financial instruments such as mutual funds have contributed significantly to the development of financial markets, and the foundation of a new financial system has been laid down in the new world order (Yagci, 2018: 18).

2.4 Reformation of the Global Financial System

In the Bretton Woods system under the fixed exchange rate, capital inflows and outflows were audited with the use of the US dollar indexed to the gold. Other countries had to use their national currencies indexed to the US dollar. The circulation of capital was limited. When necessary, capital inflows and outflows were intervened. In some countries, foreign exchange was prohibited. In this way, states were controlling foreign capital inflows and speculative transactions originating from capital. Because the fact that capital caused speculative movements, this could in turn negatively affect the financial systems of the countries and could cause depreciation in national currencies, deterioration in supply and demand balance in the financial system. Therefore, controlled capital inflows were allowed. The Bretton Woods system had actually collapsed as the US President said that the fixed exchange rate system would be abandoned. Then, with the oil crises a new monetary system had to be established. States in the European region did not want to be dependent on the US dollar. The ideas of the market should determine the value of currencies and obstacles to capital entry and exit should be removed become dominant. Therefore, the new financial system focused on free market economy. Keynesian economic policy was applied until the 1970s. With reduction in effectiveness of the fiscal policy of Keynesian in managing the economy, Milton Friedman's monetarist approach began to be adopted. Central banks began to dominate the world economy and started using monetary policy tools to support the fiscal policy. Consequently, the governments that owned money were directing the world.

With the Bretton Woods system, world currencies were indexed to US dollar and all currencies had to use US dollar. In the new system, central banks had to use the US dollar as reserve money. Because, a large part of the money circulating in the world economy was the American dollar, and had the necessary conditions for being reserve money. Therefore, the Central Bank began to use the US dollar as reserve money. The FED which manages the dollar was not only managing the monetary policy of the United States, but it was also indirectly managing all the money markets in the world. The US dollar had become a tool for the United States to raise money and generate revenue in exchange for the growth of the world economy. As time passed, the US dollar became a tool for political decisions as if they were no longer a tool to manage monetary policy. Because, about 60% of the money circulating in the world financial system was the US dollar, which made it a common currency of the world. With the transition to floating exchange rate, the value of currencies began to be determined in international markets. The interest rate decision of the Central Bank had a significant impact on the value of the currencies. With many commodities priced in US dollar, the world became a dollar addict. As the price of oil and gold was traded in US dollar, the change in dollar had a direct impact on oil and gold prices. The rise in oil price and oil crises affected the world economy negatively.

The United States national currency (US dollar) is transacted as a common currency of the world. The Federal Reserve (FED) is managing the world money market through monetary policy practices. From time to time many problems in the financial system have caused crises and in 2008 the global financial crisis emerged. As the crisis was due to the current monetary system, new currency and monetary system were needed, and the US dollar was the one that was thought to have damaged the world economy. Many negative situations in the past years led to the idea that the monetary system should be renewed. The first step of the said idea was laid one and half months after the 2008 Global Financial Crisis and the cryptocurrency "Bitcoin" was extracted for the first time as a new currency. At that time, Bitcoin and Cryptocurrencies were not paid much attention. After a few years, the term "Cryptocurrency" was considered to play an important role in the financial system.

Economists and bankers could not have predicted that we would have cryptocurrencies in our lives with the internet era which just began 20 years ago. Even though they may have guessed, they may not have thought that some developments in technology would have a negative impact on life. For example, with the introduction of the cryptocurrencies in the market during the planned period the rise in the value of the currency may lead to an increased inflation. However, limited supplied currencies such as Bitcoin has a very low inflation risk. Because in the following years, the ones who have cryptocurrency at hand cannot sale at a high price or lack of demand may force them to sell at a low price. As a consequence, devaluation in cryptocurrencies may result in significant losses (IMF, 2018: 14–16).

2.5 New Money of the Era: Electronic Money

Electronic money are those virtual currencies stored in the electronic environment which are used in businesses payments other than electronic money makers (Bilir & Cay, 2016: 23). Electronic money is extracted with the national currency units to create legal infrastructure for their use by legal regulation. Transaction is done over the Central banks extracted currency units in circulation. For example, those who want to use electronic money in the United States trade on US dollars. Electronic wallet applications have been developed for electronic money storage. The accumulated money in electronic wallets can be converted into cash or national currency under certain conditions and can also be used in real life. The field of application of electronic or digital money is increasing every year. Since there is no need for banknotes and coins for the existence of electronic money, the costs of printing money by states are decreasing. It has become easier to make shopping using electronic money like credit cards in transactions. The renewal costs may be reduced with the decrease in the use of banknotes and coins (Yuksel, 2015: 193–194).

With the development of software technology, cyber-attacks can be made on electronic money and monetary transactions can be prevented from reaching their goals, or the electronic wallet can be seized by hackers. In order to avoid these negative situations, electronic money are coded with various encryption methods that they are protected against attacks and started to be used as virtual and digital currency.

2.6 Virtual Currency and Usage: Digital Currency

Virtual currencies are money that is not valid and nothing is issued in return (Sauner, 2016: 118). The European Central Bank defines the virtual currency as digital money which is extracted and regulated by its developers without any regulation (ECB, 2012: 13). Virtual currencies can be used in the purchase of virtual goods and services extracted by virtual communities. Virtual currencies can be extracted in exchange for real money. However, this application has not been observed. Virtual currencies are not in circulation physically (Sahin, 2018: 77). The e-Gold application can be considered as an extracted virtual currency for gold, silver and platinum. However, the application is actually treating the value of precious metals as a form of currency. Therefore, while purchasing a precious metal in the application in return there is no physical delivery just e-gold is given in a virtual environment (ECB, 2012: 13).

There are structural differences between electronic money and virtual coins as digital money applications (Yuksel, 2015: 197). Unlike electronic currencies, the supply of virtual currency is not constant and can be produced by the issuers in the virtual environments to the desired amount (Yuksel, 2015: 197). They are issued or extracted by virtual groups in virtual environment. An important feature is the lack of official records of the issuers. Therefore, virtual money is not issued by financial institutions or organizations. Bitcoin, Ethereum, and XRP are examples of the virtual coins with the highest value.

Since the supply of virtual currency is not as stable as electronic money or currency, virtual money supply is determined according to the commands of the issuers. Electronic money has the same value as money in circulation. In virtual currencies, the value is determined by virtual communities. Since electronic currency has the same value as national currency of the governments, the value of the electronic money is affected in the same direction as national currency. The value of virtual currency is influenced by the supply and demand in the market and with speculative transactions. According to Ceylan, Tuzun, Ekinci, and Kahyaoglu (2018), numerous bubbles originating from speculation were detected in Bitcoin and Ethereum crypto coins. With the rapid development of technology, virtual currencies can easily be affected by speculative transactions (Dizkirici & Gokgoz, 2018: 93). Moreover, developments in internet technology facilitate the transfer of speculated virtual money.

In terms of real economy, virtual currencies are generally evaluated in three categories (Gultekin & Bulut, 2016: 84):

Closed Virtual Currency Schemes Virtual money has no connection with the real economy in this scheme. The money obtained is used for receiving virtual goods and services.

Virtual Currency Schemes with Unidirectional Flow The virtual currency can be purchased by fixed or floating exchange rate but cannot be converted into real money, and are used in purchasing virtual goods and services.

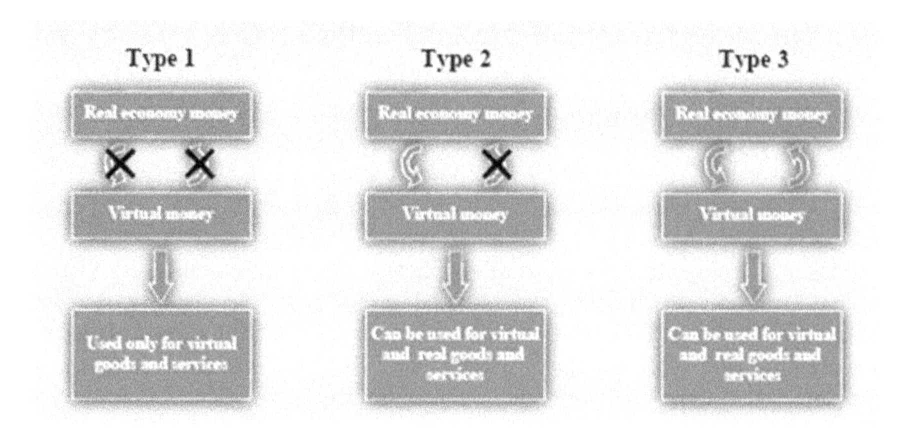

Fig. 2.1 Types of virtual currency scheme. Source: ECB (2012: 13)

Virtual Currency Schemes with Bidirectional Flow The virtual currency can be obtained by fixed or floating exchange rates and can be used as a mean of exchange in the real economy, there is not much difference from real money. Therefore, even with the low probability, virtual currency can be converted into real money.

Virtual currencies are categorized into three groups given in Fig. 2.1 (ECB, 2012: 13):

First one is the closed system virtual currency (ECB, 2012: 13). Coins in this group can only be used to play certain games on the internet. Players are members of a site to play the game and pay the registration fee in cash. Money is earned in exchange for playing. Player can use the earned virtual currency to raise the level and equipment level in the game. The virtual currency gained from the game can be used in purchasing the services offered by the virtual community (ECB, 2012: 13).

Those in the second group are virtual currencies with unidirectional flow. Virtual currencies in this group can be purchased at a particular rate in exchange with real currencies. Since its inverse is not possible, virtual currencies cannot be converted to real money. The conversion of real money into virtual currency is made according to the predefined conditions. With the virtual money earned from the system virtual goods or services can be purchased, and can even be used to purchase real goods or services with some of virtual currencies. For example, in 2009 "Facebook Credits (FB)" was issued to purchase virtual goods on Facebook platform (ECB, 2012: 14). Virtual currency is being purchased through the Facebook platform using credit cards, PayPal and other payment tools. Some people or organizations on the platform allow FB to be used purchasing real goods or services. For example, customers can shop at Nintendo stores with the "Nintendo Points" virtual currency or use for the games. It is not possible for customers to convert their remaining virtual currency back to the real money (ECB, 2012: 14).

In the third and last group, there are virtual currencies with bidirectional flow (ECB, 2012: 14). In a virtual currency system with bidirectional flow, users can purchase virtual currency at a settled exchange rate and sell them at any time. Virtual currency convertibility is considered in the same category as the real money. If you buy foreign currency in exchange for the national currency, it can be converted back to national currency when requested; you can also convert the virtual currency to real money at any time after purchase. In this case, virtual currencies can be used for real goods and services. For example, Linden Dollars (L $) is a virtual currency used in the game called second life, a virtual world in which users create characters called avatars (ECB, 2012: 14). Avatar is known as the digital representatives of the users in the game. Linden Dollars is used to purchase the goods and services needed in the second life game. In this case, Linden Dollars is purchased with real money using payment methods such as credit card and PayPal. Linden Dollars is transacted in US dollars. The value of Linden Dollars is determined automatically in the market. It is not possible to trade with different currencies. Therefore, it is necessary to have American dollar to buy Linden Dollars. Players can purchase Linden Dollars whenever they want, and can convert it back to the US dollar over the market value at any time.

2.7 Encrypted Monetary System: Crypto Currency

With the development of computer and communication technology, the volume of electronic commerce has greatly increased. The possibility of payment instruments, such as credit cards, PayPal and electronic money being exposed to cyber-attacks has created the need for virtual money to be used in transactions in a safe environment while trading online. Digital currency is considered to be a reliable currency that can be used for online shopping.

The concept of digital currency has first been suggested by David Chaum in 1982. Although David Chaum preferred the digital currency with a centralized structure, in the following years, it has turned into a system where the cryptographic transactions are done dispersedly on the network (Chaum, 1983). Brassard (1988) describes the cryptology as *the science and art allowing the data on unreliable channels to be transmitted in a fast and reliable way*. The cryptology sends the data, such as letters and digits, to the receiver within a reliable environment by encrypting them with algorithms, and deciphers the codes again to reveal the data (Gulec, Cevik, & Bahadir, 2018: 19). In this respect, the crypto currency is considered to be the most advanced version of digital currencies.

The money, in real terms, is visible and tangible. However, the digital currency is described as the money that is created within the virtual environment, can be bought and sold in the virtual environment and has certain cryptographic specifications. This is because the currency that we use in real life are processed by banks, the identity information is provided when asked during money transfers and a certain amount of service commission can be charged for transfers (Khalilov, Gundebahar, & Kurtulmuslar, 2017: 1–2). Gandal and Halaburda (2014) state that the crypto currencies have cryptographic features, in other words, that they are produced with an encryption process. Therefore, the most important feature of crypto currencies is the special encryption technique, after which the currency was named (Gandal & Halaburda, 2014: 4). It is suggested that the crypto currencies that remain their mystery, such as Bitcoin, were treated in the market as a financial instrument by performing better than other financial instruments, or they could even be used to lower the market risk during hedging transactions, just like gold (Dyhrberg, 2016: 139).

The digital currency is expected to reduce the usage of real money in daily life, allow the transfers to be made on the electronic environment without any physical exchange and contribute to preventing the use of forged money. The money transfer can be performed via mobile phones or smart tablets by logging into the account on the Internet at the moment when one desires to make payment. In order to do that, however, every business should support the digital currency. In this case, there may occur the problem of pricing the digital currency. Since the power of purchasing the national currency would increase if the money keeps raising in value, the individuals would not want to make payments with the digital currency. In this case, it may cause the digital currencies to be used as investment instruments.

The use of crypto currencies as official currencies would remove the forged money from the market. This is because it is not quite possible to end the use of forged money for the governments even though they take various measures and impose sanctions for fighting against forged money. In terms of crypto currencies, on the other hand, the said negative situation is prevented (Karaoglan, Arar, & Bilgin, 2018: 16). One of the most important features of crypto currency is to prevent the money from losing its value due to devaluation (Gulec et al., 2018: 19). With the national currencies constantly losing their values, the prestige and value of the money and the possibility of them being preferred decrease. The interest rates increase due to deficient and incorrect decisions of Central Banks on interests, and the national currencies lose their values as the foreign currencies increase in value. As the prices constantly increase, the purchasing power decreases. Crises can be experienced after a while. In this case, the confidence of people in the money decreases. Therefore, another reason for the emergence of crypto currency is the economic and financial crises going on in many countries around the world. It has also been considered as a critical problem that the Central Banks had a monopoly on the money (Srokosz and KopyĞciaĚski, 2015: 622). The failure of achieving the goals desired with the implementation of monetary and financial policies decreases the confidence in the Ministries of Treasury or Finance and the Central Bank. It is believed that the digital currencies were issued in order to avoid these negative situations caused by real money. In this way, it is considered that it is possible to prevent the devaluation, revaluation and economic crises by not depending on central authority and taking under control the decrease of money in value (Gulec et al., 2018: 19).

Crypto currencies do not possess the three main features based on the modern money theory (Icellioglu & Ozturk, 2018: 55–57). The first of those is whether the crypto currencies have the feature of exchange. Traditional currencies can easily be exchanged in the market. In order to purchase crypto currencies, national or foreign currencies are given in return. However, since there are no psychical exchanges like banknotes, crypto currencies are not exchangeable. The second feature is that the crypto currencies cannot be used as reserve currency. Therefore, while the central banks are storing foreign currencies as reserve currency, they do not see Crypto currencies as reserve currencies as they do not officially recognize them. Another feature of the currency is that it has a value measurement. Since the management of monetary policies and instruments is at the disposal of central banks, the value of modern currencies is closely related to the decisions taken applicably and correctly on the management of money. However, the value of crypto currencies is determined based on the supply and demand in the market. Supplying a small amount of crypto currency to the market may cause its market value to increase. For this reason, there can be seen speculative transactions on the prices of crypto currencies.

Crypto currencies are not used as reserve money by central banks. However, some central banks have stated that they could use crypto currencies as reserve and allow them to be exchanged with various financial instruments. The Central Bank of China has forbidden the use of Crypto currencies, Germany has accepted the use of Crypto currency for banking transactions, JP Morgen has stated that they would fire their employees in case that they buy crypto currencies, and Japan stated that the digital currencies, namely crypto currencies, can be used in shopping (Icellioglu & Ozturk, 2018: 57–58). For example, the European Central Bank has classified the crypto currencies into two categories; the ones that are not subjected to the regulations as virtual currencies, and the ones that are subjected to the regulations as electronic currency, which the Bank stated that they can be used as deposits in commercial banks (ECB, 2012: 11). Although the crypto currencies have been known as digital currencies, it creates uneasiness that they can be used as money laundering means for the money earned in illegal ways as they are not recognized as official currencies. Today, the states doing transactions in their official currencies can increase the value of the national currency, and printing money provides a seniorage income. However, crypto currencies restrict the domination of states on the money and may affect the fight against inflation negatively.

Crypto currencies are kept in digital wallets after being bought on the Internet. In this way, the customers whose identities are kept confidential can purchase the crypto coins, keep them in digital wallets and convert them into money whenever they desire. As the digital wallets work by encryption techniques in purchase and sale transactions, no information is given to third parties about the person conducting the transaction (Gulec et al., 2018: 26). Crypto currencies can be considered as a transparent currency as they have a feature of keeping accounts open to the public (Gultekin, 2017: 97). In addition, the transaction data is shared with all users after the parties have confirmed the money transfer. In this way, the data loss is avoided and altering the transaction information afterwards is prevented as the related data about the confirmation of the money transfer is shared with all users using the block chain (Ates, 2016: 356).

2.8 The Emergence of Crypto Currency: Bitcoin

The structure of crypto currency consists of several algorithms and encryption techniques. For this reason, it requires the use of advanced technological knowledge in mathematics and information systems (Gultekin & Bulut, 2016: 82). It is stated that the theoretical infrastructure of crypto currency was created by Wei Dai in 2008 (Gultekin & Bulut, 2016: 82). It was Bitcoin who has entered the market first as a crypto currency (The Economist, 2011). Developed on 1 November 2008 and generated on 3 January 2009, Bitcoin is thought to have been created by an individual nick-named Satoshi Nakamoto (Gulec et al., 2018: 22). Satoshi Nakamoto, whose real identity remains unknown but who is fluent in English language, introduced himself as a 37-years-old male living in Japan but not revealed his real identity; however, he made his name by issuing Bitcoin with the "block chain" system that crypto currencies are grounded on and with the cryptographic design that hides the personal infrastructure (Khalilov et al., 2017: 2). The crypto currency is described as a virtual currency or financial asset that is issued without being depended on the central bank or any official authority (Ozturk et al., 2018: 218). It is possible to conduct transactions with Bitcoin by establishing a direct connection between the buyer and seller at no cost without any transaction fee (Carpenter, 2016: 3). The crypto currency has been developed as an alternative payment instrument within the current monetary system (Nakamoto, 2008). Bitcoin is the currency that has the highest market value and transaction volume among more than a thousand crypto currencies (Ozturk et al., 2018: 219). According to Gandal & Halaburda (2014), Bitcoin is neither a currency nor an asset. It is stated that Purchasers prefer Bitcoin because of its value-saving or wealth-increasing functions (Icellioglu & Ozturk, 2018: 56). Over time, the expectation that Bitcoin would increase in value gives the impression as if it was an instrument of investment. Can crypto currencies be considered as financial investment instruments? The answer shall reveal itself in the coming years.

Bitcoin is known as the most widespread and valuable crypto currency among all Crypto currencies in the market. The currency of Bitcoin is "bitcoin" and its abbreviation is "BTC". The smallest unit of Bitcoin in value is "0,00000001 BCT", which is one in a million, and it was named after its inventor "satoshi" (Khalilov et al., 2017: 2; Bitcoin Wiki, 2018).

Bitcoin can be used for online shopping and can serve as a kind of e-wallet. Bitcoin transfer is possible between users having Bitcoin network. While transferring money between addresses, there is no need for revealing the identities of individuals. The flow of money is visible on the network but there is no information regarding between whom the money is transferred. For this reason, the money transfer remains confidential; and since no central structure has control and approval over it, the money transfer is realized under the initiative of individuals (Khalilov et al., 2017: 3). Therefore, the value of the currency is automatically determined based on the supply and demand of money in the free-market as Crypto currencies are not related to any country. It is stated that there can be speculative price trends due to the effects of speculative information. Even thought it was stated that Bitcoin prices were speculative in a research conducted in 2012 in the USA, the buyers did not pay any attention to this idea (Perugini & Maioli, 2014: 11). While Bitcoin price was 997$ at the beginning of 2017, it reached approximately to 20,000$ within the year and was traded at 14,165$ at the end of the year (BitcoinFiyati, 2018). As investors had predicted the price of Bitcoin would go upwards, they sold Bitcoins at its peak, which they had bought at a lower price, and the price began to go down. Therefore, the price increasing more than 20 times and then decreasing by half in a short period within the same year leads the suspicions that there might have been speculative transactions.

2.9 Block Chain: Bitcoin Example

Bitcoin consists of "Block Chain" units as all crypto currencies do. Every transaction made with Bitcoin is open to the public and recorded on a regular and timely basis, and added to the end of the chain as a block. This system prevents the previous transactions to be repeated and is used to protect them against alterations. Each node on bitcoin network is, independently of each other, stored in a block chain that contains blocks approved by that node. Containing more than one nodes in the same block chain is unanimously accepted. The rules for confirmation followed by those nodes in order to reach a consensus are named as the rules of consensus. The rules of consensus used by Bitcoin are described in advanced. Bitcoin's block chain structure is given in Fig. 2.2.

The confirmation of Bitcoin transactions are made with "mining" process. For Bitcoin transactions to be made, Bitcoin users called "miner" are informed. Users gather the transactions approved to solve Bitcoin problems in the Bitcoin chain and compete with each other to solve the algorithmic problem. The first miner solving

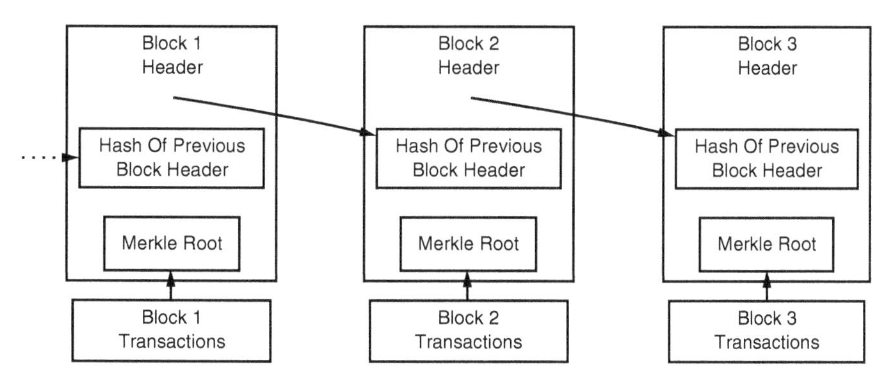

Fig. 2.2 Simplified Bitcoin block chain. Source: Bitcoin (2018)

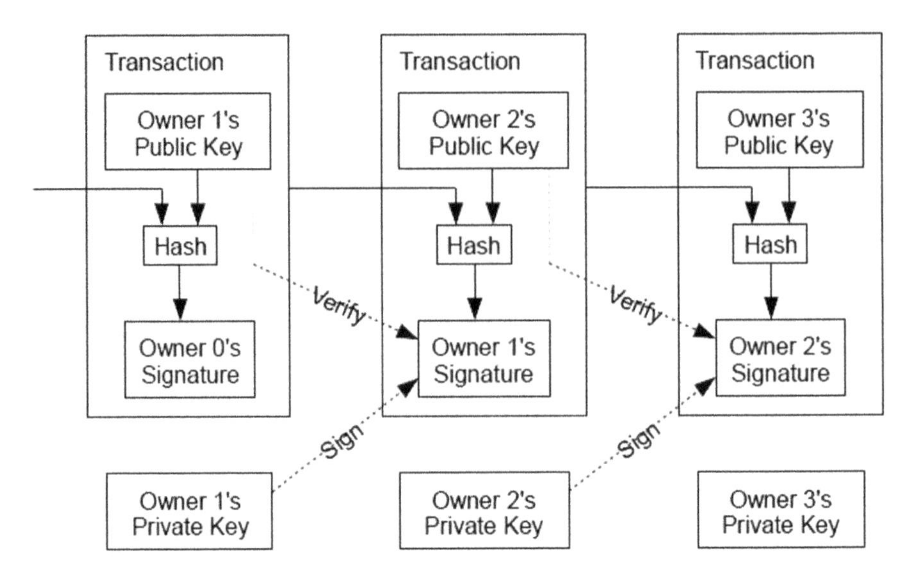

Fig. 2.3 Transaction process of Bitcoin. Source: Nakamoto (2008: 2)

the problem adds their block at the end of the chain. When the miner creates 210,000 blocks, they are entitled to the rewards; and the reward for 210,000 blocks has been 12.5 BTCs since 2016 (Bitcoin, 2018). It is expected that one block is created every 10-minutes on average (Bitcoin, 2018). The targeted number of Bitcoins is 21 million and it is estimated that the targeted number would be reached in 2140 (Khalilov et al., 2017: 3). Since 22 December 2018, there are 16,522,800 BTCs and they worth 63,381,460,800 USD (Bitcoincharts, 2018).

Figure 2.3 describes the working principle of crypto currencies. There is a digital signing process in the working principle of crypto currencies. The electronic money is transferred with signing process. The security of the operation of system is at its highest. After the user has signed for the completion of hash transaction, next users are expected to sign. Every user in the system signs it one by one and the money is added to the end of the chain as block at the end of the digital signing process.

The working principle of all crypto currencies in the market is based on the block chain principle. There are approximately 1977 Crypto currencies, which increase in number every passing day. As of 22 December 2018, the market value of crypto currencies is approximately 128 billion USD (CCMC, 2016).

2.10 Properties of Crypto Currencies

Crypto currencies have much more different features than real money. In literature, these are the features, in brief, that differentiate crypto currencies from real money:

- They do not have a physical circulation unlike real money. Crypto currencies are immovable. They do not have a circulation.
- They can be bought and sold through the Internet or ATMs.
- There is no taxation for Crypto currencies. Since the crypto currencies are not classified as revenue-generating money or capital market instruments, it is not possible for them to be excised (Hepkorucu & Genc, 2017: 49). For this reason, they are not considered as an item of income that can be subjected to declaration in terms of tax regulations (Tufek, 2017: 78–79).
- As all transactions involving crypto currencies are made with encryption technique, they are not visible to third parties.
- As they are not issued by official government institutions, they cannot be tracked by official institutions.
- Money transfers are not shared with third parties unless the concerning parties desire. The parties conducting the money transfers remain unknown.
- The transfer costs of crypto currencies are relatively lower compared to real currencies (Tufek, 2017: 78–79).
- As crypto currencies are generated with an encryption method, the level of reliability is high. When crypto currencies are bought, all users receive encrypted messages. Encryption is different in each process. Therefore, each user sees the transaction but not between whom and how it is done. It is impossible for crypto coins to be taken by others or they cannot be spent even if they are as Bitcoin belongs to the individual (Yuksel, 2015: 202).
- Crypto currencies can be bought, used and sold anytime. There are no time restrictions. For example, Bitcoin buyers can sell them in the market whenever they desire. There are no restrictions on money transfer. The transfer can easily be performed even between the most distant places of the world. As the transfer of crypto currencies are made in electronic environment, it is easy to perform transactions everywhere where there is Internet access.
- In real money transfers, the banks can see the amount and parties of the transfer. In crypto currency transfers, the parties usually hide their identities. Even though the coin transfer is known, between which it is made remains unknown. Some people may not need to hide their identities. Therefore, the parties of the transfer are visible to everyone.
- It is not possible to transfer the money to a wrong address with crypto currencies. It is because the software would recognize the wrong addresses with the encryption technique in the network and not perform the transfer (Yuksel, 2015: 202).
- As crypto currencies are stored in virtual wallets, it does not take time to store them. They can be bought 24/7, whenever desired.
- The transactions in the block chain are open to the public. They are transparent and the transaction history does not disappear since the logs cannot be deleted (Sercemeli, 2018: 43).
- Even though the crypto currencies have no physical circulation, there is a risk of them to be stolen in the virtual environment. Although crypto currencies are coded with high algorithm and encryption techniques to increases their reliability, there is a possibility that the codes of the program are deciphered in the virtual

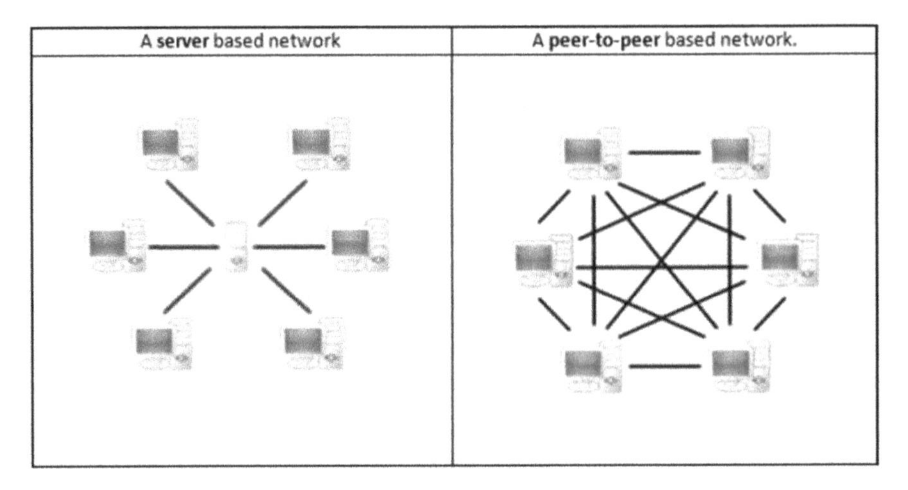

Fig. 2.4 A server and A peer-to-peer based network. Source: Gigatribe (2018)

environment. In this case, there are no extra safety measures that virtual currency users can take to secure their crypto coins.

- The transactions made with crypto coins are recorded and it is not possible to reverse the transaction made (Hepkorucu & Genc, 2017: 49). In case of incorrect transactions, the owners of crypto coins may suffer.

2.11 Working System of Crypto Currencies: Example of Bitcoin

There is the network technology in the working system of crypto currencies. Computers and networks are not connected to each other from a single center but all computers are directly connected to each other as if they are servers. In other words, while traditional network systems command the computers from a single center, the network established for crypto currencies connect all computers in the system to each other. In this way, the data is not stored at a single center but at all users. The data is shared with all users in the block chain with encryption technology (Fig. 2.4).

One of the most important features of crypto currencies is their network structures. With the networks having different algorithms, the money transfers are encrypted. The network structure of Bitcoin is called Peer-to-Peer (P2P). There are often clients and servers in their network structures. The most important feature of the network is that the whole network can easily learn all kinds of information. With peer-to-peer encryption technique, every user transfers the data to all users on the network by using their own computers as a server without using any actual server (Dulupcu, Yiyit, & Genc, 2017: 2244). In this way, the transaction speed is increased

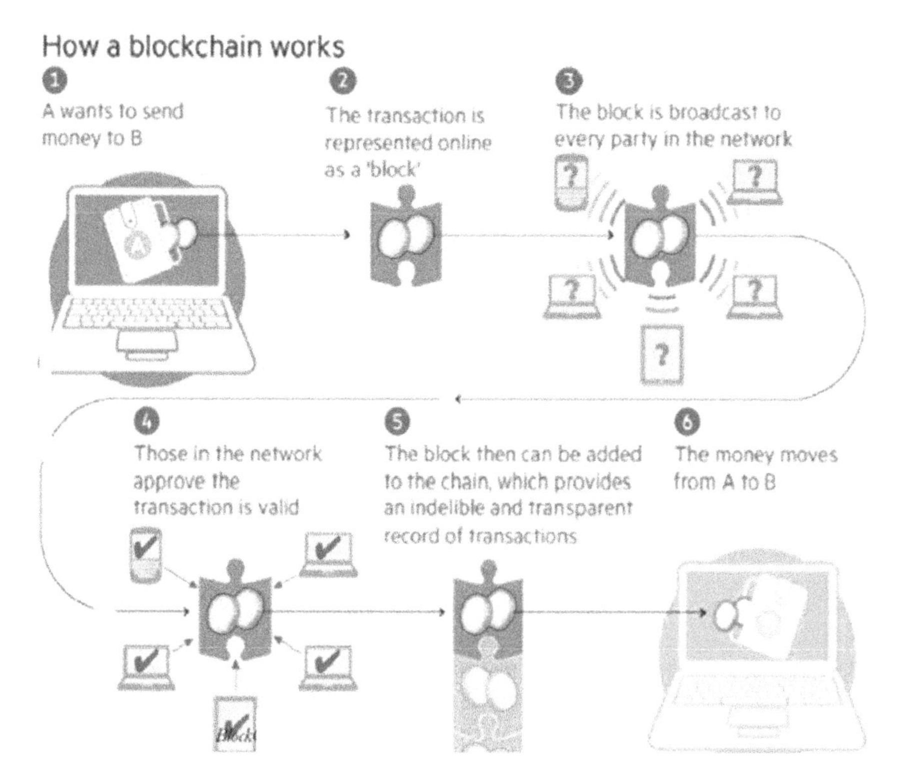

Fig. 2.5 Working system of block chain. Source: Crosbyi et al. (2016: 10)

without any obstructs in the network. Otherwise, the transactions could be delayed as there would be traffic in the network due to transactions on a server (Dulupcu et al., 2017: 2244). In this way, the possibility of tracking all computers on the network at the same time is avoided. When the coin transfer is completed, the ones on the network do not know who send how much money to where. As the identities of the ones making the transaction remain unknown and the transactions are made under encryption, the individuals can comfortably transfer coins.

The working system of Block Chain system is given in Fig. 2.5. The working system of Block Chain system given in Fig. 2.5 can be explained in brief as follows: For example, let's assume that Person A wants to send coin to Person B The transaction for money transfer represents a block in the system and registered to the system; in this way, a block is created. The block is spread to everywhere on the network via the Internet. Users on the network confirm this valid transaction. After its confirmation, the transaction is added to a block chain that is visible to everyone and cannot be deleted by anyone. In this way, the coin transfer from Person A to Person B is performed.

2.12 Crypto Currency as a Financial Instrument: Why Do Investors Buy Crypto Currency?

Investors who do not want to use their savings in the markets of money, capital or precious metals prefer crypto currency as a different means of investment. Investors buy crypto currencies with the expectation that their value would increase. As crypto currencies are generated by encryption and released to the market at certain volumes per day, the amount of their supplies is quite low. Therefore, the increase in the demand for these currencies that are in limited numbers in the market may cause the prices to increase. In this case, it is expected the prices to go down after a while when there is a supply and demand balance in the market. This is because the increase in the number of crypto currencies and the decrease in the demand for the coin may cause the coin to lose its value and the price to fall down. In this case, the fact that there are no third parties to buy the crypto coin prevents the money to be sold. The crypto coins that are bought for the purpose of investment may cause their investors to experience losses. This is owed to the fact that the crypto currency reaches to very high prices in a short notice and the base and maximum prices are not applied may cause the currency to be traded at low or high prices based on the supply and demand in the market.

Plassaras (2013) addresses that the investors want to benefit from the value-earning characteristic of the currency by purchasing the crypto coins that are in small numbers.

In the Republic of Turkey, the use of crypto currencies is allowed. However, in the statement made by the Capital Markets Board of Turkey, it is stated that the crypto currencies are outside of the Board's supervision. In addition, it is also stated that the digital assets are highly risky and speculative investments as they are issued with indistinct promises (SPK, 2018: 4).

Brière, Oosterlinck, and Szafarz (2015) addressed that Bitcoin was a massive virtual currency, that the portfolio created the risk and returns balance, and that it can be used for the diversification of portfolio. Alexander, Gasser, and Weinmayer (2015), with an opposing view, stated that Bitcoin reduces the portfolio risk and ensures the increase in returns. Therefore, although Bitcoin is a crypto currency, it is involved in portfolios as a financial investment instrument. Dyhrberg (2016) stated that Bitcoin can be used as a hedging instrument like gold against Financial Times Stock Exchange Index and U.S Dollar. Ozturk et al. (2018), in their research where they studied the possibility of using Bitcoin for diversification of portfolio and as a financial asset, addressed that there was a causality relationship from Bitcoin to gold and that it could be used like a gold, which is a safe haven, during periods where risks increase. In many studies, even though Bitcoin is not considered as a financial asset, it is stated that it can be used for diversification of portfolio. Yermack (2013) addressed that Bitcoin is not able to carry out the duties of money in classical terms, and stated that daily changes in its price show no correlation with any currencies or investment instruments and this is impractical in risk management. Baek and Elbeck (2014) compared the volatility of crypto currencies and other investment instruments, stated that this can simply be explained with only internal factors (buyers and sellers) and concluded that the market is highly speculative.

Maurer, Nelms and Swartz (2013) stated that Bitcoin, which was introduced against standard currencies and payment systems that are considered as a threat in terms of user privacy and restriction of personal liberty, has solved these problems with cryptographic protocols instead of regulatory bodies or interpersonal trust. In their research where they studied whether Crypto currencies are financial instruments, Kocoglu, Cevik, and Tanrioven (2016) have observed the relationship between Crypto currencies with high market values and exchange rates and determined that Bitcoin has a reverse directed relationship with exchange rates and financial investment instruments. It was stated that Crypto currencies, such as Bitcoin, could be used in the diversification of portfolio. Icellioglu and Ozturk (2018) has observed the relationship between crypto currencies and exchange rates, and added that the crypto currencies act independently from exchange rates. In an empirical study on Bitcoin, it was determined that Bitcoin was both a standard financial instrument and a unique investment instrument that has speculative characteristics. (Kristoufek, 2015: 1). Considering the studies in the literature, it is seen that many investors buy crypto currencies with the expectation that their prices would increase. As it is not possible to use Crypto currencies in daily life and to establish the required infrastructure for this in the following years, it is considered impossible for Crypto currencies to replace real money and it is thought that they are only bought and sold for investment purposes.

2.13 Conclusion

Allowed to be traded in the market, digital currencies not being subjected to any supervision causes a great uneasiness for investors. Investors invest by purchasing digital currencies that have no regulations. Considered as a new investment instrument among financial investment instruments, digital currencies are actually not of capital and monetary market instruments, and this causes uncertainties regarding what kind of problems would occur in the case that the said digital currencies dominate the market. As a virtual market of which number of investors and the volume of transactions increase every passing day, the crypto currencies not being subjected to regulations leaves the question of how the investors could manage a financial crisis that may be cause them suffer from damages and lose their savings unanswered. This is because the digital currencies are bought and sold in virtual markets; and since they are not subjected to any supervision, there will not be any legal actions when the investors are victimized. The reason why central banks insist on not recognizing digital currencies to avoid such situations is that they do not comply with the description of money in legal terms. However, considering that the digital currencies are not used as a medium of exchange but an instrument of investment, these currencies should be supervised.

In the research conducted by Wall Street Journal news agency based in the USA, it is addressed that crypto currencies can be used in unofficial transactions in order to launder money. It is stated that the crypto currencies are bought as a money laundering instrument, that it is possible to take the money to different countries and that the money is especially transferred to Switzerland. The lack of supervision by governments on crypto currencies makes it possible to transfer the illicit money. For this reason, many countries have banned the use of crypto currencies and some countries, including the Republic of Turkey, have not yet issued a regulation regarding crypto currencies.

Considering the electronic money with all the functions of real money, it is seen that electronic currencies have the characteristics of an exchange instrument of real money. The studies reviewed state that electronic currencies can be used for exchange. However, this does not mean that these virtual currencies can be used everywhere. This is because it is not possible for virtual currencies to be used as an instrument of exchange at places where they are not accepted. The determination of the value of virtual currencies against real money might take some time for customers while shopping. Especially when it is considered that the value of money changes instantly, it is seen that the customers and sellers race against time while deciding to trade. In this case, it is possible to see serious differences between the values one minute earlier and later. When faced with such situations, the customers may show a negative reaction. This is because the price of virtual currencies not being resistant to speculative transactions increases the volatility of the price of the money.

During purchases made with real money, the amount to be paid for goods or services is given in an instant. In terms of virtual currencies, however, it requires the software technology that is suitable for the Internet and electronic wallets to be used in transactions. For purchases, payment with virtual currencies may take some time. On the other hand, since the payment is generated by encryption method in electronic environment, no forgery is possible for both buyer and seller. In this way, the buyers and sellers are able to perform a secure trade. The real money keeps its existence as long as it's not physically damaged or the state does not disappear. The value of money may decrease in the future. However, it is not possible for it to be destroyed in physical terms. The future of virtual currencies is completely uncertain. Nobody can guarantee that even the most valuable virtual currency of the world will not disappear completely in the coming years. The fact that the legally unreliable virtual currencies are not accepted by governments reduces the possibilities of their use. In this case, they turn into investment instruments where the savings are used by considering that they would increase in value like foreign currencies. Will the crypto currency, the new financial instrument of our era, cause new financial crises in the future? We will see the answer to this question in the following years.

References

Alexander, E., Gasser, S., & Weinmayer, K. (2015). *Caveat emptor: Does Bitcoin improve portfolio diversification?*. Accessed August 05, 2018, from https://ssrn.com/abstract=2408997

Alpago, H. (2018). From Bitcoin to Selfcoin to crypto currency. *Journal of the International Scientific Researches, 3*(2), 411–428.

Ates, B. A. (2016). Crypto currencies Bitcoin and accounting. *Cankiri Karetekin University Journal of Institute of Social Sciences, 7*(1), 349–366.

Baek, C., & Elbeck, M. (2014). Bitcoins as an investment or speculative vehicle? A first look. *Applied Economics Letters, 22*(1), 30–34.

Bilir, H., & Cay, S. (2016). A relation between electronic money and financial markets. *Nigde University Academic Review of Economics and Administrative Sciences, 9*(2), 21–31.

Bitcoin. (2018). *Bitcoin developer guide*. Accessed August 05, 2018, from https://bitcoin.org/en/developer-guide#block-chain

Bitcoin Wiki. (2018). *Satoshi unit*. Accessed August 05, 2018, from https://en.bitcoin.it/wiki/Satoshi_(unit)

Bitcoincharts. (2018). *Bitcoin charts/Bitcoin network*. Accessed December 22, 2018, from http://bitcoincharts.com/bitcoin/

BitcoinFiyati. (2018). *Bitcoin 2017 yili fiyati*. Accessed August 20, 2018, from http://www.bitcoinfiyati.com/2017-yilinda-btc-fiyati-ne-kadardi.html

Bradbury, D. (2013). The problem with Bitcoin. *Computer Fraud & Security, 2013*(11), 5–8.

Brassard, G. (1988). *Modern cryptology: A tutorial*. New York, NY: Springer.

Brière, M., Oosterlinck, K., & Szafarz, A. (2015). Virtual currency, tangible return: Portfolio diversification with Bitcoin. *Journal of Asset Management, 16*, 365–373.

Carpenter, A. (2016). Portfolio diversification with Bitcoin. *Journal of Undergraduate in France, 6*, 1–27.

CCMC. (2016). *Crypto-currency market capitalizations*. Accessed August 15, 2018, from http://coinmarketcap.com/currencies/views/all

Ceylan, F. Tuzun, O., Ekinci, R., & Kahyaoglu, H. (2018). *Financial bubbles in cryptocurrencies markets: Bitcoin and ethereum*. In: 4th SCF International Conference on "Economic and Social Impacts of Globalization" and "Future of Turkey-EU Relations", 26–28 April, Nevsehir, Turkey.

Chaum, D. (1983). Blind signatures for untraceable payments. In *Advances in cryptology* (pp. 199–203). Boston, MA: Springer.

Crosbyi, M., Nachiappan, Pattanayarak, P., Verma, S., & Kalyanaraman, V. (2016). Blockchain technology: Beyond Bitcoin. *Applied Innovation Review, 2*, 7–19. Accessed from http://scet.berkeley.edu/wp-content/uploads/AIR-2016-Blockchain.pdf

Dizkirici, A. S., & Gokgoz, A. (2018). Crypto currencies and accounting of Bitcoin in Turkey. *Journal of Accounting, Finance and Auditing Studies, 4*(2), 92–105.

Dulupcu, M. A., Yiyit, M., & Genc, A. G. (2017). The rising face of the digital economy: The analysis of relationship between the value of Bitcoin and its popularity. *Suleyman Demirel University the Journal of Faculty of Economics and Administrative Sciences, 22*, 2241–2258.

Dyhrberg, A. H. (2016). Hedging capabilities of bitcoin. Is it the virtual gold? *Finance Research Letters, 16*, 139–144.

European Central Bank. (2012). *Virtual currency schemes*. Accessed September 22, 2018, from http://www.ecb.europa.eu/pub/pdf/other/virtualcurrencyschemes201210en.pdf

Gandal, N., & Halaburda, H. (2014). *Competition in the cryptocurrency market*. Bank of Canada working paper, N0: 2014-33. Ottawa: Bank of Canada.

Gigatribe. (2018). *What is the peer-to-peer (p2p)?*. Accessed September 29, 2018, from https://www.gigatribe.com/en/help-p2p-intro.

Gulec, O. F., Cevik, E., & Bahadir, N. (2018). Investigation of association between Bitcoin and financial indicators. *Kırklareli University Journal of Economics and Administrative Sciences, 7*(2), 18–37.

Gultekin, Y. (2017). Cryptocurrencies as an alternative medium of payment in tourism industry: Bitcoin. *Guncel Turizm Arastirmalari Dergisi, 1*(2), 96–113.
Gultekin, Y., & Bulut, Y. (2016). Bitcoin economy: Emerging new sectors of Bitcoin eco-system. *Adnan Menderes University Journal of Institute of Social Sciences, 3*(3), 82–92.
Hepkorucu, A., & Genc, S. (2017). Examination of Bitcoin as a financial asset and an application on unit root structure. *Osmaniye Korkut Ata University Journal of Economic and Administrative Sciences, 1*(2), 47–58.
Icellioglu, C. S., & Ozturk, M. B. E. (2018). In search of the relationship between Bitcoin and selected exchange rates: Johansen test and Granger causality test for the period 2013–2017. *Journal of Finance Letters, 109*, 51–70.
IMF. (2018). Money transformed, the future of currency in a digital world. *Finance and Development.* Accessed December 05, 2018, from https://www.imf.org/external/pubs/ft/fandd/2018/06/pdf/fd0618.pdf
Inshyn, M., Mohilevskyi, L., & Drozod, O. (2018). The issue of cryptocurrency legal regulation in Ukranie and all over the world: A comparative analysis. *Baltic Journal of Economic Studies, 4* (1), 169–174.
Karaoglan, S., Arar, T., & Bilgin, O. N. (2018). Crypto currency awareness in Turkey and motivations of businesses that accept crypto currency. *Isletme Iktisat Calismalari Dergisi, 6* (2), 15–28.
Khalilov, M.C.K., Gundebahar, M., & Kurtulmuslar, I. (2017). Bitcoin ile Dunya ve Turkiye'deki Dijital Para Calismalari Uzerine Bir Inceleme, 19. Akademik Bilisim Konferansi, 8–10 Subat, Aksaray, Turkiye. https://ab.org.tr/ab17/
Kocoglu, S., Cevik, Y. E., & Tanrioven, C. (2016). Efficiency, liquidity and volatility of Bitcoin markets. *Journal of Business Research Turk, 8*(2), 77–97.
Kristoufek, L. L. (2015). What are the main drivers of the Bitcoin price? Evidence from wavalet coherence analysis. *PLoS One, 10*(4), 1–15.
Maurer, B., Nelms, T. C., & Swartz, L. (2013). "When perhaps the real problem is money itself!": The practical materiality of Bitcoin. *Social Semiotics, 23*(2), 261–277.
Nakamoto, S. (2008). *Bitcoin: A peer-to-peer electronic cash system.* Accessed September 01, 2018, from https://bitcoin.org/bitcoin.pdf
Ozturk, M. B., Arslan, H., Kayhan, T., & Uysal, M. (2018). Bitcoin as a new hedge instrument tool: Bitconomy. *Nigde Omer Halisdemir University Academic Review of Economics and Administrative Sciences, 11*(2), 217–232.
Perugini, M. L. and Maioli, C. (2014). *Bitcoin: Between digital currency and financial commodity.* Accessed September 30, 2018, from https://ssrn.com/abstract=2526472 or https://doi.org/10.2139/ssrn.2526472
Plassaras, N. A. (2013). Regulating digital currencies: Bringing Bitcoin within the reach of the IMF. *Chicago Journal of International Law, 14*(1), 376–407.
Sahin, E. E. (2018). Kripto para Bitcoin: ARIMA ve Yapay Sinir Aglari ile Fiyat Tahmini. *Fiscaoeconomia, 2*(2), 74–92.
Sauner, B. (2016). Virtual currencies, the money market, and monetary policy. *International Advances in Economic Research, 22*, 117–130.
Sercemeli, M. (2018). Accounting and taxation of crypto currencies. *Finans-Politik ve Ekonomi Yorumlar Dergisi, 639*, 33–66.
SPK. (2018). *Sermaye Piyasası Kurulu Bulteni.* Accessed September 29, 2018, from http://www.spk.gov.tr/Bulten/Goster?year=2018&no=42
Srokosz, W., & KopyĞciaĚski, T. (2015). Legal and economic analysis of the cryptocurrencies impact on the financial system stability. *Journal of Teaching and Education, 4*(2), 619–627.
The Economist (2011). *The Economist: Virtual currency-bits and bob.* Accessed September 24, 2018, from http://www.economist.com/blogs/babbage/2011/06/virtual-currency
Tufek, B. U. (2017). *Electronic payment tools and approach of the future crypto currency.* Istanbul: Master of Business Administration, University of Bahcesehir.

Turan, Z. (2018). Crypto money, Bitcoin, blockchain, PetroGold, digital money and their usage areas. *Academic Review of Economics and Administrative Sciences, 11*(3), 1–5.
Vora, G. (2015). Crytocurrencies: Are disruptive financial innovations here? *Modern Economy, 6*, 816–832.
Yagci, M. (2018). Yukselen Finansal Teknolojilerin Ekonomi Politigi: Fintek ve Bitcoin Ornekleri. *Iktisat ve Toplum Dergisi, 88*, 17–24.
Yalciner, K. (2012). *Uluslararasi Finansman*. Ankara: Detay Yayincilik.
Yermack, D. (2013). *Is Bitcoin a real currency? An economics appraisal*. NBER Working Paper Series, No. 19747.
Yuksel, A. E. B. (2015). Looking at electonic money, virtual money, Bitcoin and Linden dollars from a legal perspective. *Istanbul Universitesi Hukuk Fakultesi Mecmuasi, 73*(2), 173–220.

Faruk Dayi is an Assistant Professor in Kastamonu University in Kastamonu (Turkey). He received a Ph.D. in Accounting-Finance at Gazi University in Ankara (Turkey). His research interests cover topics in the field of corporate finance and international finance, such as international portfolio management, working capital, financial statement analysis, and investment decisions. He took part in different national and international conferences, in several workshops and international projects.

Chapter 3
Redesigning Current Banknotes with Blockchain Infrastructure: A Model Proposal

Sabri Erdem and Derya Altun

Abstract The aim of the model proposed is to introduce a new hybrid system that can be used with less effort and requiring very few modifications in the existing structures. The model combines the advantages of both digital currency and paper-based banknote system. It uses QR Coded banknotes, digital wallets and blockchain technology to ensure, confirm and enhance the ownership of the banknotes. This model is intended to be used at the macro level by governmental policies since the money system can only be regulated by them.

3.1 Introduction

In parallel with the widespread involvement of technology in every part of daily life, trade has benefited from almost all aspects of technology. Marketplaces have now become electronic marketplaces; a number of services, such as counseling and training, have become remote services; contracts, work orders, and other bureaucratic processes, which used to require tedious paperwork, have become electronic transactions. Finally, many commercial transactions which previously used real and tangible money have turned into transactions that can be done online.

As a result of these developments, a number of innovations have emerged in both payment processes, payment instruments and commodities subject to exchange. Now reliable online payment tools such as Paypal and the evolution of banking on mobile devices have brought new issues to the agenda about monetary systems. Cryptocurrencies are now worldwide currencies, and they can be expressed in and exchanged for the currencies of countries. This situation has raised certain questions about the role and importance of money in the new economy.

Initially the world's major investors, state and reserve bank managers, and the managers of leading financial institutions had a variety of mostly pessimistic views

---

S. Erdem (✉)
Department of Business, Faculty of Business, Dokuz Eylul University, Izmir, Turkey
e-mail: sabri.erdem@deu.edu.tr

D. Altun
Department of Disabled, Izmir Metropolitan Municipality, Izmir, Turkey

© Springer Nature Switzerland AG 2019

U. Hacioglu (ed.), *Blockchain Economics and Financial Market Innovation*, Contributions to Economics, https://doi.org/10.1007/978-3-030-25275-5_3

towards cryptocurrencies. However, over the course of time, people's pessimistic attitudes towards the future of cryptocurrencies has shifted. We can now see that cryptocurrencies are accepted as a means of payment in many countries, and there is an increase in the variety of cryptocurrencies and cryptocurrency exchanges that can be used electronically in different parts of the world.

Fluctuations in cryptocurrency markets occur with both a high frequency and high amplitude. However, as cryptocurrencies become more widespread in daily life, it would not be too irrational to think that both the frequency and the magnitude of these large fluctuations will reach more reasonable levels as cryptocurrencies shift from being an investment tool to medium of payment.

Crypto coins are electronically revealed by coin mining and every coin produced is delivered through blockchain technology. Therefore, the amount of money circulated through technology in the world economy market is supplied by the blockchain that guarantees the robustness and reliability of this circulation such in transactions as shopping, dividing them into smaller amounts as they are used in purchases and payment processes, and merging them to get larger amounts.

Blockchain technology is almost impossible to break, as it uses an encryption technique to ensure the safe, non-destructive and impossible-to-steal features differentiate cryptocurrencies from traditional banknotes. Because of this crypto coins have significant advantages in electronic payment processes comes.

On the other hand, although countries continue to offer banknotes pressed in their national currencies, the amount of real banknote money in circulation is decreasing day by day. As an example, Norway, Sweden and Denmark are preparing say "No" to cash (Cointelegraph, 2016) and banknotes are rarely circulated in countries like India (Masciandaro, Cillo, Borgonovo, Caselli, & Rabitti, 2018).

It may therefore be the case that the banknote system is approaching its end. Although there are some countries, which still insist on using banknotes, there are also countries that want to take advantage of electronic payment systems (Claeys, Demertzis, & Efstathiou, 2018). Some countries do not want to give up their currency because they have strong economies and their currency is more valuable than that of other countries. For instance, although the UK is a EU member, it continued to maintain its currency and did not even enter the Eurozone.

In the course of this irresistible transformation of money, countries that want to continue using banknotes can improve their banknote systems and still make use of the benefits of crypto money and blockchain technology. In this study, we considering such an alternative means of using money. We propose a hybrid model to ensure that banknote money is safe, non-destructive and unbreakable.

In the second part of the study we discuss the history of the banknote system and the digitalization process of the money, information about the crypto and blockchain technology. The third part introduces the model proposed for the hybrid money system as a conceptual model and the fourth part includes conclusions and discussions.

3.2 Theoretical Framework

3.2.1 Historical Background

Money is considered a measurable value that mediates the purchase and sale of an asset in a given maturity and takes the form of sovereign currencies, which are generally associated with physical money, such as banknotes and coins. Money has a changeover tool, a value measure and a storage function. In the economy, like currencies and banknotes, demand deposits and credit cards are all considered money. Exchange instruments such as time deposits and government bonds are considered t money-like tools. Money and money-like instruments are also used to direct economic policies. Money, which is a powerful tool for easily transforming, storing and transporting, has also been the most important basis for the size, dynamism and complexity of trade and markets.

Commodities are the physical assets that take the value of money from a product. Commodities used as money are copper, salt, tea, pearl, ivory, cattle, iron, slaves, and cigarettes. These were used in different regions and at different times for thousands of years. Gold and silver were the most widely accepted commodity used as money of all time (SPK, 2016). The standardized coin is the most known type of money. Seashells, which were used as money before the invention of coin, were used as money in all Africa, South Asia, East Asia and Oceania for 4000 years. At the beginning of the twentieth century, British Uganda still accepted tax payments with seashells (Harari, 2014).

In his study, Harari (2014) states that the raw material of money is trust and explains how this mutual trust system works; "When the rich farmer sold his entire savings to a sack of sea shells and arrived at the new spot away, he believed that the people there would sell him brass, houses and fields for those shells." In the first types of money, the way to achieve trust was to define what was really valuable as "money." The first currency in history, known as the Sumerian barley (3000 BC), is the best example of this. When people began to trust money that was not worth anything, but was easy to store and carry, money had a breakthrough. The silver shekel, which emerged in Mesopotamia (3000 BC), was the first money that used a mutual trust system. Weights made of different metals in different places at different times were used as money. This meant the birth of coins and the first coin in history was printed in the Lydian Kingdom (640 BC). The coins with a standard weight in gold or silver have marked the amount of precious metal they contain and the signs that define the authority that guarantees money (Harari, 2014).

The Chinese were the first to use leather money (118 BC) and paper money (AD 806) (TCMB, 2019a, b) as shown in Fig. 3.1. Trade increased as a result of the development of the commodity economy, which led to higher demand for money. In this case, traders needed an easier-to-carry and maintainable currency. The first time money was used in place of commodity money was in 1023 by 16 merchant princes in Chengdu and Sichuan Province together. The world's oldest paper currency is

Fig. 3.1 First banknote in history. Source: SPK (2016)

Jiao Zi, emerged in the early Song Dynasty (I-China, 2012) in which there was a piece of paper printed with houses, trees, men, and passwords.

It seems that China was the first country that introduced fiat money (i.e., paper currency made legal tender by fiat, not convertible into coin) and the concept of the legal tender. Prior to fiat money, the history of paper money goes back to more than 1000 year ago in China, where the bills of exchange (used as money) were known as 'flying money' (Eswar, 2017).

There are also some historical examples of parallel currencies which include many gold and silver bimetallic monetary systems; one of the earliest recorded examples of a parallel currency system took place when copper and silver were circulated alongside one another in Ptolemaic Egypt in 220 B.C. (Reekmans, 1949).

After the establishment of the British Central Bank in 1694, the number of central banks increased and the use of paper money became more widespread (TCMB, 2019a, b). The reasons for the digitalization of currency has been argued for at least three decades. Tobin (1985) figured out some drawbacks to paper currencies and coins as well as facilitating and easing large payments, discreditable reasons, tax avoidance or crime. Agreeing on the drawbacks to banknotes and coins, he pointed out that they are too bulky for large legitimate transactions, awkward because it comes only in a few denominations, vulnerable to loss or theft and unsuitable for remittance by mail.

Tobin (1985) could be regarded as one of the pioneers of digital currencies, offering new transaction technology for withdrawals and payment to third parties through a computerized payments network instead of using paper currencies and coins. He proposed a computerized network system in which payments could be made at the time of purchases or settlements, or scheduled to be executed at a designated future time.

Bitcoin was first proposed by Satoshi Nakamoto in 2008 as an electronic transactions system and fully operational in January 2009. Bitcoin is a digital currency not issued by any government, bank or organization. Bitcoins rely on cryptographic protocols and a distributed network of users to be minted, stored, and transferred (Nakamoto, 2008).

3.2.2 *Digitalization of Money*

Today, there are various kinds of money in use that can be classified with respect to their issue, form and transaction. Bech and Garratt (2017) present a taxonomy of money that is based on four key properties: issuers (central bank or other); form (electronic or physical); accessibility (universal or limited); and transfer mechanism (centralized or decentralized). The taxonomy defines a Central Bank Crypto Currencies (CBCC) as an electronic form of central bank money that can be exchanged in a decentralized manner known as peer-to-peer, meaning that transactions occur directly between the payer and the payee without the need for a central intermediary. It can be shown in Fig. 3.2.

The emergence of what are frequently referred to as "digital currencies" was noted in recent reports by the Committee on Payments and Market Infrastructures (CPMI, 2015) on innovations and non-banks in retail payments. A subgroup was formed within CPMI, which has identified three key aspects relating to the development of digital currencies. The first is the assets (such as Bitcoins) featured in many digital currency schemes; the second key aspect is the way in which these digital currencies are transferred, typically via a built-in distributed ledger; and the third aspect is the variety of third-party institutions, almost exclusively non-banks, which have been active in developing and operating digital currency and distributed ledger mechanisms. These three aspects characterize the types of digital currencies discussed in CPMI's report (CPMI, 2015).

There are many different terms used to refer to intangible currency such as virtual currency, digital currency, e-money and cryptocurrency (CPMI, 2015). Contrary to

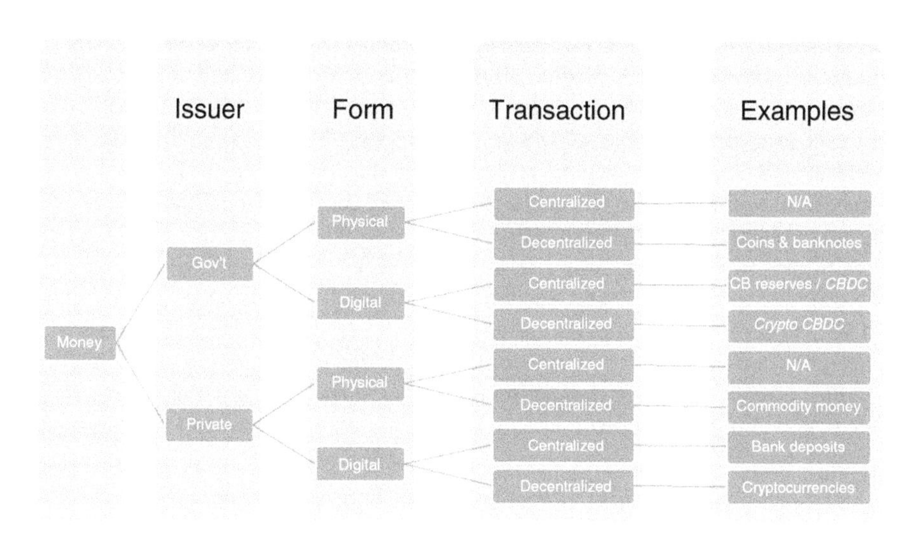

Fig. 3.2 Classification of money in use. Source: Bruegel based partly on the typology proposed by Bech and Garratt (2017). Note: CB stands for Central Bank and CDBC stands for Central Bank Digital Currency

the decentralized attribute of digital currencies like cryptocurrencies, the CPMI Report (2015) can be regarded as one of the pioneer studies that has announced the innovation of hybrid-like use of digital currency and a centralized system called "central bank digital currencies".

From the viewpoint of account holders and banknotes owners, anonymity and privacy are very important. To some extent, digital wallets could provide them privacy and anonymity. On the other hand, the digital wallet providers guarantee digital wallet owners' privacy and anonymity. This means that accounts are as private and secure as the promises of the digital wallet providers. We assume that they are fully private and anonymous (Masciandaro et al., 2018). On the other hand, according to Kahn, Rivadeneyra, and Wong (2018), record-keeping systems have tradeoffs in their level of access, privacy and security. For a given cost, there is a trilemma: no system can simultaneously have universal access, perfect security and complete privacy.

Kahn et al. (2018) has figured out that new technology as distributed ledger technology (DLT) and mobile computing have not significantly changed the tradeoffs for the specific case of providing central bank accounts to the public. Therefore some additional innovations are still needed to handle it.

Central banks have recognized that using distributed ledger technology or blockchain schemes will not be a cost-effective substitute for the core infrastructure of national payments systems (Chapman, Chiu, & Molico, 2013). The reason lies in the tradeoff between the openness of the system and the cost of verification: distributing the updating of the ledger makes it more costly to control counterfeiting (Kahn et al., 2018).

Figure 3.3 is a money flower sketched by Bech and Garratt (2017) to cover all potential types of money whether "tangible or intangible" or "centralized or peer-to-peer". Figure 3.4 shows some real and applied examples for different sub-sections of this flower. For example, Fedcoin, is an example of a retail CBCC. The concept proposed by Koning (2016) which has not been endorsed by the Federal Reserve is for the central bank to create its own cryptocurrency. CADcoin is an example for a wholesale CBCC. It is the original name for digital assets representing central bank money used in the Bank of Canada's proof of concept for a DLT-based wholesale payment system (Bech & Garratt, 2017).

According to Furche and Sojli (2018), a Central Bank can issue digital currencies with different levels of direct transactional involvement. It could act as a wholesaler and issue to commercial banks only, it could participate in the individual transactions in its digital currency either by monitoring or approving transactions, or it could offer a digital currency including transaction processing directly to any user. A financial system that simulates transactions among CBDC, commercial bank and users is shown in Fig. 3.5.

This figure includes the following steps (Furche & Sojli, 2018):

1. Payment sent from User A to user B, as a certificate issued by User A's Bank
2. User B deposits the certificate A with their Bank B

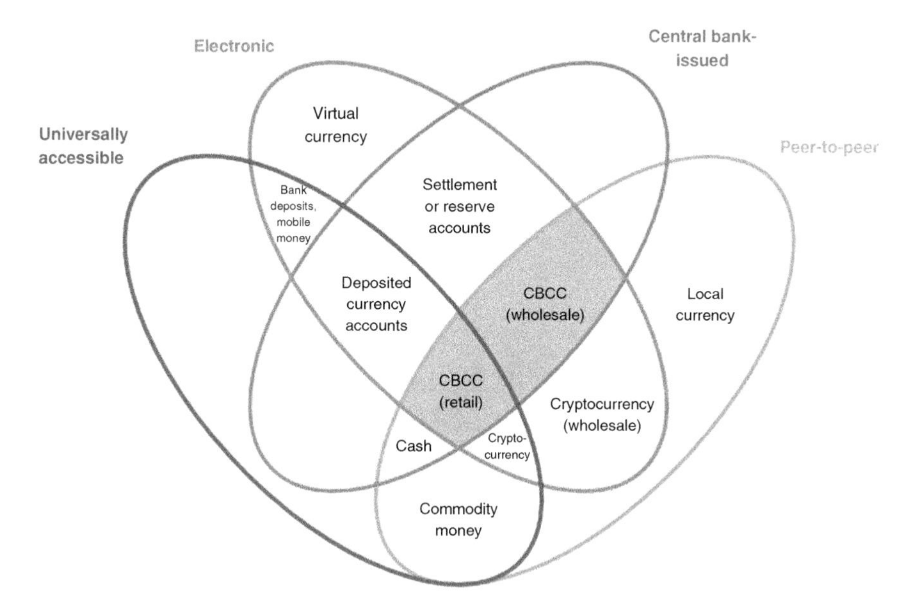

Fig. 3.3 Money flower. Source: Bech and Garratt (2017)

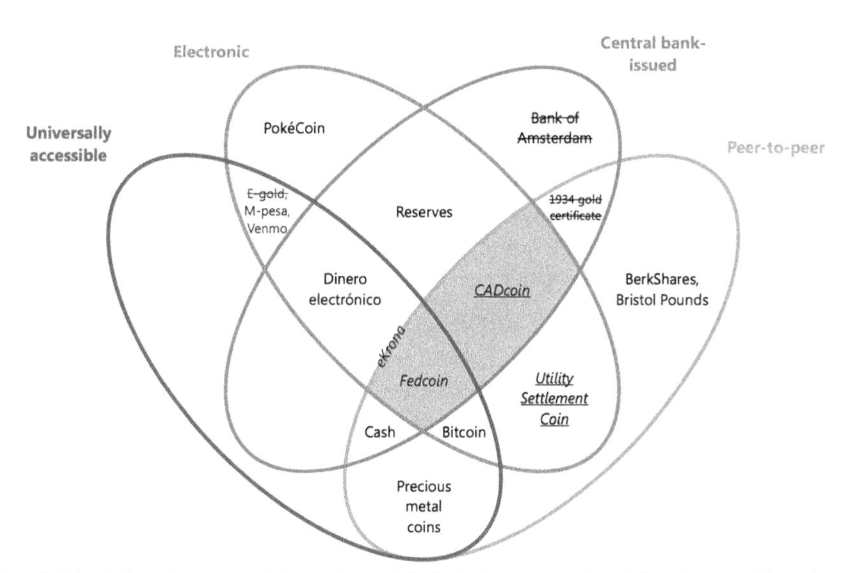

A standard font indicates that a system is in operation; an *italic* font indicates a proposal; an *italic and underlined* font indicates experimentation; a ~~strikethrough~~ font indicates a defunct company or an abandoned project.

Fig. 3.4 Money flower example. Source: Bech and Garratt (2017)

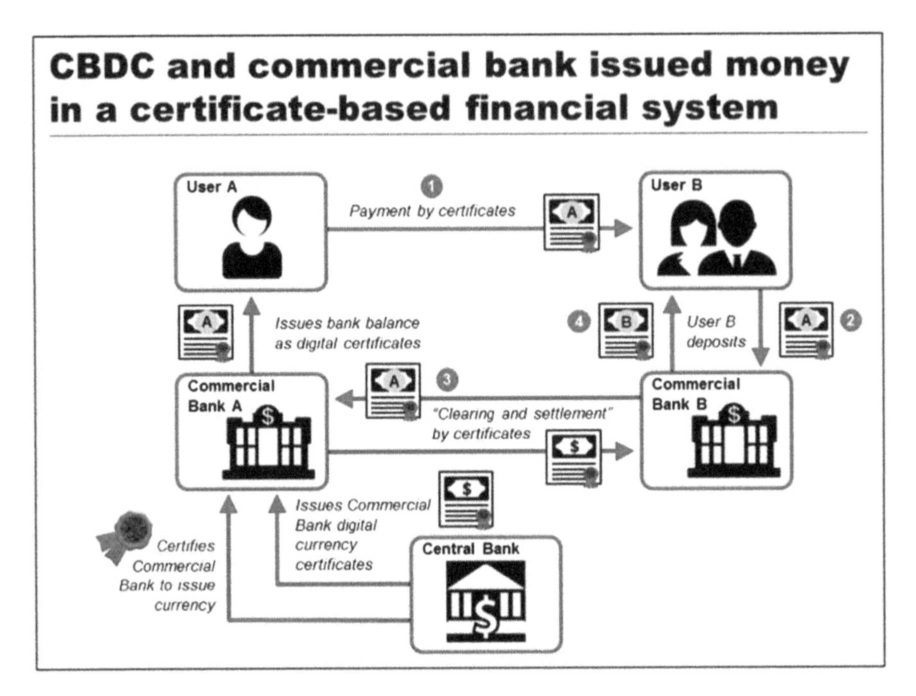

Fig. 3.5 A financial system that simulates transactions among CBDC, commercial bank and users. Source: Furche and Sojli (2018)

3. Bank B 'redeems' the certificate A with Bank A for a CB issued certificate – effecting settlement
4. User B obtains an increase in their account balance with Bank B, issued as digital certificate by Bank B. [This would occur prior to step 3 in a delayed settlement, or after step 3 in a real time settlement]

There are many studies about CBDC issues (Barrdear & Kumhof, 2016; Kahn & Wong, 2019; Niepelt, 2018; Raskin & Yermack, 2017) money on blockchain technology (Koch & Pieters, 2017; Pilkington, 2015; Walker & Luu, 2016), digital currencies and cryptocurrencies (Ali, Barrdear, Clews, & Southgate, 2014; Eken & Baloğlu, 2017; Jafari, Vo-Huu, Jabiyev, Mera, & Mirzazade, 2018; Nagpal, 2017; Peters, Panayi, & Chapelle, 2015; Raskin & Yermack, 2016; Tasca, 2015).

3.2.3 *Demand for Banknotes in Near Future*

The public utility of paper currency is increasingly disputed, given that it has been claimed (Rogoff, 2017) that paper currency has at least two important drawbacks: on the one side, it facilitates the growth of the illegal economy, with corresponding losses from missing tax revenues, without mentioning other social negative spillovers; on the other side it hampers the effectiveness of monetary policy, being the basis of the existence of the zero lower bound on the nominal interest rate (Borgonovo, Cillo, Caselli, & Masciandaro, 2018).

In 2015, per capita holdings of paper currency on GDP have been about 20 percent in Japan, 11 percent in Switzerland and in the Euro area and 8 percent in the US (Jobst & Stix, 2017). Even more puzzling, paper currency circulation has gone up in the recent years and can be observed in several and heterogeneous economies (Berentsen & Schar, 2018; Jobst & Stix, 2017), as well as inside and outside the issuing country if we are looking at a global reserve currency (Feige, 2012; Judson, 2012).

It is still a matter of fact that the electronic peer to peer currencies have been associated with the risks of money laundering (Brayans, 2014), given that the crypto currencies seem to be especially effective for conducting illegal transactions (Hendrickson, Hogan, & Luther, 2015).

According to a recent survey by Barontini and Holden (2019) with 63 central banks worldwide, many central banks in both advanced economies and emerging market economies (EMEs) are attempting to replicate wholesale payment systems using distributed ledger technology (e.g., projects Jasper, Ubin and Khokha (Bank of Canada (2018), Monetary Authority of Singapore (2018), South African Reserve Bank (2018)).

Physical cash is the main mechanism that facilitates the use of currencies, such as the USD and the Euro, as the backup to the global monetary system. As these currencies are used as a store of value and a fail-safe option outside their own country of issue or currency area, issuing CBDC to replace physical cash would jeopardize the role of these currencies in the global payments and monetary systems, a policy concern that should not be overlooked in the decision over issuing CBDC (Nabilou, 2019).

Mas (2010) claims that a new kind of "smart banknote" that can be activated or deactivated electronically (e.g., through RFID) could be created when you would like to transfer value between a banknote and your bank account, right from your mobile handset. Deactivated cash could be transported cheaply; stores could make their accumulated cash balances vanish at will by simply deactivating it; and bank customers would be able to convert their bank account into cash or vice versa anywhere, anytime.

3.2.4 The Blockchain Mechanism

Perhaps the most important difference between Bitcoin (and its variants) and past alternative currencies is the distributed timestamp ledger on which Bitcoin transactions are recorded, called the blockchain (Hileman, 2013).

A typical blockchain structure that is a base for a cryptocurrency like Bitcoin or CBDC that requires DLT is shown in Fig. 3.6. Any block with this chain structure is

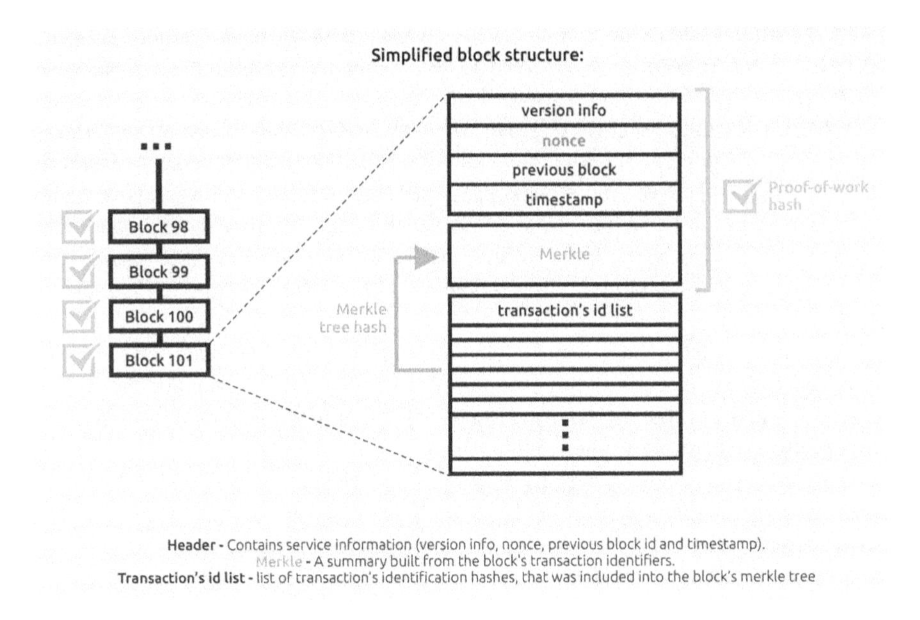

Fig. 3.6 Typical block structure in blockchain. Source: SMU (2019)

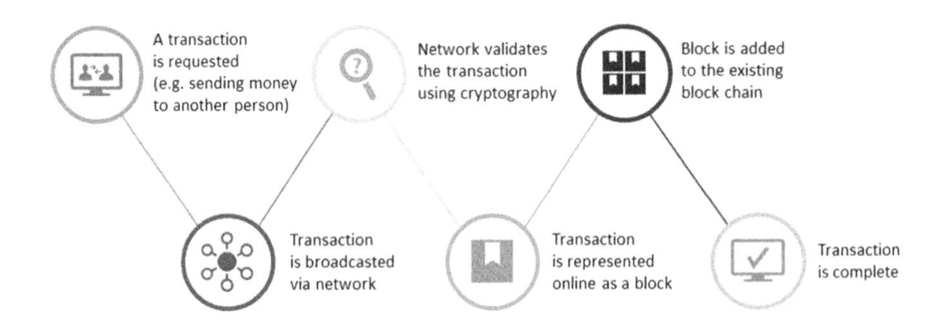

Fig. 3.7 A typical blockchain process. Source: www.infodiagram.com

identical to the other data structure with a header that points the merkle, the time stamp for creating the block and the address of the previous block as a hash-key.

A typical blockchain process for payment transactions can be summarized as in Fig. 3.7. According to this structure, any process is triggered by a transaction request upon a payment. Then, the transaction is announced to the network and all clients in

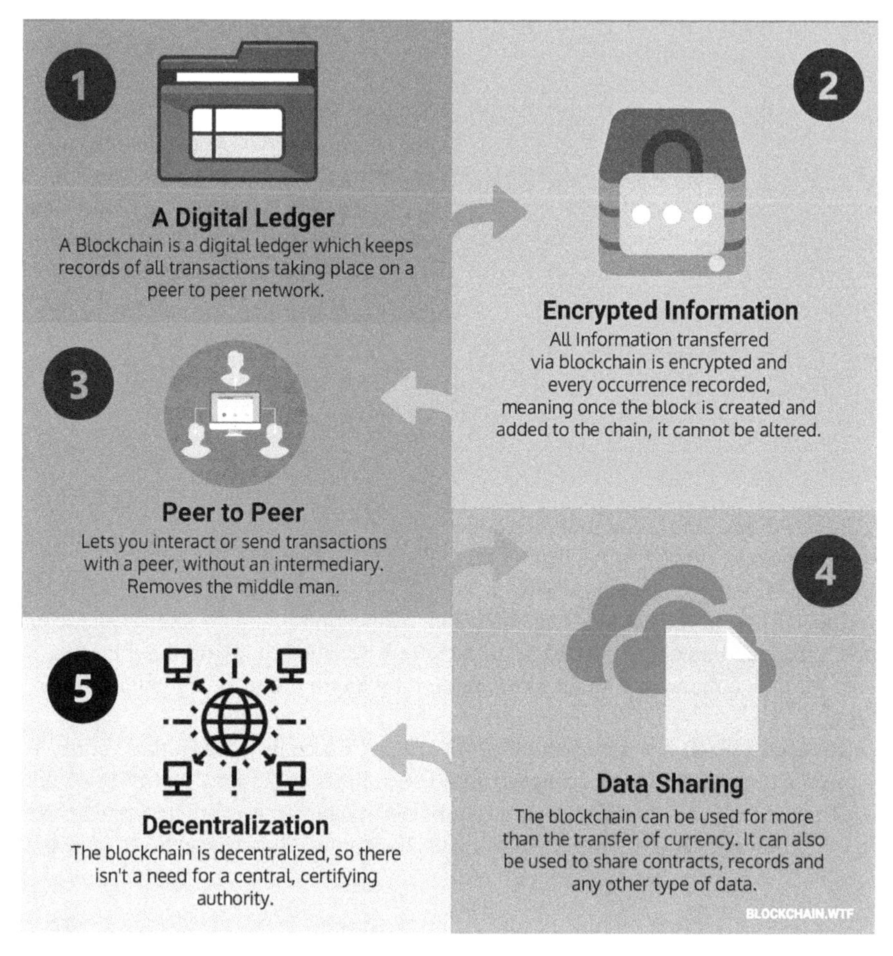

Fig. 3.8 Representation of transaction processing based on blockchain. Source: Blockchain.wtf (2019)

the network check the system and validate the transaction. After that, as a final step, a new transaction is completed by linking it to the end of the chain.

By using the blockchain structure in monetary transactions, users have a digital ledger to maintain the records of transactions, system sending/receiving money in peer-to-peer, which does not need a center for payments, and they have an encryption mechanism to protect transactions and share information about the system, transactions and contacts (Fig. 3.8).

3.2.5 Emerging the QR Coded Valuable Papers and Currencies

In Turkey, to increase the trust in check circulation, commercial banks started using QR-codes. An example can be seen in Fig. 3.9. Using the QR code people can easily use an app on their mobile phones, connecting to the web services of the bank and have the check confirmed before accepting it. In fact, there is a trust mechanism similar to the token-based system. This raises the question of using QR codes for paper currency.

Russia's Central Bank and the National Mint Goznak unveiled new 200- and 2000-ruble bills. The front side of the new banknotes are fitted with a QR code as shown in Fig. 3.10, which links to the bank's website where there is detailed information on their artistic design and security features. "The 200- and 2000-ruble bills will enter into circulation in stages after banks and businesses upgrade their technology to accommodate them" (Moscowtimes, 2017).

In 2014, the Central bank of Nigeria (CBN) introduced a feature called Quick Response Code, (QRC). It is a feature that highlights and sources all the information about the Nigerian Centenary. "This makes the note the first digital banknote in the world," CBN said (Fig. 3.11). The QRC is an application found on a smart phone or Ipad. Once the barcode on the back of the note is scanned, it shows the President's face and then comes up with all information on Nigeria's history (Premium Times, 2014).

Rossiaud (2018) has described a local currency officially used in three cities in Switzerland, named Leman, launched in 2015. Leman is the local currency of the economic life basin that develops around Lake Geneva. As can be seen in the Fig. 3.12, there is a QR-code on it, the new banknotes allow traceability, and the

Fig. 3.9 QR codes on a check in Turkey. Source: KKC (2019)

Fig. 3.10 QR code in a Ruble. Source: Moscowtimes (2017)

Fig. 3.11 QR code on a banknote issued by Central Bank of Nigeria. Source: Premium Times (2014)

blockchain makes it possible to better estimate the speed of circulation, which is an indicator of local wealth production. By using the QR Code, anyone can check its validity by scanning it through the application of the electronic Léman.

The e-Leman uses the Biletujo electronic wallet on ComChain (for "blockchain of commons"), as its infrastructure. This is based on ethereum technology, but unlike all other cryptocurrencies, there is not any mining cost. This voluntarily developed system is open to anyone, in a spirit of collaboration and consortium, to all complementary currencies that would like to benefit from this technology while collaborating in the extension and resilience of the network (Rossiaud, 2018).

Fig. 3.12 A QR coded local currency in Switzerland. Source: http://monnaie-leman.org/nouveaux-billets-en-circulation/

3.3 A Model Proposal for Banknotes Using the Blockchain Infrastructure

In our model presentation, we aim to propose a new hybrid system that can be applied with little effort and requiring very few modifications to the existing structure by combining the advantages of both a digital currency system and a paper based banknote system.

The central bank of each country issues the banknotes that are printed by that bank with a single serial number. This serial number cannot be repeated and reproduced, and it is not reused even if the banknote is somehow removed from circulation. A QR code that corresponds to the serial number and which records the monetary value of the banknote is placed in in a suitable location on the banknote.

While the banknotes' ownership during initial production is the central bank, the ownership of the banknote during circulation will be wholesaled towards banks and other financial institutions. In the meantime, money in the bank accounts of financial institutions will be kept as electronic records rather than banknotes. Banknotes will only be produced for the money needed in circulation but not for purpose of deposit. Therefore, the commercial banks and central banks will keep the banknotes for only circulation purpose, not for the deposits. It can be considered an individual ledger maintained by commercial banks on behalf of the users against any change/transaction between users as in peer-to-peer networks.

Indeed, inspired by Etherium technology, the Russian Ruble and Nigerian Naira with QR codes, and e-Leman local currency's digital wallet Biletujo and blockchain infrastructure ComChain, could be a good starting point to evolve existing banknotes in the local currencies of almost all countries especially for relatively low value transactions between users. We offer a distributed general ledger that maintains the wholesale of money transactions between financial institutions and the central bank.

Additionally we also offer a linkage between the QR code of a currency and user accounts. Here, as in POS devices, the banknote could be scanned by a QR code scanner and thus the banknote can change hands as a payment, and the money transfer of the person/institution will be confirmed with a password like authorization method.

During the payment process, a secure mobile application could verify the transactions concurrently with the physically change banknotes. In other words, ownership of the banknotes due to peer-to-peer transactions should be changed simultaneously both in physical and electronic media by means of a mobile device/ATM/handheld POS device that enables one to scan QR Codes and send the confirmation messages to move the electronic form of the money from one users' account to another's, otherwise transactions would not be fulfilled and never confirmed by buyers and sellers. In such a model, digital currency and banknotes reflect and protect each other and like a shadow they are projections of each other.

This could be called hybrid money that takes the advantages of both of these systems. Firstly, it cannot be hacked electronically because one most own the physical banknote with the unique QR code. Moreover, it is not vulnerable to physical theft and counterfeiting because nothing is electronically changed on users' accounts. This means that one would need to both hack the system electronically and steal the physical banknote at the same time to be a theft in such a hybrid system.

This hybrid structure is as easy-to-use as e-Leman like systems and ATM systems, with easy access to digital reflection of banknotes. It is also safer than ever, since users' information is still private and protected with digital wallet systems. There could be an individually organized general ledger for each user by synchronizing with the commercial banks and there is no mining and therefore no need for energy consumption. Moreover, counterfeiting and hacking could be more costly than its outcome. This means that it meets the requirements offered by Furche and Sojli (2018).

Now let's look at the same scenario from the block chain window. When the central bank picks up the bank note and picks it up to its own account, the QR code on the banknote is scanned by an automated system, adding a new block to the last chain of the blockchain. This block is processed simultaneously by the central bank, which has ownership of all banknote records. The verification of the transaction takes place through the central bank.

3.4 Discussions and Conclusions

In this study, we have presented a model proposal for a new local currency with a hybrid structure. This structure has a position between paper banknotes and digital currencies, and it utilizes the benefits of both of these types.

There are different perspectives of the proposed system from viewpoints of users, commercial banks and central banks of countries. Central banks play the wholesaler role and print only banknotes with a QR Code. There are also some aspects of central bank digital currencies as economic, politic economy and technological perspectives (Masciandaro et.al, 2018).

Commercial banks, in the proposed system, are intermediaries between users for fulfilling the transactions that must be performed both physically and electronically in daily low amounts of payment. Large payments in transactions are not held through banknotes but only allowed through the electronic transfer of money. During the withdrawal of money from commercial banks, QR coded banknotes are delivered to account owners, the corresponding digital records are entered in the owners' accounts. Counterfeiting paper currencies is prohibited, and the cost of doing this is relatively high for small quantities; therefore, banknotes could be effective for transactions of small value.

However, one would be the cost to the well-intended users of cash who lack access to electronic devices (Camera, 2017). Also, if only digital currency were to substitute cash, criminal organizations would likely respond by finding new means of payments and stores of value, possibly with worse outcomes (McAndrews, 2017).

The only problem here can be realized in the case of monetary exchange by interacting with the ATM via the applications installed on their smartphones providing access to their own accounts from the ATM devices through QR code scanning and verifying the amount of money they want to withdraw without using any card.

To some extent, anonymity is still preserved since central banks and private banks are only account holders and responsible for validating and confirming the ownership of paper currencies based on relevant electronic records through distributed general ledger among financial institutions.

Kahn et al. (2018) discuss the drawbacks of e-money like cyber-attacks; whereas in paper currency, Masciandaro et al. (2018) emphasize money laundry. This model proposal considers these drawbacks as well and resolves them by combining their power and benefits.

The hybrid money proposed through this study, takes advantages of digital currency since electronic money transfer is possible only when digital currency in the account holders deposit is available. On the other hand, circulating money as paper currencies with a unique id printed as QR code on it must be synchronized electronically as well among banknote owners' electronic records. Therefore, we should know the data as a pair of QR code and account holder transparently. It means that money in deposit account is anonymous whereas paper banknotes are transparent.

This model comes out as a substitution for money owners about how much money is to be available in circulating physically (i.e., amount of paper currency with QR codes) and electronically (i.e., amount of anonymous digital currency).

This study can be regarded as a starting point of modeling a new money for a near future, considering that since the signs recently show increasing demand for banknotes in many countries, the banknotes will still be in use for several decades along with digital ones. The model proposed here is only applicable at the macro level by rule makers.

References

Ali, R., Barrdear, J., Clews, R., & Southgate, J. (2014). The economics of digital currencies. *Bank of England Quarterly Bulletin, 54*(3), 276–286.

Barontini, C., & Holden, H. (2019). *Proceeding with caution: A survey on Central Bank digital currency* (BIS Paper No. 101).

Barrdear, J., & Kumhof, M. (2016). *The macroeconomics of Central Bank issued digital currencies* (Bank of England Working Paper No. 605). https://doi.org/10.2139/ssrn.2811208

Bech, M. L., & Garratt, R. (2017). Central bank cryptocurrencies. *BIS Quarterly Review.*

Berentsen, A., & Schar, F. (2018). A short introduction to the world of cryptocurrencies. *Federal Reserve Bank of St. Louis Review, 100*, 1–16.

Blockchain.wtf. (2019). Accessed 4/3/2019, from https://static.blockchain.wtf/wp-content/uploads/what-is-a-blockchain-infographic.png

Borgonovo, E. and Cillo, A., Caselli, S., & Masciandaro, D. (2018). *Between cash, deposit and Bitcoin: Would we like a Central Bank digital currency? Money demand and experimental economics* (BAFFI CAREFIN Centre Research Paper No. 2018-75). https://doi.org/10.2139/ssrn.3160752

Brayans, D. (2014). Bitcoin and money laundering: Mining for an effective solution. *Indiana Law Journal, 89*, 441–472.

Camera, G. (2017). A perspective on electronic alternatives to traditional currencies. *Sveriges Riksbank Economic Review, 1*, 126–148.

Chapman, J., Chiu, J., & Molico, M. (2013). A model of tiered settlement networks. *Journal of Money, Credit and Banking, 45*(2–3), 327–347.

Claeys, G., Demertzis, M., & Efstathiou, K. (2018). Cryptocurrencies and monetary policy. *Policy Contribution, 10.*

Cointelegraph. (2016). Accessed 3/3/2019, from https://cointelegraph.com/news/cash-electronic-money-scandinavia

CPMI (Committee on Payments and Market Infrastructures). (2015, November). *Digital currencies.* ISBN 978-92-9197-384-2 (print) ISBN 978-92-9197-385-9 (online).

Eken, M. H., & Baloğlu, E. (2017). Crypto currencies and their destinies in the future. *International Journal of Finance & Banking Studies, 6*(4), 2017.

Eswar, S. P. (2017). *Gaining currency: The rise of the Renminbi.* New York: Oxford University Press.

Feige, E. L. (2012). *The myth of the cashless society: How much of America's currency is overseas?* (RePEc Archive, MPRA Paper Series, n. 42169).

Furche, A., & Sojli, E. (2018). *Central bank issued digital cash.* https://doi.org/10.2139/ssrn.3213028

Harari, Y. N. (2014). *Sapiens: A brief history of humankind.* London: Random House.

Hendrickson, J. R., Hogan, T. L., & Luther, W. J. (2015). *The political economy of Bitcoin.* Mimeo.

Hileman, G. (2013). *Alternative currencies: A historical survey and taxonomy.* https://doi.org/10.2139/ssrn.2747975

I-China. (2012). Accessed 4/3/2019, from http://www.i-china.org/news.asp?type=1&id=1195

Jafari, S., Vo-Huu, T., Jabiyev, B., Mera, A., & Mirzazade F. R. (2018). *Cryptocurrency: A challenge to legal system.* https://doi.org/10.2139/ssrn.3172489.

Jobst, C., & Stix, H. (2017, October). *Is cash back? Assessing the recent increase in cash demand*? (SUERF Policy Note, n.19).

Judson, R. (2012). *Crisis and calm: Demand for U.S. currency at home and abroad from the fall of the Berlin Wall to 2011* (Board of Governors of the Federal Reserve System, International Finance Discussion Papers, n.1058).

Kahn, C. M., Rivadeneyra, F., & Wong, R. (2018). *Should the Central Bank issue e-money?* https://doi.org/10.2139/ssrn.3271654

Kahn, C. M., & Wong, T. (2019). *Should the Central Bank issue e-money?* (FRB St. Louis Working Paper No. 2019-3). http://dx.doi.org/doi.org/10.20955/wp.2019.003

KKC. (2019). Accessed 4/3/2019, from https://www.kkc.com.tr/assets/img/guides/guides-cek.png
Koch, C., & Pieters, G. C. (2017). *Blockchain technology disrupting traditional records systems*. Financial Insights – Dallas Federal Reserve Bank.
Koning, J. P. (2016). *Fedcoin: A central bank-issued cryptocurrency*. R3 report, 15.
Mas, I. (2010). Smart banknotes: A proposal for bank notes that bridge the gap between physical and electronic money. *The Futurist, 5*(1), 5.
Masciandaro, D., Cillo, A., Borgonovo, E., Caselli, S., & Rabitti, G. (2018). *Cryptocurrencies, central bank digital cash, traditional money: Does privacy matter?* (BAFFI CAREFIN Centre Research Paper No. 2018-95). https://doi.org/10.2139/ssrn.3291269
McAndrews, J. J. (2017). *The case for cash* (Asian Development Bank Institute Working Paper Series (679).
Moscowtimes. (2017). Accessed 4/3/2019, from https://www.themoscowtimes.com/2017/10/12/russia-unveils-qr-coded-200-2000-ruble-bills-a59250
Nabilou, H. (2019). Central Bank digital currencies: Preliminary legal observations. *Journal of Banking Regulation*. https://doi.org/10.2139/ssrn.3329993
Nagpal, S. (2017). *Cryptocurrency: The revolutionary future money*. https://doi.org/10.2139/ssrn.3090813
Nakamoto, S. (2008) *Bitcoin: A peer-to-peer electronic cash system report*. Accessed 4/3/2019, from www.bitcoin.org/bitcoin.pdf
Niepelt, D. (2018). *Reserves for all? Central Bank digital currency, deposits, and their (non)-equivalence* (CESifo Working Paper Series No. 7176).
Peters, G., Panayi, E., & Chapelle, A. (2015). Trends in crypto-currencies and blockchain technologies: A monetary theory and regulation perspective. *Journal of Financial Perspectives, 3*, 46. https://doi.org/10.2139/ssrn.2646618
Pilkington, M. (2015). In F. Xavier Olleros & M. Zhegu (Eds.), *Blockchain technology: Principles and applications research handbook on digital transformations*. Cheltenham: Edward Elgar.
Premium Times. (2014). Accessed 4/3/2019, from https://www.premiumtimesng.com/business/business-news/171066-nigeria-unveils-new-digital-n100-paper-note.html
Raskin, M. & Yermack, D. (2016). *Digital currencies, decentralized ledgers, and the future of central banking* (No. w22238). National Bureau of Economic Research. https://www.nber.org/papers/w22238
Raskin, M. & Yermack, D. (2017). Digital currencies, decentralized ledgers, and the future of central banking. In P. Conti-Brown & R. Lastra (Eds.), *Research handbook on central banking*. Cheltenham: Edward Elgar.
Reekmans, T. (1949). Economic and social repercussions of the ptolemaic copper inflation. *Chronique d'Egypte, 24*(48), 324–342.
Rogoff, K. (2017). Dealing with monetary paralysis at the zero bound. *Journal of Economic Perspectives, 31*(3), 47–66.
Rossiaud. (2018). *E-leman: a local blockchain currency in Switzerland and beyond*. Accessed 4/3/2019, from http://www.ripess.eu/e-leman-a-local-blockchain-currency-in-switzerland-and-beyond
SMU. (2019). Accessed 4/3/2019, from https://skbi.smu.edu.sg/sites/default/files/skbife/pdf/BitCoin-Seminar2.pdf
SPK. (2016). *Sermaye Piyasası Kurulu, Araştırma Dairesi: Kripto-para Bitcoin*. Accessed 4/3/2019, from http://www.spk.gov.tr/siteapps/yayin/yayingoster/1130
Tasca, P. (2015). *Digital currencies: Principles, trends, opportunities, and risks*. https://doi.org/10.2139/ssrn.2657598
TCMB. (2019a). Accessed 1/3/2019, from https://www.tcmb.gov.tr/wps/wcm/connect/d189b219-fe71-40bf-9754-6a5f7d0a65eb/KagitParaTarihce.pdf?MOD=AJPERES&CVID=
TCMB. (2019b). *Kağıt Paranın Tarihi*. Accessed 4/3/2019, from https://www.tcmb.gov.tr/wps/wcm/connect/d189b219-fe71-40bf-9754-6a5f7d0a65eb/KagitParaTarihce.pdf?MOD=AJPERES&CVID=

Tobin, J (1985). *Financial innovation and deregulation in perspective* (Cowles Foundation Papers, no 635). http://www.imes.boj.or.jp/research/papers/english/me3-2-3.pdf
Walker, M., & Luu, J. (2016). *Blockchain and the nature of money*. https://doi.org/10.2139/ssrn.2939042

Sabri Erdem is a full-time professor at Dokuz Eylul University, Faculty of Business, Business Administration Department, and Chair of Quantitative Methods Division. He is Industrial Engineer since 1996 and holds two MSc degrees in both Computer Engineering and Business Administration. He received a Ph.D. in Computer Engineering in 2007. He published many articles in national and international journals and has many proceedings at the national and international symposium and congress and partner of three patents about automated medication dispensing machines. Main interest areas are optimization algorithms, nature-inspired algorithms, intelligent systems, data mining, healthcare information technologies, healthcare materials management, data envelopment analysis, and structural equation modeling.

Derya Altun is a feminist economist. She graduated from the DEU Faculty of Economics and Administrative Sciences in 1999. She has received a master degree in General Economics in 2008. She is currently a Ph.D. candidate at Ege University, Economics Department. Her master thesis about efficiency and productivity in NGOs, and her Ph.D. thesis on gender budgeting in local governments. She had been working as a Ph.D. researcher at the University of Essex, 3 months, in 2012. Gender, NGOs, local governments, migration, and development are the academic fields of her study. She is a member of local and national NGOs. Currently, she is working as an economist at Izmir Metropolitan Municipality.

Chapter 4
Initial Coin Offerings: Tokens as Innovative Financial Assets

Saman Adhami and Giancarlo Giudici

Abstract In this chapter we describe the phenomenon of Initial Coin Offerings (ICOs), i.e. unregulated offerings of digital tokens, built on the innovative blockchain technology, as to provide a means to collect finance for a project on the Internet, disintermediating any external platform, payment agent or professional investor. ICO tokens allow the access to platform services, may serve as cryptocurrencies, or grant profit rights; they are traded on electronic exchanges and represent a new financial asset. We highlight the issues emerged with respect to information asymmetries and moral hazard and we review the nascent empirical literature exploring the ICO token market.

4.1 Introduction

An 'initial coin offering', also called 'token offering, is essentially a funding event for a blockchain-based company or project: the company receives financing in the form of cryptocurrencies and rewards contributors with digital 'tokens'. This transaction occurs through the codification of a 'smart contract', which in its simplest form autonomously enforces the exchange between cryptocurrencies (received by entrepreneurs) and tokens (received by contributors). The issued tokens can serve multiple purposes but the primary one is to grant pledgers (as well as developers) access to the services offered by the company. Tokens are often traded on electronic unregulated exchanges after the offering: therefore, they can be considered as a liquid innovative financial asset.

S. Adhami (✉)
Department of Finance, Accounting and Statistics, Vienna Graduate School of Finance, Wien, Austria
e-mail: saman.adhami@vgsf.ac.at

G. Giudici
School of Management, Politecnico di Milano, Milan, Italy
e-mail: giancarlo.giudici@polimi.it

© Springer Nature Switzerland AG 2019

U. Hacioglu (ed.), *Blockchain Economics and Financial Market Innovation*, Contributions to Economics, https://doi.org/10.1007/978-3-030-25275-5_4

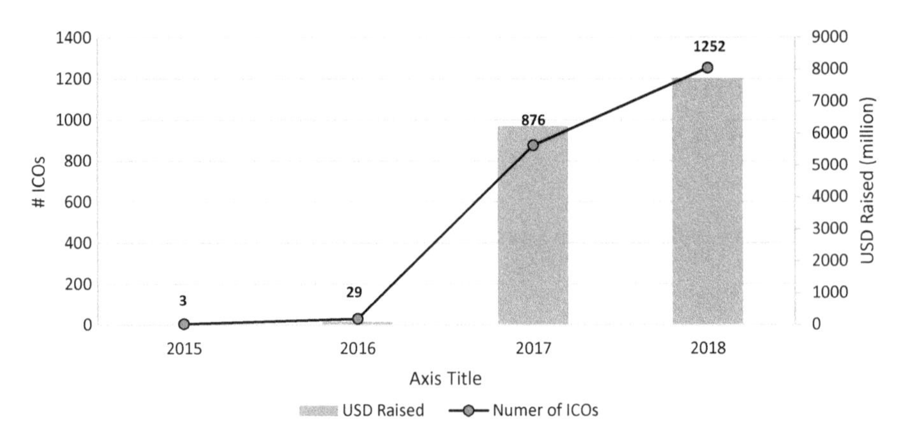

Fig. 4.1 ICO flow over time. Source: www.icodata.io

ICOs were born during the infancy of the blockchain revolution (the first ever ICO took place in 2013) and thus they were initially targeted to a small public of people, in many cases computer programmers or crypto-enthusiasts. As the audience was very restricted, the amounts of funding raised were also not exceptionally high. What made ICOs popular among entrepreneurs was their negligible transaction cost at the time. The direct transaction between contributors and ventures eliminated the need and costs of financial intermediaries, while the absence of regulations removed all compliance costs. Only recently relevant portions of market participants started to disentangle the value of the blockchain as a technology from cryptocurrencies, which are just one of the products or applications of such technology. The rise of the Bitcoin price in 2017 and the success of ICOs in 2018 nurtured a hype in the interest of practitioners and scholars in the topic. Figure 4.1 provides a quick snapshot of ICO volumes over time.[1]

From the last quarter of 2017, some countries started to scrutinize ICOs; others (e.g. China and South Korea) banned them, and some (e.g. Switzerland, Singapore and the Baltic republics) set themselves to become the hotspot of high-tech ventures by providing ICO-specific legal provisions that are either very attractive for entrepreneurs or, simply, very transparent and clear-cut. With the number of ICOs steadily increasing worldwide, being able to raise the target amount of funding became much tougher and significant initial investments in the ICO structuring and marketing were done by proponents. For instance, ICO project Friendz disclosed that they had to use about $2 million of VC financing to be able to successfully conduct a fundraising of about $21 million.

[1]The data is provided by www.icodata.io

4.2 The Structure of an ICO

Momtaz (2018) dissects the ICO process into four major steps: (1) ICO structuring and marketing, (2) pre-ICO, (3) ICO or main crowd sale event, and (4) token listing. A preliminary 'zero' step can also be identified in which the project promoters must assess whether an ICO fits with the strategy and needs of the business. Lipusch (2018) stresses that the ICO decision is fundamentally dependent on two dimensions, namely the willingness to disclose sensitive information and the necessity of a tokenized platform for the business plan. Indeed, a significantly high level of disclosure is necessary for successful public token sales, including the sharing of relevant coding efforts, which act as quality signal (Adhami, Giudici, & Martinazzi, 2018). Moreover, there must be an economic justification on why the business would need a token. For instance, MedicalChain is a decentralized platform that securely stores patients' medical data that medical experts can access by using the MedToken, which are also rewarded to networks operators (or 'miners') for the maintenance and update of the ledger. Possible use cases for the tokens are research institutions paying individuals for accessing their health data used in trials, or patients paying physicians for telemedicine consultations, which generate token circulation within the system. Market participants considered such storage and use of sensitive data valuable, leading MedicalChain to collect about $24 million though a one-day ICO in February 2018.

The first step is the set up of all the details of the campaign, which include the creation of tokens with specific bundle of rights attached to them, the setting of the timeline of the liquidity events, the targeted audience (e.g. developers, qualified or institutional investors, the public at large), the pricing of the tokens (which generally follows an increasing path over time reflecting the early bird discounts), the content of the offer documentation (the main piece being the 'white paper'), the legal framework of reference (e.g. for placing tokens to US citizens without triggering security law requirements, the Howey test is currently used as discriminator), and the business plan for the venture (product definition, target market, projected financials, team and roadmap). Moreover, the social media tools for the marketing activity must be put in place, which include channels of direct contact between the promoters and the public (e.g. Telegram or Slack), main forums of discussion of the trends in the crypto-world (e.g. BitcoinTalk, TokenMarket, etc.) and active accounts on major traditional social media, i.e. Facebook, Twitter and LinkedIn. Moreover, some online ICO data aggregators provide rating services for tokens using panels of supposed experts (e.g. ICOBench), both for upcoming ICOs and for some of the concluded ones that are actively trading. This adds to the lucrative business of advisory services to ICOs that have been gaining momentum as the popularity of this form of financing increased in 2018.

The second step in the process is the pre-ICO or presale, which entails the sale of a smaller proportion of the issued tokens (at a much lower prices than the following rounds of the main ICO event) to either a selected group of contributors or to the public at large. While a clear benefit of a successful presale is the collection of funds

Table 4.1 Major descriptive features of ICOs

Country of origin	Number	%	Core team size (average)
US	165	17.6	6.76
Russia	107	11.4	7.54
Decentralized	90	9.6	6.07
UK	59	6.3	6.05
Singapore	48	5.1	8.91
Switzerland	38	4.1	13.09
Canada	30	3.2	6.42
Australia	24	2.6	8.53
Others/NA	374	40	6.99
Blockchain adopted			
Ethereum	744	79.6	7.62
Own blockchain	78	8.3	5.80
Waves	46	4.9	5.55
Bitcoin	18	1.9	5.92
Others/NA	49	5.2	7.27
ICO stage			
Initial	810	86.6	7.33
Follow-up	13	1.4	8.25
Presale	108	11.6	6.52
NA	4	0.4	7
Bonuses			
Early bird	543	58.1	7.26
Early bird + major contributions	52	5.6	10.45
Other	48	5.1	9.09
None/NA	292	31.2	6.46
Code availability			
Yes	550	58.8	8.15
No	385	41.2	5.88

Source: analysis of proprietary sample of ICO occurred from 2015 to 2017

to be used for marketing the main sale event (which is especially crucial for non-VC-backed projects), there are also indirect benefits such as a test for the market demand of tokens (with a role comparable to that of book-building for an IPO) or the attraction of notable contributors to the project's cause (e.g. star entrepreneurs, entertainment celebrities or crypto-gurus), creating a goodwill that is spent during the proper ICO. Howell, Niessner, and Yermack (2018) highlight some similarities between the pre-ICO and the convertible notes used by business angels to fund a startup, as they also facilitate investment in a very high-risk stage by allowing for large discounts. Obviously, an unsuccessful presale would delay or preempt the ICO. Up to December 2017, according to our database, 50.43% of ICOs organized a presale event. Table 4.1 presents some statistics regarding the main features of ICO fundraising events occurred from the onset of the phenomenon until the end of 2017.

During the third phase, starting from the predefined opening time for the call for contributions, Ethers, Bitcoins and other accepted cryptocurrencies can be sent to a public digital wallet address. Depending on the coding of the smart contract, tokens are sent to the wallets of the contributors, according to the exchange rate awarded by the timing of the contribution (or the size of the contribution, if there are large-contribution bonuses), which can happen immediately or after the crowd sale is concluded. There is generally a theoretical window for the crowd sale, typically of 1 month, but the actual duration of the campaign depends also on token demand, which is out of the will of project promoters. In fact, a crowd sale can end as soon as the maximum target (called 'hard-cap') is reached, with this limit being sometimes hidden from the public as a device to avoid market manipulation by large contributors. Moreover, a crowd sale may be ended by project promoters in response to scarce demand. There are also some ICOs with a long open window for fund collection, called 'ongoing'. For instance, the EOS offering closed after 1 year, collecting a record amount of $4 billion. An ICO is deemed to have been 'successful' if the funds raised reach or surpass the minimum target stated by promoters ('soft-cap'). Many ICOs reached funding success in less than a day, but a significant proportion also failed in reaching their minimum target (18.48% as of the end of 2017), in which cases either the project continues with the collected funds (despite below their ideal target) or the cryptocurrencies are sent back to the contributors. For the cases of fraudulent projects, the funds are never reimbursed.

Conditional on a successful offering, the last step of an ICO process is the listing of the token on an unregulated secondary digital exchange. These exchanges can be either centralized and managed by a private organization, or decentralized and managed through automated match-making. While centralized exchanges allow higher trading volumes and permit the conversion of crypto-assets to fiat currency (which adds some valuable flexibility for retail investors willing to participate in these markets), they are more exposed to hostile hacker attacks. Given the large availability and competition of crypto-exchanges, listing requirements vary a lot and surely are not nearly as stringent as those of regulated exchanges. If the ICO-backed project is in trouble, or due to very large drop in price, the token will be delisted, and will become immediately illiquid. In analogy to lockup periods applicable to insiders' shares after an IPO, there are frequent vesting periods embedded in the smart contract of tokens in order to align founders' objectives to those of the external token-holders. Sometimes, these lockups are linked to certain milestones in the project development roadmap.

Tokens are built on a blockchain. Most projects build their tokens on popular existing blockchains such as Ethereum or Waves, with a predefined set of smart contracts which make the issuance of token quick and relatively cheap for developers, while also granting a high level of standardization. Still, some projects, especially during the first phase of the ICO hype, created their own ex-novo blockchains, which allowed them to have a higher level of customization of the smart contract features and gave rise to opportunities for other applications to build on their platform, thus promising faster diffusion. In general, if the project entails the creation of a token that can also act as a full-fledged currency, then project promoters must build their own blockchain. Otherwise, as in most cases, if the project is meant to be a decentralized app needing just an 'internal' currency for its functioning, then existing blockchains can be used to create the token as a derivative product. By the end of 2017, most ICO projects relied on the ETH blockchain (79.61%) while only about 8% developed their own distributed ledger system. Adhami et al. (2018) isolate five roles or rights that can be attached to a token, either singularly or as different bundles. These are: token acting as a currency, tokens granting access to a service, token granting voting rights, token granting profit sharing rights, and tokens granting contribution rights. As stressed above, currency tokens require an ad-hoc blockchain, whose development cost is justified only if it fits the business plan of the project, since given the plethora of cryptocurrencies available on the Internet, the likelihood of a new-entry currency-token of becoming a major means of payment is negligible. Tokens serving as 'internal' currency on the platform to be developed (or already in place) are the most common, and are commonly called 'utility' tokens, while, if the service access right is coupled with profit sharing, they are more precisely defined 'security' tokens. Utility tokens can be used to buy a physical product or a service offered by the company, but the most common idea is that tokens grant exclusive access to a digital service (typically videogames, cloud computing or other platforms). Tokens can also grant different levels of governance rights to holders, ranging from control power over the venture in the case of 'decentralized autonomous organizations' (DAOs), to a very limited power of securing blocks in a proof-of-stake blockchain. In reality, it is rare that investors have rights comparable to those of common shareholders in a traditional listed company, since in most of the cases where voting rights are present, they regard only few decisions and are not exercisable on a regular basis, such as through pre-scheduled shareholder meetings. Profit rights, on the other hand, when present, are more similar in their structure to those of traditional equity securities: according to predetermined deadlines, they pay token-holders with dividends or token-buybacks proportionally to their holdings. As a consequence of this functional resemblance to traditional securities, these types of tokens are under constant scrutiny by national security regulatory agencies. Finally, a less frequent occurrence is the embedding of contribution rights in the tokens, which allow the holders to play an active role in the growth of the project, generally as a developer of the platform or of applications built on it. The aim of the tokens is to allow collaboration between the core team and external developers or influencers, who will receive proportional rewards in exchange for their active role in the expansion or diffusion of the project. The contribution token, thus, acts in many cases as a 'status' signal or it allows access to some internal information of the project, such as the real-time complete source code. Figure 4.2 presents the possible classification and aggregation of the five primary token rights according to dimensions reflecting different stakes already existing in traditional finance and the needs of regulators analyzing this phenomenon.

Regardless of the rights attached to a token, which we recall are mutually non-exclusive, the supply of token is generally limited before the presale. The monetary policy of tokens is called 'tokenomics' and it is thoroughly discussed by

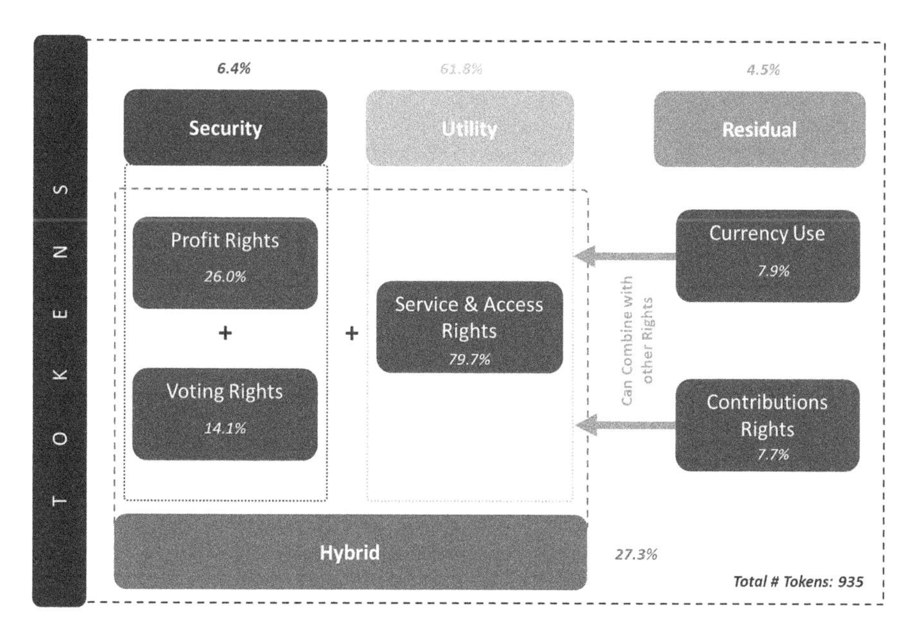

Fig. 4.2 ICO token segmentation. Source: analysis of proprietary sample of ICO occurred from 2015 to 2017

Cong, Li, and Wang (2018), but one simple principle dictates that a limited supply of tokens is necessary to avoid spiraling inflation risk and systematic devaluations. Nevertheless, a scarce supply of tokens that are only usable within a single closed marketplace would hinder the long-term viability of the venture's business model, which could eventually create deflationary pressure for the prices of tokens with a strict fixed-supply regime. Finally, it is very common by project promoters to retain a portion of the tokens in a treasury and use them to reward future collaborators and advisors, or to smooth down operations within the ecosystem of utility tokens, especially in the case of spikes in demand and shortage of available tokens to be spent on the platform. Moreover, if further ICO rounds are planned, larger proportions of tokens are not distributed. On average, for the ICOs concluded by 2017 and tracked in our research sample, the average token retention rate was 39.90% for utility tokens and 28.90% for security tokens.

Usually, tokens are priced prior to the ICO by project promoters in terms of a conversion rate vis-à-vis a major cryptocurrency, which being itself volatile makes the fiat or dollar money price also volatile. These tokens are offered on a first-come first-served basis, and generally present an early bird discount scheme. However, other types of pricing and price discovery systems are not that unusual as the promoters have total flexibility in the structuring of the sale. For instance, the Polkadot ICO (conducted during October 2017) was set up as a spend-all second-price Dutch auction: the price of the token decreased at a predetermined schedule and the auction closed the moment the sum of the offers received at a given price were sufficient to cover all the available token supply.

4.3 Critical Issues in the ICO Universe

The emergence of a new technology and a novel avenue for entrepreneurial financing clearly create some frictions that need to be addressed in order for entrepreneurs and investors to fully benefit from financial innovation. In the following sections, we discuss the most pressing challenges affecting the functioning and attractiveness of ICOs.

4.3.1 Agency Issues and Governance Solutions

Currently, lacking a standard regulation, the major governance challenge for the ICO ecosystem is the imbalance of power between ICO proponents (i.e. the entrepreneurs and insiders of the project) and the token holders (i.e. the contributors of funding in the token offering). At the moment, the imbalance of power is totally in favor of entrepreneurs as in the typical ICO they gain total control of the funds raised at the end of a successful ICO, i.e. one that reaches its minimum target. The need for governance improvements is so dire, as to protect token buyers from scams and project mismanaging the funds, that market participants have proposed different solutions to agency problems while awaiting a probable regulatory response. Any device that is able to involve investors in the management of the collected funds would certainly increase the attractiveness of ICOs. One of such devices is the DAICO, which merges some of the benefits of decentralized autonomous organizations (DAOs) with the ICO fundraising mechanism. As any type of improvement of governance in the blockchain, the DAICO relies on the specific smart contract created for the tokens being offered. Unlike usual ICO tokens, where the smart contract only regulates the inflow of funds in exchange for tokens (based upon all possible specification of the sale such as KYC-based, auction type, minimum and maximum caps, and so on), the DAICO contract remains operational in a ‘tap mode’; similarly to ‘stage financing’ in venture capital (Gompers & Lerner, 2000), the money is kept in an escrow account and released in different tranches, according to the milestones decided in the initial business plan. If contributors do not approve the management conduct or are disappointed by the project performance, they can vote to shut down the DAICO and get their cryptocurrencies back, i.e. determining the self-destruction of the contract. DAICOs allow investors to permanently exert a strong control over the managers as they can autonomously shut down the project at any time and therefore force the agent-managers to continuously engage with contributors and follow the roadmap proposed at the time of ICO inception or require approval of new plans.

Another solution to the moral hazard issue arising after the ICO is the project of the Coin Governance System (CGS), a venture registered in Spain proposing a product meant to protect ICO investors and be a managerial tool for current and future ICO companies. Once an ICO assisted by the CGS scheme is completed, the raised capital will be held in escrow and the CGS will gradually release the money to the ICO launcher, similarly to the DAICO-type smart contract. In case a token holder thinks the project is not being executed properly, she can submit a claim to the CGS by depositing a certain number of tokens. If the claim receives enough support from other ICO investors, it will be handled by a decentralized judge: the community of 'CGS Arbiters', who vote on the petition. There are two possible outcomes for this vote: "OK" or "KO". If the outcome of the voting is "OK", the CGS smart contract continues releasing money to the ICO launchers as before the claim. If the outcome of the voting is "KO", the CGS smart contract will enter into "withdrawal mode" and ICO token holders can withdraw remaining money by sending back their tokens. The CGS arbiters act as a decentralized judicial system which is made up by the CGS community (anyone holding CGS tokens) and is incentivized to vote correctly as they earn CGS tokens depending on the outcome. The votes are confidential, and each arbiter puts at stake a number of CGS tokens by entering the voting process with the hopes of earning more CGS tokens if the vote they cast ("OK" or "KO") coincides with the vote of the majority. This type of mechanism is known as the "prisoner's dilemma" with the possible results ("OK" or "KO") that are neutral for the arbiters, whose incentives depends on the voting of the majority and not on the result itself. According to the CGS founders this type of voting mechanism is optimal to oversee the behavior of ICO promoters and it is a sign of their goodwill to proceed with the ICO. However, using CGS will have a cost for the newly developed venture and add decision-making uncertainty for the entrepreneurs as they rely on a majority voting to solve disputes and shut down the business which is very susceptible to manipulation through online trends and rumors.

In a certain way, the CGS system is just a DAICO contract that instead of giving total veto and dissolution powers to token holders, it elects online participants in its network as arbiters. This adds a further level of intermediation between the principal (the token holders) and the agents (the entrepreneurs) without necessarily granting a greater expertise vis-à-vis the autonomous decisions of investors. This is in-line with the philosophy of the crowd-consensus on the blockchain to build trust in the community, but it may be suboptimal for specific ICO investors as the crowd, in the example of the CGS, is potentially neutral with respect to the outcomes and might not act in the best interest of token holders. On the other hand, with the general DAICO contract, ICO promoters could, depending on market circumstances during their ICO window, simply increase their token retention rate as to have a perennial majority and avoid any chance of self-destruction vote by crowd-investors. In this scenario, the election of external parties as collective judges for the issue raised also by a minority of unsatisfied token holders can be very impactful in terms of protecting the actual providers of funds.

With the purpose of preventing other outright bans on ICOs (such as those in China and South Korea), the non-profit ICO Governance Foundation has been pushing for the self-regulation within the ICO market by promoting its IGF-1 filing form, namely a voluntary standard disclosure protocol tailored for ICOs. While ICO promoters can signal their quality through the use of this protocol (and therefore also raise more capital), investors can gain higher quality information and therefore protect themselves. Moreover, the foundation envisages also access to the "high-quality" ICO investment opportunities by institutions once they can meet their custodial requirements also thanks to this disclosure standard. The ICO Governance Foundation justifies the working of the disclosure mechanism in this way in the FAQ section of their website: "The registration process establishes your identity and produces a time stamp along with your disclosures. If you use your disclosures to raise money and they turn out to be false, you are in most jurisdictions guilty of securities fraud and can be subject to civil and criminal penalties, and your IGF-1 filing can be used as evidence in those proceedings". Among the sections included in the IGF-1 there are many of the variables identified by the recent financial literature as critical to assess the blockchain venture, such as entity structuring, key management and beneficial ownership, information about supervisors or advisors, jurisdiction, whitepaper and so on. Notably, key missing information regarding the project after the ICO's conclusion are crystalized because after filling these public forms will become more of a commitment. These include: timing and structure of employee compensation, liquidation process and procedure of the collected cryptocurrencies, reporting obligations, frequency and degree of details. Since July 2018, five ventures have filed their information through this system within the due diligence for their ICO. It is clear that the quality of ICO disclosures and the structuring of the related venture will be greatly improved with these types of self-regulated provisions.

4.3.2 The Regulatory Framework

There is significant regulatory uncertainty regarding both cryptocurrencies and the ICO process and, thus, around the status of tokens. Chohan (2017) supports that tokens have systematically higher level of risk with respect to cryptocurrencies as they do not simply operate in a trust-less setting of distributed ledgers, but rather depend on the operation of the company that created them (i.e. they are exposed to a sizeable idiosyncratic risk). With the rise in popularity of token offerings, different stances were taken by national regulators vis-à-vis the ICO mechanism for fundraising. Some banned it outright such as China, but most, with different time lags, provided guidelines on how existing security laws applied to ICOs and tokens.

In the US, the Securities and Exchange Commission, after a thorough analysis of the ICO of 'The DAO', concluded that tokens built with profit and voting rights are to be considered investment contracts as they meet the three conditions of the Howey Test. After that, it created a special cyber-unit that started investigating all alleged misconducts regarding ICOs, intervening in many instances where either federal security laws were broken, or outright frauds were committed. For instance, in December 2017 the Plexcoin ICO was stopped with accusations of both being a scam and selling not registered securities to the public. Formally, there is a rigid position from the SEC, with Chairman Jay Clayton stating in November 2017 that he has "*yet to see an ICO that doesn't have a sufficient number of hallmarks of a security*".[2] To avoid regulatory scrutiny, many ICO promoters started running Know-you-customer (KYC) measures to identify and register potential contributors to their upcoming crowd sales. This identification of investors allows both to discern retail from institutional and qualified investors, and to target the token sale to the appropriate target. A common example is forbidding US citizens to invest in the ICO to avoid SEC investigations. Still, ICO promoters may exploit the routes set out by the various exemptions and the Title III of the JOBS Act of 2012 and avoid compliance with most securities requirements, other than anti-fraud provisions, if the instrument qualifies and the promoter applies for an exemption to registration. Thus, tokens may be issued under Regulation D (the private-placement-to-accredited-investors), Regulation S (the offshore-offers-and-sales-safe- harbor), Regulation A (small-and-additional-issues-of-securities), and Regulation Crowdfunding exemptions rather than facing compliance with initial public offering requirements. Still, there are important limitations and legal requirements for each route, which generate significant compliance costs with respect to the 'wild west' era for ICOs which ruled until the 3Q2017.

Many legal scholars do not support the currently strict stance of the SEC. For instance, Crosser (2018) states that, despite the SEC's determination to treat all ICOs as sales of securities, courts should exercise restraint and hold that ICOs can and sometimes should be viewed as the sale of digital consumer assets or commodities rather than investment contracts. Given the rationale behind the existence of a pure utility token (i.e. its use within a decentralized ecosystem), Crosser suggests that these tokens should be regulated more like the sale of commodities, or assets like software licenses, gift cards, or gambling chips.

As there is not yet a comprehensive communitarian regulation regarding crypto finance in the EU, most national authorities are left to their own in providing guidelines. Hacker and Thomale (2018) reveal that also some hybrid forms of tokens (i.e. having both the utility component and one of the ownership rights embedded) are subject to EU securities regulation. The UK FCA and the German BaFin have provided similar signals as the SEC, defining the rights that if attached to tokens make them financial securities. The French AMF, after long consultations, took a bolder stance and started developing a new ad-hoc legislation for ICOs, with specific focus on providing potential investors with the minimum quality and quantity of information before the start of crowd sales.

In May 2018 the Russian government approved an extensive legislative document regarding cryptocurrencies and tokens, which earned them the status of 'property'. In practice, this made the conduction of an ICO in Russia impossible and pushed many Russian fintech startups out of the country, similarly to what happened in China.

On the other side of the spectrum, many smaller countries and microstates have established themselves as crypto-havens with the goal of becoming the next hubs of tech startups. Examples include Switzerland, Singapore, Estonia, Liechtenstein,

---

[2]https://www.wsj.com/articles/sec-chief-fires-warning-shot-against-coin-offerings-1510247148

Malta, and Gibraltar. For instance, in Estonia there are no VAT taxes on cryptocurrencies and ICOs, and there is a cheap and fast online process for incorporating and registering ventures. Both the Estonian authorities and the Swiss ones established a case by case assessment of ICOs to determine if security law applies, providing clear guidance in terms of the documentation to be presented. Moreover, so far, their attitude has been very conducive to this fundraising mechanism, as it allowed exemption from security registration for all properly defined utility tokens. The tax-shelter considerations should not be understated as they have been so far crucial for the investment momentum in cryptocurrencies and tokens. In particular, selling Bitcoins and other cryptocurrencies for tokens has become an attractive diversification mechanism for hoarders of such coins, especially due to their historical surge in price and the amount of taxable wealth they create if cashed in.

Overall, most jurisdictions currently apply pre-existing securities law to tokens issued for an ICO with the differential treatment and conduciveness for this type of crowd sale depending mostly on the flexibility with which such laws are enforced and the effort made by national authorities to ease the bureaucratic procedure for blockchain startups.

4.4 ICOs Among Other Financing Sources

Other than an assonance with IPOs, ICOs share few of their features. Firstly, in both instances a public offering, with a predefined duration and a non-negotiable price, is made by the company. Moreover, after the conclusion of the offering, shares or tokens are listed on at least one secondary exchange, which makes them potentially liquid financial assets. Furthermore, both financing mechanisms have an important informative component for managers: IPOs allow the company to learn about their market value (Subrahmanyam & Titman, 1999; Bertoni & Giudici, 2014), while ICOs are a relevant early signal of consumer demand (Catalini & Gans, 2018). Still, a major difference between these two types of offerings is the object of the offering itself: IPO investors are allocated exclusively securities while ICO contributors receive tokens, which are a bundle of rights that not very often contains dividend and voting rights. In most cases, a utility-token is offered, i.e. a right to access the service of the company, which as we stressed earlier is generally not on the market and will be financed itself through the ICO. Other novelties of token offerings are the absence of an underwriter and the much lighter compliance and disclosure requirements (if any, depending on the jurisdiction). Finally, these two offerings target companies in different stages of their lifecycle. Maybe in the future it will be more common to see late-stage, high-growth companies opt for an ICO rather than an IPO, but today ICOs are for the large part a tool for venture financing. In this regard, ICOs seem to be an alternative to VC financing for project promoters. Accessing VC funds is generally a complex endeavor as it entails in-depth screening by the institutional investor and it requires a minimum viable product, networking and extensive negotiations. Moreover, VCs are cumbersome minority partners for the entrepreneurs as they limit their flexibility. The objective of VC managers is to actively support the ventures in their portfolio; they sign complex agreements with the entrepreneurial team, aimed at protecting and monitoring their investment, through a number of clauses (veto and control power, pre-emption rights, representation in the board of directors, tag along and drag along options). In contrast, ICO investors are dispersed and have no opportunity to discuss any provision with the proponents.

Token offerings may also occur alongside the involvement of professional investors in a venture (Arslanian, Diemers, Dobrauz, McNamara, & Wohlgemuth, 2018). Howell et al. (2018) find that VC-backed ICOs more easily collect money in exchange for digital tokens; yet there is no evidence of lower default rates in the short run.

The presence of an online campaign open to retail investors and the use of social media for the marketing of the event are structural features common to both crowdfunding and ICOs. Depending on the type of token issued, ICOs can be compared to different types of crowdfunding: tokens granting ownership rights ('security tokens') have close similarities with equity crowdfunding, whereas 'utility tokens' share major similarities to reward-based crowdfunding. The fact that crowdfunding campaign promoters receive fiat money while ICO entrepreneurs collect crypto money does not truly account as a striking difference *per se*, but the types of businesses that employed ICOs are systematically different. These are businesses that rely on the blockchain as ecosystem for their product to be developed and later used, which is a step further of being 'just' high-tech companies. On the other hand, the bulk of crowdfunded companies more frequently deploy physical products or traditional services. A second fundamental difference is that, in crowdfunding, an intermediary exists, namely the crowdfunding platform, which scrutinizes the projects, selects the most promising ones and assists the entrepreneurial team in the marketing activity, hosting the project in its infrastructure in exchange for a fee on the collected capital. ICO projects can autonomously organize their own platform for the offering and adopts the blockchain infrastructure. In the case of crowd-investing (equity and lending) the platform is in charge of complying with the regulation and is supervised by public authorities (Giudici, 2015), while ICOs are mostly unregulated.

4.5 An Overview of Empirical Studies on ICOs

The nascent financial literature on ICOs focuses on four areas of empirical investigation: (1) the determinants of success of an ICO campaign; (2) the underpricing of ICO tokens; (3) short-term and medium-term performance of token returns; (4) the economics of cryptocurrencies underlying ICO pricing. In the following we review the existing contributions on the topics above.

4.5.1 Determinants of ICO Success

The earliest dimension of empirical investigation of the ICO phenomenon relies on signaling theory (Spence, 1973) and tests the effect on fundraising success (or on the amount of funds raised) of different quality-related signals that ventures can send to potential investors. Signals are observable actions providing information regarding unobservable attributes of high-quality projects, and therefore should have the capability of dissipating, in part, the asymmetric information existing between potential contributors and the venture. Signals must not only be observable but also costly for 'sellers', or more precisely, they must be more costly to produce for low-quality 'sellers' than for high-quality ones. This condition would allow a separating equilibrium, as described by Bergh, Connelly, Ketchen Jr, and Shannon (2014), so that only high-quality sellers would end up signaling themselves and allow to avoid the generation of a "market for lemons" where, due to extreme uncertainty, buyers and sellers of securities cannot trade efficiently (Akerlof, 1970).

The presence of signaling devices addressed to potential investors is already established in the crowdfunding literature, despite its very short history. Among the features of a campaign and of a project that can induce trust within investors we find: the founder and team characteristics (Ahlers, Cumming, Günther, & Schweizer, 2015; Block, Hornuf, & Moritz, 2018; Piva & Rossi-Lamastra, 2018), the quality and content of the pitch or prospectus (Ahlers et al., 2015; Block et al., 2018), the geographical localization of the project (Hornuf & Schmitt, 2016), the duration of the campaign (Lukkarinen, Teich, Wallenius, & Wallenius, 2016) and the social media presence of the project (Block et al., 2018; Lukkarinen et al., 2016).

In the context of ICOs, the absence of intermediation, the lack of regulation and the higher degree of complexity of the projects, given the innovativeness of the blockchain concept coupled with digital services provision or fintech, make signaling even more crucial for the funding.

The first empirical analysis of ICOs was conducted by Adhami et al. (2018), who determined the main elements of an ICO that were able to significantly affect the probability of fundraising success. They found that the presence of publicly available coding effort, a presale event and a voluntary choice of a jurisdiction of reference are all valid signals that encourage larger volumes of investments in the ICO project. Amsden and Schweizer (2018) confirm that the presence of shared coding repositories for the projects positively affects the probability of success, while adding that social media presence (measured in terms of Telegram groups), lower token retention rate and larger teams are also conducive to higher fundraising success. Giudici and Adhami (2019) study the human capital of ICO teams and the effectiveness of their governance signals. They find that the size of the core team, the size of the advisory network and the token retention rate are all significantly and positively correlated with ICO success. On the other hand, these studies find no signaling effect in terms of the number and type of past professional roles of team members, the presence of a white paper, and the underlying volatility or return of the cryptocurrency market underlying each respective project (mostly, in terms of the Ethereum blockchain and the Ether).

While Adhami et al. (2018) only use a binary variable indicating whether the ICO reached its minimum funding target or not as the dependent variable, Giudici and Adhami (2019) and Amsden and Schweizer (2018) also consider amounts raised and whether or not the token was listed as further proxies of ICO 'success'. As listing only occurs after an ICO managed to collect at least the 'soft cap' level of funds, and entails further steps toward making the token liquid, it can be considered as a more restrictive measure of overall success of the project. Still, there are many ICOs that do not list their tokens in the short term despite having raised sufficient funds, which would exclude them from the sample. As most market practitioners do, also scholars use the aggregator CoinMarketCap.com as main source for assessing if a token is listed and to gauge its average market price.

Fisch (2019) studies the determinants of the amount of funds raised and gives further confirmation of the fact that publicly available high-quality source codes are effective signals for investors, which also indicates the relatively high-level of technical wisdom for the average ICO investor to be able to interpret and value such signal (Connelly, Certo, Ireland, & Reutzel, 2011). Perhaps due to the practice of sharing coding efforts freely or the jurisdictional complexity of registering intangible assets for such global projects, the author finds that patents do not have signaling power, which instead have been a key driver of business angel and VC investments (Block, De Vries, Schumann, & Sandner, 2014).

Finally, Lee, Li, and Shin (2018) specifically focus on the role of independent experts and analysts in the ICO industry. They analyze the contents of ICO documents and rate the projects, reducing information asymmetry (Sehra, Smith, & Gomes, 2017). As stressed earlier, the main online hub for ICO ratings is ICOBench.com; using its data and one-to-five ratings, Lee et al. (2018) find that the probability of a successful fundraising increases by 19.8% for any 1% increase in the average analyst rating, after controlling for other project features.

4.5.2 *The Underpricing of Tokens*

Ritter and Welch (2002) and Ljungqvist (2007) discuss the underpricing of IPO shares (i.e. the first-day return of IPO shares after the admission to trading, compared to the offer price) analyzing the reasons why the shares of newly listed companies are issued at a discount. This can be related to a number of reasons: information asymmetry, signaling, risk aversion, information gathering from the market, bribery. In ICOs, it is more complex to define the token initial underpricing due to the possibility that tokens are allocated at different prices during different phases of the offering. A possible solution is to compute an average of the prices, weighted by the volumes of the different tranches. Another problem is that during the offering the exchange rate of cryptocurrencies against major fiat currencies may vary, this increasing or decreasing the underpricing in US dollars. Adhami et al. (2018) and

Table 4.2 Statistics on ICO tokens

	Mean	Median	Sample size
Token supply (million)	220,562	100	808
% distributed tokens	62%	66%	794
First day trading volume (USD)	2,647,677	55,391	398
First day underpricing	745%	19%	366

Source: analysis of proprietary sample of ICO occurred from 2015 to 2017. Trading data are obtained from CoinMarketCap.com, as of July 2018

Lee et al. (2018) use proxies to measure the actual offering price, whereas Momtaz (2018) simply considers the underpricing as the percentage difference between the first day opening price of the tokens on the secondary market and the closing price on the same day, but clearly this is a biased measure of the ICO underpricing. Adhami et al. (2018) find that for the earliest 140 successfully listed tokens, the average first-day underpricing level was huge, at about 920%, which surely contributed to the speculative pressure in the market and the runups to token sales. Lee et al. (2018), using a more recent sample of ICOs, find that the average underpricing of tokens is 158%, which is still impressively higher compared to the average IPO underpricing in the US, measured by Jay Ritter[3] (an average of 18%). Still, in both the previous studies, the median ICO underpricing was about 24%, showing that the distribution of underpricing (and thus of the "money left on the table" by ICO entrepreneurs) is very highly skewed to the right. Benedetti and Kostovetsky (2018) analyze the performance of 609 listed tokens collected from CoinMarketCap.com and decompose the underpricing into offering-to-first-day-open return, on average 246% (median rate of return of 21%), and first-day-opening-to-close, on average 273% (median rate of return of 29%). Indeed, Momtaz (2018) finds a significant but much smaller mean underpricing for ICOs, ranging from 6.8 to 8.2%, due to the bias mentioned before. Nevertheless, the findings are all compatible with Momtaz's 'market liquidity hypothesis for ICOs', which states that project promoters have strong incentives to underprice their tokens to increase both market liquidity and the number of users in their platform. The value of a platform increases with the number of users (Catalini & Gans, 2018), this providing an incentive to attract new adopters. Therefore, we posit that the significantly large underpricing of ICO tokens is a reward for the high initial risk taken by crowd sale participants but is designed also to attract new market participants.

Table 4.2 presents the statistics about the initial token underpricing based on our hand-collected sample of 922 ICOs concluded up to December 2017, according to the methodology by Adhami et al. (2018). The mean value is 745% and the median value is equal to 19%.

So far, only few statistically significant determinants of the ICO underpricing have been found, and they include: the quality of the management team (Momtaz, 2018), previous work experiences of the CEO (Momtaz, 2018), if the ERC20

[3]https://site.warrington.ufl.edu/ritter/ipo-data/

technical standard is used to build the token (Momtaz, 2018), the fundraising volume (Benedetti & Kostovetsky, 2018), the bonuses offered during the campaign, like early birds or bounties (Lee et al., 2018). Marketing costs and bonuses can be compared to the promotional costs of an IPO, which have been found to counter-balance IPO underpricing (Habib & Ljungqvist, 2001). There is clearly much room for further exploratory studies in this field.

4.5.3 The Aftermarket Performance of Tokens

Another possible future research stream on ICOs is the dynamics of token prices after the offering. The IPO literature shows that the return of newly listed shares is generally poor in the medium and long run (Ritter, 1991). Lee et al. (2018) find similar results in the ICO market; they compute the token buy-and-hold returns on a 3, 6 and 12-month basis and find on average negative values (−35%, −37% and − 125% respectively). These are excess returns calculated with respect to a value-weighted index of Bitcoin and Ether, the main benchmark of the crypto market. Longer time series of token prices will be available in the future and it will be possible to engage in more significant analyses. Combining figures about token returns in the short and long run, we underline that many ICO pledgers initially earned significant profits, but suffered losses afterwards, with a number of tokens that have been delisted from electronic exchanges. The most significant determinant of the token return performance found by Lee et al. (2018) is that positive analyst coverage of a token is able to predict positive 3-month and 12-month returns, supposedly because of the informative due diligence performed by experts that focuses on the team and long-term vision of the project.

After talks with practitioners in the market, we underline that token markets are neither regulated nor supervised by public authorities, and transparency is very poor. Token prices may be easily manipulated; therefore, we must be very careful in interpreting the data. Using our dataset of ICOs concluded by the end of 2017, Table 4.3 presents the cumulative returns of tokens in different time periods after the ICO. Tokens de-listed as of July 2018 were not included in this sample, this introducing a survivorship bias. Nonetheless, absolute returns are significantly negative in all the periods observed, and adjusting for the return of the Ether cryptocurrency only worsens the performance, as confirmed by previous empirical works.

Studying a related but different topic, Masiak, Block, Masiak, Neuenkirch, and Pielen (2018), still relying on the literature of IPOs (precisely, Lowry & Schwert, 2002), have developed a VAR model to study market cycles for ICOs. They show that shocks to ICO volumes are persistent and that shocks in Bitcoin and Ether prices have a substantial and positive effect on ICO volumes, and higher ICO volumes cause lower cryptocurrency prices. While these results shed some light on the interaction between cryptocurrency trends and ICOs, they may also be useful for

Table 4.3 ICO token performance

Days		Median (%)	Mean (%)	Sample size
30	Return	−15	−13	319
	Adj return	−27	−28	
60	Return	−6	−9	318
	Adj return	−37	−39	
90	Return	−11	−11	315
	Adj return	−40	−48	
120	Return	−17	−16	309
	Adj return	−55	−58	
150	Return	−13	−18	302
	Adj return	−65	−64	
180	Return	−18	−17	289
	Adj return	−72	−71	
210	Return	−1	0	259
	Adj return	−78	−80	

Source: analysis of proprietary sample of ICO occurred from 2015 to 2017. Trading data are obtained from CoinMarketCap.com as of July 2018

project promoters in order to better time their crowd sales, as the aggregate volume of ICO funds is linked to these macro-like shocks.

4.5.4 *Token Valuation*

There is no consensus on how cryptocurrencies should be valued, or in other words what are the fundamental drivers of their prices. The pricing of ICO tokens is even more complex, as not only they are dependent on venture-specific factors but are also, to some extent, derivatives of native cryptocurrencies (at least the bulk of the tokens, i.e. those without their own specific blockchain). Rephrasing our previous definition, a token is a unit of value created by an organization to fuel its business model by favoring user interaction with its product while facilitating the distribution of benefits to all its stakeholders. From this point of view, token transactions within the venture ecosystem are a key driver of token value.

Cong et al. (2018) present a dynamic model of a digital economy with endogenous user adoption and native tokens that facilitate transactions and business operations. Their model is a continuous-time asset pricing model, with a continuum of agents that are heterogeneous in their transactional needs. In such a setting, agents consider a two-step decision on (1) whether to pay a participation cost to join the platform, and if so, (2) how many tokens to hold, this depending on both blockchain trade surplus and the expected future token price. The token pricing formula is the solution to an ordinary differential equation with boundary conditions that rule out bubbles, in line with our goal of valuing tokens based on platform fundamentals. The basic dynamics depend on the expectation of token price appreciation, that leads more agents to join the platform, allowing tokens to capitalize future user adoption, eventually enhancing welfare and reducing user-base volatility. Various extensions to the basic model are also presented in the paper.

Another relevant economic modelling effort for token valuation has been pursued by Sockin and Xiong (2018). They introduce a crypto asset developed to facilitate transactions of certain goods or services in a platform, thus serving as a membership fee for households to join the marketplace and as a fee for miners to provide transaction services on the platform. The crypto asset price has then to clear such sub-markets and results in either no equilibrium or two equilibria, determining a scenario where it is sub-optimal for ICO promoters to commit ex-ante to fully disclose all the information about platform fundamentals.

Both the above-mentioned models have attractive features, and present some of the complex dynamics at play when a single token price must channel the different expectations of heterogenous agents in context of multiplicity of the roles or uses of the token itself. However, no practical solution to the imminent pricing concerns of investors has been proposed at the moment, with each analyst using different market prices as alleged drivers of token values and mostly relying on a judgement on the variegated and volatile news regarding ICO projects and the crypto world in general.

An attempt has been made by Rasskazova and Koroleva (2018) who propose a bottom-up approach to simulate the value of tokens, which includes steps such as the estimation of the tokens in circulation, the growth rate of the project, and applying a basic CAPM model. Still, they do not show systematic tests on a representative sample of the prediction power of this piece-wise model. Once operative, similarly to the regular revenue disclosures by listed companies, blockchain ventures should provide a public ledger of all the transactions occurring on their platform, as this is the key driver of token valuation. Surely, there are other factors affecting changes in ICO token values (e.g. increase of liquidity on secondary exchanges, hoarding of tokens by suppliers and investors, or 'burning' of a small fraction of tokens used on the platform by project managers), but the fundamentals of a (utility) token are still related to its core use within the platform, i.e. aimed at the service provided by the organization that issued them. However, contrary to IPOs where a company exhibits a certain value of revenues prior to the issuance of public shares of stock, ICOs must be pursued using only forecasts of future token transaction volumes, subject to the offering success. This makes the valuation of tokens prior to an ICO highly arbitrary, and – to some extent – very similar to the pricing of a complex option.

4.6 Conclusions

The new industry of token offerings is challenging the models of venture financing and offers new financial assets (ICO tokens) to be added in the investors' portfolio. The blockchain technology, through the development of 'smart contracts' has the potential to 'digitalize' a number of provisions commonly adopted in corporate finance (raising money) and asset management with evident advantages for entrepreneurs and investors. Blockchain finance should be rapidly regulated, as to enrich the opportunities for entrepreneurs to raise money and to bring transparency and protection for investors. Academic researchers are spending many efforts in the field, which will provide valuable insights to practitioners and regulators.

References

Adhami, S., Giudici, G., & Martinazzi, S. (2018). Why do businesses go crypto? An empirical analysis of Initial Coin Offerings. *Journal of Economics and Business, 100*, 64–75.

Ahlers, G. K., Cumming, D., Günther, C., & Schweizer, D. (2015). Signaling in equity crowdfunding. *Entrepreneurship Theory and Practice, 39*(4), 955–980.

Akerlof, G. A. (1970). The market for "lemons": Quality uncertainty and the market mechanism. *The Quarterly Journal of Economics, 84*(3), 488–500.

Amsden, R., & Schweizer, D. (2018). *Are blockchain crowdsales the new 'gold rush'? Success determinants of initial coin offerings* (Working Paper).

Arslanian, H., Diemers, D., Dobrauz, D., McNamara, G., & Wohlgemuth, L. (2018). *Initial coin offerings: A strategic perspective*. Pwc.

Benedetti, H., & Kostovetsky, L. (2018). *Digital tulips? Returns to investors in initial coin offerings*. Working paper.

Bergh, D. D., Connelly, B. L., Ketchen Jr., D. J., & Shannon, L. M. (2014). Signalling theory and equilibrium in strategic management research: An assessment and a research agenda. *Journal of Management Studies, 51*(8), 1334–1360.

Bertoni, F., & Giudici, G. (2014). The strategic reallocation of IPO shares. *Journal of Banking & Finance, 39*, 211–222.

Block, J. H., De Vries, G., Schumann, J. H., & Sandner, P. (2014). Trademarks and venture capital valuation. *Journal of Business Venturing, 29*(4), 525–542.

Block, J., Hornuf, L., & Moritz, A. (2018). Which updates during an equity crowdfunding campaign increase crowd participation? *Small Business Economics, 50*(1), 3–27.

Catalini, C., & Gans, J. S. (2018). *Initial coin offerings and the value of crypto tokens (No. w24418)*. Cambridge, MA: National Bureau of Economic Research.

Chohan, U. (2017). *Initial coin offerings (ICOs): Risks, regulation, and accountability* (Working Paper).

Cong, L. W., Li, Y., & Wang, N. (2018). *Tokenomics: Dynamic adoption and valuation* (Working Paper).

Connelly, B. L., Certo, S. T., Ireland, R. D., & Reutzel, C. R. (2011). Signaling theory: A review and assessment. *Journal of Management, 37*(1), 39–67.

Crosser, N. (2018). *Initial coin offerings as investment contracts: Are blockchain utility tokens securities?* (Working Paper).

Fisch, C. (2019). Initial coin offerings (ICOs) to finance new ventures. *Journal of Business Venturing, 34*(1), 1–22.

Giudici, G. (2015). Equity crowdfunding of an entrepreneurial activity. In D. B. Audretsch, E. E. Lehmann, M. Meoli, & S. Vismara (Eds.), *University Evolution, Entrepreneurial Activity and Regional Competitiveness*. Berlin, Heidelberg: Springer International Publishing.

Giudici, G., & Adhami, S. (2019). The impact of governance signals on ICO fundraising success. *Journal of Industrial and Business Economics, 46*(2), 283–312.

Gompers, P., & Lerner, J. (2000). *The venture capital cycle*. Cambridge, MA: MIT.

Habib, M. A., & Ljungqvist, A. P. (2001). Underpricing and entrepreneurial wealth losses in IPOs: Theory and evidence. *The Review of Financial Studies, 14*(2), 433–458.

Hacker, P., & Thomale, C. (2018). Crypto-securities regulation: ICOs, token sales and cryptocurrencies under EU financial law. *European Company and Financial Law Review, 15* (4), 645–696.
Hornuf, L., & Schmitt, M. (2016). *Does a local bias exist in equity crowdfunding?* (Working Paper).
Howell, S. T., Niessner, M., & Yermack, D. (2018). *Initial coin offerings: Financing growth with cryptocurrency token sales (No. w24774)*. Cambridge, MA: National Bureau of Economic Research.
Lee, J., Li, T., & Shin, D. (2018). *The wisdom of crowds and information cascades in fintech: Evidence from initial coin offerings* (Working Paper).
Lipusch, N. (2018). *Initial coin offerings: A paradigm shift in funding disruptive innovation* (Working Paper).
Ljungqvist, A. (2007). IPO underpricing. In *Handbook of empirical corporate Finance* (pp. 375–422). London: Elsevier.
Lowry, M., & Schwert, G. W. (2002). IPO market cycles: Bubbles or sequential learning? *Journal of Finance, 57*(3), 1171–1200.
Lukkarinen, A., Teich, J. E., Wallenius, H., & Wallenius, J. (2016). Success drivers of online equity crowdfunding campaigns. *Decision Support Systems, 87*, 26–38.
Masiak, C., Block, J. H., Masiak, T., Neuenkirch, M., & Pielen, K. (2018). *The market cycles of ICOs, Bitcoin, and Ether* (Working Paper).
Momtaz, P. P. (2018). *Initial coin offerings* (Working Paper).
Piva, E., & Rossi-Lamastra, C. (2018). Human capital signals and entrepreneurs' success in equity crowdfunding. *Small Business Economics, 51*(3), 667–686.
Rasskazova, A., & Koroleva, E. (2018, October). Investment simulation model for estimating the future value of tokens. In *2018 Eleventh International Conference "Management of Large-scale System Development" (MLSD)* (pp. 1–5). Prague: IEEE.
Ritter, J. R. (1991). The long-run performances of initial public offerings. *Journal of Finance, 46*, 3–27.
Ritter, J. R., & Welch, I. (2002). A review of IPO activity, pricing, and allocations. *Journal of Finance, 57*(4), 1795–1828.
Sehra, A., Smith, P., & Gomes, P. (2017). *Economics of initial coin offerings*. London: Allen & Overy.
Sockin, M., & Xiong, W. (2018). *A model of cryptocurrencies* (Working Paper).
Spence, M. (1973). Job market signaling. *The Quarterly Journal of Economics, 87*(3), 355–374.
Subrahmanyam, A., & Titman, S. (1999). The going-public decision and the development of financial markets. *The Journal of Finance, 54*(3), 1045–1082.

Saman Adhami is a Ph.D. student at the Vienna Graduate School of Finance. He is affiliated both with the Department of Finance, Accounting & Statistics at WU and with the Department of Finance at the University of Vienna. His research focus is on entrepreneurial finance with notable interest in fintech, crowdfunding, and initial coin offerings. In the past, he was teaching assistant of corporate finance at Bocconi University, where he also graduated from the Master of Science in Finance.

Giancarlo Giudici is an associate professor of corporate finance at Politecnico di Milano. He belongs to the faculty of MIP Graduate School of Business where he teaches Finance. He wrote several publications on domestic and international journals on the topics of entrepreneurship, corporate financing, listings and IPOs, venture capital and crowdfunding. He is the director of the Italian Observatories on Mini-Bond and Crowdinvesting at Politecnico di Milano, School of Management. He led several projects financed by public and private entities on the topics of competitiveness and firm financing. He is an adjunct professor at the Ton Duc Thang University in Ho Chi Minh City (Vietnam).

Chapter 5
The Blockchain–Sustainability Nexus: Can This New Technology Enhance Social, Environmental and Economic Sustainability?

Semen Son-Turan

Abstract With the rise and fall of the prominence of Bitcoin, blockchain technology, which provides public online ledgers used for the verification and recording of transactions, has started to become the center of attention for diverse parties in the global financial system. This chapter explores the nature of blockchain and discusses how it may contribute to, or obstruct, sustainability. To this end, first, blockchain technology is introduced. Next, a short discussion on sustainability is presented, including how it is defined, measured, reported, and understood in theoretical frameworks. After that, the 2015 United Nations Sustainable Development Goals are briefly explained. This is followed by a systematic literature review, which highlights the scarcity of literature linking blockchain to sustainability. Finally, the author offers her own reflections on the potential of blockchain to revolutionize the financial services industry and weighs up the pros and cons vis-a-vis sustainable development.

5.1 Introduction

First developed for the Bitcoin cryptocurrency, blockchain uses a decentralized transaction and data management technology. It relies on distributed ledger technology, which offers a consensus validation mechanism through a network of computers that facilitates peer-to-peer transactions. Through this process, the need for an intermediary or a centralized authority to update and maintain the information generated by transactions is eliminated, as each transaction is validated through a chain of transactions and is added as a new "block" to an already existing chain. In general, this cannot be changed or removed. The name "blockchain" was coined to encapsulate this concept (Michael, Cohn, & Butcher, 2018). Within this system,

S. Son-Turan (✉)
Department of Business Administration, MEF University, Istanbul, Turkey
e-mail: sons@mef.edu.tr

© Springer Nature Switzerland AG 2019

U. Hacioglu (ed.), *Blockchain Economics and Financial Market Innovation*, Contributions to Economics, https://doi.org/10.1007/978-3-030-25275-5_5

blockchain is: distributed, in the sense that it runs on computers that are provided by volunteers worldwide; public, since it resides on the network in lieu of a single institution, meaning everyone can view it at any time; and encrypted, as it uses public and private keys (Tapscott & Tapscott, 2016).

To create a flow of transactions using blockchain, first, the sender creates a smart contract by entering an encryption code in order to create a new block. Next, after being validated by all computers in the network, this new block gets added to the existing chain of blocks. Finally, the transaction is settled (Fig. 5.1). Thus, by its very nature, blockchain technology is a tamper-proof record of information, as it is accessible to everyone online.

Tapscott and Tapscott (2016) describe blockchain as an ingeniously simple, revolutionary protocol that allows transactions to be simultaneously anonymous and secure by providing a tamper-proof public ledger of value. However, they believe the potential of blockchain is much greater than merely driving digital currencies. In fact, they go as far as predicting that it may eventually record everything of value to humankind. And we are starting to see this happen. Since 2008, when blockchain was invented as a support for Bitcoin, it has increasingly been gaining prominence in other sectors (Yli-Huumo, Ko, Choi, Park, and Smolander (2016). These include food supply chains (Tian, 2016), healthcare (Mettler, 2016), smart cities (Biswas & Muthukkumarasamy, 2016), financial services (Trautman, 2016), medical data access (Azaria, Ekblaw, Vieira, & Lippman, 2016), sustainable development (Nguyen, 2016), and law enforcement (De Filippi & Hassan, 2018).

This chapter explores the nature of blockchain and discusses how it may contribute to, or obstruct sustainability within its three dimensions: social, environmental and economic. Therefore, the research questions become: (1) Can blockchain revolutionize the financial services industry? and, (2) is blockchain useful for, or detrimental to, the advancement of sustainable development? To answer these

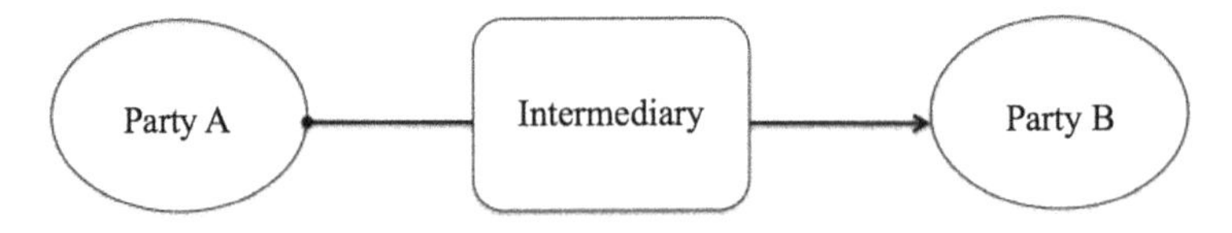

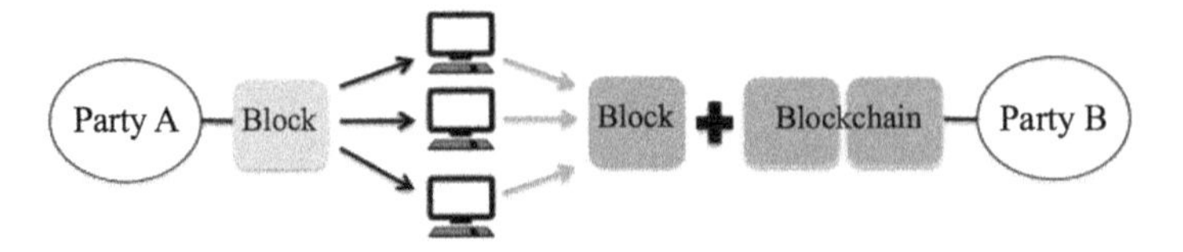

Fig. 5.1 Pre and post-blockchain technology "BT" transaction flow

questions, first blockchain technology has been introduced. Next, a brief overview of theoretical frameworks for sustainability reporting is provided. After that, the United Nations Sustainable Development Goals are briefly expatiated on. Subsequently, a systematic literature review is provided, highlighting the scarcity of literature linking blockchain to sustainability. Finally, the author offers her own thoughts and reflections on the potential of blockchain to revolutionize the financial services industry, while also contemplating the pros and cons of blockchain vis-a-vis sustainable development.

5.2 Sustainability

5.2.1 History

Environmental problems, or what we would today refer to as sustainability problems, such as deforestation and the salinization and loss of fertility of soil, go as far back as the ancient Egyptian, Mesopotamian, Greek, and Roman civilizations (Du Pisani, 2006). However, global recognition of the growing negative externalities, such as the generation of waste and ecological damage on a massive scale, associated with the excessive consumption of finite natural resources exacerbated by an unbridled population growth as well as a myriad of problems like climate change, desertification, land degradation, food security, and greenhouse gas fluxes in terrestrial ecosystems, did not arrive until the second half of the twentieth century.

1972 marks the beginning of a concerted global effort towards solving problems related to environmental sustainability through the UN Conference on the Human Environment held in Stockholm. Another milestone in the history of sustainability is the UN release of the Brundtland Report in 1987 where the first definition of sustainable development (SD) was coined: “SD is development that meets the needs of the present without compromising the ability of future generations to meet their own needs”. What followed was a series of conferences, summits, resolutions and declarations that culminated in the 17 United Nations (UN) Sustainable Development Goals (SDGs) and Agenda 2030, which were unanimously agreed by all UN member states in 2015. Therefore, during this time period, sustainability has evolved from being a rather inconsequential matter to becoming a mainstream concern for many stakeholders.

While used interchangeably, sustainable corporate social responsibility (CSR), environmental social governance (ESG), social responsibility, triple bottom line, and social reporting all refer to the same phenomenon: the non-financial disclosure by institutions that clearly show (i.e. quantify) how much harm they inflict upon, or good they do, to their communities, the environment and the economy.

Deegan, 2010 provides a comprehensive definition of social reporting, which he defines as the provision of information about the performance of an organization in relation to its interaction with its physical and social environment. This definition includes, among others: (1) interaction with the local community; (2) level of support for community projects; (3) level of support for developing countries; (4) health and safety record; (5) training, employment and education programs; and, (6) environmental performance.

5.2.2 Reporting and Theoretical Perspectives

As the world has watched the concept of sustainable development unfold on national, corporate, and individual levels, regulators and accounting professionals have been somewhat perplexed about its measurement, assessment and reporting. Sustainability reporting is a relatively new term that began appearing in the literature in the 1990s under names such as "environmental reporting", "social accounting", and "corporate social responsibility" (Lozano et al., 2015). As with the definition of SD and the confusion surrounding its nature, scholars have proposed various methods of measuring and reporting it.

Over the years, an increasing number of national and international institutional groups have convened and cooperated to discuss how non-financial information should be reported and to what extent. Among others, the Social Accounting Standards Board (SASB) (https://www.sasb.org), the Global Reporting Initiative (GRI) (https://www.globalreporting.org/), and the International Integrated Reporting Council (IIRC) (https://integratedreporting.org/) have paved the way in developing social accounting standards and related disclosures for organizations.

Currently, around the world, sustainability accounting is not enforced in the majority of jurisdictions. However, in Europe, the debate on whether sustainability reporting is necessary or not has been put to rest through the introduction of the groundbreaking EU Directive on the disclosure of non-financial and diversity information (Directive 2014/95/EU), which aims at greater business transparency and accountability on social and environmental issues.

According to a report by the World Business Council for Sustainable Development (WBSCD) and its global partners, since the 1992 Rio Earth Summit the number of sustainability reporting requirements—the provisions which specify either mandatory or voluntary disclosure requirements of specific non-financial information—has increased more than tenfold. Consequently, there are now over 1000 reporting requirements that have been introduced by various national and supranational bodies.

While, as indicated above, sustainability reporting, with some regional exceptions, is still voluntary in nature, theoretical perspectives have been adopted as to why it has become so widespread. Among these, stakeholder theory and legitimacy theory stand out as potentially viable explanations. According to stakeholder theory, a corporation's continued existence requires the support of its stakeholders, which means that a corporation must adjust its activities to gain that approval. The more powerful the stakeholders, the more the company must adapt. Due to this, social disclosure is seen as part of the dialogue between a company and its stakeholders (Gray, Kouhy, & Lavers, 1995). On the other hand, legitimacy theory is probably the most widely used theory to explain non-financial reporting (Campbell, Craven, & Shrives, 2003). Legitimacy theory has an advantage over other theories in that it provides disclosing strategies that may be empirically tested, which organizations can adopt to legitimize their existence (Gray et al., 1995).

5.2.3 Corporate Adoption of Sustainability Standards

When directives exist that are mandatory, such as Directive 2014/95/EU of the European Parliament and of the Council, corporations are bound to follow the outlined sustainability targets. However, it is at the discretion of corporations whether or not they wish to follow the targets of other organizations. Despite this, it has been seen that corporations, regardless of their origin and location, have been adopting these standards. In fact, they have been adopting them, at a faster rate and on a more comprehensive level than non-profit institutions. One reason for this may be the fact that "corporation" has long been associated with capitalistic and pragmatic practices that serve, in essence, the selfish pursuits of the owner. These pursuits are reliant on running legitimized operating practices and pleasing stakeholders. This is probably why corporations have been leading the way in closely following the evolution of sustainability. Standards serve as criteria and, in the corporate world, it is these criteria that affect reputation and stakeholder pressure.

Many other organizations have sprung up which are establishing frameworks for reporting sustainability-oriented activities. These include the Carbon Project (https://www.globalcarbonproject.org/), the GRI, provisional standards by the SASB, and Integrated Reporting Standards. However, despite recent interest, the sustainability targets laid out by these organizations are still not being met by most corporations.

5.2.4 The United Nations Sustainable Development Goals and Agenda 2030

As has been discussed, since 1972, numerous organizations have been developing sustainability targets for voluntary uptake by corporations. However, 2015 saw a leap forward in the development of sustainability targets when the United Nations declared 17 Sustainable Development Goals and Agenda 2030, which were unanimously agreed by all UN member states. The SDGs comprise a global agenda to end poverty, hunger and inequality, and to effectively tackle climate change by 2030. The SDGs contain 17 core goals which are geared towards establishing a sustainable and prosperous world. As an appendage to this agenda, the UN, in particular, urges the private sector and impact investors to contribute towards the achievement of these 17 goals, which include: (1) No Poverty; (2) Zero Hunger; (3) Good Health and Well-being; (4) Quality Education; (5) Gender Equality; (6) Clean Water and

Sanitation; (7) Affordable and Clean Energy; (8) Decent Work and Economic Growth; (9) Industry, Innovation and Infrastructure; (10) Reduced Inequality—within and among countries; (11) Sustainable Cities and Communities; (12) Responsible Consumption and Production; (13) Climate Action; (14) Life Below Water; (15) Life on Land; (16) Peace and Justice Strong Institutions; and (17) Partnerships to Achieve the Goal.

5.3 Blockchain and Governmental Projects

To move closer to the goals of Agenda 2030, several governmental projects are already using blockchain technology. For instance, the World Food Programme's "Blockchain for Zero Hunger" is taking the first steps to harness blockchain technology to enhance their ability to provide effective, efficient assistance to the people they serve. So far, pilot projects with Syrian refugees in Pakistan and Jordan have been successfully implemented whereby blockchain is being used to record every transaction that occurs at a certain retailer; this ensures greater security and privacy for the refugees and also allows for improved reconciliation and a significant reduction of transaction fees.[1]

5.4 Literature Review

In this section, a systemic literature review is provided regarding blockchain and sustainability. A "systematic review is a specific methodology that locates existing studies, selects and evaluates contributions, analyses and synthesizes data, and reports the evidence in such a way that allows reasonably clear conclusions to be reached about what is and is not known" (Denyer & Tranfield, 2009: 671). Due to its wide coverage, the ISI Web of Science database ("WoS") is used for this review. 2008 is used as the starting point for the review, as this is the year that Satoshi Nakamoto invented blockchain to serve as the public transaction ledger to support Bitcoin (Nakamoto, 2008).

To conduct this literature review, first the selection of databases and time range was determined. Next, the search terms were chosen at the author's discretion. Finally, papers deemed irrelevant as they lie outside the social sciences discipline were eliminated (Fig. 5.2).

Thus, using a reductionist approach, the output of the search was limited solely to relevant papers that specifically focus on the given search terms. Additionally, non-contextual studies, such as those referring to sustainability in the "continuity" sense, were eliminated. Table 5.1 presents the search findings.

[1]https://innovation.wfp.org/project/building-blocks

1. Selection of Database: Web of Science Database (2008 - December 2018)

2. Selection of Search Terms:

"Blockchain and sustainable development"

"Blockchain and sustainable development goals"

"Blockchain and sustainability reporting"

"Blockchain and SDGs"

3. Reduction of output to relevant papers

Fig. 5.2 Systematic literature review representation

Table 5.1 Output from WoS

Search terms	Search results
Blockchain and sustainability	Giungato, P., Rana, R., Tarabella, A., & Tricase, C. (2017). Current trends in sustainability of bitcoins and related blockchain technology. *Sustainability*, *9*(12), 2214.
Blockchain and sustainable development	Nguyen, Q. K. (2016, November). Blockchain -a financial technology for future sustainable development. In *Green Technology and Sustainable Development (GTSD), International Conference on* (pp. 51–54). IEEE.

As can be seen, the search resulted in only one paper and one conference proceeding published over the past 2 years. However, both touch upon the relation of sustainability and blockchain and are, in that aspect, unique. Giungato et al. (2017) seek to define and evaluate the literature concerned with the sustainability of bitcoin, considering the environmental impacts, social issues and economic aspects and, in particular, focus on the current trends in exploring the sustainability concepts related to bitcoin diffusion. They conclude that bitcoin is a living organism in the digital ecosystem and therefore can be environmentally sustainable since it requires few natural resources (e.g., fossil fuels) to sustain and maintain the exchange system of value in comparison with other payment or banking circuits (like credit cards). Nguyen (2016), on the other hand, attempts to synthesize and analyze available information with a focus on the role of blockchain as a financial tool that can potentially play an important role in the sustainable development of the global economy.

Since the WoS search returned too few results, a Google Scholar search was added, which resulted in 29 additional papers (Table 5.2).

The Google Scholar search shows that while the topic is relatively more addressed in conference proceedings and book chapters, the relation between blockchain technology and sustainability refers predominantly to the ecological dimension of sustainability, such as green supply chains (articles 15, 18, 21, 23), carbon emissions (article 19) and manufacturing (articles 16, 17) and architecture/ smart cities (articles 20, 22), respectively. Interdisciplinary manuscripts on finance, blockchain and sustainability are almost non-existent. As a publication outlet, the journal, titled "Sustainability", stands out with respect to high quality manuscripts on various kinds of sustainability-related issues.

Table 5.2 Output from Google Scholar search

Search terms	Search results
Blockchain and sustainability	1. Giungato, P., Rana, R., Tarabella, A., & Tricase, C. (2017). Current trends in sustainability of bitcoins and related blockchain technology. *Sustainability*, *9*(12), 2214. 2. Chapron, G. (2017). The environment needs cryptogovernance. *Nature*, *545*(7655). 3. Sulkowski, A. J. (2018). Blockchain, Law, and Business Supply Chains: The Need for Governance and Legal Frameworks to Achieve Sustainability. 4. Vranken, H. (2017). Sustainability of bitcoin and blockchains. *Current opinion in environmental sustainability*, *28*, 1–9. 5. Truby, J. (2018). Decarbonizing Bitcoin: Law and policy choices for reducing the energy consumption of Blockchain technologies and digital currencies. *Energy research & social science*. 6. Wu, J., & Tran, N. (2018). Application of Blockchain Technology in Sustainable Energy Systems: An Overview. *Sustainability*, *10*(9), 3067. 7. Cocco, L., Pinna, A., & Marchesi, M. (2017). Banking on blockchain: Costs savings thanks to the blockchain technology. *Future Internet*, *9*(3), 25. 8. Park, L. W., Lee, S., & Chang, H. (2018). A Sustainable Home Energy Prosumer-Chain Methodology with Energy Tags over the Blockchain. *Sustainability*, *10*(3), 658. 9. Adams, R., Kewell, B., & Parry, G. (2018). Blockchain for good? Digital ledger technology and sustainable development goals. In *Handbook of sustainability and social science research* (pp. 127–140). Springer, Cham. 10. Delliere, E., & Grange, C. (2018). Understanding and Measuring the Ecological Sustainability of the Blockchain Technology. Thirty Ninth International Conference on Information Systems, San Francisco 2018. Retrieved from: https://aisel.aisnet.org/cgi/viewcontent.cgi?article=1121&context=icis201 11. Wehner, N. (2018). Sustainability Certification Goes

(continued)

Table 5.2 (continued)

Search terms	Search results
	Blockchain. *IIIEE Master Thesis*. 12. Poberezhna, A. (2018). Addressing Water Sustainability With Blockchain Technology and Green Finance. In *Transforming Climate Finance and Green Investment with Blockchains* (pp. 189–196). Academic Press. 13. Nikolakis, W., John, L., & Krishnan, H. (2018). How Blockchain Can Shape Sustainable Global Value Chains: An Evidence, Verifiability, and Enforceability (EVE) Framework. *Sustainability*, *10*(11), 3926. 14. Lee, H. L., Lin, Y. P., Petway, J., Settele, J., & Lien, W. Y. (2018). Consumption-Based Blockchain Accounting of Telecoupled Global Land Resource Debtors and Creditors. 15. Kouhizadeh, M., & Sarkis, J. (2018). Blockchain Practices, Potentials, and Perspectives in Greening Supply Chains. *Sustainability*, *10*(10), 3652. 16. Fu, B., Shu, Z., & Liu, X. (2018). Blockchain Enhanced Emission Trading Framework in Fashion Apparel Manufacturing Industry. *Sustainability*, *10*(4), 1105. 17. Ko, T., Lee, J., & Ryu, D. (2018). Blockchain Technology and Manufacturing Industry: Real-Time Transparency and Cost Savings. *Sustainability*, *10*(11), 4274. 18. Thiruchelvam, V., Mughisha, A. S., Shahpasand, M., & Bamiah, M. (2018). Blockchain-based Technology in the Coffee Supply Chain Trade: Case of Burundi Coffee. *Journal of Telecommunication, Electronic and Computer Engineering (JTEC)*, *10* (3-2), 121–125. 19. Mao, D., Hao, Z., Wang, F., & Li, H. (2018). Innovative Blockchain-Based Approach for Sustainable and Credible Environment in Food Trade: A Case Study in Shandong Province, China. *Sustainability*, *10*(9), 3149. 20. Orecchini, F., Santiangeli, A., Zuccari, F., Pieroni, A., & Suppa, T. (2018, October). Blockchain Technology in Smart City: A New Opportunity for Smart Environment and Smart Mobility. In *International Conference on Intelligent Computing & Optimization* (pp. 346–354). Springer, Cham. 21. Saberi, S., Kouhizadeh, M., Sarkis, J., & Shen, L. (2018). Blockchain technology and its relationships to sustainable supply chain management. *International Journal of Production Research*, 1–19. 22. Sun, J., Yan, J., & Zhang, K. Z. (2016). Blockchain-based sharing services: What blockchain technology can contribute to smart cities. *Financial Innovation*, *2*(1), 26. 23. Ahmed, S., & ten Broek, N. (2017). Food supply: Blockchain could boost food security. *Nature*, *550*(7674),
Blockchain and sustainable development	1. Nguyen, Q. K. (2016, November). Blockchain-a financial technology for future sustainable development. In *Green Technology and Sustainable Development (GTSD), International Conference on* (pp. 51-54). IEEE. 2. Alcamo, J. (2003). *Ecosystems and human well-being: a framework for assessment* (p. 245p). Island Press, Washington,

(continued)

Table 5.2 (continued)

Search terms	Search results
	DC, USA. 3. Tian, F. (2016, June). An agri-food supply chain traceability system for China based on RFID & blockchain technology. In *Service Systems and Service Management (ICSSSM), 2016 13th International Conference on* (pp. 1–6). IEEE. 4. Cocco, L., Pinna, A., & Marchesi, M. (2017). Banking on blockchain: Costs savings thanks to the blockchain technology. *Future Internet*, *9*(3), 25. 5. Chapron, G. (2017). The environment needs cryptogovernance. *Nature*, *545*(7655). 6. Rocamora, A. E., & Amellina, A. (2018). Blockchain Applications and the Sustainable Development Goals-Analysis of Blockchain Technology's Potential in Creating a Sustainable Future. Kanagawa, Japan: Institute for Global Environmental Strategies.

5.5 The Potential of Blockchain Technology to Revolutionize the Financial Services Industry, Sustainability Reporting and Sustainable Development

5.5.1 How Blockchain Can Revolutionize the Financial Services Industry

Trust in financial services has been widely lost due to several controversial developments, in particular with regard to events leading up to the financial (mortgage) crisis of 2008 (Son-Turan, 2017). However, blockchain has the potential to restore confidence in the finance sector and even add more value to it due to its ability to hold an accurate global ledger of the ESG performance of companies, thereby presenting to stakeholders a complete and more accurate picture of the organization. Blockchain's decentralized structure has the potential to provide more transparency to reporting and performance management. It can also provide a solution to the infamous principal-agent problem that is pervasive in financial markets and institutions by controlling and hindering managers from acting in their own interest. Furthermore, since verification of data is required by all stakeholders, from customers to auditors, and since clear reporting is made possible through blockchain technology, the market value of a public firm may be more truthfully reflected. This, in turn, will make markets more efficient and trustworthy. In particular, historically less efficient developing markets who have difficulties attracting foreign investors and direct investments may witness an inflow of hard currencies with blockchain-based less ambivalent market mechanisms in place, as violations of law will no longer go unnoticed. Moreover, blockchain may also act as an early warning system of, for instance, liquidity crises or fraudulent activity, thereby serving as an oversight system that may prevent potential financial crises.

5.5.2 *How Blockchain Is Useful for the Advancement of Sustainable Development*

There are many potential benefits arising from blockchain regarding sustainability. One benefit is reduced costs. The results of Cocco et al.'s (2017) study show that if the disadvantages of the Bitcoin system are overcome to allow blockchain technology to be implemented, it may be possible to handle financial processes in a more efficient way than under the current system. They believe that if banks are able to save on costs, these savings can then be used to promote economic growth and accelerate the development of green technologies. Another benefit is increased transparency. If corporations are more transparent, this may lead to greater pressure from stakeholders regarding sustainability-related issues. It is possible that such stakeholder pressure may drive more companies of all sizes, even start-ups, to adopt more sustainability-oriented approaches and sustainability reporting standards. A further benefit of the transparent nature of blockchain is that it has the ability to harness behavioral economic theory, and this in turn may empower consumers to call upon corporations to move toward sustainability. Behavioral economic theory recognizes the limits of human cognition and willpower. One aspect of this is herd behavior. Herd behavior causes individuals to think and act instinctively like the crowd they associate with, which is triggered through transparent transactions—the very essence of blockchain. Blockchain technology is dependent upon herd behavior, as to operate it requires millions of volunteers globally to provide the network on which the blockchain technology exists. As each new computer is added to the network, every volunteer sees a diminished cost for their share of the infrastructure.

While, as was seen previously, the transparency inherent in blockchain may lead shareholders to push for corporations to move towards sustainability, the same can be seen with the general public. This is particularly true of millennials. Millennials, the "green generation", are said to be very sensitive toward the economic, social and environmental impacts of their actions. Accordingly, top sustainability purchasing drivers for participants aged 15–20 (Generation Z) of a global online study[2] indicate that this group is willing to pay more if products: (1) come from a brand they trust (72%), (2) Are known for its health and wellness benefits (70%), (3) are made from fresh, natural and/or organic ingredients (69%) and, (4) are from a company known for being environmentally friendly. That humanity is moving towards a more socially conscious generation is evidenced by a study carried out by Goldman Sachs, in which Millennials are determined to "have a disproportionate desire for their investment decisions to reflect their social, political and environmental values". Further this group of people is also "more likely to accept a lower return or a higher risk related to an investment if it's in a company that has a positive impact on society

[2]https://www.nielsen.com/eu/en/insights/reports/2015/the-sustainability-imperative.html

and the environment".[3] Therefore, the green and tech-savvy generation will become a major stakeholder demanding sustainable investment products offered through reliable and transparent networks.

However, stakeholder pressure and pressure from the general public regarding sustainability will only happen if there is transparent and standardized reporting from corporations. Yet, to date, no harmonization, standardization, or common conceptual framework for sustainability reporting exists, which is currently causing legitimacy issues. This can be resolved if transactions are translated into a uniform language, by way of smart contracts facilitated through blockchain technology, as this will enable practitioners and policy makers to move closer to a unified sustainability reporting framework. Therefore, much needs to be done to move to a system of smart contracts enabled through blockchain that are enforceable without reliance on state intervention.

5.5.3 How Blockchain Is Detrimental to the Advancement of Sustainability

Despite the benefits, blockchain technology may also have drawbacks for sustainability. From a societal perspective, the integration of this technology into our lives and workplace may result in the potential loss of jobs. For decades, automation has already been replacing manual labor, and now scientists working on artificial intelligence are contemplating how managerial tasks can be outsourced to robots. Another disadvantage of blockchain technology is the sizable amount of energy it utilizes and the enormous carbon footprint it produces. Environmental activists, calling blockchain technology an "experimental concept", not yet a decade old and with no tangible intrinsic value, are purportedly aghast at the amount of global power consumption required to sustain the system (McGirk, 2018). The same source reports that by 2020, the Bitcoin network alone could use as much electricity as the entire world does today (McGirk, 2018). Evidence for this is already being seen. For example, the cryptocurrency, Bitcoin, uses an energy-intensive "mining process", which purportedly consumes nearly the same amount of energy as Ireland every year- 2.55 Gigawatts (Martineau, 2018). These factors are of concern if blockchain technology is to be considered part of the sustainability solution. However, scientists have recognized this problem and are currently searching for solutions to go "crypto-clean" (Kugler, 2018). In fact, measures are constantly being implemented to minimize the negative externalities not only peculiar to blockchain but to technology in general.

[3]https://medium.com/unleash-lab/un-sdgs-the-world-s-largest-intergenerational-wealth-transfer-f5a032b8542e

5.5.4 Solutions for Overcoming the Disadvantages of Blockchain for Sustainability

Some solutions are currently being investigated to alleviate the disadvantages of blockchain regarding sustainability. One solution is to use cryptographic purpose-driven tokens, issued by smart contracts, that can be used to fulfill societal goals or encourage governments and institutions to act sustainably. This approach is already being used by the SolarCoin Foundation who reward solar energy producers with blockchain-based digital tokens at the rate of one SolarCoin (SLR) per Megawatt-Hour (MWh) of solar energy produced (https://solarcoin.org) if they provide proof of carbon footprint reduction. Another solution has been put forward by Eikmanns (2018), who proposes four applications of blockchain that may promote future ecological and societal sustainability. These are as follows. First, blockchain may be applied in order to achieve efficiency increases through enabling the direct financing of sustainable projects and by cutting out middle men. Second, blockchain may be used to track resources; there is transparency and trust in blockchain-based corporate provided information, as transactions are infinitely and openly recorded and cannot be altered. Third, resource pricing systems, implemented through cap and trade or Pigovian taxes, could benefit from the use of blockchain and smart contracts. Fourth, complementary currency systems incentivizing individuals and corporations to act in a sustainable manner could be implemented; for example, offering reward tokens as proof of undertaking ecologically friendly behavior, such as planting trees or recycling.

5.5.5 Organizations and Corporations Using Blockchain Toward Sustainability

A number of organizations are currently implementing blockchain-based solutions towards sustainability. One example is the platform Global Fishing Watch who is, at present, harnessing big data and recognizing the potential for blockchain to crack down on the $23 billion annual global cost of illegal fishing. According to WWF and strategic partner BCG, it is believed these new technologies have the potential to secure sustainability across the ocean economy—a $2.5 trillion market—(Verberne, 2018). Another example is the UN Blockchain Multi-UN Agency Platform, which was founded by a group of UN employees. Their vision is for every single person on earth to have an ID and get access to education, health and other social services, and for the 2.5 billion unbanked people to be included in the global financial system, [...]".[4] It is likely that the UN's interest in such an approach is driven by the desire to

[4]https://un-blockchain.org

make UN services resilient, so that they could survive even if the UN is dissolved.[5] A further example is The World Blockchain Organization (WBO), registered with the UN Department of Economic and Social Affairs. The WBO is encouraging the implementation of a Global Code of Ethics for Blockchain to maximize blockchain technology's socio-economic contribution whilst minimizing its possible negative impacts. It is committed to promoting blockchain technology as an instrument in achieving the UN SDGs.[6] Another example is The European Blockchain Foundation. This foundation alone hosted 175 Blockchain related events in 2018.[7] Moreover, although primarily among practitioners, conferences focusing solely on blockchain and finance are gaining in popularity.[8]

A number of corporations are also implementing blockchain-based solutions towards sustainability. For example, technology market leaders like IBM, already have in place a broad range of blockchain banking solutions that enhance banking experiences for customers by condensing transactions, removing manual processes, and reducing friction in day-to-day trade finance, digital identities and cross-border payments.[9] It is not only the financial services sector that is undergoing change. Consulting firms are also seeing the potential of blockchain to transform financial services. Some firms are now offering new ranges of consulting, such as partnership support and blockchain applications workshops[10] as well as encouraging banks to "see the bigger picture and work together—and with non-banks—to help define the backbone that can underpin a universally accepted, ubiquitous global payment system that can transform how banks execute transactions".[11]

While blockchain is viewed as a game-changer, experts do not consider it to be a replacement for capital markets. Instead it is believed that this technology will offer the opportunity to fundamentally re-architect processes–driving blockchain from experimentation to mainstream adoption across multiple business applications, such as settlement optimization, client onboarding KYC/AML, standard settlement instructions, collateral management, regulatory audit and reporting, and a host of others.

[5]https://medium.com/%C3%B5petfoundation/world-blockchain-organization-united-nations-blockchain-foundation-endorses-opet-foundation-dc97d8c26ce8

[6]http://www.unwbo.org/#whoweare

[7]https://www.europeanblockchainfoundation.org

[8]http://blockchain.fintecnet.com/

[9]https://www.ibm.com/downloads/cas/DKQPWYXN

[10]https://www.capgemini.com/service/blockchain-solutions-for-banking-financial-services/

[11]https://www.accenture.com/in-en/insight-blockchain-technology-how-banks-building-real-time

5.6 Conclusion

As has been seen, while still in its infancy, blockchain technology has already started to revolutionize the financial services industry by triggering new business models, modes of transaction and consumer habits. For traditional institutions, blockchain may prove to be a destructive force. The fact that banks, are currently investing more in blockchain technology than tech companies is a testament to them viewing blockchain as an existential threat to their business.[12] However, for financial service providers, blockchain has marked a historical turning point. And for new ventures, entrepreneurs, and start-ups, blockchain is providing a way to democratize the business world. This is because it provides low barriers to entry, boosts the speed, efficiency and verification of transactions, and makes those transactions immutable and transparent. Thus, what was rhetoric only a couple of years ago has swiftly become reality. And, judging by the speed with which this new technology has entered our lives, it is sure to broaden its sphere of influence. However, as has also been seen, the question of whether or not blockchain will contribute to sustainable development or pose a major threat to societal, economical, and ecological subsistence is still open for debate.

Various supra-national organizations, such as the UN, IMF, OECD and The World Bank, have acknowledged that, regarding sustainability, something needs to be done quickly and on a universal scale to stop the various dangers and threats to human existence. Thus, the main goal of every act, be it on a personal, company or governmental level, needs to be, first and foremost, to strive to meet the aims of at least one of the 17 UN SDGs. The same is true of financial service providers. Many actors in the finance sector have embraced sustainability on a wide scale, which is evident in their comprehensive sustainability reports published on their websites and also made available through sustainability initiatives such as the GRI database.[13]

There is no question, therefore, that blockchain is revolutionizing the industry with every bitcoin that is being mined and each smart contract being settled. How much and to what is extent it will contribute to the advancement of sustainable development still remains to be seen. However, it is looking like the pros will outweigh the cons. Traditional industries and state mechanisms have brought the world to the point where trust in most industries, led by the financial services industry, is lost, natural resources are at the verge of depletion and plans for expeditions to Mars are underway. Blockchain technology, if its negative externalities are controlled, may be the light at the end of the tunnel for a sustainable future, efficient and transparent markets, and the mindful green generation.

---

[12]https://www.euromoney.com/article/b19711r9ycx8xk/banks-investing-more-in-blockchain-than-tech-companies

[13]http://database.globalreporting.org/

References

Azaria, A., Ekblaw, A., Vieira, T., & Lippman, A. (2016, August). Medrec: Using blockchain for medical data access and permission management. In *Open and Big Data (OBD), International Conference on* (pp. 25–30). IEEE.

Biswas, K., & Muthukkumarasamy, V. (2016, December). Securing smart cities using blockchain technology. In *High Performance Computing and Communications; IEEE 14th International Conference on Smart City; IEEE 2nd International Conference on Data Science and Systems (HPCC/SmartCity/DSS), 2016 IEEE 18th International Conference on* (pp. 1392–1393). IEEE.

Campbell, D., Craven, B., & Shrives, P. (2003). Voluntary social reporting in three FTSE sectors: A comment on perception and legitimacy. *Accounting, Auditing & Accountability Journal, 16*(4), 558–581.

Cocco, L., Pinna, A., & Marchesi, M. (2017). Banking on blockchain: Costs savings thanks to the blockchain technology. *Future Internet, 9*(3), 25.

De Filippi, P., & Hassan, S. (2018). *Blockchain technology as a regulatory technology: From code is law to law is code.* arXiv preprint arXiv:1801.02507.

Deegan, C. (2010). Organizational legitimacy as a motive for sustainability reporting. In *Sustainability accounting and accountability* (pp. 146–168). New York: Routledge.

Denyer, D., & Tranfield, D. (2009). Producing a systematic review. In D. Buchanan & A. Bryman (Eds.), *The Sage handbook of organizational research methods* (pp. 671–689). Thousand Oaks, CA: Sage Publications Inc.

Du Pisani, J. A. (2006). Sustainable development–historical roots of the concept. *Environmental Sciences, 3*(2), 83–96.

Eikmanns, B. C. (2018). Blockchain: Proposition of a new and sustainable macroeconomic system. FSBC working paper. Retrieved from: http://www.explore-ip.com/2018_Blockchain-and-Sustainability.pdf

Giungato, P., Rana, R., Tarabella, A., & Tricase, C. (2017). Current trends in sustainability of bitcoins and related blockchain technology. *Sustainability, 9*(12), 2214.

Gray, R., Kouhy, R., & Lavers, S. (1995). Corporate social and environmental reporting: A review of the literature and a longitudinal study of UK disclosure. *Accounting, Auditing & Accountability Journal, 8*(2), 47–77.

Kugler, L. (2018). Why cryptocurrencies use so much energy—And what to do about it. *Communications of the ACM, 61*(7), 15–17. Retrieved from: https://cacm.acm.org/magazines/2018/7/229045-why-cryptocurrencies-use-so-much-energyand-what-to-do-about-it/fulltext

Lozano, R., Ceulemans, K., Alonso-Almeida, M., Huisingh, D., Lozano, F. J., Waas, T., et al. (2015). A review of commitment and implementation of sustainable development in higher education: Results from a worldwide survey. *Journal of Cleaner Production, 108*, 1–18.

Martineau, P. (2018). Bitcoin is consuming as much energy as the country of Ireland. In *The future*. Retrieved from: https://theoutline.com/post/4561/bitcoin-is-consuming-as-much-energy-as-the-country-of-ireland?zd=1&zi=7tusqufp

McGirk, J. (2018). Is blockchain energy use sustainable? *Chinadialogue*. Retrieved from: https://www.chinadialogue.net/article/show/single/en/10606-Is-blockchain-energy-use-sustainable-.

Mettler, M. (2016, September). Blockchain technology in healthcare: The revolution starts here. In *e-Health Networking, Applications and Services (Healthcom), 2016 IEEE 18th International Conference on* (pp. 1–3). IEEE.

Michael, J., Cohn, A., & Butcher, J. R. (2018). BlockChain technology. *The Journal*. Retrieved from: https://www.steptoe.com/images/content/1/7/v3/171269/LIT-FebMar18-Feature-Blockchain.pdf

Nakamoto, S. (2008). Bitcoin: A peer-to-peer electronic cash system.

Nguyen, Q. K. (2016, November). *Blockchain-a financial technology for future sustainable development.* In Green Technology and Sustainable Development (GTSD), International Conference on (pp. 51–54). IEEE.

Son-Turan, S. (2017). Emerging trends in the post-regulatory environment: The importance of instilling trust. In *Risk management, strategic thinking and leadership in the financial services industry* (pp. 345–354). Cham: Springer.

Tapscott, D., & Tapscott, A. (2016). *Blockchain revolution: How the technology behind bitcoin is changing money, business, and the world.* New York: Penguin.

Tian, F. (2016, June). *An agri-food supply chain traceability system for China based on RFID & blockchain technology.* In Service Systems and Service Management (ICSSSM), 2016 13th International Conference on (pp. 1–6). IEEE.

Trautman, L. J. (2016). *Is disruptive blockchain technology the future of financial services*? 69 The Consumer Finance Law Quarterly Report 232 (2016). Retrieved from https://papers.ssrn.com/sol3/papers.cfm?abstract_id=2786186.

Verberne, J. (2018). How can blockchain serve society? In *World Economic Forum.* Retrieved from: https://www.weforum.org/agenda/2018/02/blockchain-ocean-fishing-sustainable-risk-environment/

Yli-Huumo, J., Ko, D., Choi, S., Park, S., & Smolander, K. (2016). Where is current research on blockchain technology?—A systematic review. *PLoS One, 11*(10), e0163477.

Semen Son-Turan received her BA degree from Boğaziçi University, Istanbul, Turkey, (1998). She completed her MBA at The Pennsylvania State University (Penn State), USA (2000) and holds a Ph.D. degree from Yaşar University, Izmir, Turkey (2014). She earned the title of “Associate Professor of Finance” in 2017 from YÖK (Council of Higher Education of Turkey). After her graduate education, Dr. Son-Turan took various corporate and managerial positions at companies such as Ford Otosan, Rehau, İş Asset Management, Ashmore Asset Management and ultimately established her own consulting company. Dr. Son-Turan has been teaching finance and accounting courses at US, German and Turkish institutions for over twenty years. She holds the capital market activities advanced level license and various other professional certifications. Her work has been published in acclaimed international books and journals. She joined MEF University’s business administration department as a full-time faculty member in 2014 and is still working there. Her research interests include behavioral finance, higher education finance, financial innovation, and sustainable finance.

Part II
Crypto-Currency Investment Strategies and Crypto-Markets

Chapter 6
Herding Behaviour in Cryptocurrency Market: CSSD and CSAD Analysis

Gülüzar Kurt Gümüş, Yusuf Gümüş, and Ayşegül Çimen

Abstract Cryptocurrencies get substantial attention of investors as recently created innovations. This chapter focuses on herding behaviour in the cryptocurrency market, considering CSSD and CSAD approaches for cryptocurrencies in the CCI 30 Index. The analysis focuses on the cryptocurrency index and cryptocurrencies, which have existed since the arbitrarily set starting date of the index. In addition to the CCI 30 Index, as a proxy for market, Bitcoin, Litecoin, Stellar, Monero, Dogecoin and Dash are used for empirical analysis. To the best of the author's knowledge, the CCI 30 Index is used for the first time as a proxy for market return. Despite the growing literature on cryptocurrencies, there is still a gap in herding behaviour in the cryptocurrency market. Results indicate no evidence of herding behaviour in the cryptocurrency market in both CSSD and CSAD approaches. The findings of both approaches are in line with the findings of the previous literature regarding the herding behaviour in cryptocurrencies such as Bouri et al. (Financ Res Lett 29:216–221, 2018) and Vidal-Tomás et al. (Financ Res Lett doi: 10.1016/j.frl.2018.09.008, 2018).

G. Kurt Gümüş (✉)
Department of International Business and Trade, Faculty of Business, Dokuz Eylül University, İzmir, Turkey
e-mail: guluzar.kurt@deu.edu.tr

Y. Gümüş
Department of Tourism Management, Reha Midilli Foça Tourism Faculty, Dokuz Eylül University, İzmir, Turkey
e-mail: yusuf.gumus@deu.edu.tr

A. Çimen
Department of Business Administration, Faculty of Economics and Administrative Sciences, Dokuz Eylül University, İzmir, Turkey
e-mail: aysegul.cimen@deu.edu.tr

© Springer Nature Switzerland AG 2019

U. Hacioglu (ed.), *Blockchain Economics and Financial Market Innovation*, Contributions to Economics, https://doi.org/10.1007/978-3-030-25275-5_6

6.1 Introduction

Technological developments affect all sides of the economy. Those improvements and explorations reshape the financial system as well. An article published in 2008 by Nakamoto also mentions these types of improvements. The author discussed the possibility of an electronic payment system between two willing parties to transact directly without a trusted third party. This system is called a blockchain and the medium of exchange is called crypto currency.

Bitcoin was the first cryptocurrency generated in that system. Today, there are almost 2000 cryptocurrencies (CoinMarketCap). Their total market value is $104.542.780.889 as of February 4, 2019. Bitcoin exists as the first mover advantage and has 58% market share according to the total market capitalization. XRP and Etherium are the closest followers with 12 and 11% market share respectively.

This chapter intends to identify and analyze the herding behaviour in the cryptocurrency market, taking CCI 30 Index and cryptocurrencies of that index into account for the January 1, 2015 and December 31, 2018 historical periods. The study contributes to the cryptocurrency market literature by considering herding behaviour of a larger sample and by using the first generated cryptocurrency index with the longest time period.

This paper is organized as follows. Section 6.2 describes the theoretical framework and terminology used throughout this paper as well as the previous studies. Section 6.3 describes data and research methodology used for empirical analysis. Sections 6.4 indicates the findings of the analysis and finally Sect. 6.5 provides the conclusion of this paper.

6.2 Theoretical Framework

Herding is defined as the result of a clear intent by investors to mimic other investors' behaviours (Bikhchandani & Sharma, 2000). For imitating another, an investor must be aware of and be affected by other investors' positions. For instance, the investor would invest without having information about other investors' actions, but does not invest when learning that other investors' don't invest. Thus, investors may herd on a wrong investment decision, if they are influenced by other investors' decisions.

One of the main reasons of following the others rather than using their own beliefs is the uncertainty. Owing to evolutionary reasons, people tend to imitate others in the case of uncertainty according to the socioeconomic theory. Parker and Prechter (2005) state that, people unconsciously follow and imitate others, which is known as herd behavior in finance literature. By imitating others, traders think that others know better than they do, so they keep imitating.

Emerging markets differ from developed markets in terms of depth of financial markets and variety of financial instruments, as well as the differences in regulations.

Kremer (2010) states that in developing countries uncertainty is higher than in developed ones, due to less developed regulatory frameworks. Namely, due to the uncertainty, traders may follow others with a hope of having higher returns.

Herding behaviour has taken place in financial markets since the first financial crisis known as Tulipmania in the seventeenth century followed by the 2008 subprime mortgage crisis and the dot-com bubble of the 2000s (Bouri, Gupta, & Roubaud, 2018).

Herd behaviour is common among all types of investors (institutional or individual), and generally causes high market volatility and instability in the markets (Spyrou, 2013). Causes of herding include inferring information from previous investors' actions, reacting to newly arrived important information, protecting reputation, and irrationality of the investors.

Avery and Zemsky (1998) stated that herding caused by informational cascade is not possible in case of plain information structures and price mechanisms, however herd may exist if complicated information structures, and uncertainty in asset value and information are assumed. Technology itself and also the ambiguity related to the blockchain system and crypto currencies are the rationality behind searching herding behaviour in the cryptocurrency market. Intrinsically, the system in and of itself causes participants to perceive as if there is uncertainty. From causes of herding behaviour perspective, the complicated structure of the blockchain system and differences among average cryptocurrency traders' information levels may explain the possibility of the existence of herding in the cryptocurrency market.

Cryptocurrencies are financial instruments currently introduced to the financial markets. For this reason, this gap has attracted the attention of researchers. The recent studies mostly focus on market efficiency and price dynamics (Corbet, Lucey, & Yarovaya, 2018a).

Considering the extreme speculative nature of cryptocurrencies, Bitcoin being the largest currency in cryptocurrency market, makes the cryptocurrency market volatile (Baur, Hong, & Lee, 2018). This extreme volatility in the market might result in herding behaviour. For this reason, traders of cryptocurrencies are not as sensitive as the traders in the financial markets (Bouri et al., 2018).

Sapuric and Kokkinaki (2014) investigates the relationship and volatility between Bitcoin and six major currencies whereas Cheung, Roca, and Su (2015) empirically analyzes the bubbles in the Bitcoin market, using the Phillips et al. methodology. With the help of this method, the authors find a number of short-lived bubbles, and three large bubbles lasting from 66 to 106 days. The occurrence of these bubbles is coincided with some major events that take place in the Bitcoin market, the most significant of these being the demise of the Mt. Gox exchange.

Zhang, Wang, Li, and Shen (2018) focuses on statistical characteristics of the cryptocurrencies return based on the existence of heavy tails, volatility clustering, leverage effects and the presence of a power-law correlation between price and volume. Chuen and Deng (2017) implemented some statistical methods such as ARIMA, GARCH and EGARCH modelling to the CRIX indices family in order to find out the volatility clustering phenomenon and the presence of fat tails.

Urquhart (2017) investigates the efficiency of Bitcoin returns from August 2010 to July 2016 by implementing various tests for randomness such as Ljung-Box, Runs, Bartels, Automatic variance test, BDS, R/S Hurst tests. The study finds that returns are significantly inefficient over the analyzed period. Then, the period is divided into two equal sub-samples. The tests reveal efficiency of returns in the sub-samples, suggesting that Bitcoin may be moving towards becoming more efficient.

Nadarajah and Chu (2017) test the efficiency in Bitcoin in USD from August 1, 2010 and July 31, 2016. Data is analyzed in three periods: the full period from the first of August 2010 to 31st of July 2016; the subsample period from the first of August 2010 to 31st of July 2013 and the subsample period from the first of August 2013 to 31st of July 2016. Eight different tests are implemented on the data to find out the efficiency of the Bitcoin market.

Kristoufek (2015) researches the main drivers of the price and price formation of Bitcoin by using utilized wavelets methodology. Although Bitcoin is assumed as a speculative financial instrument, the findings reveal that usage in trade, money supply and price level are the factors that play a role in Bitcoin price in the long term.

In the financial markets, the traders can either behave rationally or irrationally. When the traders behave rationally, assumptions of asset pricing models are proved. On contrary, if the investors behave irrationally and imitate others rather than using their own beliefs based on the information, herd behaviour occurs. The existence of herd behaviour in the financial markets means that the assumptions of Efficient Market Hypothesis are disagreed upon (Caparrelli, D'Arcangelis, & Cassuto, 2004; Fama, 1965; Lao & Singh, 2011).

Some papers have analyzed the relationship between digital currencies in the cryptocurrency market. For instance, Ciaian and Rajcaniova (2018) examines the interdependencies between Bitcoin and 16 digital currencies from 2013 to 2016. Findings show that Bitcoin and altcoin markets are interdependent. In the short term, the Bitcoin-altcoin price relationship is significantly stronger than in the long run.

Gandal and Halaburda (2016) investigate the daily price (i.e., exchange rate) data in the analysis from 2 May 2013 to 1 July 2014 between Bitcoin and 7 altcoins. The study examines how the prices of cryptocurrencies change in time by applying a reinforcement effect and a substitution effect. Findings show positive correlations between the cryptocurrencies.

Osterrieder and Lorenz (2017) analyze an extreme value analysis of the returns of Bitcoin. Study focuses on the risk properties of the Bitcoin exchange rate versus USD. The Data set is from September 2013 to September 2016 for Bitcoin and the G10 currencies. Empirical findings show Bitcoin returns are much more volatile, much riskier and exhibit heavier tail behaviour than the traditional currencies.

Corbet, Meegan, Larkin, Lucey, and Yarovaya (2018b) examines the return and volatility transmission among three Bitcoin, Ripple and Litecoin, and gold, bond, equities and the global volatility index (VIX). Findings indicate the major cryptocurrencies Bitcoin, Ripple and Lite are interconnected whereas the cryptocurrencies are isolated from other markets. The Bitcoin price can affect the levels of Ripple and Lite and cryptocurrencies may offer diversification benefits for investors in the short term period.

Due to being a new financial instrument, studies related to herding behaviour in the cryptocurrencies market is limited. For instance, Pele and Mazurencu-Marinescu-Pele (2019) investigate the herd behaviour in cryptocurrencies market, especially in Bitcoin by using Metcalfe's law in Bitcoin evaluation, however in the long-run, validity of Metcalfe's law for Bitcoin is debatable.

Vidal-Tomás, Ibáñez, and Farinós (2018) examines the herding in the cryptocurrency market with a dataset of 65 digital currencies from 1 January 2015 to 31 December 2017. Both cross sectional deviations of returns (CSSD) and cross sectional absolute deviation of returns (CSAD) approaches are used in the empirical analysis. The findings indicate that based on both approaches, there is no evidence of herd behaviour in the cryptocurrencies market, showing that the extreme price movements are explained by rational asset pricing models.

Poyser (2018) studies the empirical herding model based on Chang, Cheng, and Khorana (2000) methodology, and developed the model for both under asymmetric and symmetric conditions and the existence of different herding regimes by implementing the Markov-Switching approach. First 100 leading cryptocurrencies are analyzed for the study.

Bouri et al. (2018) examines the existence of herding behaviour in the cryptocurrency market. Cross-sectional absolute standard deviations (CSAD) approach is implemented on 14 leading cryptocurrencies. Based on the CSAD approach, existence of herding cannot be found. As Balcilar, Demirer, and Hammoudeh (2013) has mentioned, the parameters are assumed to be constant over time, which might result in misleading conclusions. For this reason, Bai and Perron (2003), tests are applied for structural breaks. In addition, Stavroyiannis and Babalos (2017) a time varying approach is implemented for a rolling window of 250 observations. Significant herding is found in some rolling windows.

6.3 Data and Methodology

6.3.1 Methodology

Herding behaviour is commonly discussed in financial markets especially for stock markets with different methodologies and definitions (Bikhchandani & Sharma, 2000; Chang et al., 2000; Chiang, Li, Tan, & Nelling, 2013; Demirer & Kutan, 2006; Olsen, 1996). The studies focusing on herding behaviour in cryptocurrencies are conducted by Poyser (2018), and Bouri et al. (2018) by following the similar methodologies of the herding literature.

Many models are generated to measure herd behaviour: Lakonishok, Shleifer and Vishny Measurement (developed by Lakonishok, Shleifer, & Vishny, 1991); Cross Sectional Volatility of Stocks (developed by Christie & Huang, 1995, and Chang et al., 2000); and Beta Herding (developed by Hwang & Salmon, 2004).

This chapter uses Christie and Huang (1995) methodology to detect herd behaviour in cryptocurrency market.

$$\mathrm{CSSD_t} = \sqrt{\frac{\sum_{i=1}^{N}\left(R_{i,t} - R_{m,t}\right)^2}{N-1}} \tag{6.1}$$

$CSSD_t$ stands for the cross-sectional standard deviation of stock return rates from the market return rate in period t. $R_{i,t}$ shows the return rate on i for time t and $R_{m,t}$ shows the return on the market portfolio in time t. N is the number of cryptocurrencies for the selected period.

Christie and Huang (1995) analysed herd behaviour under extreme market conditions with the following regression formula:

$$\mathrm{CSSD_t} = \alpha + \beta_1 D_t^L + \beta_2 D_t^U + \varepsilon_t \tag{6.2}$$

Cryptocurrency daily return is calculated from the following formula:

$$R_{i,t} = \frac{P_{i,t} - P_{i,t-1}}{P_{i,t-1}} \tag{6.3}$$

In Eq. (6.3), $P_{i,t}$ is the closing price of cryptocurrency i on day t and $P_{i,t-1}$ is the closing price of cryptocurrency i on the previous day (t-1). For the 1% upper and lower tails, 15 days of the highest and lowest returns are taken to justify the stress in tails. For the 5% upper and lower tails, 73 days of the highest and lowest returns are taken to show the stress in tails.

In addition to the CCSD methodology, cross-sectional absolute standard deviations (CSAD) is also applied to the same dataset.

$$\mathrm{CSAD_t} = \alpha_0 + \alpha_1 \left|R_{m,t}\right| + \alpha_2 R_{m,t}^2 + \varepsilon_t \tag{6.4}$$

Herding is assumed to be absent if $\alpha_1 > 0$ and $\alpha_2 = 0$. On the contrary, if herding is present, $\alpha_2 < 0$.

6.3.2 Data

The Crypto Currencies (CCI 30) Index is rule based and is delineated to gauge the size and movement of the cryptocurrency market. The index tracks the 30 largest cryptocurrencies by market capitalization. Main characteristics of the index are being diversified, being replicable, being transparent, providing detailed coverage of the whole blockchain sector and presenting the beyond compare risk-adjusted performance figure. The CCI 30 index was started on Jan. 1st, 2015. Constituents of

Table 6.1 Constituents of the crypto currencies index

Bitcoin	NEO
Ethereum	Ethereum Classic
XRP	NEM
EOS	Zcash
Litecoin	Waves
Bitcoin Cash	Tezos
Stellar	VeChain
TRON	Ontology
Binance Coin	Dogecoin
Cardano	Bitcoin Gold
Bitcoin SV	Qtum
Monero	OmiseGO
IOTA	Basic Attention Token
Dash	Zilliqa
Maker	0x

the index are listed in Table 6.1. Cryptocurrencies in the index are put in an order according to their market capitalization, which indicates that Bitcoin has the highest market capitalization, and on the other hand 0x is the last currency with the lowest market capitalization.

Constituents are selected by considering their adjusted market capitalization. Adjusted market capitalization regards volatility as a destabilizing factor in index composition. The index employs an exponentially weighted moving average of the market capitalization to smooth the volatility and achieve the most accurate market capitalization values.

The index value is calculated with the formula below:

$$I_t = \sum_{j=1}^{30} W_j \frac{P_j(t)}{P_j(0)} \tag{6.5}$$

Where I_t is the index value at time t, W_j is the weight of the jth name in the index, and P_j is the price of the jth name as a function of time.

Data covers the same period as the index. Daily price data of the cryptocurrencies in the index are gathered from CoinMarketCap. High price fluctuations are noticed on analysed cryptocurrencies (Fig. 6.1).

6.4 Results

Prior to running the regression analyses, preliminary tests should be applied. These preliminary tests include checking the normality with the Jarque-Bera test, testing serial correlation analysed with the Breusch-Godfrey Serial Correlation LM Test and White test is implemented for heteroscedasticity.

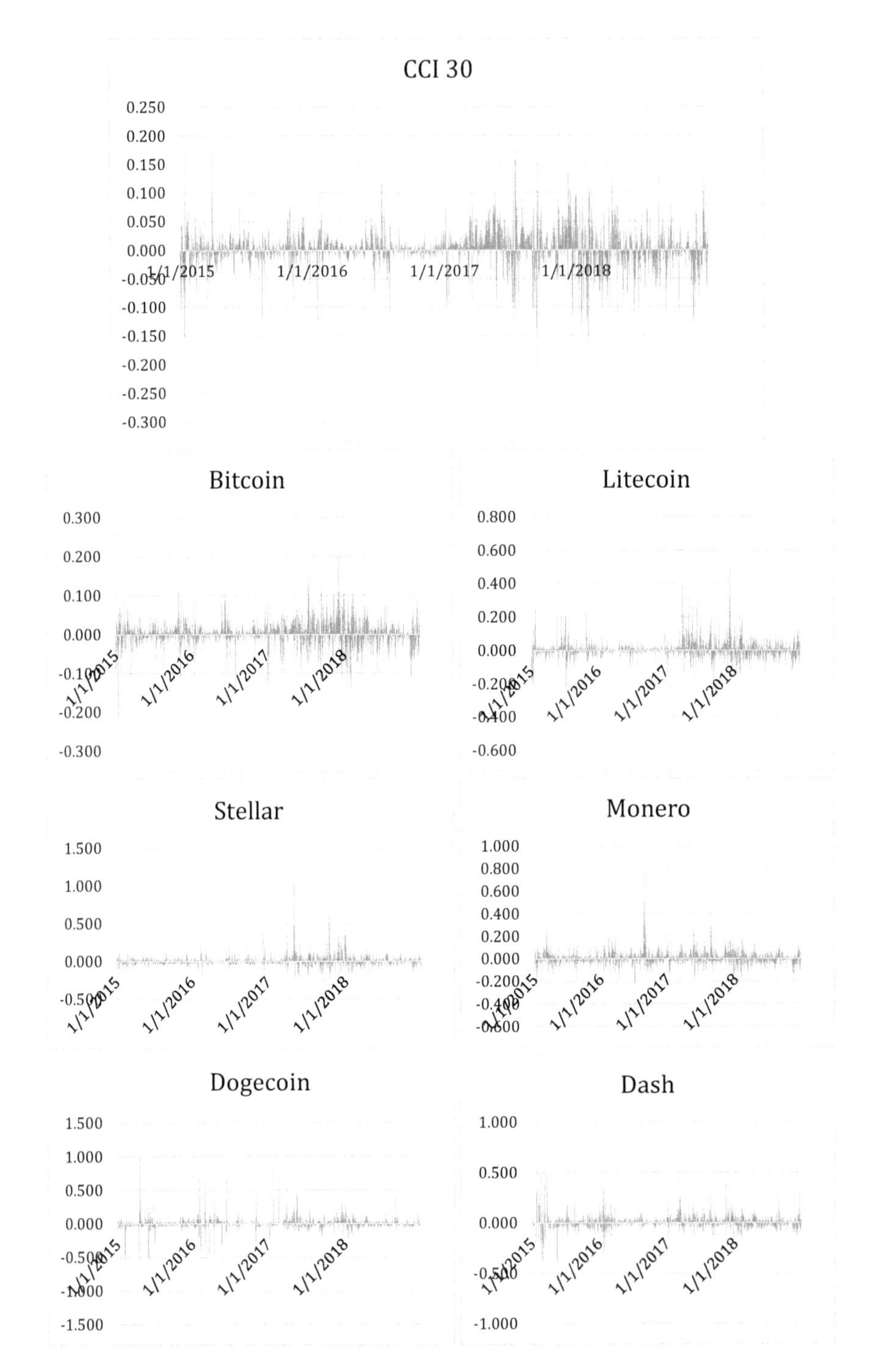

Fig. 6.1 CCI 30 Index and constituents daily return

Table 6.2 Regression results of daily CSSD on market returns

	1% Criterion	5% Criterion
Variables	Coefficients	Coefficients
Included Observations	1459	1459
α	0.062317 (0.0000)	0.059374 (0.0000)
D_t^L (β_1)	0.071806 (0.0000)	0.032540 (0.0000)
D_t^U (β_2)	0.078289 (0.0000)	0.057126 (0.0000)
F-statistic	26.20895	46.44711

Table 6.3 Regression results of daily CSAD on market return

Variables	Coefficients
Included Observations	1459
α	0.036052 (0.0000)
α_1 ($\lvert R_{m,t} \rvert$)	0.379745 (0.0010)
$\alpha_2 \left(R^2_{m,t}\right)$	0.151303 (0.8591)
F-statistic	33.51159

After the preliminary analysis, regression Eq. (6.2) was run to find out the presence of herd behaviour cryptocurrency market.

Table 6.2 shows the results of estimated coefficients for CSSD of returns during the period Jan 1, 2015 and December 31, 2018. The data is consisted of 1459 daily return for 6 cryptocurrencies that were in market during the given period. The (β_1) coefficient indicates the change in the amount of return dispersion given that cryptocurrency return is in the lowest 1 and 5% return, which is mentioned as lower market stress. On the other hand, the (β_2) coefficient shows the change in the amount of return dispersion given that cryptocurrency return is in the highest 1 and 5% return, which is also mentioned as upper market stress. The lowest and highest 1 and 5% refer to the extreme price movement days that lie in the upper and lower tails of the market return distribution.

Table 6.2 indicates the β_1 and β_2 coefficients of the regression analysis for CSSD. Negative value of β_1 is assumed as a proxy of herd behaviour existence. On the contrary, positive β_1 and β_2 coefficients are predicted as rational asset pricing models. According to Table 6.1, β_1 coefficient is not negative, but statistically significant in both 1 and 5% extreme tails. Findings show that there is no evidence of herd behaviour in the cryptocurrency market.

Table 6.3 shows the regression results for CSAD method. A positive and statistically significant α_1 coefficient shows that CSAD returns on cryptocurrencies is an increasing function of absolute value of markets returns.

Herding is assumed to be absent if $\alpha_1 > 0$ and $\alpha_2 = 0$. On the contrary, if herding is present, $\alpha_2 < 0$. Based on the findings of Table 6.2, α_2 is positive, which means that with the CSAD method herding is not present in the cryptocurrency market.

6.5 Conclusion

Behavioral issues in the field of finance has gained importance in the last few decades as the traders started to behave irrationally and disagree with the rational asset pricing models. Herding behavior is just one of the behavioral finance topics which has gotten the attention of researchers especially for stock markets. Cryptocurrencies are the other trending topic in financial markets. This chapter focuses on herding behaviour in the cryptocurrency market, considering CSSD and CSAD approach for cryptocurrencies in CCI 30 Index.

CSSD results indicate that during the period from January 1, 2015 and December 31, 2018, there is no evidence of herd behaviour in the cryptocurrency market when CCI 30 index is taken as market portfolio for both 1 and 5% extreme tails.

CSAD findings also show that, α_2 coefficient which is accepted as a proxy for herding behaviour is positive, but not statistically significant. Overall, given the findings of both methodologies, there is not statistically significant evidence that shows the presence of herd behaviour in the cryptocurrency market.

The findings of the both approaches are in line with the findings of the previous literature regarding the herding behaviour in cryptocurrencies. Bouri et al. (2018) and Vidal-Tomás et al. (2018) also investigates the herding behaviour in cryptocurrencies and in both studies empirical findings indicate the absence of herd behaviour in CSAD and CSSD methodologies, which can be interpreted as extreme price movements are explained by rational asset pricing models, just like the findings of this paper.

References

Avery, C., & Zemsky, P. (1998, September). Multidimensional uncertainty and herd behavior in financial markets. *The American Economic Review, 88*(4), 724–748.

Bai, J., & Perron, P. (2003). Computation and analysis of multiple structural change models. *Journal of Applied Econometrics, 18*(1), 1–22.

Balcilar, M., Demirer, R., & Hammoudeh, S. (2013). Investor herds and regime-switching: Evidence from Gulf Arab stock markets. *Journal of International Financial Markets, Institutions and Money, 23*, 295–321.

Baur, D. G., Hong, K., & Lee, A. D. (2018). Bitcoin: Medium of exchange or speculative assets? *Journal of International Financial Markets, Institutions and Money, 54*, 177–189.

Bikhchandani, S., & Sharma, S. (2000). Herd behaviour in financial markets. *IMF Staff papers, 47* (3), 279–310.

Bouri, E., Gupta, R., & Roubaud, D. (2018). Herding behaviour in cryptocurrencies. *Finance Research Letters, 29*, 216–221.

Caparrelli, F., D'Arcangelis, A. M., & Cassuto, A. (2004). Herding in the Italian stock market: A case of behavioural finance. *The Journal of Behavioural Finance, 5*(4), 222–230.

Chang, E. C., Cheng, J. W., & Khorana, A. (2000). An examination of herd behaviour in equity markets: An international perspective. *Journal of Banking & Finance, 24*(10), 1651–1679.

Cheung, A., Roca, E., & Su, J. J. (2015). Crypto-currency bubbles: An application of the Phillips–Shi–Yu (2013) methodology on Mt. Gox bitcoin prices. *Applied Economics, 47*(23), 2348–2358.

Chiang, T. C., Li, J., Tan, L., & Nelling, E. (2013). Dynamic herding behaviour in Pacific-Basin markets: Evidence and implications. *Multinational Finance Journal, 17*(3/4), 165–200.

Christie, W. G., & Huang, R. D. (1995). Following the pied piper: Do individual returns herd around the market? *Financial Analysts Journal, 51*(4), 31–37.

Chuen, D. L. K., & Deng, R. H. (2017). *Handbook of blockchain, digital finance, and inclusion: cryptocurrency, fintech, insurtech, regulation, Chinatech, mobile security, and distributed ledger*. London: Academic.

Ciaian, P., & Rajcaniova, M. (2018). Virtual relationships: Short-and long-run evidence from BitCoin and altcoin markets. *Journal of International Financial Markets, Institutions and Money, 52*, 173–195.

Corbet, S., Lucey, B., & Yarovaya, L. (2018a). Datestamping the Bitcoin and ethereum bubbles. *Finance Research Letters, 26*, 81–88.

Corbet, S., Meegan, A., Larkin, C., Lucey, B., & Yarovaya, L. (2018b). Exploring the dynamic relationships between cryptocurrencies and other financial assets. *Economics Letters, 165*, 28–34.

Demirer, R., & Kutan, A. M. (2006). Does herding behaviour exist in Chinese stock markets? *Journal of international Financial markets, institutions and money, 16*(2), 123–142.

Fama, E. F. (1965). The behaviour of stock-market prices. *The journal of Business, 38*(1), 34–105.

Gandal, N., & Halaburda, H. (2016). Can we predict the winner in a market with network effects? Competition in cryptocurrency market. *Games, 7*(3), 16.

Hwang, S., & Salmon, M. (2004). Market stress and herding. *Journal of Empirical Finance, 11*(4), 585–616.

Kremer, S. (2010). *Herding of Institutional Traders. SFB 649 Discussion Paper*. Berlin, Germany: Humboldt University.

Kristoufek, L. (2015). What are the main drivers of the Bitcoin price? Evidence from wavelet coherence analysis. *PloS one, 10*(4), e0123923.

Lakonishok, J., Shleifer, A., & Vishny, R. W. (1991). *Do institutional investors destabilize stock prices? Evidence on herding and feedback trading (No. w3846)*. National bureau of economic research.

Lao, P., & Singh, H. (2011). Herding behaviour in the Chinese and Indian stock markets. *Journal of Asian economics, 22*(6), 495–506.

Nadarajah, S., & Chu, J. (2017). On the inefficiency of Bitcoin. *Economics Letters, 150*, 6–9.

Olsen, R. A. (1996). Implications of herding behaviour for earnings estimation, risk assessment, and stock returns. *Financial Analysts Journal, 52*(4), 37–41.

Osterrieder, J., & Lorenz, J. (2017). A statistical risk assessment of Bitcoin and its extreme tail behaviour. *Annals of Financial Economics, 12*(01), 1750003.

Parker, W. D., & Prechter, R. R. (2005). *Herding: An interdisciplinary integrative review from a socionomic perspective*. Available from SSRN: https://ssrn.com/abstract=2009898 or https://doi.org/10.2139/ssrn.2009898

Pele, D. T., & Mazurencu-Marinescu-Pele, M. (2019). *Metcalfe's law and herding behaviour in the cryptocurrencies market* (No. 2019–16). Economics Discussion Papers.

Poyser, O. (2018). *Herding behaviour in cryptocurrency markets*. arXiv preprint arXiv:1806.11348.

Sapuric, S., & Kokkinaki, A. (2014, May). Bitcoin is volatile! Isn't that right?. In *International Conference on Business Information Systems* (pp. 255–265). Springer, Cham.

Spyrou, S. (2013). Herding in financial markets: A review of the literature. *Review of Behavioural Finance, 5*(2), 175–194.

Stavroyiannis, S., & Babalos, V. (2017). Herding, faith-based investments and the global financial crisis: Empirical evidence from static and dynamic models. *Journal of Behavioural Finance, 18* (4), 478–489.

Urquhart, A. (2017). Price clustering in Bitcoin. *Economics Letters, 159*, 145–148.
Vidal-Tomás, D., Ibáñez, A. M., & Farinós, J. E. (2018). Herding in the cryptocurrency market: CSSD and CSAD approaches. *Finance Research Letters.*
Zhang, W., Wang, P., Li, X., & Shen, D. (2018). Some stylized facts of the cryptocurrency market. *Applied Economics, 50*(55), 5950–5965.

Gülüzar Kurt Gümüş is a Professor of Finance at Dokuz Eylul University Department of International Business, Izmir-Turkey. Dr. Kurt Gumus has an MBA from Dokuz Eylul University (2004) and a Ph.D. in Business from Dokuz Eylul University (2007). Her research interests lie in responsible investment, corporate finance, sustainability in finance, capital markets, and investment, and international finance. She has taught Responsible Investment, International Financial Management, Corporate Finance courses, among others, at both graduate and undergraduate levels.

Yusuf Gümüş is a Professor of Accounting at Dokuz Eylul University Department of Tourism Management, Izmir-Turkey. Dr. Gumus has an MBA from Dumlupınar University (2002) and a Ph. D. in Business from Dokuz Eylul University (2007). His research interests lie in financial statement analysis, financial accounting, cost and managerial accounting, and international accounting standards. He has taught Financial Accounting, Cost Accounting, Financial Statement Analysis, and Managerial Accounting courses, among others, at both graduate and undergraduate levels.

Ayşegül Çimen after graduating from Dokuz Eylul University, Department of Business Administration in 2009, she received her MSc degree in Finance at Dokuz Eylül University in 2012. She was awarded the Jean Monnet Scholarship Programme funded by the European Union and received her MSc degree in Investment Management from Cranfield University, the UK which was ranked 8th in the field of finance by Financial Times in 2014. Cimen has successfully completed her Ph.D. and was awarded the title of Doctor of Philosophy in the Department of Business Administration majoring in the field of finance, with a thesis entitled "Underpricing in IPOs" in 2017. She has been working at the Faculty of Economics and Administrative Sciences, Dokuz Eylul University and has articles and papers published in the field of finance.

Chapter 7
News Sentiment and Cryptocurrency Volatility

Serkan Cankaya, Elcin Aykac Alp, and Mefule Findikci

Abstract The cryptocurrency market has shown remarkable growth in the last decade, resulting in heightened interest in research on several aspects of cryptocurrencies. The drastic price fluctuations have attracted attention from investors, but they have also raised concerns from national regulatory institutions. Several studies are conducted to understand the factors and the dynamics of its value formation. It is becoming more important to be able to value cryptocurrencies as an investor and as part of the process to legitimize them as a financial asset. This study aims to contribute to this field of research by examining the relationship between cryptocurrency's volatile returns and the effects of different types of news on selected cryptocurrencies. This paper categorizes the news about cryptocurrencies and determines the effect of news from each category on the return structure of each cryptocurrency. By using 1054 news sources, 22 categories are created, and a clustering analysis is used to set these categories into six groups. These groups are modelized in proper ARCH family models, which are created for different cryptocurrencies to analyze the effect on volatility. The results show that different cryptocurrencies react differently to various news categories. News about regulations from national authorities exhibit a significant effect on all selected cryptocurrencies.

S. Cankaya (✉) · M. Findikci
Graduate School of Finance, Istanbul Commerce University, Beyoglu, Istanbul, Turkey
e-mail: scankaya@ticaret.edu.tr; mfindikci@ticaret.edu.tr

E. Aykac Alp
Department of Economics, School of Business, Istanbul Commerce University, Beyoglu, Istanbul, Turkey
e-mail: ealp@ticaret.edu.tr

© Springer Nature Switzerland AG 2019

U. Hacioglu (ed.), *Blockchain Economics and Financial Market Innovation*, Contributions to Economics, https://doi.org/10.1007/978-3-030-25275-5_7

7.1 Introduction

The acceptance of cryptocurrencies as a financial instrument has been an ongoing debate for a considerable time. The analysis of several cryptocurrencies attracted attention for several reasons, such as the instrument's innovative structure, the impressive price and volume development, the attractiveness for investors, and the differing views about their legality from regulatory authorities. The increased interest in cryptocurrencies can be better understood when the recent developments in fintech industry are analyzed.

Cryptocurrencies can simply be defined as a digital asset designed to function as a medium of exchange based on cryptography technology. There are plenty of cryptocurrencies worldwide, and the major ones included in our analysis are Bitcoin, Monero, Ethereum, Litecoin, Ripple, and Zcash. As indicated by Farell (2015), research on the cryptocurrency industry is still limited and mostly focused on Bitcoin. This study aims to examine the relationship between cryptocurrency return volatility and the effects of various types of news on selected cryptocurrencies. We aim to provide a broad perspective by examining other major cryptocurrencies. In particular, we follow Auer and Claessens' (2018) work for the classification of various cryptocurrency return reactions to news about regulatory actions. Their work is extended by including non-regulatory news variables. The cryptocurrencies are assumed to function outside of national regulations, but their valuations and volumes respond considerably to news about regulatory actions. There is no unified approach to regulations about cryptocurrencies. Hughes and Middlebrook's (2015) study explain competing models for regulatory options for individuals and institutions that accept cryptocurrencies as payment. The threats and opportunities of cryptocurrencies and the future of these instruments is an ongoing debate for financial practitioners and academics. Amanzholova and Teslya (2018) claim that cryptocurrencies provide anonymity, remove intermediaries, and present speculative benefits. They also list the potentials threats under two headings: threats to state, country, and society and threats to cryptocurrency network users. The decentralized structure and anonymity make it almost impossible to track the origin of transactions. The free and mostly unregulated structure forms a basis for speculative and fictive transactions and eventually for high volatility. An increasing amount of cryptocurrency usage can pose a potential risk for national governments in terms of their monetary authorities and encourage unlawful tax evasion and the trade of illegal goods and services. These kinds of potential threats increase the regulatory concerns about cryptocurrency transactions. News about regulatory actions taken by governments, positive and negative, impact cryptocurrency value. As Carrera (2018) suggests, the unproven nature of these instruments requires a unique valuation approach to understand the price formation behavior for each cryptocurrency. Carrera's study also claims that value of each coin relies more heavily on sentiment than on the traditional asset price formation. Several studies investigate the price formation of cryptocurrencies. Ciaian, Rajcaniova, and Kancs (2016) mostly uses a neoclassical pricing model to examine the factors affecting Bitcoin price, including

Bitcoin supply and demand market forces, macrofinancial development, and Bitcoin attractiveness for investors as a measure of public sentiment. Given their results, additional measures to capture the sentiment could prove useful. Studies use various methods to measure sentiment. Several studies analyze the relationship between cryptocurrency returns and sentiment indicators. Most of the work concentrated on Twitter data and Bitcoin price as a sentiment measure. Xie, Chen, and Hu (2017) found no significant relationship between sentiment and Bitcoin price. Other studies, like Colianni, Rosales, and Signorotti (2016) and Carrera (2018), found a positive relationship between sentiment Bitcoin price. Carrera (2018) underlines the importance of successful sentiment categorization.

For methodology, we started by quantifying the news used in the study as dummy variables. The news was categorized according to the content structure, and 22 categories were obtained as a result. These categories were organized into six groups via a clustering analysis. In the next step, autoregressive conditionally heteroscedastic (ARCH) models were tested by adding the news groups to ARMA models. Then, the models that remove the ARCH effect for each cryptocurrency series are determined. We investigated several ARCH family type models' ability to explain the volatility of various cryptocurrency returns.

The findings of this study aim to contribute to a better understanding of the dynamics of cryptocurrencies and their potential to compete with fiat money. We believe it is necessary for investors and regulatory authorities to understand and adapt to the new financial market structure, whether you believe that cryptocurrencies are the evolution of the financial system or that they are a huge threat for the world economy.

The rest of the study is structured as follows. Section 7.2 describes the data set used in the empirical analysis and outlines the econometric approach used in the analysis. Section 7.3 details the step-by-step construction of the model and presents our results. Section 7.4 concludes the study.

7.2 Data and Methodology

This study's analysis is based on news and on data about the price of cryptocurrencies from January 1, 2014 to January 18, 2018. The news data in the study include the announcements, notifications, and regulations for cryptocurrencies. The cryptocurrencies selected in this study are Bitcoin, Monero, Ethereum, Litecoin, Ripple, and Zcash. Because the cryptocurrencies have varying initial release dates, the data in the sample takes the first trading day of each cryptocurrency except Bitcoin (Monero: 01/31/2015; Ethereum: 03/11/2016; Litecoin: 08/15/2016). The price data for cryptocurrencies are extracted from investing.com, and the news data are from ccn.com.

In the literature on the effect of news on financial asset prices, the news effect is usually analyzed using event study and ARCH models (Birz & Lott, 2011; Campbell & Hentschel, 1992; Chan, 2003; Gidofalvi & Elkan, 2001; Hanlon & Slemrod,

2009; Vega, 2006). Moreover, the literature that analyzed the structure of the price of the cryptocurrencies and the volatility used event study and ARCH models (Auer & Claessens, 2018; Chu, Chan, Nadarajah, & Osterrieder, 2017; Corbet & Katsiampa, 2018; Corbet, Larkin, Lucey, Meegan, & Yarovaya, 2018; Dyhrberg, 2016; Hayes, 2017; Katsiampa, 2017). In this study, ARCH models by Engle (1982) are used to analyze the possibility of estimating the series variance at a given time by allowing the conditional variance to change over time while also accepting the constant unconditional variance. Bollerslev (1986) improved these models to form generalized autoregressive conditionally heteroscedastic (GARCH) models that use autoregressive and moving averages to model conditional variance. In these models, the parameters in the variance equation have a nonnegative condition and ignore the symmetric effect. Nelson (1991), in response to the symmetric assumption's weakness, revealed exponential GARCH (EGARCH) models with a conditional variance formulation that successfully captured the asymmetric effect in the conditional variance. As an example, an EGARCH (1,1) is shown in (7.1).

$$\log h_t = \alpha_0 + \delta_1 \frac{|u_{t-1}|}{\sqrt{h_{t-1}}} + \gamma_1 \frac{u_{t-1}}{\sqrt{h_{t-1}}} + a_1 \log h_{t-1} \tag{7.1}$$

The coefficient y_1 in the equation measures the asymmetry or the leverage effect. If coefficient y_1 is statistically significant and negative, negative shocks will cause more volatility than positive shocks. Another model that accounts for asymmetry's effect is the threshold GARCH, improved by Glosten, Jagannathan, and Runkle (1993). This model's objective is to measure asymmetric information with dummy variables added to the GARCH model. As an example, a TGARCH (1,1) is shown in (7.2).

$$\mathrm{h_t} = \alpha_0 + \alpha_1 \mathrm{u}_{t-1}^2 + \gamma_1 \mathrm{u}_{t-1}^2 \mathrm{I}_{t-1} + \beta_1 \mathrm{h}_{t-1} \tag{7.2}$$

The coefficient y_1 in the equation is statistically significant and positive to represent asymmetric information and to indicate the leverage effect. In this case, negative shocks increased volatility (Bildirici, Alp, Ersin, & Bozoklu, 2010).

The return series of the cryptocurrencies were calculated with Eq. (7.3). The unit root tests are performed to check the stationarity. Then, the variables' ARMA structure is obtained. An ARCH-LM test is used to determine if ARMA models have an ARCH effect.

$$return_t = \frac{price_t - price_{t-1}}{price_{t-1}} x100 \tag{7.3}$$

The news data used in the study are quantified using dummy variables. The news data used to determine the price effects of news are categorized according to the content structure. We obtained 22 categories from the news. These categories were grouped using a clustering analysis. ARCH models were tested by adding the news

Table 7.1 Variable definitions

Variable	Definition
Cryptocurrency	
Bitcoin	Return of Bitcoin price
Monero	Return of Monero price
Ethereum	Return of Ethereum price
Litecoin	Return of Litecoin price
Dummy variables	
News	"1" news about cryptocurrencies; "0" otherwise
AML	"1" news on the evaluation of cryptocurrencies within context of antimoney laundering; "0" otherwise
BAN	"1" news about strict cryptocurrency bans or blocking; "0" otherwise
CFTC	"1" news about the commodity futures trading commission (CFTC) or its members have made statements about the cryptocurrencies; "0" otherwise
Crime	"1" news about transactions of cryptocurrency is a crime; "0" otherwise
Currency/Not Currency	"1" news about cryptocurrencies are considered to be a currency or not currency; "0" otherwise.
ETF	"1" news about exchange traded funds; "0" otherwise.
Exchange	"1" news on exchange of cryptocurrencies; "0" otherwise.
General	"1" news on general evaluations, comments, or announcements about cryptocurrencies; "0" otherwise.
ICO	"1" news about Initial Coin Offerings (ICO); "0" otherwise
Law	"1" news about laws or law proposals for cryptocurrencies; "0" otherwise.
Legal	"1" news about legalization of cryptocurrencies; "0" otherwise.
License	"1" news about establishing and operating cryptocurrency licenses; "0" otherwise.
Mining/Production Of Money	"1" news on announcements from countries or institutions about mining or money production of cryptocurrency; "0" otherwise.
Price	"1" news on cryptocurrency prices; "0" otherwise.
Regulation	"1" news about past or future legal regulations related to cryptocurrencies; "0" otherwise.
Restrict	"1" news about restrictions on cryptocurrencies; "0" otherwise.
Risk	1″ news about cryptocurrency risks; "0" otherwise.
SEC	"1" news that the securities and exchange commission (SEC) or its members have made statements about the cryptocurrencies; "0" otherwise
Security	"1" news that cryptocurrencies are considered a security; "0" otherwise.
Tax	"1" news about cryptocurrency taxation; "0" otherwise.
Terror	"1" news on the use of cryptocurrencies as part of terrorist financing; "0" otherwise.
Warning	"1" news on warnings about cryptocurrencies; "0" otherwise.

groups to ARMA models. Then the coin series models that best eliminate the ARCH effect are determined. The variable definitions used in the models are listed in Table 7.1.

Table 7.2 Variable definitions

		Bitcoin	Monero	Ethereum	Litecoin
Augmented Dickey-Fuller test statistic		−18.303	−4.815	−11.628	−9.713
Test critical values	1% level	−3.434	−3.435	−3.436	−3.438
	5% level	−2.863	−2.863	−2.864	−2.865
	10% level	−2.568	−2.568	−2.568	−2.568

Table 7.3 Cluster of news

Cluster	1	2	3	4	5	6
Variable	License Minning/ Currency Produc- tion Restrict Law Risk	Genaral CFTC Security Currency/No Cur- rency Crime Tax	Terror AML	SEC ETF	Ban Legal Exchange Price	ICO Warn Regulation

The Augmented Dickey-Fuller unit root test was performed to test the stationarity of the cryptocurrency returns. As a result of the test, we determined that the Augmented Dickey-Fuller test statistic of the cryptocurrencies is more negative than all the test critical values. Hence, the return series were stationary (Table 7.2).

7.3 Empirical Results

7.3.1 Grouping News Via Cluster Analysis

The study includes 1054 news items from January 1, 2014 to January 18, 2018. In the first step, the news items are divided into 22 categories according to the content. These categories are classified using the Ward technique of hierarchical clustering analysis. This technique utilizes the total deviation squares, which are based on the average distance from the observation in the middle of a cluster to other observations (Table 7.3).

As a result of the analysis, 22 news categories are grouped into six clusters, and those clusters become the benchmark of our ARCH family models. Using these groups, ARCH models are examined without any specification error.

Table 7.4 The most appropriate ARMA (p, q) model

	Bitcoin	Monero	Ethereum	Litecoin
	ARMA(4,4)	ARMA(2,2)	ARMA(1,1)	ARMA(2,2)
C	0.148***	1.535***	0.440***	0.479**
AR(1)	−0.290*	1.115*	0.980*	−1.194*
AR(2)	0.227*	−0.138***		−0.858*
AR(3)	−0.378*			
AR(4)	−0.899*			
MA(1)	0.280*	−1.419*	−0.964*	1.207*
MA(2)	−0.253*	0.465*		0.825*
MA(3)	0.409*			
MA(4)	0.899*			
R^2	0.015	0.104	0.007	0.011
Adjusted R^2	0.011	0.101	0.006	0.007
Akaike info criterion	5.521	8.575	6.523	6.827
Schwarz criterion	5.548	8.593	6.537	6.855
Hannan-Quinn criter.	5.531	8.581	6.528	6.838

It refers to the significance level of *%1 **%5 ***%10

7.3.2 *ARCH(p,q) Models*

7.3.2.1 ARMA Models for Cryptocurrencies

Several ARMA (p, q) models were tested to obtain the most appropriate ARMA (p, q) model. The most appropriate model was selected according to Akaike, Schwarz, and Hannan-Quinn's information criteria. The ARMA (4,4) model is the most appropriate model for Bitcoin's return variable. The ARMA (2,2) model is the most appropriate model for Monero's and Litecoin's return variables. The ARMA (1,1) model is the most appropriate model for Ethereum's return variable.

The ARMA model created from this point is tested with the ARCH-LM Test, and the results are shown in Table 7.4. In the ARCH-LM test, the null hypotheses are defined as "There is no ARCH effect in the model." and the alternative hypothesis is defined as "There is ARCH effect in the model."

According to the ARCH-LM test results, the null hypothesis is rejected because the value of $T * (R^2)$ is greater than the chi-square table value with 5 degrees of freedom. Therefore, the presence of the fifth-degree ARCH effect in the models' residuals is determined. ARMA models for Ripple and Zcash cryptocurrencies were excluded in much as they did not have an ARCH effect (Table 7.5).

7.3.2.2 The Effect of News on Cryptocurrencies

In this section, ARCH models are examined to eliminate the ARCH effect in the most appropriate model for various cryptocurrencies, and the effects of news on

Table 7.5 ARCH-LM test

Bitcoin	F-statistic	28.74491	Prob. F(5,1828)	0.000
	Obs*R^2	133.6854	Prob. Chi-Square(5)	0.000
Monero	F-statistic	46.511	Prob. F(5,1436)	0.000
	Obs*R^2	200.978	Prob. Chi-Square(5)	0.000
Ethereum	F-statistic	21.984	Prob. F(5,1032)	0.000
	Obs*R^2	99.916	Prob. Chi-Square(5)	0.000
Litecoin	F-statistic	2.668	Prob. F(5,864)	0.021
	Obs*R^2	13.227	Prob. Chi-Square(5)	0.021

cryptocurrencies' volatility are examined. The variables' statistical significance in the model and the information's criteria were examined to select the model with the best representation among the ARCH models.

From all evaluations, EGARCH (1,1) is selected as the most appropriate model for Bitcoin's return variable, TGARCH (1,1) is selected as the most appropriate model for Monero's return variable, GARCH (1,1) is selected as the most appropriate model for Ethereum's return variable, and TGARCH (1,1) is selected as the most appropriate model for Litecoin's return variable (Table 7.6).

The results of the EGARCH (1,1) model for Bitcoin's return shows that all coefficients except the constant term's coefficient are statistically significant. In the variance equation, negative volatility is more effective on conditional variance due to the negative statistical significance of coefficient γ for Bitcoin's return in the EGARCH models. Therefore, negative news leads to more volatility. Moreover, it is determined that the news has an effect on the volatility of Bitcoin's return, and the news increased volatility by 0.0876.

After examining the results of the TGARCH (1,1) model for Monero's return, we see that all coefficients except for the constant term's coefficient are statistically significant. As coefficient γ is not positive and denotes the asymmetry in the model, there is no leverage effect. In other words, there is no evidence that negative and positive news lead to different reactions. In addition, the news items are found to be effective on Monero's return volatility, and decreased volatility by 4.99. After examining the results of the GARCH (1,1) model for Ethereum's return, we see that all coefficients except for the constant term's coefficient are statistically significant. The volatility parameters of the variance equation in the GARCH (p, q) models are not negative. We determined that the news items are effective on Ethereum's return volatility, and the volatility decreased by 0.684. After examining the results of the TGARCH (1,1) model for Litecoin's return, we see that all coefficients except for the constant term's coefficient are statistically significant. As coefficient γ is not positive and denotes the asymmetry in the model, there is no leverage effect. Therefore, there is no evidence that positive or negative news differ in terms of effect. Furthermore, it is found that the news are effective on Litecoin's return volatility and showed a 1.72 increase.

Table 7.6 Cryptocurrencies ARCH family models of news

	Bitcoin		Monero		Ethereum		Litecoin	
	EGARCH(1,1)		TGARCH(1,1)		GARCH(1,1)		TGARCH(1,1)	
C	0.043	0.847	0.407	0.079	0.154	0.456	0.194	0.354
AR(1)	1.889	0.000	−1.507	0.000	0.918	0.000	0.121	0.000
AR(2)	−1.384	0.000	−0.986	0.000			−0.989	0.000
AR(3)	0.199	0.000						
AR(4)	0.293	0.000						
MA(1)	−1.874	0.000	1.514	0.000	−0.906	0.000	−0.122	0.000
MA(2)	1.324	0.000	0.998	0.000			0.995	0.000
MA(3)	−0.106	0.000						
MA(4)	−0.337	0.000						
α_0	−0.124	0.000	4.693	0.000	4.192	0.000	0.123	0.014
a_1			0.149	0.000	0.162	0.000	0.016	0.000
δ_1	0.348	0.000						
$\boldsymbol{\gamma_1}$	**−0.034**	0,001	−0.070	0.000			−0.030	0.000
β_1	0.936	0.000	0.866	0.000	0.743	0.000	0.975	0.000
News	0.088	0.000	−4.999	0.000	−0.684	0.376	1.720	0.000
R^2	0.007		0.032		0.003		0.004	
Adjusted R^2	0.003		0.030		0.001		0.000	
Log likelihood	−4777.695		−5352.549		−3325.173		−2863.324	
Akaike info criterion	5.211		7.412		6.390		6.568	
Schwarz criterion	5.253		7.448		6.423		6.622	
Hannan-Quinn criter.	5.227		7.426		6.402		6.588	

$GARCH(p,q): h_t = \alpha_0 + \sum_{i=1}^{p} a_i u_{t-i}^2 + \sum_{j=1}^{q} \beta_j h_{t-j}$

$EGARCH(1,1):\ \log h_t = \alpha_0 + \delta_1 \frac{|u_{t-1}|}{\sqrt{h_{t-1}}} + \gamma_1 \frac{u_{t-1}}{\sqrt{h_{t-1}}} + a_1 \log h_{t-1}$

$TGARCH(1,1): \mathrm{h_t} = \alpha_0 + \alpha_1 \mathrm{u}_{t-1}^2 + \gamma_1 \mathrm{u}_{t-1}^2 \mathrm{I}_{t-1} + \beta_1 \mathrm{h}_{t-1}$

7.3.2.3 The Reaction of Cryptocurrencies to News Groups

Among the models created for cryptocurrencies and for the first news group, EGARCH (1,1) is selected for Bitcoin, Monero, and Litecoin; GARCH (1,1) is selected for Ethereum (Table 7.7).

After examining the models' results, we see that all coefficients except for the constant term's coefficient are statistically significant. In the variance equation, negative volatility in the model is more effective on conditional variance due to the negative statistical significance of coefficient γ for the returns of Bitcoin and Monero cryptocurrencies in the EGARCH models. Therefore, negative news leads to more volatility. However, coefficient γ shows the calculated asymmetry for the return of Litecoin and is positive, demonstrating that negative and positive news did not lead to different reactions because of the lack of leverage effect on the models. In the Ethereum return model, the variance equation of the GARCH (1,1) model shows that the values related to volatility are not negative.

When the models are examined in terms of news types for the first group (with the "License" variable), the effect of Bitcoin, Ethereum, and Litecoin on return volatility is statistically significant. Licensing-related news reduced the volatility of Bitcoin return and increased Ethereum's and Litecoin's returns. The effect of news about Mining/Currency Production and Risk on the return volatility of Monero, Ethereum, and Litecoin were statistically significant. News on Mining/Currency Production increased Monero's and Ethereum's return volatility while reducing Litecoin's return volatility. News on Risk reduced the volatility of Monero and Ethereum while increasing Litecoin's return volatility. The effect of Restrict on the return volatility of Bitcoin, Monero, and Litecoin was statistically significant. News related to restrictions of cryptocurrencies increased return volatility of Bitcoin and Monero and reduced Litecoin's return volatility. The news about Law was statistically significant for all four cryptocurrencies. News of the law increased the Bitcoin, Monero and Litecoin volatility of return while reducing Ethereum's return volatility. The most appropriate model for cryptocurrency returns and the second news group was EGARCH (1,1) for Bitcoin, Monero, and Litecoin and TGARCH (1,1) for Ethereum (Table 7.8).

After examining the results of the models, we see that all coefficients except the constant term's coefficient are statistically significant. In the variance equation, negative volatility in the model is more effective on conditional variance due to the negative and statistical significance of coefficient γ for Bitcoin's return in the EGARCH models. However, there is no leverage effect due to the model established for Monero's and Litecoin's returns. The model obtained for Bitcoin shows that negative news causes more volatility. However, the models for Monero and Litecoin do not provide different reactions for negative and positive news. The effect of the Currency/No Currency and Tax variables on the return volatility of Bitcoin, Monero, Ethereum, and Litecoin were statistically significant. News on whether or not cryptocurrencies are fiat money increased the return volatility of Bitcoin, Monero, and Litecoin but reduced Ethereum's return volatility. Tax-related news reduced the

Table 7.7 Effects of the first group news on the volatility of cryptocurrencies

	Bitcoin		Monero		Ethereum		Litecoin	
	EGARCH(1,1)		EGARCH(1,1)		GARCH(1,1)		EGARCH(1,1)	
	Coefficient	Prob.	Coefficient	Prob.	Coefficient	Prob.	Coefficient	Prob.
C	0.083	0.218	−2.053	0.284	−0.175	0.726	0.057	0.654
AR(1)	−0.380	0.001	0.979	0.000	0.992	0.000	−0.939	0.000
AR(2)	0.524	0.000	0.020	0.000			−0.959	0.000
AR(3)	−0.444	0.000						
AR(4)	−0.789	0.000						
MA(1)	0.395	0.000	−1.229	0.000	−0.983	0.000	0.927	0.000
MA(2)	−0.533	0.000	0.237	0.000			0.956	0.000
MA(3)	0.468	0.000						
MA(4)	0.827	0.000						
α_0	−0.119	0.000	0.279	0.000	2.599	0.000	0.045	0.000
a_1					0.133	0.000		
δ_1	0.387	0.000	0.508	0.000			−0.033	0.000
$\boldsymbol{\gamma_1}$	**−0.035**	0.001	**−0.074**	0.000			0.075	0.000
β_1	0.934	0.000	0.854	0.000	0.795	0.000	0.989	0.000
Licence	**−0.325**	0.002	0.463	0.101	**35.922**	0.019	**1.284**	0.000
Mining/Production of Money	0.022	0.869	**0.321**	0.017	**18.305**	0.018	**−0.073**	0.050
Restrict	**0.410**	0.001	**0.740**	0.000	6.509	0.154	**−0.352**	0.000
Law	**0.131**	0.077	**0.692**	0.000	**−8.053**	0.000	**0.507**	0.000
Risk	0.188	0.163	**−0.647**	0.026	**−6.994**	0.007	**0.172**	0.006
R^2	0.09259		0.09259		0.005		−0.000635	
Adjusted R^2	0.090072		0.090072		0.002768		−0.005235	
Log likelihood	−5475.176		−5475.18		−3308.08		−2749.455	
Akaike info criterion	7.586975		7.586975		6.36448		6.31647	

(continued)

Table 7.7 (continued)

	Bitcoin		Monero		Ethereum		Litecoin	
	EGARCH(1,1)		EGARCH(1,1)		GARCH(1,1)		EGARCH(1,1)	
	Coefficient	Prob.	Coefficient	Prob.	Coefficient	Prob.	Coefficient	Prob.
Schwarz criterion	7.638033		7.638033		6.416684		6.392857	
Hannan-Quinn criter.	7.60603		7.60603		6.384281		6.34569	

$GARCH(p,q): h_t = \alpha_0 + \sum_{i=1}^{p} a_i u_{t-i}^2 + \sum_{j=1}^{q} \beta_j h_{t-j}$

$EGARCH(1,1): \log h_t = \alpha_0 + \delta_1 \frac{|u_{t-1}|}{\sqrt{h_{t-1}}} + \gamma_1 \frac{u_{t-1}}{\sqrt{h_{t-1}}} + a_1 \log h_{t-1}$

$TGARCH(1,1): \mathrm{h_t} = \alpha_0 + \alpha_1 \mathrm{u}_{t-1}^2 + \gamma_1 \mathrm{u}_{t-1}^2 \mathrm{I}_{t-1} + \beta_1 \mathrm{h}_{t-1}$

Table 7.8 Effects of the second group news on the volatility of cryptocurrencies

	Bitcoin		Monero		Ethereum		Litecoin	
	EGARCH(1,1)		EGARCH(1,1)		TGARCH(1,1)		EGARCH(1,1)	
C	0.053	0.446	0.328	0.087	−0.203	0.779	0.234	0.227
AR(1)	−0.196	0.000	0.792	0.000	0.994	0.000	−1.819	0.000
AR(2)	0.482	0.000	−0.981	0.000			−0.835	0.000
AR(3)	−0.723	0.000						
AR(4)	−0.406	0.000						
MA(1)	0.212	0.000	−0.815	0.000	−0.985	0.000	1.780	0.000
MA(2)	−0.509	0.000	0.999	0.000			0.794	0.000
MA(3)	0.734	0.000						
MA(4)	0.457	0.000						
α_0	−0.112	0.000	−0.085	0.000	2.830	0.000	0.059	0.000
a_1					0.134	0.000		
δ_1	0.363	0.000	0.228	0.000			0.113	0.000
$\boldsymbol{\gamma_1}$	**−0.036**	0.000	0.046	0.000	0.013	0.631	0.114	0.000
β_1	0.942	0.000	0.988	0.000	0.804	0.000	0.972	0.000
Genaral	0.031	0.358	−0.007	0.864	−0.752	0.722	**−0.359**	0.000
CFTC	0.039	0.711	−0.136	0.117	0.974	0.829	**0.427**	0.000
Security	−0.052	0.797	**−0.529**	0.000	0.416	0.949	**−0.865**	0.000
Currency/ No Currency	**0.298**	0.003	**0.412**	0.000	**−9.305**	0.006	**0.990**	0.000
Crime	−0.111	0.336	**−0.227**	0.016	−1.449	0.862	**−0.428**	0.008
Tax	−0.153	0.036	**−0.517**	0.000	**−9.255**	0.000	**−0.418**	0.000
R^2	0.006		0.016		0.005		−0.002	
Adjusted R^2	0.002		0.013		0.003		−0.006	
Log likelihood	−4778.195		−5340.406		−3317.268		−2819.073	
Akaike info criterion	5.217		7.402		6.386		6.478	

(continued)

Table 7.8 (continued)

	Bitcoin		Monero		Ethereum		Litecoin	
	EGARCH(1,1)		EGARCH(1,1)		TGARCH(1,1)		EGARCH(1,1)	
Schwarz criterion	5.274		7.457		6.448		6.560	
Hannan-Quinn criter.	5.238		7.422		6.409		6.509	

$GARCH(p,q): h_t = \alpha_0 + \sum_{i=1}^{p} a_i u_{t-i}^2 + \sum_{j=1}^{q} \beta_j h_{t-j}$

$EGARCH(1,1): \log h_t = \alpha_0 + \delta_1 \frac{|u_{t-1}|}{\sqrt{h_{t-1}}} + \gamma_1 \frac{u_{t-1}}{\sqrt{h_{t-1}}} + a_1 \log h_{t-1}$

$TGARCH(1,1): \mathrm{h_t} = \alpha_0 + \alpha_1 \mathrm{u}_{t-1}^2 + \gamma_1 \mathrm{u}_{t-1}^2 \mathrm{I}_{t-1} + \beta_1 \mathrm{h}_{t-1}$

return volatility of four cryptocurrencies. The effect of Security and Crime variables on the return volatility of Monero and Litecoin was statistically significant, and both variables decreased the return volatility of the cryptocurrencies. The CFTC's explanations and general crypto money news were statistically significant only on Litecoin's return volatility. While general announcements reduce volatility, the CFTC's explanations increased volatility.

The most appropriate model for cryptocurrencies' returns and the third news group was EGARCH (1,1) for Bitcoin, Monero, and Litecoin and GARCH (1,1) for Ethereum (Table 7.9).

When the models' results are examined, all coefficients except for the constant term's coefficient are statistically significant in the average equation of the models.

In the variance equation of the Bitcoin model, coefficient γ, which shows the asymmetric structure in EGARCH models, is negative and statistically significant. Thus, the negative news effect caused more volatility in the Bitcoin model. In the model for Monero's and Litecoin's volatility, however, there is no leverage effect because coefficient γ is positive. The negative or positive news did not cause different reactions for Monero and Litecoin.

The most appropriate model for cryptocurrencies' return and the fourth news group was EGARCH (1,1) for Bitcoin and Litecoin, TGARCH (1,1) for Ethereum, and GARCH (2,2) for Monero (Table 7.10).

After examining the models' results, we see that all coefficients except for the constant term's coefficient are statistically significant in the average equation of the models. In the variance equation, the negative volatility in the model has more effect on conditional variance because coefficient γ shows the asymmetric structure in EGARCH models, and it is negative and statistically significant for Bitcoin's return volatility model. In the model obtained for Bitcoin, negative news led to more volatility. In the GARCH (2,2) variance equation established for Monero's return volatility, the coefficients related to volatility are nonnegative.

The effect of the ETF variable on Bitcoin's and Litecoin's return volatility is statistically significant and caused an increase in volatility. The effects of the SEC variable on the return volatility of Monero, Ethereum, and Litecoin are statistically significant. News on SEC announcements reduced return volatility for Monero but increased return volatility for Ethereum and Litecoin.

The most appropriate model for cryptocurrency returns and the second news group is EGARCH (1,1) for Bitcoin, Ethereum, and Litecoin and GARCH (1,1) for Monero (Table 7.11).

When the models' results are examined, all coefficients except for the constant term's coefficient are statistically significant in the average equation of the models. In the variance equation of Bitcoin's return volatility model, negative volatility had less effect on conditional variance because the coefficient of asymmetric structure in EGARCH models was negative and statistically significant. This shows that negative news leads to more volatility. In the GARCH (1,1) variance equation for Monero's return volatility, the volatility coefficients satisfy the nonnegative condition.

Table 7.9 Effects of the third group news on the volatility of cryptocurrencies

	Bitcoin		Monero		Ethereum		Litecoin	
	EGARCH(1,1)		EGARCH(1,1)		GARCH(1,1)		EGARCH(1,1)	
C	0.045	0.542	0.455	0.086	−1.810	0.725	0.272	0.234
AR(1)	0.618	0.000	−1.875	0.000	0.999	0.000	0.280	0.000
AR(2)	0.135	0.000	−0.879	0.000			−0.915	0.000
AR(3)	−0.939	0.000						
AR(4)	0.532	0.000						
MA(1)	−0.568	0.000	1.838	0.000	−0.996	0.000	−0.260	0.000
MA(2)	−0.168	0.000	0.842	0.000			0.961	0.000
MA(3)	0.929	0.000						
MA(4)	−0.468	0.000						
α_0	−0.110	0.000	−0.079	0.000	3.128	0.000	0.108	0.000
a_1					0.184	0.000		
δ_1	0.373	0.000	0.221	0.000			0.156	0.000
γ_1	**−0.039**	0.000	0.067	0.000			0.130	0.000
β_1	0.941	0.000	0.983	0.000	0.749	0.000	0.950	0.000
Terror	−0.295	0.104	**−0.500**	0.000	**−57.078**	0.000	**0.591**	0.058
AML	−0.182	0.184	**0.766**	0.000	**41.538**	0.000	**−1.855**	0.000
R^2	0.001		−0.017		0.000		−0.002	
Adjusted R^2	−0.003		−0.020		−0.002		−0.007	
Log likelihood	−4776.909		−5382.257		−3311.671		−2852.696	
Akaike info criterion	5.211		7.454		6.366		6.546	
Schwarz criterion	5.256		7.495		6.404		6.606	
Hannan-Quinn criter.	5.228		7.469		6.380		6.569	

$GARCH(p,q): h_t = \alpha_0 + \sum_{i=1}^{p} a_i u_{t-i}^2 + \sum_{j=1}^{q} \beta_j h_{t-j}$

$EGARCH(1,1): \log h_t = \alpha_0 + \delta_1 \frac{|u_{t-1}|}{\sqrt{h_{t-1}}} + \gamma_1 \frac{u_{t-1}}{\sqrt{h_{t-1}}} + a_1 \log h_{t-1}$

$TGARCH(1,1): h_t = \alpha_0 + \alpha_1 u_{t-1}^2 + \gamma_1 u_{t-1}^2 I_{t-1} + \beta_1 h_{t-1}$

Table 7.10 Effects of the fourth group news on the volatility of cryptocurrencies

	Bitcoin		Monero		Ethereum		Litecoin	
	EGARCH(1,1)		GARCH(2,2)		TGARCH(1,1)		EGARCH(1,1)	
C	0.060	0.416	0.202	0.272	−0.100	0.875	0.067	0.709
AR(1)	0.700	0.000	−0.709	0.000	0.994	0.000	−0.946	0.000
AR(2)	0.058	0.000	−0.886	0.000			−0.964	0.000
AR(3)	−0.993	0.000						
AR(4)	0.638	0.000						
MA(1)	−0.650	0.000	0.639	0.000	−0.985	0.000	0.944	0.000
MA(2)	−0.086	0.000	0.894	0.000			0.966	0.000
MA(3)	0.986	0.000						
MA(4)	−0.574	0.000						
α_0	−0.124	0.000	6.161	0.000	2.235	0.000	−0.029	0.044
a_1			0.049	0.024	0.119	0.000		
a_2			0.248	0.000				
δ_1	0.385	0.000					0.149	0.000
$\boldsymbol{\gamma_1}$	**−0.037**	0.001			0.009	0.708	0.046	0.000
β_1	0.938	0.000	0.372	0.000	0.812	0.000	0.972	0.000
β_2			0.333	0.000				
SEC	−0.016	0.889	**−27.463**	0.000	**7.480**	0.073	**0.613**	0.000
ETF	**0.255**	0.012	0.090	0.954	3.831	0.242	**0.207**	0.002
R^2	−0.003		0.016		0.005		−0.002	
Adjusted R^2	−0.007		0.013		0.004		−0.007	
Log likelihood	−4777.009		−5342.206		−3322.007		−2780.584	
Akaike info criterion	5.212		7.400		6.387		6.381	
Schwarz criterion	5.257		7.444		6.430		6.441	
Hannan-Quinn criter.	5.228		7.417		6.404		6.404	

$GARCH(p,q): h_t = \alpha_0 + \sum_{i=1}^{p} a_i u_{t-i}^2 + \sum_{j=1}^{q} \beta_j h_{t-j}$

$EGARCH(1,1): \log h_t = \alpha_0 + \delta_1 \frac{|u_{t-1}|}{\sqrt{h_{t-1}}} + \gamma_1 \frac{u_{t-1}}{\sqrt{h_{t-1}}} + a_1 \log h_{t-1}$

$TGARCH(1,1): h_t = \alpha_0 + \alpha_1 u_{t-1}^2 + \gamma_1 u_{t-1}^2 I_{t-1} + \beta_1 h_{t-1}$

Table 7.11 Effects of the fifth group news on the volatility of cryptocurrencies

	Bitcoin		Monero		Ethereum		Litecoin	
	EGARCH(1,1)		GARCH(1,1)		EGARCH(1,1)		EGARCH(1,1)	
C	0.031	0.930	0.070	0.702	0.122	0.480	0.091	0.640
AR(1)	−0.088	0.018	−1.200	0.000	−0.884	0.000	0.518	0.000
AR(2)	1.617	0.000	−0.215	0.000			−0.967	0.000
AR(3)	0.139	0.000						
AR(4)	−0.669	0.000						
MA(1)	0.112	0.005	1.053	0.000	0.899	0.000	−0.502	0.000
MA(2)	−1.616	0.000	0.053	0.027			0.996	0.000
MA(3)	−0.164	0.000						
MA(4)	0.675	0.000						
α_0	−0.096	0.000	3.387	0.000	0.111	0.009	0.007	0.725
a_1			0.139	0.000				
δ_1	0.340	0.000			0.214	0.000	0.177	0.000
γ_1	**−0.032**	0.003			0.011	0.485	0.032	0.049
β_1	0.932	0.000	0.856	0.000	0.924	0.000	0.952	0.000
BAN	**0.499**	0.000	2.644	0.292	0.107	0.118	0.045	0.337
LEGAL	**−0.396**	0.000	**−12.517**	0.000	**−0.112**	0.092	**0.423**	0.000
EXCHANGE	**0.097**	0.038	**−3.619**	0.015	**0.141**	0.006	**0.364**	0.000
PRICE	**0.184**	0.053	**−7.904**	0.004	**−0.449**	0.000	**−0.554**	0.000
R^2	0.005		0.016		0.001		0.014	
Adjusted R^2	0.000		0.013		−0.001		0.009	
Log likelihood	−4752.167		−5336.624		−3319.371		−2799.956	

Akaike info criterion	5.187		7.393		6.386		6.430	
Schwarz criterion	5.238		7.436		6.438		6.501	
Hannan-Quinn criter.	5.206		7.409		6.406		6.457	

$$GARCH(p,q): h_t = \alpha_0 + \sum_{i=1}^{p} a_i u_{t-i}^2 + \sum_{j=1}^{q} \beta_j h_{t-j}$$

$$EGARCH(1,1): \log h_t = \alpha_0 + \delta_1 \frac{|u_{t-1}|}{\sqrt{h_{t-1}}} + \gamma_1 \frac{u_{t-1}}{\sqrt{h_{t-1}}} + a_1 \log h_{t-1}$$

$$TGARCH(1,1): \mathrm{h_t} = \alpha_0 + \alpha_1 \mathrm{u}_{t-1}^2 + \gamma_1 \mathrm{u}_{t-1}^2 \mathrm{I}_{t-1} + \beta_1 \mathrm{h}_{t-1}$$

The effect of the Ban variable expressing the news about prohibitions on Bitcoin's return volatility is statistically significant. This news increases the volatility. The effects of the Legal, Exchange, and Price variables on the four cryptocurrencies are statistically significant. Positive news about the legality of the cryptocurrencies decreased the return volatility of Bitcoin, Monero, and Ethereum but increased the Litecoin's return volatility.

The Exchange variable refers to the exchange of cryptocurrencies to Bitcoin and the conversion of cryptocurrency to other types of financial instruments, such as fiat money. Bitcoin, Ethereum, and Litecoin increased in return volatility, and Monero decreased in return volatility. News on cryptocurrency prices (Price) had an increasing effect on the volatility of Bitcoin's return. However, Monero, Ethereum, and Litecoin showed a reduction in return volatility.

The most appropriate model for cryptocurrencies' returns and the second news group is TGARCH (1,1) for Bitcoin, EGARCH (1,1) for Monero and Ethereum, and GARCH (1,2) for Litecoin (Table 7.12).

The model results show that all coefficients except for the constant term's coefficient are statistically significant in the average equation of the models. In the model's variance equation for the volatility of Bitcoin's return, coefficient γ for asymmetric structure in TGARCH models is positive and statistically significant. This shows the leverage effect in the model that negative news causes more volatility. In the GARCH (1,2) variance equation established for Litecoin's return volatility, the volatility coefficients satisfy the nonnegative condition.

In the sixth group, the effect of ICO and news on Bitcoin's and Litecoin's return volatility is found to be statistically significant. While the news on initial public offerings raised the volatility of Bitcoin returns, Litecoin's return volatility was reduced. The effects of the Warning variable that represents warnings about the use of cryptocurrency on Bitcoin's, Ethereum's, and Litecoin's return volatility was statistically significant and increased the return volatility. The effect of the Regulation variable on the return volatility of Bitcoin, Monero, and Ethereum was statistically significant. News about regulation decreased the return volatility of Bitcoin and Monero; however, Ethereum showed an increase.

7.3.2.4 ARCH-LM Test Results for Models

The ARCH-LM test is used to determine the effect of news on cryptocurrencies. The results showed that the ARCH effect was eliminated at the 5% statistical significance level (Chi-square tail probability >0.05) (Table 7.13).

Table 7.12 Effects of the sixth group news on the volatility of cryptocurrencies

	Bitcoin		Monero		Ethereum		Litecoin	
	TGARCH(1,1)		EGARCH(1,1)		EGARCH(1,1)		GARCH(1,2)	
C	0.084	0.288	−0.417	0.180	0.192	0.266	0.144	0.592
AR(1)	0.385	0.000	1.696	0.000	−0.864	0.000	−1.869	0.000
AR(2)	−0.801	0.000	−0.698	0.000			−0.932	0.000
AR(3)	0.123	0.000						
AR(4)	0.207	0.000						
MA(1)	−0.328	0.000	−1.791	0.000	0.879	0.000	1.859	0.000
MA(2)	0.814	0.000	0.793	0.000			0.930	0.000
MA(3)	−0.071	0.000						
MA(4)	−0.202	0.000						
α_0	0.583	0.000	−0.093	0.000	0.100	0.017	7.149	0.000
a_1	0.172	0.000					0.100	0.000
δ_1			0.317	0.000	0.214	0.000		
γ_1	**0.091**	0.000	0.047	0.000	0.018	0.251		
β_1	0.766	0.000	0.977	0.000	0.923	0.000	0.268	0.034
β_2							0.512	0.000
ICO	**6.042**	0.018	**−0.123**	0.069	−0.056	0.416	**−19.413**	0.000
Warnining	**4.522**	0.006	0.128	0.298	**0.278**	0.082	**24.370**	0.000
Reglation	**−0.933**	0.000	**−0.121**	0.000	**0.093**	0.034	−0.493	0.854
R^2	0.008		0.042		0.002		0.003	
Adjusted R^2	0.004		0.040		0.000		−0.002	
Log likelihood	−4766.225		−5379.456		−3321.452		−2906.754	
Akaike info criterion	5.201		7.452		6.388		6.671	
Schwarz criterion	5.249		7.496		6.436		6.737	
Hannan-Quinn criter.	5.219		7.468		6.406		6.696	

$GARCH(p,q): h_t = \alpha_0 + \sum_{i=1}^{p} a_i u_{t-i}^2 + \sum_{j=1}^{q} \beta_j h_{t-j}$

$EGARCH(1,1): \log h_t = \alpha_0 + \delta_1 \frac{|u_{t-1}|}{\sqrt{h_{t-1}}} + \gamma_1 \frac{u_{t-1}}{\sqrt{h_{t-1}}} + a_1 \log h_{t-1}$

$TGARCH(1,1): h_t = \alpha_0 + \alpha_1 u_{t-1}^2 + \gamma_1 u_{t-1}^2 I_{t-1} + \beta_1 h_{t-1}$

Table 7.13 ARCH-LM test results

Group 1	EGARCH(1,1)		EGARCH(1,1)		GARCH(1,1)		EGARCH(1,1)	
	F-statistic	Obs∗R-squared	F-statistic	Obs∗R-squared	F-statistic	Obs∗R-squared	F-statistic	Obs∗R-squared
	1.489	7.441	1.981	9.878	0.159	0.797	0.326	1.639
	Prob. F (5,1828)	Prob. Chi-Square (5)	Prob. F (5,1436)	Prob. Chi-Square (5)	Prob. F (5,1032)	Prob. Chi-Square (5)	Prob. F (5,864)	Prob. Chi-Square (5)
	0.190	0.190	0.079	0.079	0.977	0.977	0.897	0.897
Group 2	EGARCH(1,1)		EGARCH(1,1)		TARCH(1,1)		EGARCH(1,1)	
	F-statistic	Obs∗R-squared	F-statistic	Obs∗R-squared	F-statistic	Obs∗R-squared	F-statistic	Obs∗R-squared
	1.727	8.623	0.590	2.954	0.272	1.367	0.073	0.369
	Prob. F (5,1828)	Prob. Chi-Square (5)	Prob. F (5,1436)	Prob. Chi-Square (5)	Prob. F (5,1032)	Prob. Chi-Square (5)	Prob. F (5,864)	Prob. Chi-Square (5)
	0.125	0.125	0.708	0.707	0.928	0.928	0.996	0.996
Group 3	EGARCH(1,1)		EGARCH(1,1)		GARCH(1,1)		EGARCH(1,1)	
	F-statistic	Obs∗R-squared	F-statistic	Obs∗R-squared	F-statistic	Obs∗R-squared	F-statistic	Obs∗R-squared
	1.685	8.411	0.548	2.746	0.259	1.301	0.035	0.178
	Prob. F (5,1828)	Prob. Chi-Square (5)	Prob. F (5,1436)	Prob. Chi-Square (5)	Prob. F (5,1032)	Prob. Chi-Square (5)	Prob. F (5,864)	Prob. Chi-Square (5)
	0.135	0.135	0.740	0.739	0.935	0.935	0.999	0.999
Group 4	EGARCH(1,1)		GARCH(2,2)		TGARCH(1,1)		EGARCH(1,1)	
	F-statistic	Obs∗R-squared	F-statistic	Obs∗R-squared	F-statistic	Obs∗R-squared	F-statistic	Obs∗R-squared
	1.746	8.716	0.444	2.226	0.465	2.333	0.295	1.485
	Prob. F (5,1828)	Prob. Chi-Square (5)	Prob. F (5,1436)	Prob. Chi-Square (5)	Prob. F (5,1032)	Prob. Chi-Square (5)	Prob. F (5,864)	Prob. Chi-Square (5)
	0.121	0.121	0.818	0.817	0.803	0.801	0.916	0.915

Group 5	EGARCH(1,1)		GARCH(1,1)		EGARCH(1,1)		EGARCH(1,1)	
	F-statistic	Obs∗R-squared	F-statistic	Obs∗R-squared	F-statistic	Obs∗R-squared	F-statistic	Obs∗R-squared
	1.407	7.031	1.951	9.730	0.567	2.842	0.143	0.719
	Prob. F (5,1828)	Prob. Chi-Square (5)	Prob. F (5,1436)	Prob. Chi-Square (5)	Prob. F (5,1032)	Prob. Chi-Square (5)	Prob. F (5,864)	Prob. Chi-Square (5)
	0.219	0.218	0.083	0.083	0.726	0.724	0.982	0.982
Group 6	TGARCH(1,1)		EGARCH(1,1)		EGARCH(1,1)		GARCH(1,2)	
	F-statistic	Obs∗R-squared	F-statistic	Obs∗R-squared	F-statistic	Obs∗R-squared	F-statistic	Obs∗R-squared
	1.510	7.543	1.311	6.551	0.557	2.792	0.014	0.071
	Prob. F (5,1828)	Prob. Chi-Square (5)	Prob. F (5,1436)	Prob. Chi-Square (5)	Prob. F (5,1032)	Prob. Chi-Square (5)	Prob. F (5,864)	Prob. Chi-Square (5)
	0.184	0.183	0.257	0.256	0.733	0.732	1.000	1.000

7.4 Conclusion

In this paper, we examined the relationship between cryptocurrency return volatility and the effects of various types of news on selected cryptocurrencies. Previous studies drew attention mostly to the links between Bitcoin and news. Following Auer and Claessens (2018), we used 1054 news items and created 22 categories. We extended the previous study by employing a clustering analysis to form six groups. We also included several other major cryptocurrencies besides Bitcoin. These groups were added to the best ARCH family models created for various cryptocurrencies to analyze the effect on volatility. The results show that different cryptocurrencies react differently to various news categories.

We examined six news groups, and the first group contains five news groups: License, Mining/Currency Production, Restrict, Law, and Risk. In line with previous literature, one of the most significant findings of this news group is the Law category, which showed consistent findings for all coins in the analysis. We observed significant effects for the Mining/Currency Production and Risk categories on Monero's, Ethereum's, and Litecoin's return volatilities. The Restrict category shows a similar effect on Bitcoin, Monero, and Litecoin, and License shows a similar effect on Bitcoin, Ethereum, and Litecoin. The second group includes six news groups: General, CFTC, Security, Currency/Not Currency, Crime and Tax. News about the Currency/Not Currency and Tax categories had a significant effect on all four cryptocurrencies. The Security and Crime categories only had an effect on Monero and Litecoin. The General news and CTFC announcements only had an effect on Litecoin. The third group covers only two news categories: Terror and AML. Both categories did not have any statistically significant effect on Bitcoin's return volatility. All of the other coins' return volatility was affected by Terror and AML news. The fourth group contains two news categories: SEC and ETF. Significance was present in the ETF category for Bitcoin and Litecoin. Significance was present in the SEC category for Monero, Ethereum, and Litecoin. The fifth group consists of four categories: Ban, Legal, Exchange, and Price. The Ban category, which includes regulatory measures that ban and block cryptocurrencies, has an effect only on Bitcoin's return volatility. All remaining categories show a significant effect on the rest of the coins. The sixth group has three categories: ICO, Warning, and Regulation. News related to the ICO category showed a significant effect on Bitcoin and Litecoin, the Warning category on Bitcoin, Ethereum, and Litecoin, and the Regulation category on Bitcoin's, Monero's, and Ethereum's return volatility.

Finally, despite the free and borderless structure of cryptocurrencies, news about regulations or potential actions and all other news have a certain level of effect on cryptocurrency market valuations. Nonfundamental factors, such as public sentiment, should be included in cryptocurrency valuations. This paper seeks to narrow the gap between traditional and behavioral asset pricing by understanding the dynamics of cryptocurrency price formation with a focus on the effect of news on various coins.

References

Amanzholova, B. A., & Teslya, P. N. (2018, October). Threats and opportunities of cryptocurrency technologies. In *2018 XIV International Scientific-Technical Conference on Actual Problems of Electronics Instrument Engineering* (APEIE) (pp. 335–339). IEEE.

Auer, R., & Claessens, S. (2018, September). Regulating cryptocurrencies: Assessing market reactions. *BIS Quarterly Review*.

Bildirici, M. E., Alp, E. A., Ersin, Ö. Ö., & Bozoklu, Ü. (2010). *İktisatta kullanılan doğrusal olmayan zaman serisi yöntemleri*. Türkmen Kitabevi.

Birz, G., & Lott Jr., J. R. (2011). The effect of macroeconomic news on stock returns: New evidence from newspaper coverage. *Journal of Banking & Finance, 35*(11), 2791–2800.

Bollerslev, T. (1986). Generalized autoregressive conditional heteroskedasticity. *Journal of Econometrics, 31*(3), 307–327.

Campbell, J. Y., & Hentschel, L. (1992). No news is good news: An asymmetric model of changing volatility in stock returns. *Journal of Financial Economics, 31*(3), 281–318.

Carrera, B. P. (2018). *Effect of sentiment on bitcoin price formation* (pp. 1–49). North Carolina: Duke University Durham.

Chan, W. S. (2003). Stock price reaction to news and no-news: Drift and reversal after headlines. *Journal of Financial Economics, 70*(2), 223–260.

Chu, J., Chan, S., Nadarajah, S., & Osterrieder, J. (2017). GARCH modelling of cryptocurrencies. *Journal of Risk and Financial Management, 10*(4), 17.

Ciaian, P., Rajcaniova, M., & Kancs, D. A. (2016). The economics of BitCoin price formation. *Applied Economics, 48*(19), 1799–1815.

Colianni, S., Rosales, S., & Signorotti, M. (2016). Algorithmic trading of cryptocurrency based on Twitter sentiment analysis. *SSRN Electronic Journal.*

Corbet, S., & Katsiampa, P. (2018). Asymmetric mean reversion of bitcoin price returns. *International Review of Financial Analysis.* https://doi.org/10.1016/j.irfa.2018.10.004

Corbet, S., Larkin, C. J., Lucey, B. M., Meegan, A., & Yarovaya, L. (2018). The volatility generating effects of macroeconomic news on cryptocurrency returns. (March 16, 2018). Retrieved from https://ssrn.com/abstract=3141986 or https://doi.org/10.2139/ssrn.3141986

Dyhrberg, A. H. (2016). Bitcoin, gold and the dollar–a GARCH volatility analysis. *Finance Research Letters, 16*, 85–92.

Engle, R. F. (1982). Autoregressive conditional heteroscedasticity with estimates of the variance of United Kingdom inflation. *Econometrica: Journal of the Econometric Society, 50*, 987–1007.

Farell, R. (2015). An analysis of the cryptocurrency industry.

Gidofalvi, G., & Elkan, C. (2001). *Using news articles to predict stock price movements.* San Diego: Department of Computer Science and Engineering, University of California.

Glosten, L. R., Jagannathan, R., & Runkle, D. E. (1993). On the relation between the expected value and the volatility of the nominal excess return on stocks. *The Journal of Finance, 48*(5), 1779–1801.

Hanlon, M., & Slemrod, J. (2009). What does tax aggressiveness signal? Evidence from stock price reactions to news about tax shelter involvement. *Journal of Public Economics, 93*(1–2), 126–141.

Hayes, A. S. (2017). Cryptocurrency value formation: An empirical study leading to a cost of production model for valuing bitcoin. *Telematics and Informatics, 34*(7), 1308–1321.

Hughes, S. J., & Middlebrook, S. T. (2015). Advancing a framework for regulating cryptocurrency payments intermediaries. *Yale Journal on Regulation, 32*, 495.

Katsiampa, P. (2017). Volatility estimation for bitcoin: A comparison of GARCH models. *Economics Letters, 158*, 3–6.

Nelson, D. B. (1991). Conditional Heteroskedasticity in asset returns: A new approach. *Econometrica, 59*(2), 347.

Vega, C. (2006). Stock price reaction to public and private information. *Journal of Financial Economics, 82*(1), 103–133.

Xie, P., Chen, H., & Hu, Y. J. (2017). *Network structure and predictive power of social media for the bitcoin market* (Georgia Tech Scheller College of Business Research Paper No. 17-5).

Serkan Cankaya is an associate professor of Banking & Finance at Istanbul Commerce University and has been the dean of the graduate school of finance since 2017. He received an MBA in Finance from the University of West Georgia and a Ph.D. in Banking and Finance from Kadir Has University, where he held an administrative and a research position at the Graduate School of Social Sciences. His primary research interests include behavioral finance and market volatility. He has taught a range of courses for graduate and undergraduate students, including Behavioral Finance, Investment Theory, and Portfolio Management and Financial Mathematics.

Elcin Aykac Alp is a Professor in Economics at Istanbul Commerce University in Istanbul (Turkey). She has graduated from the Econometrics Department of Marmara University. She holds a master's degree in Econometrics at the Institute of Social Sciences of the same University. She received a Ph.D. in Economics at Yildiz Technical University. Her research interests cover topics in the field of time series econometrics, applied econometrics, and macroeconomic analysis. Her publications have appeared in several academic journals such as; Applied Economics, The Journal of Energy and Development, Expert Systems with Applications, Applied Econometrics and International Development, International Journal of Applied Econometrics and Quantitative Studies.

Mefule Findikci is a researcher of the Rectorate Office at Istanbul Commerce University in Istanbul (Turkey). She has graduated from the Statistics Department of İstanbul Commerce University with an honorary degree in 2013. She has her post-graduate degree in the Department of Economics at the Institute of Social Sciences of the same University in 2015. Her thesis subject is "The Analysis of the Relationship Between League Performance and The Shares of the Publicly-Traded Sports Clubs in the Stock Market." Currently, she is pursuing two PhDs in the department of Econometrics in Marmara University and the Financial Economics Program of Graduate School of Finance in İstanbul Commerce University. Her research interests focus on statistics, econometrics, finance and sports' economy.

Chapter 8
Bitcoin Market Price Analysis and an Empirical Comparison with Main *Currencies, Commodities, Securities and Altcoins*

Burak Pirgaip, Burcu Dinçergök, and Şüheda Haşlak

Abstract The purpose of this study is to analyze Bitcoin (BTC) market prices and to answer the question of whether there is a relationship between BTC and other asset prices, where other assets include currencies, commodities, securities and altcoins. In the empirical part, we evaluate the lead-lag relationships among each type of asset. Consequently, we compare BTC with major currencies and stock exchanges of the U.S., the EU, the U.K. and Japan (USD-SPX, EUR-DAX, GBP-FTSE and JPY-NIK), with currencies and stock exchanges of the U.S., the U.K., Russia, Venezuela and China where BTC is actively traded (USD-SPX, GBP-FTSE, RUB-MOEX, VEF-IBVC and YUAN- SSCE), with major commodities (GOLD and OIL) and with major altcoins (ETH, XRP and LTC) on a daily basis for the period spanning from 2010.07 to 2018.12. We employ Johansen co-integration, Granger causality, impulse response functions and forecast error variance decomposition analyses in our study. Our results show that BTC does not have a long-run relationship with any asset type, but that it has a short-run relationship with gold and especially altcoins, which are both significant and bidirectional. While BTC and altcoins are closely interrelated with each other, BTC price variation is mostly borne by its own prices in all cases.

B. Pirgaip (✉)
Department of Business Administration, Hacettepe University, Ankara, Turkey
e-mail: burakpirgaip@hacettepe.edu.tr

B. Dinçergök
Department of Business, Atilim University, Ankara, Turkey
e-mail: burcu.dincergok@atilim.edu.tr

Ş. Haşlak
Graduate School of Social Sciences, Cankaya University, Ankara, Turkey

© Springer Nature Switzerland AG 2019

U. Hacioglu (ed.), *Blockchain Economics and Financial Market Innovation*, Contributions to Economics, https://doi.org/10.1007/978-3-030-25275-5_8

8.1 Introduction

Bitcoin (BTC), as the leading cryptocurrency, has recently been one of the highly debated topics in finance circles. The appearance of BTC concept dates back to late 2008[1] when an anonymous figure called Satoshi Nakamoto (2008) issued a whitepaper titled "Bitcoin: A Peer-to-Peer Electronic Cash System". Despite the fact that this seminal publication failed to attract immediate attention, investors who lost trust in the financial system following the global financial crisis started to resort to BTC as a part of an independent economy (Bouri, Molnár, Azzi, Roubaud, & Hagfors, 2017). This environment of uncertainty also served to increase the profitability of BTC both during and after the crisis (Weber, 2014).

The total number of BTC in circulation reached approximately 17.45 million in December 2018, which is triple the 5.8 million figure seen in the first quarter of 2011 (Statista, 2019). In addition the market capitalization of BTC is currently around 63 billion USD (Bitcoinity, 2019). Given these striking values, BTC has not only become one of the most prominent investment options for investors, but has also begun to be accepted as a mode of payment in some companies, and even in some countries. It is worth emphasizing that BTC markets such as Bittrex, Binance, Poloniex, HADAX and Kraken facilitate BTC trading in exchange for other currencies. Hence, in sectors as diverse as general merchandise (e.g. Overstock.com), computer (e.g. Microsoft), web (e.g. Shopify), travel (e.g. CheapAir.com), and food (e.g. PizzaForCoins), enterprises are increasingly accepting BTC transactions for their goods or services in today's business environment (Moreau, 2019). When countries are considered, it is possible to observe that while some jurisdictions in the world are still contentious (e.g. China, India, Russia) or hostile (e.g. Iceland, Bolivia, Ecuador) toward BTC, most of them are permissive (e.g. G7 countries, i.e. Canada, France, Germany, Italy, Japan, the UK and the US) of BTC (BitLegal, 2017). Apart from its legitimacy, however, market practice shows that about 70% of all trading occurs in the U.S., Russia, the U.K., Venezuela, and China in order of volume (Alford, 2018).

The use of BTC continues to increase because of its value as a fast, cheap, secure and convenient payment mechanism, although this growing popularity is overshadowed by a pretty volatile history. After following a stable path between 2009 and 2012, the price of BTC has experienced a dramatic increase from $13.00 in late 2012 to reach its highest level to about $20,000[2] in late 2017. The current BTC price is surprisingly close to $3600 meaning that BTC has lost almost 80% of its value from its all-time high levels. Moreover, although there is great consensus among policy-makers in recognizing BTC as "legal tender", there is ambiguity in whether it should be classified as *"a currency"*, *"a commodity"* or *"a security"*.[3]

[1]BTC scheme is operationalized in January 2009 when the very first transfer of BTC was occurred.

[2]Exact price was $19.783,21 and exact date was 17th December 2017 (Shome, 2019).

[3]A fourth classification indeed exists. The Internal Revenue Service in the U.S. declared BTC *"as property"* for tax purposes in 2014 (IRS, 2014).

Table 8.1 BTC classification puzzle

BTC	Currency	Commodity	Security
Why	Tradable and exchangeable	A basic good utilized in commercial transactions	Fungible and negotiable
		Interchangeable with the same type of other commodities	Promise to recover a financial value in future
		Scarce, finite supply, and inherent value (similar to gold)	Investment with a reasonable expectation of profit
Why not?	No centralized governing authority	No use value	Not secured by a third party
	No ties back to any jurisdiction		No managerial effort of the seller
	Not generated as and when required		
	Total units is limited (21 mn)		
	Lacks relative stability	Bears a significant counterparty risk	

Source: Compiled from various sources by authors

For instance, Australia treats BTC *"just like money"* (Australian Government, 2017), the U.S. Commodity Futures Trading Commission classifies BTC as *"a commodity"* (CFTC, 2018), and Malaysia considers BTC offerings (i.e., initial coin offerings, namely ICO) to comply with *"securities"* laws in the country (SCM, 2019).

The classification issue is crucial in clarifying how to protect investors in a given market, and yet this issue is far from being resolved. Conceptually speaking, BTC shares some similarities with all three types of assets. It is designed to perform the common functions of fiat currencies such as being a medium of exchange, a unit of account and a store of value. In addition, just like a commodity, BTC supply is limited by the protocol design, it is independent of government control and regulations, and it has a decentralized nature. Despite the similarities, BTC also possesses some significant differences. The following table (Table 8.1) shows the main reasons why (or why not) BTC may be regarded as a currency, a commodity and a security:

As long as the fundamental question of whether BTC should be regarded as a currency, a commodity or a security remains, regulatory guidance concerning BTC will be absent, in turn creating uncertainty for market participants (Mandjee, 2015).

This confusion over classification occupies a major place in the literature. Scholars have tried to determine which asset type BTC belongs to by examining its financial features and statistical properties (e.g. Baur, Hong, & Lee, 2018), arriving, in general, at inconclusive results. These results are discussed at length in the next section.

Within this framework, our study aims to contribute to this discussion by analyzing the behavior of BTC prices vis-à-vis that of other currencies, commodities, and securities. Our main motivation is to provide new insight into the general understanding of BTC as well as to shed light on the curious classification case of

BTC in the market. To this end, we include major alternative digital currencies, also known as altcoins, in our analyses in order to infer the price variation among cryptocurrencies. We investigate the reasons behind BTC price changes and the lead-lag relationship between BTC and other assets by means of different methodologies that include Johansen co-integration, Granger causality, impulse response functions and forecast error variance decomposition analyses. Our results show that Bitcoin does not have a long-run relationship with any other asset type, but that it has a short-run relationship with gold, and especially altcoins, that are significant and bidirectional. Further, BTC and altcoins are closely interrelated with each other when compared to other assets. On the other hand, BTC price variation is stimulated mainly by its own price.

Section 8.2 briefly reviews the literature. Section 8.3 presents the data and methodology. Section 8.4 presents the main findings. Finally, Sect. 8.5 concludes.

8.2 Literature Review

There is a vast amount of research analyzing BTC price behavior with respect to other investment alternatives. Buchholz, Delaney, Warren, and Parker (2012) assert that the interplay between BTC's supply and demand is one of the most important determinants of BTC price in the long run. Van Wijk (2013) investigates the financial factors that have a potential effect on the value of BTC using an error correction model on a 3-year data that extends from 19.07.2010 to 13.06.2013 and finds that the Dow Jones Index significantly affects BTC both in the short-term and in the long-term, whereas the euro-dollar parity and the WTI oil price have a significant long run effect only.[4] Glaser, Zimmermann, Haferkorn, Weber, and Siering (2014) underline the fact that users of BTC are not primarily interested in an alternative transaction system but rather they seek to participate in an alternative investment vehicle.

Yermack (2015) addresses the currency criteria of BTC and concludes that BTC does not appear to have unit of account nor does it have store of value features due to the large risk that BTC imposes on its holders. This excessive risk arises from its volatility and low correlation with other currencies. Eisl, Gasser, and Weinmayer (2015), by adopting a Conditional Value-at-Risk framework that does not require normally distributed returns, report that BTC should be an investment option for optimal portfolios due to its low correlations with currencies, stocks, bonds or commodities. Selgin (2015) and Baek and Elbeck (2015) refer to BTC as synthetic commodity money that lacks nonmonetary value and is naturally scarce and speculative where volatility is internally driven. Cheah and Fry (2015) and Cheung,

[4]Other variables concerning the stock markets, exchange rates and oil prices are FTSE 100-Nikkei 225, yen-dollar parity, and Brent oil-the UBS Bloomberg Constant Maturity Commodity Index of Oil, respectively.

Roca, and Su (2015) also argue that BTC exhibits speculative bubbles. According to Kristoufek (2015), the speculative feature of BTC makes it a unique asset possessing the properties of a standard financial asset.

Ciaian, Rajcaniova, and Kancs (2016) analyze BTC price formation and find that both supply-demand and BTC attractiveness for investors and users significantly affect BTC prices with variation over time. Dyhrberg (2016) questions the hedging capabilities of BTC in comparison with gold over the 2010–2015 period. The results of the asymmetric GARCH methodology show that BTC has some of the hedging capabilities of gold since it can be used as a hedge against stocks in the FTSE and against the U.S. dollar. Hence, BTC can be included in portfolios to mitigate the harmful effects of sudden shocks. Urquhart (2016) finds that BTC market is significantly inefficient that may be in the process of becoming an efficient market. He ties this finding to the relatively eccentric position of BTC as a new investment asset. By contrast, Nadarajah and Chu (2017) show that a power transformation of BTC returns can be weakly efficient.

Bouri, Azzi, and Dyhrberg (2017) study price return and changes in volatility. Based on an asymmetric-GARCH framework, their study concludes that, for the pre-crash period, BTC has a safe- haven property similar to that of traditional assets such as gold, but it turns out that this safe-haven feature disappears in the post-crash period. Baur and Dimpfl (2017) find that BTC has up to 30 times more volatility than major currencies such as the U.S. dollar, the Euro and the Japanese yen. In this regard, they assert that its high volatility prevents it from performing common currency functions efficiently. The excess volatility of BTC is considered in various studies, such as Hencic and Gouriéroux (2015), Sapuric and Kokkinaki (2014), Briere, Oosterlinck, and Szafarz (2015), Bouoiyour and Selmi (2015) and Katsiampa (2017).

Estrada (2017) analyses the Granger causality relationship between BTC price and volatility and S&P 500, VIX, and Blockchain Google Trend time series. The results reveal that while there is no Granger causality between BTC price and S&P 500 and VIX, BTC price Granger causes BTC Google trends. The author also shows that there is a unidirectional causality running from BTC volatility to S&P 500, while its causal relationship with VIX is a bidirectional one.

Employing a dynamic conditional correlation model on daily and weekly data over the 2011–2015 period, Bouri et al. (2017) analyze the diversification, hedging and safe haven properties of BTC.[5] Correlations between the returns to BTC and the returns of major stock indices (U.S, UK, Germany, Japan, China, World, Europe and Asia Pacific), commodity index, corporate bond index, gold spot price, and the U.S. dollar index show that BTC is an effective diversifier against the movements in other asset types. Additionally, BTC provides a strong hedge against movements in

[5]The authors define "diversifier" as an asset, which has low positive correlation with other assets; a weak (strong) hedge as an asset, which has no (negative) correlation with other assets; and, a weak (strong) safe haven as an asset, which has no (negative) correlation with other assets in times of stress.

Japanese, Asia Pacific and Chinese stock markets, but acts as a safe haven only in Asia Pacific and Chinese cases.

Chan, Chu, Nadarajah, and Osterrieder (2017) analyze the statistical features of BTC with other major cryptocurrencies such as Dash, Dogecoin, Litecoin, MaidSafeCoin, Monero, and Ripple. They find that cryptocurrencies are not normally distributed in that they exhibit heavy tails. As for BTC, the results reveal that the generalized hyperbolic distribution gives the best fit.

Panagiotidis, Stengos, and Vravosinos (2018) follow least absolute shrinkage and selection operator (LASSO) methodology in order to analyze the determinants of BTC returns over the period of 2010–2017.Search intensity, gold returns and policy uncertainty are among the most significant potential drivers of BTC returns. The results also show that all the uncertainty indices and all the exchange rates considered as well as interest rates, gold and oil prices have a positive impact on BTC returns, while the results for the stock markets are mixed.

Klein, Thu, and Walther (2018) analyze the time-varying correlations of BTC, gold and financial markets using the BEKK-GARCH model on daily data between 2011 and 2017. The authors find that although gold provides a hedge against financial markets, BTC cannot act as a hedge or as a safe haven for equity investments.[6] A portfolio-based test of hedging property also supports these results.

Baur et al. (2018) report the correlations of BTC with S&P 500 index, gold, silver, crude oil index, gas index, corporate bond index, high yield corporate bond index and foreign exchanges over the 2010–2015 period. The authors state that BTC is uncorrelated with other assets and these results are robust even when the effects of financial crisis were considered. These results suggest that while BTC offers great benefits for diversification, it is neither a safe haven nor a reliable hedging instrument.

8.3 Data and Methodology

We investigate whether BTC's price behavior is related to the price behavior of various currencies, commodities and securities. Relying on the extant literature we expect to provide support for the hypothesis that BTC has a greater tendency towards becoming a unique asset type rather than converging to the behavior of other assets. In this context, we compare BTC with major currencies (i.e., the U.S. dollar (USD), the euro (EUR), the British pound (GBP) and the Japanese yen (JPY)), with major commodities (i.e., gold (GOLD) and oil (OIL)), and with major stock exchanges (i.e., S&P 500 (SPX), FTSE 100 (FTSE), NIKKEI 225 (NIK), and DAX (DAX)). We also include the currencies and stock exchanges of Russia (RUB and MOEX), Venezuela (VEF and IBVC) and China (YUAN and SSEC), as BTC is intensively used in these markets. Lastly, we enhance our study with major altcoins Etherium

[6]The results are valid for S&P 500 and MSCI World indices.

Table 8.2 Variables used in the study

Variables		Proxy
	BITCOIN	BTC/USD
Currencies	USD	DXY
	EURO	EUR/USD
	GBP	GBP/USD
	JPY	JPY/USD
	RUB (Russian Ruble)	RUB/USD
	VEF (Venezuelan Bolivar)	VEF/USD
	YUAN (Chinese Yuan Renminbi)	YUAN/USD
Commodities	GOLD	GOLD/USD
	OIL	OIL/USD
Securities	S&P 500	SPX
	FTSE 100	FTSE/USD
	DAX	DAX/USD
	NIKKEI 225	NIK/USD
	MOEX (Moscow stock exchange)	MOEX/USD
	IBVC (Caracas stock exchange)	IBVC/USD
	SSEC (Shanghai stock exchange)	SSEC/USD
Altcoins	ETH	ETH/USD
	XRP	XRP/USD
	LTC	LTC/USD

(ETH/USD), Ripple (XRP/USD) and Litecoin (LTC/USD) so as to evaluate cryptocurrencies as a separate asset class.

Table 8.2 displays a brief description of the variables used in this study. All variables are converted into USD and are available on a daily basis for the period between 2010:07 and 2018:12. In order to compare BTC with USD itself, we use the U.S. Dollar Index (DXY). Data are obtained from Bloomberg.

The price levels of these assets, denominated in U.S. dollars, are graphically represented for the period 16.07.2010–31.12.2018 in Fig. 8.1.

We follow Johansen co-integration and Granger causality methodologies in our study.[7] These tests require an examination of whether or not the variables are stationary. We use Augmented Dickey-Fuller (ADF) (Dickey & Fuller, 1979) and Philips-Perron (PP) (Phillips & Perron, 1988) unit root tests to examine the stationarity of our series. The relevant hypotheses are as follows:

H_0 Series contains a unit root (they are not generated by a stationary process)

H_1 Series does not contain a unit root (they are generated by a stationary process)

The equations used in the ADF (8.1) and PP (8.2) tests are provided below:

[7]We rely on the (A)kaike (I)nformation (C)riterion in lag selection throughout our study.

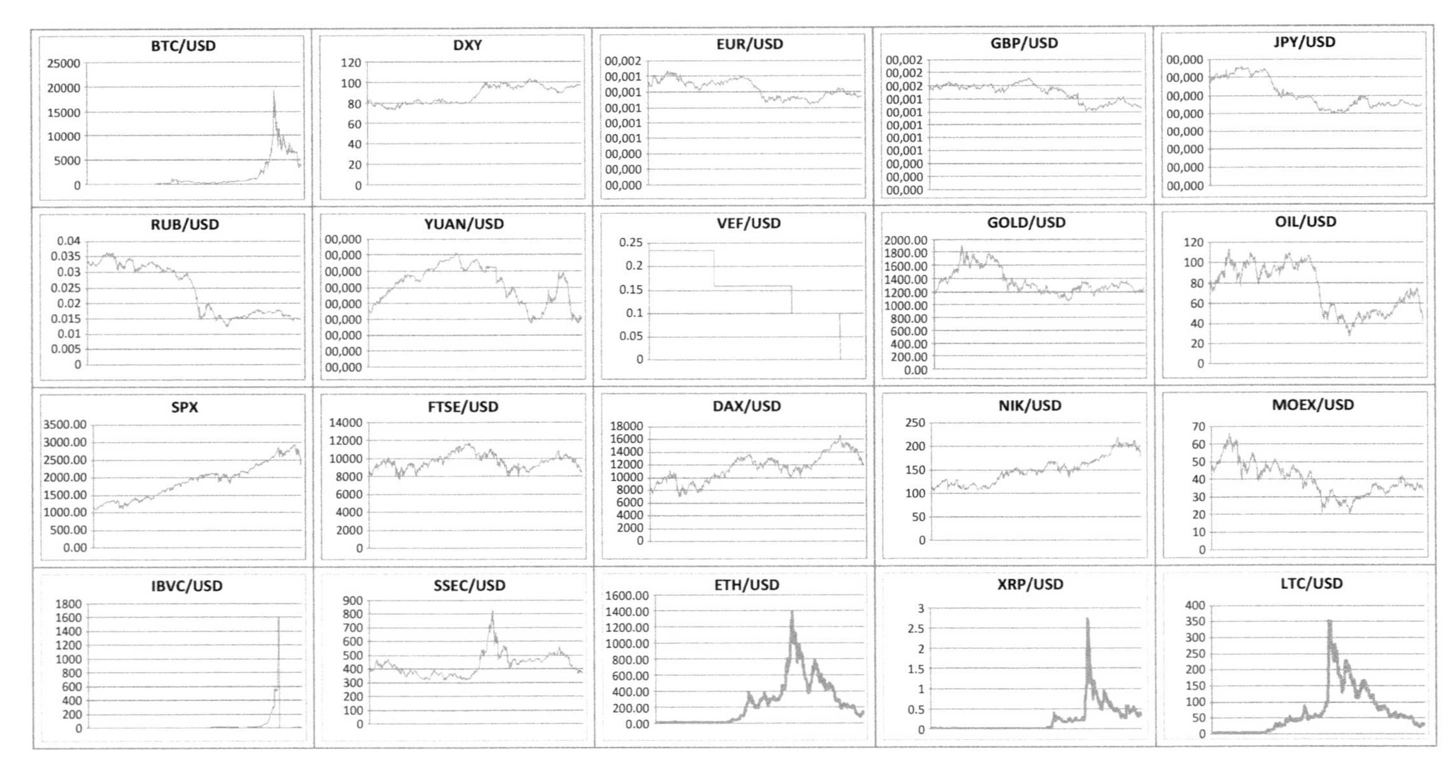

Fig. 8.1 Price levels of various asset types in U.S. dollars (16.07.2010–31.12.2018) (ETH/USD, XRP/USD and LTC/USD data starts from 10.03.2016, 31.01.2015 and 24.08.2016, respectively). Source: Bloomberg

$$\Delta Y_t = \beta_0 + \beta_1 t + \delta Y_{t-1} + \sum_{i=1}^{m} \beta_i \Delta Y_{t-i} + u_t \tag{8.1}$$

$$\Delta Y_t = \alpha_0 + \alpha_1 (t - T/2) + \alpha_2 Y_{t-1} \sum_{i=1}^{m} \Delta Y_{t-i} + \varepsilon_t \tag{8.2}$$

Unless the unit root tests yield the same order of stationarity, it is necessary to iterate with the first differences of the series. The reason why the series should have the same order of stationarity is so that the analyses of the co-integration relationship can be processed further.

In this regard, we investigate the presence of a long-run linear relationship between the series by employing the Johansen (1988) and the Johansen and Juselius (1990) standard co-integration methodology. These tests provide trace and maximum eigenvalue likelihood ratios in order to make inferences. The following hypotheses will be examined with the Johansen co-integration test.

H_0 There is no co-integration relationship between the variables.

H_1 There is a co-integration relationship between the variables.

In the presence of co-integration, the causality relationship is detected by Granger causality methodology applied in accordance with the vector error correction model (VECM). In the absence of co-integration, however, the causality relationship is tested with the standard Granger causality methodology. Standard equations regarding the VECM are as follows:

$$\Delta X_t = a_0 + \sum_{i=1}^{a} a_i \Delta Y_{t-i} + \sum_{i=1}^{b} \beta_i \Delta X_{t-i} + \lambda EC_{t-1} + u_{xt} \tag{8.3}$$

$$\Delta Y_t = \beta_0 + \sum_{i=1}^{a} a_i \Delta Y_{t-i} + \sum_{i=1}^{b} \beta_i \Delta X_{t-i} + \lambda EC_{t-1} + u_{yt} \tag{8.4}$$

In Eqs. (8.3) and (8.4), α and β are the coefficients of interest, the a and b above the operators indicate selected lag lengths, EC_{t-1} is the error correction term, and X and Y are the independent and dependent variables respectively.

Standard Granger causality methodology is employed by means of the following model so as to determine the short-term link between the variables of interest (Granger, 1969):

$$Y_t = \alpha_0 + \sum_{i=1}^{k1} \alpha_i Y_{t-i} + \sum_{i=1}^{k2} \beta_i X_{t-i} + \varepsilon_t \tag{8.5}$$

$$X_t = \chi_0 + \sum_{i=1}^{k3} \chi_i X_{t-i} + \sum_{i=1}^{k4} \delta_i Y_{t-i} + v_t \tag{8.6}$$

As it is clear from (8.5) (and (8.6)), Y (X) is regressed on its own lagged values and on the lagged values of X (Y). In the equations, α, β, χ and δ are the coefficients and the "k_i"s indicate the selected lag lengths. Error terms are assumed to be independent from each other (Granger, 1969). For instance, if the lagged values of X (Y) are jointly zero, then there is a one-way or unidirectional causality running from Y (X) as the independent variable to X (Y) as the dependent variable. When lagged values of both X and Y are statistically significant, then there is a two-way or bidirectional Granger causality relationship between the variables.

Lastly, we employ impulse response function (IRF) and forecast error variance decomposition (FEVD) analyses. IRF can be used to estimate the effects of an exogenous shock to a single variable on the dynamic paths of all of the variables of the system. Therefore, we further characterize the IRF between the variable sequences to analyze the short-term dynamic relationship between them. IRF measures the dynamic marginal effects of each unexpected shock on the variables over a given period. Variance decompositions, however, investigate how important each of the shocks is as a component of the overall variance of each variable over that period. FEVD basically separates forecast error variance into components attributed to each of these sources.

8.4 Empirical Findings

We transform each price data into their natural logarithms and report the descriptive statistics in Table 8.3.

The mean and median figures in Table 8.3 imply that BTC is one of the most valuable assets, although the value of gold and major stock exchanges surpasses it in value. BTC has the largest maximum value among all assets, which shows that it can reach very high price levels over time. It has the highest range and standard deviation as well as a relatively high coefficient of variation.

It is also one of the most negatively skewed of all the assets. In terms of kurtosis, BTC is less peaked, but is close to the levels similar to that of the normal distribution. Although we do not report it here for the sake of brevity, the log returns of the price data also show a similar pattern. In this respect, BTC brings in significant returns in exchange for a relatively high risk.[8] Figure 8.2 below portrays the normal distribution plot for BTC prices. We can conclude that BTC is not normally distributed.

Table 8.4 displays the correlation matrix of the variables used in this study. Panel A of Table 8.4 shows that BTC has a positive significant correlation with DXY only,

[8]The mean return of BTC is the highest among other asset classes and the variation in returns shows high levels of risk. Data is available on request.

Table 8.3 Descriptive statistics for log transformed price levels of each asset type in U.S. dollar

	Mean	Med.	Max	Min	St. Dev.	CV	Skew.	Kurt.	Obs.
BTC/USD	4.94	5.79	9.87	−3.01	3.04	0.62	−0.70	2.81	2.207
DXY	4.47	4.45	4.64	4.29	0.10	0.02	0.02	1.49	2.207
EUR/USD	0.21	0.22	0.39	0.04	0.09	0.43	−0.07	1.68	2.207
GBP/USD	0.40	0.44	0.54	0.19	0.09	0.23	−0.64	2.03	2.207
JPY/USD	−4.60	−4.64	−4.33	−4.83	0.15	−0.03	0.47	1.78	2.207
RUB/USD	−3.78	−3.69	−3.31	−4.41	0.34	−0.09	−0.06	1.23	2.207
VEF/USD	−1.64	−1.84	0.00	−6.34	0.66	−0.40	0.89	7.75	2.207
YUAN/USD	−1.86	−1.85	−1.80	−1.94	0.04	−0.02	−0.40	2.00	2.207
GOLD/USD	7.20	7.17	7.55	6.96	0.13	0.02	0.73	2.57	2.207
OIL/USD	4.25	4.33	4.73	3.28	0.34	0.08	−0.43	1.92	2.207
SPX	7.52	7.58	7.98	6.95	0.27	0.04	−0.19	1.90	2.207
FTSE/USD	9.17	9.17	9.38	8.94	0.09	0.01	0.14	2.43	2.207
DAX/USD	9.34	9.36	9.72	8.84	0.20	0.02	−0.34	2.36	2.207
NIK/USD	4.99	5.00	5.39	4.65	0.19	0.04	0.11	2.03	2.207
MOEX/USD	3.65	3.64	4.20	3.00	0.24	0.07	−0.11	2.49	2.120
IBVC/USD	0.52	0.57	7.38	−2.75	2.22	4.27	0.45	2.97	2.054
SSEC/USD	6.05	6.05	6.72	5.75	0.19	0.03	0.72	3.53	1.996
ETH/USD	4.52	5.20	7.23	1.96	1.70	0.38	−0.24	1.41	733
XRP/USD	−3.21	−4.51	1.00	−6.10	2.06	−0.64	0.36	1.38	1.021
LTC/USD	3.48	3.94	5.86	1.25	1.40	0.40	−0.46	1.89	614

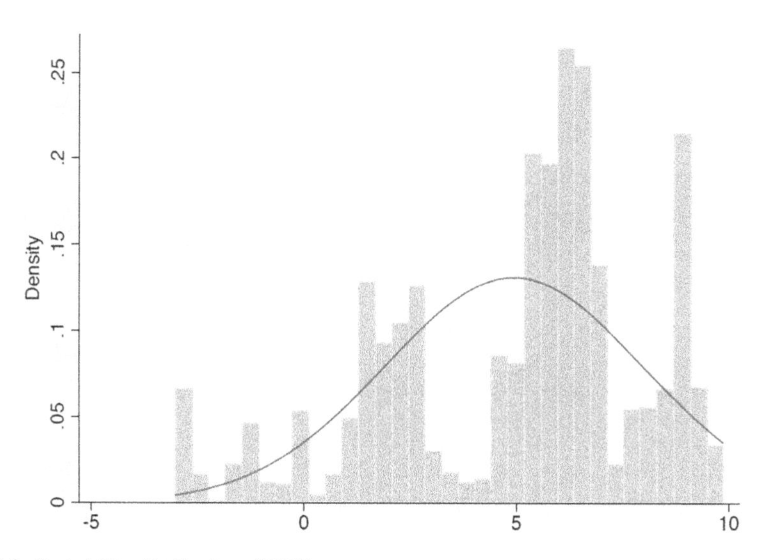

Fig. 8.2 Probability distribution of BTC

Table 8.4 Correlation matrix of variables

Panel A—Correlations between BTC and currencies

	BTC	DXY	EUR/USD	GBP/USD	JPY/USD	RUB/USD	VEF/USD	YUAN/USD
BTC	1.00							
DXY	0.69***	1.00						
EUR/USD	−0.56***	−0.98***	1.00					
GBP/USD	−0.60***	−0.80**	0.76***	1.00				
JPY/USD	−0.80***	−0.84***	0.74***	0.51***	1.00			
RUB/USD	−0.75***	−0.95***	0.90***	0.75***	0.84*	1.00		
VEF/USD	0.01	−0.14***	0.17***	0.00	0.16***	0.04**	1.00	
YUAN/USD	−0.10***	−0.48***	0.48***	0.75***	0.12***	0.44***	−0.05**	1.00

Panel B—Correlations between BTC and commodities

	BTC	GOLD/USD	OIL/USD
BTC	1.00		
GOLD/USD	−0.54***	1.00	
OIL/USD	−0.50***	0.63***	1.00

Panel C—Correlations between BTC and stock exchanges

	BTC	SPX	FTSE/USD	DAX/USD	NIK/USD	MOEX/USD	IBVC/USD	SSEC/USD
BTC	1.00							
SPX	0.95***	1.00						
FTSE/USD	0.39***	0.33***	1.00					
DAX/USD	0.88***	0.91***	0.61***	1.00				
NIK/USD	0.90***	0.97***	0.34***	0.92***	1.00			
MOEX/USD	−0.62***	−0.68***	−0.16***	−0.48***	−0.58***	1.00		

IBVC/USD	0.77***	0.83***	0.16***	0.67***	0.76***	−0.72	1.00	
SSEC/USD	0.28***	0.46***	0.04**	0.38***	0.42***	−0.54***	0.48***	1.00

Panel D—Correlations between BTC and Altcoins

	BTC	ETH/USD	XRP/USD	LTC/USD
BTC	1.00			
ETH/USD	0.96***	1.00		
XRP/USD	0.94***	0.97***	1.00	
LTC/USD	0.97***	0.98***	0.97***	1.00

*,** and *** indicate statistical significance at the 10%, 5% and 1% level, respectively

whereas its correlation with other currencies is negative. The highest correlation is with JPY/USD (−80%) and the lowest correlation is with YUAN/USD (−10%). In Panel B, we observe that the correlation of BTC with commodities is significantly negative at around −50%. Panel C shows that, unlike commodities, the correlation of BTC with stock markets is significantly positive, except for the Russian stock exchange, MOEX/USD. The highest correlation exists with SPX (95%) and the lowest correlation exists with SSEC/USD (28%). Panel D basically reveals that the correlation of BTC with altcoins is far higher than the correlations with other assets. The average correlation between BTC and other cryptocurrencies is about 95.6% and significantly positive.

We then employ mean and variance comparison tests as well, which are not reported here. These tests suggest that all of the series are significantly different from each other from a parametric point of view, i.e., in mean and variance terms. More importantly, the size of the variance (standard deviation) of BTC prices with respect to the variances of other asset classes is both economically and statistically significant.

The results in Table 8.5 give the ADF and PP test results for our variables.

A quick examination of Table 8.5 clearly reveals that all of the variables are non-stationary at level. However, when we convert them into first-differenced variables, they become stationary. They are all integrated of the same order, i.e. I(1). Hence, we apply the Johansen co- integration test and VECM for BTC with other assets. Table 8.6 shows the results of the co- integration tests.

In line with the results given in Table 8.5, we cannot reject the null hypothesis that there is no co-integration among our variables in the currency market (Panel A), in the commodity market (Panel B), in the stock markets (Panel C) or in the altcoins market (Panel D). More specifically, there are no co-integrating vectors and in the long run they do not move together. We thus employ Granger causality tests in order to infer the short-run relationship between our variables. First-differenced data[9] is used in order to reflect stationarity in the tests. Table 8.7 presents the Granger causality test results.

Panel A of Table 8.7 shows that there is no Granger causality between BTC and exchange rates in any of the currency markets. Unreported results, on the other hand, support the fact that information is incorporated more quickly in DXY and GBP/USD in the major currencies market, and RUB/USD and YUAN/USD in the markets where BTC is actively traded. In this case, we can easily argue that BTC does not carry any currency information in its price formation and vice versa. In Panel B, however, we observe significant bidirectional Granger causality between BTC/USD and GOLD/USD. Panel C also portrays a significant relationship between BTC/USD and DAX/USD in major stock markets and between BTC/USD and SPX and FTSE/USD in the markets where BTC is actively traded. Interestingly, BTC/USD Granger causes SPX in the major currency markets as well. These results

[9]First differenced data is in effect the daily returns generated by each asset. As a result, the analysis is on returns rather than price levels.

Table 8.5 Unit root tests

	ADF (PP)					
	Level			First difference		
	Lags	Constant	Constant and trend	Lags	Constant	Constant and trend
BTC/USD	4	−2.76* (−2.92**)	−2.43 (−2.49)	3	−22.32*** (−51.52***)	−22.42*** (−51.62***)
DXY	2	−0.95 (−0.97)	−2.41 (−2.40)	1	−24.45*** (−49.73***)	−24.44*** (−49.72***)
EUR/USD	1	−1.35 (−1.37)	−2.24 (−2.26)	0	−47.75*** (−47.74***)	−47.74*** (−47.74***)
GBP/USD	1	−0.87 (−0.89)	−2.58 (−2.56)	0	−47.47*** (−47.47***)	−47.48*** (−47.48***)
JPY/USD	2	−0.99 (−1.10)	−1.41 (−1.48)	1	−54.77*** (−54.77***)	−54.76*** (−54.76***)
RUB/USD	12	−0.61 (−0.49)	−1.75 (−1.71)	11	−12.45*** (−47.07***)	−12.45*** (−47.06***)
VEF/USD	1	−2.73* (−2.73)*	−2.65 (−2.65)	0	−46.95*** (−46.95***)	−46.97*** (−46.97***)
YUAN/USD	10	−1.09 (−0.90)	−2.02 (−1.89)	9	−12.90*** (−46.96***)	−13.07*** (−47.05***)
GOLD/USD	1	−1.84 (−1.84)	−2.82 (−2.77)	0	−47.41*** (−47.41***)	−47.41*** (−47.41***)
OIL/USD	1	−0.98 (−1.01)	−1.98 (−2.02)	0	−48.90*** (−48.90***)	−48.90*** (−48.90***)
SPX	6	−1.49 (−1.66)	−2.74 (−3.19*)	5	−20.66*** (−48.46***)	−20.68*** (−48.47***)
FTSE/USD	6	−2.72* (−3.16**)	−2.58 (−3.02)	5	−21.46*** (−44.56***)	−21.49*** (−44.57***)
DAX/USD	6	−1.99 (−2.13)	−2.22 (−2.47)	5	−20.69*** (−45.02***)	−20.71*** (−45.02***)
NIK/USD	6	−1.38 (−1.45)	−3.01 (−3.40*)	5	−21.00*** (−56.18***)	−21.00*** (−56.18***)
MOEX/USD	3	−1.74 (−1.75)	−2.34 (−2.40)	2	−27.08*** (−44.58***)	−27.07*** (−44.57***)
IBCV/USD	9	−1.95 (−2.12)	−2.49 (−2.87)	8	−5.48*** (−42.08***)	−5.45*** (−42.09***)
SSEC/USD	8	−1.15 (−1.97)	−1.40 (−1.89)	7	−14.55*** (−42.70***)	−14.54*** (−42.70***)
ETH/USD	1	−1.18 (−1.18)	−0.12 (−0.09)	0	−26.22*** (−26.22***)	−26.27*** (−26.27***)
XRP/USD	11	−0.61 (−0.79)	−2.22 (−2.42)	10	−10.94*** (−36.38***)	−10.95*** (−36.37***)
LTC/USD	1	−1.54 (−1.53)	−0.12 (−0.13)	0	−24.87*** (−24.87***)	−25.01*** (−25.01***)

*, ** and *** indicate statistical significance at the 10%, 5% and 1% level, respectively

Table 8.6 Co-integration results (Trace and Max-Eigen Statistics)

Panel A—BTC and Currencies					
Major currencies (DXY, EUR, GBP, JPY)					
	H_0	H_1	**Eigenvalue**	**Trace statistic**	**1%**
0	$r = 0$	$r = 1$	–	63.89*	76.07
1	$r \leq 1$	$r = 2$	0.01	33.23	54.46
2	$r \leq 2$	$r = 3$	0.01	12.93	35.65
	H_0	H_1	**Eigenvalue**	**Max-Eigen statistics**	**1%**
0	$r = 0$	$r = 1$	–	30.66	38.77
1	$r \leq 1$	$r = 2$	0.01	20.30	32.24
2	$r \leq 2$	$r = 3$	0.01	8.32	25.52
Active country currencies (DXY, GBP, RUB, VEF, YUAN)					
	H_0	H_1	**Eigenvalue**	**Trace statistic**	**1%**
0	$r = 0$	$r = 1$	–	91.89*	103.18
1	$r \leq 1$	$r = 2$	0.02	53.20	76.07
2	$r \leq 2$	$r = 3$	0.01	35.00	54.46
	H_0	H_1	**Eigenvalue**	**Max-Eigen statistics**	**1%**
0	$r = 0$	$r = 1$	–	38.69	45.10
1	$r \leq 1$	$r = 2$	0.02	18.20	38.77
2	$r \leq 2$	$r = 3$	0.01	14.65	32.24
Panel B—BTC and commodities					
	H_0	H_1	**Eigenvalue**	**Trace statistic**	**1%**
0	$r = 0$	$r = 1$	–	25.66*	35.65
1	$r \leq 1$	$r = 2$	0.01	10.90	20.04
2	$r \leq 2$	$r = 3$	0.00	3.50	6.65
	H_0	H_1	**Eigenvalue**	**Max-Eigen statistics**	**1%**
0	$r = 0$	$r = 1$	–	14.76	25.52
1	$r \leq 1$	$r = 2$	0.01	7.40	18.63
2	$r \leq 2$	$r = 3$	0.00	3.50	6.65
Panel C—BTC and stock exchanges					
Major stock exchanges (SPX, FTSE, DAX, NIK)					
	H_0	H_1	**Eigenvalue**	**Trace statistic**	**1%**
0	$r = 0$	$r = 1$	–	73.99*	76.07
1	$r \leq 1$	$r = 2$	0.01	42.62	54.46
2	$r \leq 2$	$r = 3$	0.01	22.16	35.65
	H_0	H_1	**Eigenvalue**	**Max-Eigen statistics**	**1%**
0	$r = 0$	$r = 1$	–	31.38	38.77
1	$r \leq 1$	$r = 2$	0.01	20.46	32.24
2	$r \leq 2$	$r = 3$	0.01	12.42	25.52
Active country stock exchanges (SPX, FTSE, MOEX, IBVC, SSEC)					
	H_0	H_1	**Eigenvalue**	**Trace statistic**	**1%**
0	$r = 0$	$r = 1$	–	75.24*	103.18
1	$r \leq 1$	$r = 2$	0.01	46.58	76.07
2	$r \leq 2$	$r = 3$	0.01	28.86	54.46

(continued)

Table 8.6 (continued)

	H_0	H_1	Eigenvalue	Max-Eigen statistics	1%
0	r = 0	r = 1	–	28.66	45.10
1	r ≤ 1	r = 2	0.01	17.72	38.77
2	r ≤ 2	r = 3	0.01	14.15	32.24
Panel D—BTC and Altcoins					
	H_0	H_1	Eigenvalue	Trace statistic	1%
0	r = 0	r = 1	–	42.22*	54.46
1	r ≤ 1	r = 2	0.05	13.25	35.65
2	r ≤ 2	r = 3	0.01	6.47	20.04
	H_0	H_1	Eigenvalue	Max-Eigen statistics	1%
0	r = 0	r = 1	–	28.97	32.24
1	r ≤ 1	r = 2	0.05	6.78	25.52
2	r ≤ 2	r = 3	0.01	3.92	18.63

indicate that BTC may act as an important information tool for stock market investments. Lastly, Panel D gives evidence that cryptocurrencies are closely interrelated since we come up with significant bidirectional Granger causalities between each cryptocurrency in general. Namely, BTC/USD-ETH/USD and BTC/USD-XRP/USD are significant bidirectional Granger causal pairs, while Granger causality runs only from BTC/USD to LTC/USD. This reveals that BTC still has greater dominance over altcoins.

In a nutshell, our Granger causality analyses emphasize that BTC price information may have a considerable impact on gold, major stock indices and altcoins and this impact is not without response from these markets. Nonetheless, Granger causality tests do not give any idea regarding the direction of the relationship. In this sense, we introduce the outcomes of IRF and FEVD analyses in conjunction with each other.

In IRF analyses, we observe the shock of a one unit standard deviation. Afterwards, we apply the variance decomposition to the forecast error of the equation where BTC price is the dependent or independent variable. The observation period is arbitrarily selected as 20. IRF graphs are displayed in Fig. 8.3 and FEVD results are presented in Table 8.8.

In Panel A of Fig. 8.3, responses to a shock in one currency on BTC are displayed on the left hand side and responses to a shock in BTC on other currencies are displayed on the right hand side. In line with our Granger causality test results regarding BTC and other currencies, very few linkages between the series are established. BTC price responses to the shocks to other currencies are very small and they disappear in a very short period of time. In other words, the initial impact of a one unit standard deviation in currency prices is generally negative but only for a while. On the other hand, the responses to shocks to BTC are more apparent in economic terms. An increase in the shock to BTC causes a short series of declines in other currencies, which die out after about five periods.

Table 8.7 Granger causality tests

Panel A—BTC and currencies

Major currencies (DXY, EUR, GBP, JPY)

Variables	Excluded	Chi²
BTC/USD	DXY	1.99
	EUR/USD	1.85
	GBP/USD	2.53
	JPY/USD	1.56
	ALL	13.56
DXY	BTC/USD	2.11
EUR/USD		2.33
GBP/USD		5.18
JPY/USD		2.41

Active country currencies (DXY, GBP, RUB, VEF, YUAN)

Variables	Excluded	Chi²
BTC/USD	DXY	1.47
	GBP/USD	1.32
	RUB/USD	1.29
	VEF/USD	0.79
	YUAN/USD	0.23
	ALL	4.24
DXY	BTC/USD	0.64
GBP/USD		3.04
RUB/USD		0.84
VEF/USD		1.40
YUAN/USD		1.56

Panel B—BTC and commodities

Variables	Excluded	Chi²
BTC/USD	GOLD/USD	24.85***
	OIL/USD	2.50
	ALL	25.90***
GOLD/USD	BTC/USD	17.90***
OIL/USD		0.69

Panel D—BTC and Altcoins

Variables	Excluded	Chi²
BTC/USD	ETH/USD	10.60**
	XRP/USD	14.53***
	LTC/USD	3.33
	ALL	28.50***
ETH/USD	BTC/USD	46.98***
XRP/USD		14.54***
LTC/USD		54.22***

Panel C—BTC and stock exchanges

Major stock exchanges (SPX, FTSE, DAX, NIK)

Variables	Excluded	Chi²
BTC/USD	SPX	7.71
	FTSE/USD	8.64
	DAX/USD	12.89**
	NIK/USD	5.65
	ALL	36.22**
SPX	BTC/USD	11.66**
FTSE/USD		9.66*
DAX/USD		28.35***
NIK/USD		2.26

Active country stock exchanges (SPX, FTSE, MOEX, IBVC, SSEC)

Variables	Excluded	Chi²
BTC/USD	SPX	10.60**
	FTSE/USD	10.17**
	MOEX/USD	2.86
	IBVC/USD	1.62
	SSEC/USD	5.38
	ALL	25.83
SPX	BTC/USD	9.96**
FTSE/USD		10.08**
MOEX/USD		8.59*
IBVC/USD		0.16
SSEC/USD		0.89

*, ** and *** indicate statistical significance at the 10%, 5% and 1% level, respectively

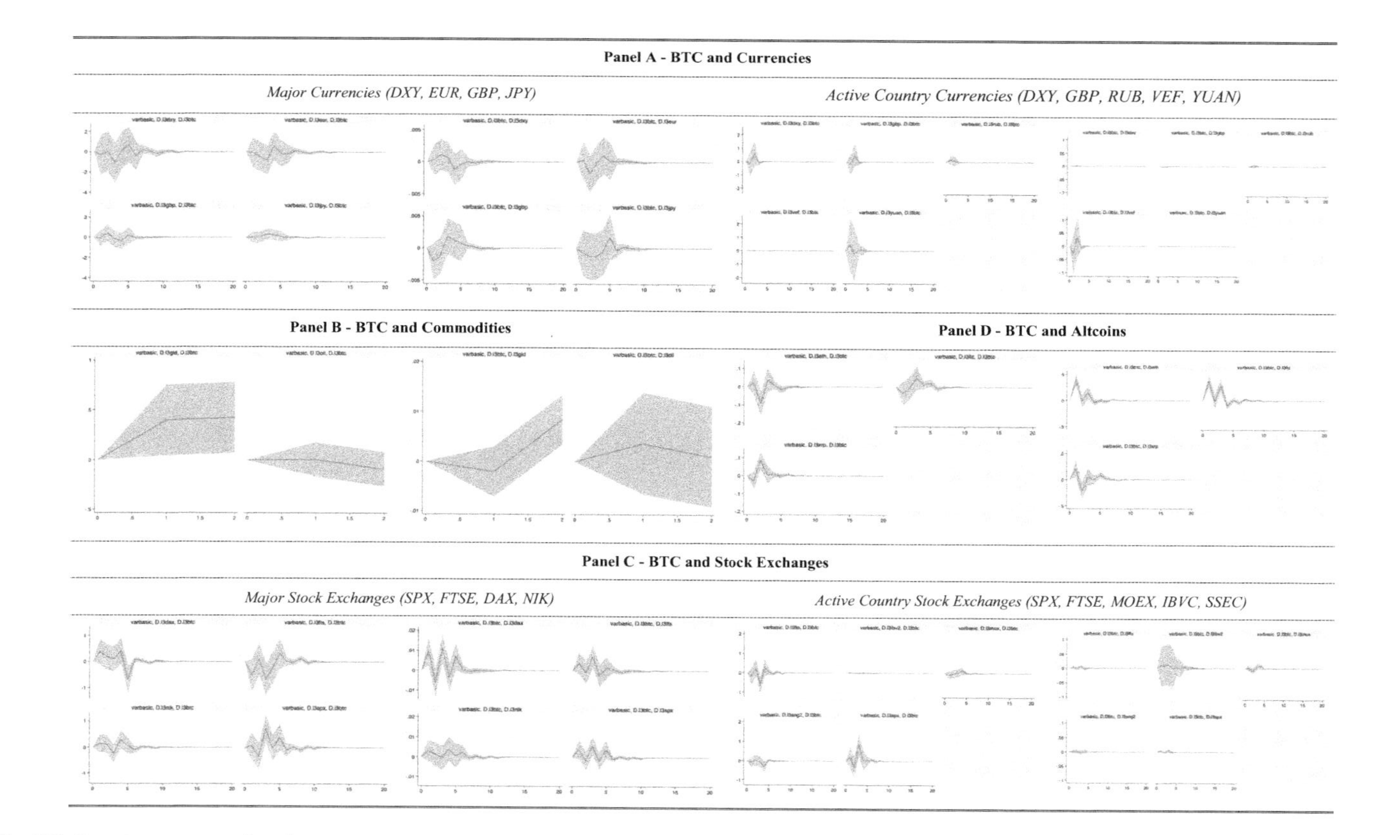

Fig. 8.3 Impulse response functions

Table 8.8 FEVD ratios

Panel A—BTC and currencies						
Major currencies (DXY, EUR, GBP, JPY)						
BTC/USD	DXY	EUR/USD	GBP/USD	JPY/USD		Total
99.38%	0.26%	0.13%	0.12%	0.11%		100.00%
	0.19%	0.25%	0.42%	0.15%		
Active country currencies (DXY, GBP, RUB, VEF, YUAN)						
BTC/USD	DXY	GBP/USD	RUB/USD	VEF/USD	YUAN/USD	Total
99.80%	0.02%	0.07%	0.06%	0.04%	0.01%	100.00%
	0.16%	0.39%	0.03%	0.07%	0.22%	
Panel B—BTC and commodities						
BTC/USD	GOLD/USD	OIL/USD				Total
98.97%	0.92%	0.11				100.00%
	0.79%	0.10%				
Panel C—BTC and stock exchanges						
Major stock exchanges (SPX, FTSE, DAX, NIK)						
BTC/USD	SPX	FTSE/USD	DAX/USD	NIK/USD		Total
98.29%	0.20%	0.61%	0.61%	0.29%		100.00%
	0.60%	0.43%	1.16%	0.22%		
Active country stock exchanges (SPX, FTSE, MOEX, IBVC, SSEC)						
BTC/USD	SPX	FTSE/USD	MOEX/USD	IBVC/USD	SSEC/USD	Total
98.74%	0.22%	0.56%	0.13%	0.08%	0.27%	100.00%
	0.50%	0.38%	0.42%	0.02%	0.04%	
Panel D—BTC and Altcoins						
BTC/USD	ETH/USD	XRP/USD	LTC/USD			Total
95.60%	1.34%	2.41%	0.65			100.00%
	14.57%	7.24%	19.66%			

In Panel B, impulses and responses of BTC and commodities are again presented in dual form. On the left hand side, we observe a positive initial impact of a one unit standard deviation in gold on BTC and on the right hand side we are presented with a similar pattern, but only after an initial immediate decrease in BTC prices.

In Panel C of Fig. 8.3, responses to a shock in one stock market index (BTC) on BTC (one stock market index) are reported. Just like currencies, although few linkages exist between series, BTC price responses to the shocks to other stock market indices are generally negative, disappearing after about five periods. It is almost the same in the case of stock market index responses to BTC price shocks. An increase in the shock to BTC causes a short series of decreases in other indices but not for a long period of time.

In Panel D, we can see a positive response to shocks, which is not long-lasting. In the altcoins world, BTC price responses to the shocks to other cryptocurrencies are, by and large, positive yet disappear quickly. The reverse is true as well.

In order to find out how important each of the shocks is as a component of the overall variance of each of the variables over time, we tabulate the FEVD results in Table 8.8.

In each panel in Table 8.8, the first row shows the variance contribution rate of each variable on BTC prices that total to 100%, while the second row shows the individual variance contribution rate of BTC on each variable. As can be seen from Table 8.8, almost 100% of the contribution comes from BTC's own standard error (e.g. 99.38% with respect to major currencies or 98.97% with respect to commodities). This outcome indicates that the transmission effect of BTC's own price is so large that the impact of other assets has minor influence on BTC price, which is consistent with the IRF analyses presented in the previous section. According to Panel A, the variance contribution rates of DXY, EUR/USD, GBP/USD and JPY/USD on BTC prices are 0.26%, 0.13%, 0.12%, and 0.11% (first row); while the variance contribution rates of BTC on these currencies are 0.19%, 0.25%, 0.42%, and 0.15%, respectively (second row). Among the other assets, DXY makes the largest contribution to the price variation of BTC, while BTC makes the largest contribution on the price variation of GBP/USD. Unreported results demonstrate that DXY is the most prominent contributory variable to the variance of GBP/USD prices with a contribution rate of 40%. We obtain similar results for currency markets where BTC is actively traded as well. Indeed, variance contribution rates of DXY, GBP/USD, RUB/USD, VEF/USD and YUAN/USD on BTC prices are 0.02%, 0.07%, 0.06%, 0.04% and 0.01% respectively (first row). These are far lower than the variance contribution ratios associated with major currency markets. The results are robust to various currency market variations. Similarly, it is on the GBP/USD exchange rate (39%) that BTC imposes the greatest variance contribution among other currencies (second row).

In Panel B, the interaction between BTC and GOLD/USD is easily seen. Though relevant figures are not that much significant, the variance contribution rate of GOLD/USD (0.92%) on BTC prices is higher than the ones pertaining to currencies (first row). This is also the case for the variance contribution rate of BTC (0.79%) on GOLD/USD (second row). In this regard, it is possible to infer that, although BTC price variation emerges mainly from its own prices, GOLD/USD seems to be the most effective instrument that contributes to BTC price formation when compared to currencies. On the other hand, the mutual contribution of BTC and OIL/USD is relatively low at around 0.10%.

Panel C, shows the variance contribution rates of major stock market indices and indices of stock markets where BTC is actively traded on BTC prices (first row) as well as the variance contribution rates of BTC on these market indices (second row). The figures show that variance contribution rates of SPX, FTSE/USD, DAX/USD and NIK/USD on BTC prices are 0.20%, 0.61%, 0.61%, and 0.29%, and the variance contribution rates of SPX, FTSE/USD, MOEX/USD, IBVC/USD and SSEC/USD on BTC prices are 0.22%, 0.56%, 0.13%, 0.08% and 0.27% (first rows). On the other hand, the variance contribution rates of BTC on SPX, FTSE/USD, DAX/USD and NIK/USD are 0.60%, 0.43%, 1.16%, and 0.22%, and the variance contribution rates of BTC on SPX, FTSE/USD, MOEX/USD, IBVC/USD and SSEC/USD are 0.50%, 0.38%, 0.42%, 0.02% and 0.04% (second rows). In addition, BTC's own variance contribution is 98.29% in major stock markets and 98.74% in BTC-active stock markets, which are both lower than its contribution in currency (99.38% and 99.80%) and commodity (98.97%) markets. These results reveal the fact that the price information is incorporated more quickly in stock markets than in the currency and commodity markets. It is also interesting to consider that while the relative variance contribution of GOLD/USD on BTC is still the highest, the opposite is not true any more, because the variance contribution rate of BTC on DAX/USD is 1.16%, which is about 1.5 times higher than the variance contribution rate of BTC on GOLD/USD (0.79%).

Panel D, lastly, gives the variance contribution ratios of cryptocurrencies on BTC and vice versa. As can be clearly seen, these ratios are higher than the ones we discussed above, as BTC's own variance contribution is 95.60% which is the lowest level. In line with that, the variance contribution rates of ETH/USD, XRP/USD and LTC/USD on BTC prices are 1.34%, 2.41% and 0.65% (first row). Altcoins, when compared to other asset types, contribute more to the BTC price variation. The variance contribution rates of BTC, however, are remarkable, since they are 14.57%, 7.24%, and 19.66% (second row). In brief, BTC has the greatest variance contributions on altcoins prices.

Overall, the results show that neither BTC prices nor other asset prices have significant predictive power on each other and its price is predominantly determined by its own supply and demand (Buchholz et al., 2012; Luther & White, 2014; Wu, Pandey, & DBA, 2014; Zhu, Dickinson, & Li, 2017). This feature of BTC carries resemblance to a currency, while it also has highly volatile price behavior like a commodity (Bianchi, 2018; Dyhrberg, 2016). In this respect, BTC forms a unique speculative asset (Baur et al., 2018; Kristoufek, 2015).

8.5 Conclusion

Cryptocurrencies, in particular BTC, have gained attention around the world as both a mode of payment and a means of investment. The former feature of BTC relies on its fast, secure and low cost transactional mechanism, while the latter feature derives its value from its expected high profitability. In this respect, companies in institutional terms, countries in governmental terms and investors in individual terms are continuously opening some room for BTC in their trade-, policy- and portfolio-related facilities.

Nevertheless, the very question of how BTC should be classified has been one of the most controversial issues among both policymakers and scholars. Is BTC a currency, a commodity, a security or a unique type of asset? Supporters of BTC as a currency point to its tradability, exchangeability and negotiability; supporters of BTC as a commodity point to its scarcity, finite supply and inherent value, and supporters of BTC as a security point to its fungibility and profit recoverability. These arguments come along with dissenting opinions.

In this study, we aimed to contribute to the clarification of this unsolved puzzle by exploring the price relationships between BTC and other types of assets. We employed Johansen co-integration, Granger causality, impulse response functions and forecast error variance decomposition analyses. Our results showed that BTC does not have a long-run relationship with any kind of assets, but it has a short-run relationship with gold, securities and especially altcoins, which is significantly bidirectional in most cases. Moreover, BTC, be it an impulse or a response variable, is in close association with altcoins prices. Overall, however, BTC price variation is stimulated by its own prices in that its supply and demand functions are the main determinants of BTC prices rather than other assets. As a consequence, due to the fact that neither BTC prices nor other asset prices have significant predictive power over each other, BTC seems to form a unique asset that could be classified separately from currencies, securities and commodities.

There are at least three policy implications. First, regulatory bodies should treat cryptocurrencies, e.g. BTC, as a unique form of asset rather than classifying it within traditional currencies, commodities or securities, since the classification confusion existing in the markets would lead international investors to trade only in specific markets where cryptocurrencies are classified in line with their interests. Due to this fact, not all investors would be protected similarly throughout the world markets. Second, if accepted as a different asset type, financial institutions and especially institutional investors would be able to hold them in their portfolios with respect to new investment criteria or new capital adequacy restrictions assigned particularly for cryptocurrencies. Third, tax issues should be handled in a way as not to create regulatory arbitrage among markets where cryptocurrencies are treated differently. Acknowledging cryptocurrencies as a unique asset type would remove such diversities in practice as well.

References

Alford, T. (2018, August 10). *Bitcoin adoption: Trading volume by country*. Retrieved January 17, 2019, from TotalCrypto.io: https://totalcrypto.io/bitcoin-adoption-trading-volume-country/

Australian Government. (2017). *Backing innovation and FinTech*. Canberra: Australian Government.

Baek, C., & Elbeck, M. (2015). Bitcoins as an investment or speculative vehicle? A first look. *Applied Economics Letters, 22*(1), 30–34.

Baur, D. G., & Dimpfl, T. (2017). Realized bitcoin volatility. *SSRN, 2949754*, 1–26.

Baur, D., Hong, K., & Lee, A. (2018). Bitcoin: Medium of exchange or speculative assets? *Journal of International Financial Markets, Institutions and Money, 54*, 177–189.

Bianchi, D. (2018, June 6). *Cryptocurrencies as an asset class? An empirical assessment. An empirical assessment* (WBS Finance Group Research Paper)

Bitcoinity. (2019). *Bitcoin market capitalization*. Retrieved January 22, 2019, from data.bitcoinity.org: https://data.bitcoinity.org/markets/market_cap/30d/USD?t=l

BitLegal. (2017). *BitLegal. "Index"*. Retrieved January 16, 2019, from BitLegal: http://bitlegal.io/list.php

Bouoiyour, J., & Selmi, R. (2015). What does Bitcoin look like? *Annals of Economics & Finance, 16*(2), 449–492.

Bouri, E., Azzi, G., & Dyhrberg, A. H. (2017). On the return-volatility relationship in the Bitcoin market around the price crash of 2013, Economics: The open-access, Open-Assessment E-Journal, ISSN 1864-6042 (Vol. 11, Iss. 2017-2, pp. 1–16). Kiel: Kiel Institute for the World Economy (IfW).

Bouri, E., Molnár, P., Azzi, G., Roubaud, D., & Hagfors, L. I. (2017). On the hedge and safe haven properties of Bitcoin: Is it really more than a diversifier? *Finance Research Letters, 20*, 192–198.

Briere, M., Oosterlinck, K., & Szafarz, A. (2015). Virtual currency, tangible return: Portfolio diversification with bitcoin. *Journal of Asset Management, 16*(6), 365–373.

Buchholz, M., Delaney, J., Warren, J., & Parker, J. (2012). Bits and bets, information, price volatility, and demand for Bitcoin. *Economics, 312*.

CFTC. (2018). *Bitcoin basics*. Chicago: CFTC.

Chan, S., Chu, J., Nadarajah, S., & Osterrieder, J. (2017). A statistical analysis of cryptocurrencies. *Journal of Risk and Financial Management, 10*(2), 12.

Cheah, E. T., & Fry, J. (2015). Speculative bubbles in Bitcoin markets? An empirical investigation into the fundamental value of Bitcoin. *Economics Letters, 130*, 32–36.

Cheung, A., Roca, E., & Su, J. J. (2015). Crypto-currency bubbles: An application of the Phillips–Shi–Yu (2013) methodology on Mt. Gox bitcoin prices. *Applied Economics, 47*(23), 2348–2358.

Ciaian, P., Rajcaniova, M., & Kancs, D. A. (2016). The economics of BitCoin price formation. *Applied Economics, 48*(19), 1799–1815.

Dickey, D. A., & Fuller, W. A. (1979). Distribution of the estimators for autoregressive time series with a unit root. *Journal of the American Statistical Association, 74*(366a), 427–431.

Dyhrberg, A. H. (2016). Hedging capabilities of bitcoin. Is it the virtual gold? *Finance Research Letters, 16*, 139–144.

Eisl, A., Gasser, S., & Weinmayer, K. (2015). Caveat emptor: Does bitcoin improve portfolio diversification?

Estrada, J. C. S. (2017). *Analyzing bitcoin price volatility*. Berkeley: University of California.

Glaser, F., Zimmermann, K., Haferkorn, M., Weber, M., & Siering, M. (2014). Bitcoin-asset or currency? Revealing users' hidden intentions.

Granger, C. W. (1969). Investigating causal relations by econometric models and cross-spectral methods. *Econometrica: Journal of the Econometric Society, 37*, 424–438.

Hencic, A., & Gouriéroux, C. (2015). Noncausal autoregressive model in application to bitcoin/USD exchange rates. In *Econometrics of risk* (pp. 17–40). Cham: Springer.

IRS. (2014). *Notice 2014–21*. Washington, DC: Internal Revenue Service.

Johansen, S. (1988). Statistical analysis of cointegration vectors. *Journal of Economic Dynamics and Control, 12*(2–3), 231–254.

Johansen, S., & Juselius, K. (1990). Maximum likelihood estimation and inference on cointegration—With applications to the demand for money. *Oxford Bulletin of Economics and Statistics, 52*(2), 169–210.

Katsiampa, P. (2017). Volatility estimation for Bitcoin: A comparison of GARCH models. *Economics Letters, 158*, 3–6.

Klein, T., Thu, H. P., & Walther, T. (2018). Bitcoin is not the new gold–a comparison of volatility, correlation, and portfolio performance. *International Review of Financial Analysis, 59*, 105–116.
Kristoufek, L. (2015). What are the main drivers of the bitcoin price? Evidence from wavelet coherence analysis. *PLoS One, 10*(4), e0123923.
Luther, W., & White, L. (2014). Can bitcoin become a major currency?
Mandjee, T. (2015). Bitcoin, its legal classification and its regulatory framework. *Journal of Business & Securities Law, 15*(2), 157.
Moreau, E. (2019). *15 major retailers and services that accept bitcoin*. Retrieved January 15, 2019, from Lifewire: https://www.lifewire.com/big-sites-that-accept-bitcoin-payments-3485965
Nadarajah, S., & Chu, J. (2017). On the inefficiency of bitcoin. *Economics Letters, 150*, 6–9.
Nakamoto, S. (2008). *Bitcoin: A peer-to-peer electronic cash system*.
Panagiotidis, T., Stengos, T., & Vravosinos, O. (2018). On the determinants of bitcoin returns: A LASSO approach. *Finance Research Letters., 27*, 235–240.
Phillips, P. C., & Perron, P. (1988). Testing for a unit root in time series regression. *Biometrika, 75* (2), 335–346.
Sapuric, S., & Kokkinaki, A. (2014, May). Bitcoin is volatile! Isn't that right? In *International Conference on Business Information Systems* (pp. 255–265). Cham: Springer.
SCM. (2019, January 14). *SC to regulate offering and trading of digital assets*. Retrieved January 18, 2019, from Securities Commission Malaysia: https://www.sc.com.my/news/media-releases-and-announcements/sc-to-regulate-offering-and-trading-of-digital-assets
Selgin, G. (2015). Synthetic commodity money. *Journal of Financial Stability, 17*, 92–99.
Shome, R. (2019). What was the highest bitcoin price in history? Retrieved January 22, 2019, from BTCWIRES: https://www.btcwires.com/analysis/what-was-the-highest-bitcoin-price-in-history/
Statista. (2019). *Number of Bitcoins in circulation worldwide from 1st quarter 2011 to 4th quarter 2018 (in millions)*. Retrieved January 22, 2019, from The Statistics Portal: https://www.statista.com/statistics/247280/number-of-bitcoins-in-circulation/
Urquhart, A. (2016). The inefficiency of Bitcoin. *Economics Letters, 148*, 80–82.
Van Wijk, D. (2013). In Erasmus Universiteit Rotterdam (Ed.), *What can be expected from the BitCoin*.
Weber, B. (2014). Bitcoin and the legitimacy crisis of money. *Cambridge Journal of Economics, 40* (1), 17–41.
Wu, C. Y., Pandey, V. K., & DBA, C. (2014). The value of bitcoin in enhancing the efficiency of an investor's portfolio. *Journal of Financial Planning, 27*(9), 44–52.
Yermack, D. (2015). Is Bitcoin a real currency? An economic appraisal. In *Handbook of digital currency* (pp. 31–43).
Zhu, Y., Dickinson, D., & Li, J. (2017). Analysis on the influence factors of Bitcoin's price based on VEC model. *Financial Innovation, 3*(1), 3.

Burak Pirgaip was graduated from Middle East Technical University (METU) BA in 2001. He has earned his Master's and Ph.D. degrees from Gazi University (BA/Accounting and Finance) and Hacettepe University (BA) in 2004 and 2014, respectively. He had the opportunity to visit the University of Hull, the UK in 2013 as a Jean Monnet Scholar. Burak PİRGAİP resigned from Capital Markets Board of Turkey where he had worked for since 2002 and started his academic career. He acted as a full faculty member in Cankaya University and a part-time faculty member in METU and Izmir University of Economics between 2016 and 2018. He is an Associate Professor of Finance at Hacettepe University as of 23.10.2019. He is also an expert in many fields concerning capital markets especially in real estate finance and investments, corporate finance and portfolio management.

Burcu Dinçergök is an Assistant Professor in Finance at Atılım University Business Department in Ankara (Turkey). She holds a master's degree in Accounting and Finance from Hacettepe University, Turkey and a Ph.D. in Accounting and Finance at Gazi University, Turkey. Her research interest covers topics such as capital structure, working capital management, dividend policies, mergers and acquisitions, financial development and economic growth relations, bitcoin, and energy finance. She participated as a researcher in the Scientific and Technological Research Council of Turkey's funded project entitled "Financial Development, Capital Accumulation, Productivity and Growth: Turkish Experience". She took part in different international conferences. She teaches financial management, investment analysis, and investment project analysis courses both at undergraduate and graduate programs.

Şüheda Haşlak was graduated from TOBB ETU BA in 2013. She has earned her Master's degree from Cankaya University (BA/Financial Economics) in 2018. Her master's thesis was about "Analysis of Bitcoin Market Volatility" which was supervised by Burak Pirgaip in majority, but the official procedure was completed under the supervision of Assoc. Prof. Dr. Elif Öznur Acar due to his resignation.

Chapter 9
The Causal Relationship Between Returns and Trading Volume in Cryptocurrency Markets: Recursive Evolving Approach

Efe Caglar Cagli

Abstract This chapter examines the time-varying causal relationship between trading volume and returns in cryptocurrency markets. The chapter employs a novel Granger causality framework based on a recursive evolving window procedure. The procedures allow detecting changes in causal relationships among time series by considering potential conditional heteroskedasticity and structural shifts through recursive subsampling. The chapter analyzes the return-volume relationship for Bitcoin and seven other altcoins: Dash, Ethereum, Litecoin, Nem, Stellar, Monero, and Ripple. The results suggest rejecting the null hypothesis of no causality, indicating bi-directional causality between trading volume and returns for Bitcoin and the altcoins except Nem and Stellar. The findings also highlight that the causal relations in cryptocurrency markets are subject to change over time. The chapter may conclude that trading volume has predictive power on returns in cryptocurrency markets, implying potential benefits of constructing volume-based trading strategies for investors and considering trading volume information in developing pricing models to determine the fundamental value of the cryptocurrencies.

9.1 Introduction

The price-volume relation has received much attention from investors and academic research within the last half century. Investigating the causal relationship between price and trading volume is a crucial task given its importance of providing (1) insight into the structure of financial markets to better understand how information is disseminated, (2) understanding of the empirical distribution of speculative prices, (3) implications for pricing formations in the futures markets, and (4) valuable inputs for the tests in event studies where price and volume information are incorporated (Karpoff, 1987, p. 109: 110).

E. C. Cagli (✉)
Faculty of Business, Dokuz Eylul University, Izmir, Turkey
e-mail: efe.cagli@deu.edu.tr

© Springer Nature Switzerland AG 2019

U. Hacioglu (ed.), *Blockchain Economics and Financial Market Innovation*, Contributions to Economics, https://doi.org/10.1007/978-3-030-25275-5_9

Academics have proposed theoretical explanations regarding the relationship between price changes and volume: Mixture of Distribution Hypothesis (Clark, 1973; Crouch, 1970; Epps & Epps, 1976; Harris, 1986; Tauchen & Pitts, 1983; Ying, 1966), Sequential Information Arrival Hypothesis (Copeland, 1976, 1977; Jennings, Starks, & Fellingham, 1981), models of tax related motives (Lakonishok & Smidt, 1989), noise trader models (De Long, Shleifer, Summers, & Waldmann, 1990), models of information endowment (He & Wang, 1995; Kyle, 1985; Llorente, Michaely, Saar, & Wang, 2002), models of the relative precision of the aggregate private information (Schneider, 2009), and models of trade based on differences of opinion (M. Harris & Raviv, 1993; Kandel & Pearson, 1995). Karpoff (1987), Hiemstra and Jones (1994), Gebka (2012) and Gebka and Wohar (2013) not only provide systematic explanations of the theoretical models but also document empirical studies predominantly on equity markets- examining the hypotheses derived from these models.

Other studies have empirically examined the return-volume relation in various markets including those of cryptocurrencies (see, Balcilar, Bouri, Gupta, and Roubaud (2017) and references therein). The chapter presents a literature review in the following section. Bitcoin, the largest cryptocurrency, has recorded exorbitant price increases since its inception; accordingly, the literature on the economics of cryptocurrencies has intensified, testing the finance theories on the data of Bitcoin and the alternative cryptocurrencies, referred to as 'altcoins'. Investigating the price-volume relation in cryptocurrency markets is of similar importance in other asset markets; Balcilar et al. (2017) highlight the lack of an asset pricing model to determine the fundamental value of the cryptocurrencies as investors construct trading strategies largely based on technical analysis.

This chapter investigates the causal relationship between trading volume and returns in cryptocurrency markets. The chapter contributes to the literature in the following ways. First, a dataset that consists of not only Bitcoin but also seven other altcoins: Dash (DSH), Ethereum (ETH), Litecoin (LTC), Nem (XEM), Stellar (XLM), Monero (XMR), and Ripple (XRP) is analyzed. Understanding the return-volume relation of Bitcoin along with those of altcoins may provide broader insight into the structure of cryptocurrency markets. The sample period, between December 27, 2013 and September 3, 2018, covers significant news and market events, such as the MtGox Bitcoin Exchange hack in 2014, the Bitcoin hard forks in 2017–18, the Bitcoin legislation in Japan in April 2017, the statement of People's Bank of China in January 2017, and so forth. Second, the Granger causality framework consists of the Shi, Phillips, and Hurn (2018) model, which is a novel method that has not been previously employed in cryptocurrency markets. The framework allows for detecting and dating changes in Granger causal relationships based on a recursive evolving window procedure. The results obtained from the recursive evolving window procedure are compared to those obtained from both the forward expanding (Thoma, 1994) and rolling window procedures (Swanson, 1998). Third, the results show that trading volume has a predictive power on the returns in most of the considered cryptocurrencies, proving useful information for future research on developing appropriate asset pricing models and profitable trading strategies in such a speculative market. The findings also highlight the importance of testing

Granger causality through the recursive evolving window procedure that detects causal changes over time.

The remainder of the chapter is organized as follows. Section 9.2 documents previous research on cryptocurrency markets. Section 9.3 explains the econometric framework. Section 9.4 presents the data and reports the empirical results. Finally, Sect. 9.5 concludes the chapter.

9.2 Literature Review

Numerous studies have contributed towards the debate on the economics of cryptocurrencies, complementing the studies approaching cryptocurrencies from a technical (Becker et al., 2013; Böhme, Christin, Edelman, & Moore, 2015; Sadeghi, 2013) and a legal (Plassaras, 2013) perspective.

Rogojanu and Badea (2014) document the advantages of using Bitcoin and the disadvantages of Bitcoin circulation, highlight the obstacles faced by Bitcoin in the economic environment and the authors state that Bitcoin is like digital gold (Rogojanu & Badea, 2014, p. 112). Glaser, Zimmermann, Haferkorn, Weber, and Siering (2014) report that users consider Bitcoin as an investment tool, rather than a payment system. Dwyer (2015) discusses the demand for digital currency, properties of Bitcoin as a digital currency, the price and volatility of Bitcoin on cryptocurrency exchanges. Kristoufek (2015) states that Bitcoin has the characteristics of both a standard financial asset and a speculative one (Kristoufek, 2015, p. 14). However, Yermack (2015) documents Bitcoin's weaknesses as a currency and concludes that Bitcoin is a speculative investment tool rather than a currency (Yermack, 2015, p. 42).

Brandvold, Molnár, Vagstad, and Andreas Valstad (2015) report that the prices of Bitcoin and altcoins vary on different exchanges and examine the contributions of cryptocurrency exchanges to price discovery process of Bitcoin. Their conclusion that the information shares of exchanges are dynamic and tend to evolve over time is later confirmed by Pagnottoni, Dimpfl, and Baur (2018). Baur and Dimpfl (2018) examine the price discovery between spot and futures prices of Bitcoin and their findings show that spot market dominates the futures market, which is contrary to those of Kapar and Olmo (2019) who report that price discovery takes place in the futures market, rather than the spot market.

Employing a battery of tests to test the efficient market hypothesis: Urquhart (2016) finds that Bitcoin prices become more efficient in the latter sample period, August 2013 to July 2016; and the findings are supported by those of Brauneis and Mestel (2018), Bariviera (2017), Vidal-Tomás and Ibañez (2018) and Sensoy (2018). Nadarajah and Chu (2017) extend the study of Urquhart (2016) by transforming Bitcoin returns and the findings indicate that modified Bitcoin returns are informationally efficient. Tiwari, Jana, Das, and Roubaud (2018) test Bitcoin's price efficiency by means of a battery of long-range dependence tests and evidence that Bitcoin market is largely efficient. On the contrary, Jiang, Nie, and Ruan (2018) present results, implying Bitcoin market has a long-term memory and has become more inefficient over sample period. Al-Yahyaee, Mensi, and Yoon (2018) report that Bitcoin market is the least efficient one among the gold, stock, and currency markets.

The rapid information flow and drastic price changes in the cryptocurrency markets have led researchers to investigate the presence of bubbles. Garcia, Tessone, Mavrodiev, and Perony (2014) and Kristoufek (2015) provide early evidence of bubbles in the Bitcoin market. Cheung, Roca, and Su (2015) employ the bubble detection framework of Phillips, Shi, and Yu (2015) and report a number of significant bubbles in sample period of 2010–2014. Cheah and Fry (2015) confirm the existence of the bubbles and note that the fundamental price of Bitcoin is zero (Cheah & Fry, 2015, p. 35). Applying the methods originated from statistical physics, Fry and Cheah (2016) find a negative bubble in the Bitcoin and Ripple markets. Cagli (2018) concludes that Bitcoin and six large altcoins exhibit explosive behavior. However, Corbet, Lucey, and Yarovaya (2018) find inconsistent results with those of the previous studies, stating that Bitcoin and Ethereum markets do not have a persistent bubble.

Ciaian, Rajcaniova, and Kancs (2016) conduct a study to investigate the determinants of Bitcoin prices. Their results show that both the "market forces of Bitcoin supply and demand" and "Bitcoin attractiveness for investors" play an important role in explaining Bitcoin prices, supporting the early evidence by Kristoufek (2013) and Bouoiyour and Refk (2015). However, the results suggest that global macro-financial development does not drive Bitcoin prices, inconsistent with those of van Wijk (2013). Panagiotidis, Stengos, and Vravosinos (2018) investigate the predictive power of 21 on Bitcoin returns; internet trends and gold returns are found to be the most important predictors of Bitcoin returns. Demir, Gozgor, Lau, and Vigne (2018) provide evidence that the "economic policy uncertainty index" is a successful predictor of the daily Bitcoin returns, suggesting that Bitcoin can be considered for hedging during a bear market.

Brière, Oosterlinck, and Szafarz (2015) report that including Bitcoin in a well-diversified portfolio offers significant diversification benefits, owing to the low correlations between Bitcoin and other assets. Dyhrberg (2016) documents that gold and Bitcoin have similar hedging capabilities. Bouri, Molnár, Azzi, Roubaud, and Hagfors (2017), Bouri, Gupta, Tiwari, and Roubaud (2017), Bouri, Jalkh, Molnár, and Roubaud (2017), Corbet, Meegan, Larkin, Lucey, and Yarovaya (2018), and Feng, Wang, and Zhang (2018) report that Bitcoin as a market leader is an effective diversifier and a hedge for various asset markets. Conducting rolling window analyses, Shahzad, Bouri, Roubaud, Kristoufek, and Lucey (2019) evidence that safe-haven role of Bitcoin is weak and time-varying. Beneki, Koulis, Kyriazis, and Papadamou (2019) find that diversifying capabilities of both Bitcoin and Ethereum decline over time.

Balcilar et al. (2017) investigate the causal relation between trading volume and Bitcoin returns and volatility, employing a non-parametric causality-in- quantiles test. Their results suggest that volume is a successful predictor of Bitcoin returns during the period 2011–2016 with some exception of bear and bull market regimes, but not of Bitcoin volatility. Bouri, Lau, Lucey, and Roubaud (2018) employ a copula-quantile causality approach on the daily data of Bitcoin and six altcoins; the findings show that volume Granger causes returns of all cryptocurrencies. El Alaoui, Bouri, and Roubaud (2018) conduct the multifractal detrended cross-correlations analysis and provide evidence of nonlinear interaction between Bitcoin returns and changes in volume. Koutmos (2018) use number of Bitcoin transactions and number of unique addresses, respectively, as a proxy for transaction activity and bi-variate VAR results suggest bidirectional linkages between transaction activity and Bitcoin returns. Aalborg, Molnár, and de Vries (2018) report that Google searches and transaction volume in the Bitcoin network have a predictive power on the trading volume of Bitcoin; however, Bitcoin returns cannot be predicted from any of the considered variables, volume, VIX index, and internet trends. Kokkinaki, Sapuric, and Georgiou (2019) estimate EGARCH models over the period 2010–2017, and find a significant relationship between trading volume and Bitcoin returns before the MtGox Bitcoin Exchange hack in early 2014, and between volume and volatility after 2013. Stosic, Stosic, Ludermir, and Stosic (2019) employ the multifractal detrended fluctuation analysis on the daily data of 50 cryptocurrencies and conclude that trading volume and returns follow different dynamics; returns are more complex than the changes in trading volume.

Based on the above explanations, the evidence is inconclusive. Several studies reveal that the behavior of prices and causal return-volume relations are subject to change over time. In this regard, the chapter attempts to fill the gap in the literature by employing a novel time-varying Granger causality testing framework on a large dataset of cryptocurrencies.

9.3 Methodology

This chapter employs the dynamic Granger causality testing procedures, as developed by (Shi et al., 2018) based on a recursive evolving window. The procedures essentially modify the *Wald* statistics obtained from the standard Granger causality tests and allow for detecting and dating the changes in causal relations.

The following unrestricted vector-autoregression (VAR(p)) model is estimated for employing a standard Granger causality test:

$$\mathbf{y}_t = \Pi \mathbf{x}_t + \varepsilon_t, \tag{9.1}$$

where $t = 1, \ldots, T$, $\mathbf{y}_t = (y_{1t}, y_{2t})'$, $\mathbf{x}_t = \left(1, \mathbf{y}'_{t-1}, \mathbf{y}'_{t-2}, \ldots, \mathbf{y}'_{t-p}\right)'$, $\Pi_{2 \times (2p+1)} = [\beta_0, \beta_0, \ldots, \beta_p]$; p is the optimal lag length determined by the Schwarz (Bayesian) Information Criterion (BIC). Wald test statistics (W) following χ^2 distribution, with p degrees of freedom, is calculated as follows (Shi et al., 2018, p. 968):

$$W = \left[\mathbf{R}\text{vec}\left(\hat{\Pi}\right)\right]'\left[\mathbf{R}\left(\hat{\Omega} \otimes \left(\mathbf{X}'\mathbf{X}\right)^{-1}\right)\mathbf{R}'\right]^{-1}\left[\mathbf{R}\text{vec}\left(\hat{\Pi}\right)\right], \tag{9.2}$$

where $\hat{\Pi}$ is the OLS estimator of Π, $\hat{\Omega} = T^{-1}\sum_{t=1}^{T}\hat{\varepsilon}_t\hat{\varepsilon}'_t$ with $\hat{\varepsilon}_t = \mathbf{y}_t - \hat{\Pi}\mathbf{x}_t$; $\text{vec}\left(\hat{\Pi}\right)$ is the $2(2p + 1) \times 1$ coefficients of $\hat{\Pi}$; $\mathbf{R}$ is the $p \times 2(2p + 1)$ selection matrix; $\mathbf{X}' = [\mathbf{x}_1, \ldots, \mathbf{x}_T]$ is the observation matrix of the regressors in Eq. (9.1). By imposing zero restriction on the coefficients ($H_0 : \beta_1 = \beta_2 = \ldots = \beta_p = 0$), the chapter tests the null hypothesis that y_{2t} does not Granger cause y_{1t} in Eq. (9.1).

The standard Granger causality test uses information from the whole sample period and thus is not capable of considering structural changes in economic policies, regulations, or socio-economic circumstances, which cause the test to lose power. Shi et al. (2018) propose a novel recursive evolving window procedure that allows detecting the changes in causal relationships, overcoming the shortcomings of the standard Granger causality test. The procedure is an extension of both the forward expanding window procedure of Thoma (1994) and the rolling window procedure of Swanson (1998); and it has its roots in the work of Phillips et al. (2015), which develops a framework for detecting and dating multiple financial bubbles in real-time.

Different from the standard Granger causality test, *Wald* test statistics are estimated across subsamples determined by the procedures to capture the impact of structural changes on the causal relationships. Figure 9.1 illustrates the subsampling processes and the window widths of the procedures. In Fig. 9.1, r is the observation of interest; r_0 is the minimum window size; r_1 and r_2 are the starting and terminal points of the sequence of regressions, respectively; and r_w is the estimation window width. The chapter refers the reader to Shi et al. (2018) for more detailed explanations on the subsampling process of the procedures.

In the recursive evolving window procedure, the analysis obtains modified *Wald* test statistics (*MWald*) for each subsample regression and estimate sup *Wald* (SW_r) as follows (Shi et al., 2018, p. 969):

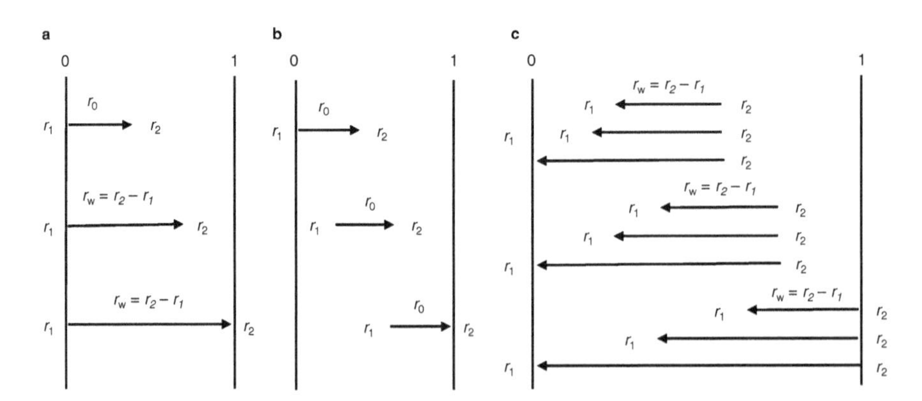

Fig. 9.1 Sample sequence for the procedures

$$SW_r(r_0) = \sup_{(r_1, r_2) \in \Lambda_0, r_2 = r} \{W_{r_2}(r_1)\} \tag{9.3}$$

where $\Lambda_0 = \{(r_1, r_2) : 0 < r_0 + r_1 \leq r_2 \leq 1, \text{and}\, 0 \leq r_1 \leq 1 - r_0\}$. Different from the recursive evolving window procedure, the forward expanding window has starting point fixed ($r_1 = 0$) and sets $r = r_2$, and the rolling window procedure has fixed window width and window initialization (Shi et al., 2018, p. 969).

The following three crossing time equations determine the origination (r_e) and termination (r_f) dates in the causal relationships for the forward expanding window, rolling window, and recursive evolving window procedures, respectively (Shi et al., 2018, p. 969):

$$\hat{r}_e = \inf_{r \in [r_0, 1]} \{r : W_r(0) > cv\};\ \hat{r}_f = \inf_{r \in [\hat{r}_e, 1]} \{r : W_r(0) < cv\} \tag{9.4}$$

$$\hat{r}_e = \inf_{r \in [r_0, 1]} \{r : W_r(f - f_0) > cv\};\ \hat{r}_f = \inf_{r \in [\hat{r}_e, 1]} \{r : W_r(f - f_0) < cv\} \tag{9.5}$$

$$\hat{r}_e = \inf_{r \in [r_0, 1]} \{r : SW_r(r_0) > scv\};\ \hat{r}_f = \inf_{r \in [\hat{r}_e, 1]} \{r : SW_r(r_0) < scv\} \tag{9.6}$$

where cv and scv are the sequences of the bootstrapped critical values of the W_f and SW_r statistics, respectively. The proposed *Wald* (W) and sup *Wald* (SW_r) statistics are heteroskedasticity consistent and are employed to consider the well-known phenomenon of conditional heteroskedasticity in financial time-series (Shi et al., 2018, p. 967).

9.4 Data and Empirical Results

9.4.1 Data

The data, obtained from Coinmarketcap.com, consist of the daily price and trading volume series for Bitcoin and seven altcoins, Dash (DSH), Ethereum (ETH), Litecoin (LTC), Nem (XEM), Stellar (XLM), Monero (XMR), and Ripple (XRP). These cryptocurrencies are large, actively traded, and have adequate data for the econometric framework employed. Their total market value was about 182 billion USD, representing approximately 77% of the total cryptocurrency market capitalization as of September 2018 (Coinmarketcap.com). The sample period varies across cryptocurrencies with December 27, 2013 being the earliest starting point and September 3, 2018 the latest ending point.

Table 9.1 reports the sample period for each cryptocurrency along with some descriptive statistics, and Augmented-Dickey (1979) (*ADF*) unit root test statistics relating to the natural logarithm (log) of price, continuously compounded (Δlog) return, log volume, and detrended volume series of the cryptocurrencies. Following the work of Balcilar et al. (2017), Gallant, Rossi, and Tauchen (1992), and Gebka

Table 9.1 Descriptive and unit root statistics

Series	Average	Std. Dev.	Skewness	Kurtosis	ADF
(a) Bitcoin (BTC), sample period: 12/27/2013–09/03/2018					
Price (log)	6.872	1.266	0.809	2.280	0.310
Return (Δlog, %)	0.134	3.982	−0.420	8.671	−41.262[a]
Volume (log)	18.831	2.217	0.690	2.093	−0.945
Det. Volume	−1.7E-15	0.711	0.265	3.060	−4.138[a]
(b) Dash (DSH), sample period: 02/14/2014–09/03/2018					
Price (log)	2.800	2.125	0.546	1.863	−1.328
Return (Δlog, %)	0.383	8.291	3.027	44.232	−41.257[a]
Volume (log)	14.135	3.046	0.418	1.653	−0.998
Det. Volume	−6.2E-15	1.142	0.597	3.120	−3.976[a]
(c) Ethereum (ETH), sample period: 08/07/2015–09/03/2018					
Price (log)	3.639	2.326	−0.168	1.717	−0.796
Return (Δlog, %)	0.414	7.977	−3.524	68.582	−36.024[a]
Volume (log)	18.080	2.941	−0.246	1.856	−1.260
Det. Volume	4.5E-15	1.146	−0.047	2.327	−3.119[b]
(d) Litecoin (LTC), sample period: 12/27/2013–09/03/2018					
Price (log)	2.315	1.487	0.743	2.198	−0.448
Return (Δlog, %)	0.060	5.959	0.562	16.311	−41.470[a]
Volume (log)	16.329	2.312	0.726	2.001	−1.824
Det. Volume	5.9E-15	1.111	0.519	3.050	−4.212[a]
(e) Nem (XEM), sample period: 04/01/2015–09/03/2018					
Price (log)	−4.698	3.049	−0.065	1.573	−0.760
Return (Δlog, %)	0.492	9.045	1.882	18.954	−28.197[a]
Volume (log)	12.689	3.740	−0.153	1.828	−1.338
Det. Volume	4.6E-15	1.516	0.054	2.869	−4.251[a]
(f) Stellar (XLM), sample period: 08/05/2014 09/03/2018					
Price (log)	−4.822	1.923	1.082	2.570	−0.148
Return (Δlog, %)	0.304	8.273	1.952	17.562	−35.773[a]
Volume (log)	12.758	3.576	0.515	1.852	−1.396
Det. Volume	−2.0E-15	1.511	0.432	3.229	−4.849[a]
(g) Monero (XMR), sample period: 05/21/2014–09/03/2018					
Price (log)	1.760	2.349	0.380	1.639	−0.060
Return (Δlog, %)	0.284	7.628	0.659	8.770	−39.260[a]
Volume (log)	13.809	3.122	0.175	1.536	−1.319
Det. Volume	3.5E-15	1.352	0.091	2.402	−3.726[a]
(h) Ripple (XRP), sample period: 12/27/2013–09/03/2018					
Price (log)	−3.715	1.877	0.944	2.282	−0.161
Return (Δlog, %)	0.147	7.117	2.404	40.941	−26.813[a]
Volume (log)	15.037	3.107	0.724	2.161	−1.481
Det. Volume	4.5E-15	1.306	0.319	3.023	−3.833[a]

Note: log and Δlog denote natural logarithm and natural logarithmic-difference, respectively. Return series is the first difference of the log-price series. Det. Volume is the detrended volume. *Std. Dev.* is the standard deviation. *ADF* is the Augmented Dickey-Fuller (Dickey & Fuller, 1979) unit root test statistics. [a] and [b] denote rejecting the null hypothesis of unit root at 1% and 5% significance levels, respectively and Wohar (2013), the study calculates detrended volume by obtaining the residuals from the following equation:

$$vol_t = \alpha_0 + \beta_1(\gamma/T) + \beta_2(\gamma/T)^2 + \varepsilon_t \tag{9.7}$$

where vol_t is the natural logarithm of volume traded, α_0 is intercept, γ is trend and T is the sample size, and ε_t is the error term. The rationale behind detrending volume data is to control the linear and nonlinear deterministic time-trends.

According to Table 9.1, Bitcoin is found to be the least volatile cryptocurrency in the sample; Nem provides the highest average return and its returns have the highest standard deviation among all. While the return series of Bitcoin and Ethereum are negatively skewed implying a higher probability of having negative returns, those of other altcoins are positively skewed indicating a heavier tail of large positive values. *Kurtosis* statistics suggest that the return series for all cryptocurrencies are leptokurtic, implying the distribution of returns have fat tails and are non-normal. Based on the *ADF* statistics, the null hypothesis of unit root for the log-price and log-volume series for all cryptocurrencies cannot be rejected. However, the return and detrended volume series of all cryptocurrencies are found to be stationary, integrated of order zero ($I(0)$), as the null hypothesis of unit root at the 5% level, or better is rejected. The return and detrended volume series for all cryptocurrencies throughout the subsequent analysis are employed since the Granger Causality framework requires stationarity of time-series.

9.4.2 Empirical Results

The study tests the causal relationship between returns and volume and reports the estimation results in Figs. 9.2 to 9.9. Findings illustrate the time-varying modified Wald (*MWald*) statistic sequence (solid lines) along with their bootstrapped 5% critical value sequence (dashed lines) under the figures. The study tests the null hypothesis of no causality between variables and rejects the null hypothesis when the *MWald* statistics exceed the 5% critical value sequence. The two columns of the figures show the results for Granger causality from returns to volume, and from volume to returns, respectively. The three rows show estimation results by the forward expanding, rolling, and recursive evolving window procedures, respectively.

Panels a and b of Fig. 9.2 show that *MWald* test statistics obtained from forward expanding procedures are always below the 5% critical value sequence, suggesting no causal relationship between Bitcoin returns and volume, in either direction. However, the estimation results by the rolling and recursive evolving window procedures reported in Panels c to f of Fig. 9.2 suggest rejecting the null hypothesis of no causality between the variables at the 5% level and indicate bi-directional causality between Bitcoin returns and volume. In Panel c of Fig. 9.2, rolling window

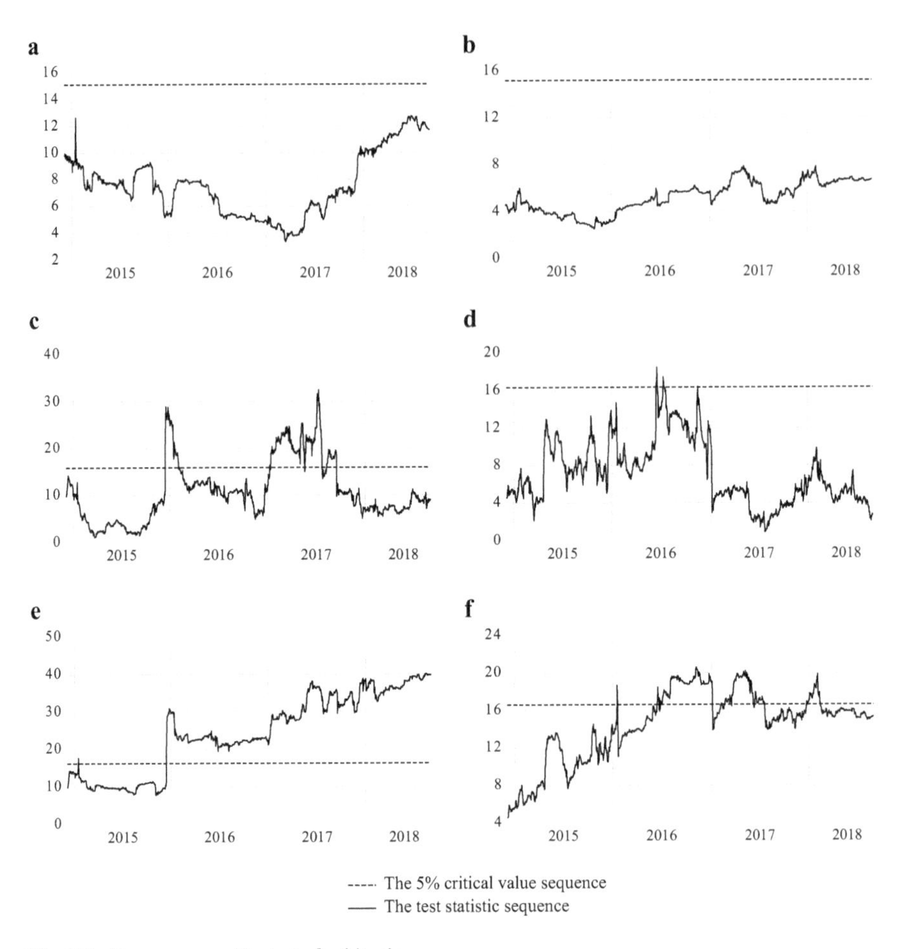

Fig. 9.2 Granger causality tests for bitcoin

procedure detects three episodes of Granger causality from Bitcoin returns to volume; the first lasts 52 days between December 2015 and February 2016; the second lasts 190 days between January 2017 and July 2017; and the third lasts 38 days between August 2017 and September 2017. Panel d of Fig. 9.2 shows that the rolling window procedure detects three short episodes of Granger causality from volume to Bitcoin returns; all episodes occur in 2016 and last 10 days in total. Panel e of Fig. 9.2 illustrates that the recursive evolving window procedure detects two episodes of Granger causality from Bitcoin returns to volume; the first lasts 2 days in January 2015; and the second starts in December 2015 and continues until the end of the sample period, for a total of 994 days. Panel f of Fig. 9.2 shows four episodes of Granger causality from volume to Bitcoin returns; the first lasts 3 days in January 2016; the second lasts 201 days between June 2016 and January 2017; the third lasting 145 days between February 2017 and July 2017; and the fourth lasts 50 days between December 2017 and February 2018.

Fig. 9.3 Granger causality tests for dash

The results reported in Panels a to d of Fig. 9.3 suggest unidirectional Granger causality from Dash returns to volume at the 5% level. The null hypothesis of no causality from volume to returns cannot be rejected since the *MWald* statistics do not exceed the 5% critical value sequences in Panels b and d of Fig. 9.3. In Panel a of Fig. 9.3, the forward expanding procedure detects two episodes of Granger causality from Dash returns to volume, in the later periods of the sample; the first lasts 71 days between July 2017 and September 2017; and the second starts in November 2017 and lasts 296 days, until the end of the sample period. Panel c of Fig. 9.3 illustrates three episodes of Granger causality from Dash returns to volume, clustered at the beginning of the sample period; the first lasts 179 days between July 2015 and January 2016; the second lasts 18 days between January 2016 and February 2016; and the third lasts 43 days between August 2016 and October 2016. Different from the results obtained from the previous procedures, those obtained from the recursive evolving window procedure suggest bi-directional Granger causality between Dash

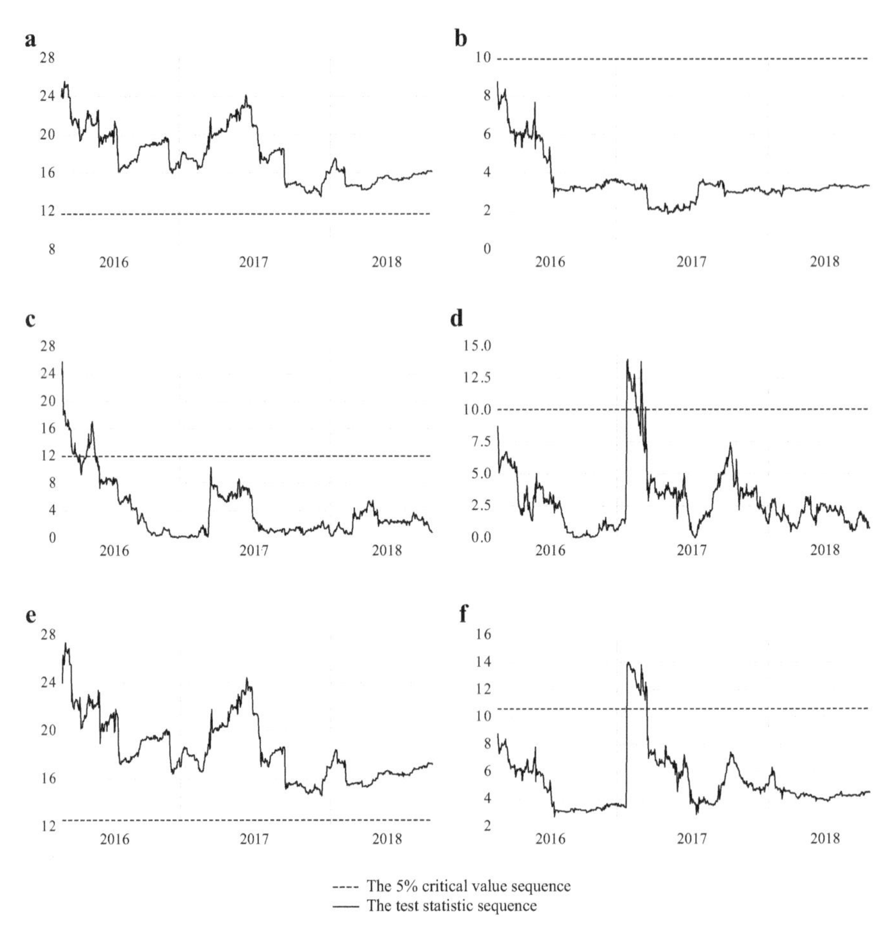

Fig. 9.4 Granger causality tests for Ethereum

returns and volume at the 5% level, reported in Panels e and f of Fig. 9.3. The chapter evidences the episode of Granger causality from Dash returns to volume, starting at the same point of time detected by the rolling window procedure, July 2015, and lasting until the end of the sample period. The recursive evolving window procedure detects three main episodes of Granger causality from volume to Dash returns; the first lasts 12 days between July 2015 and August 2015; the second lasts 80 days between November 2015 and February 2016; and the third lasts 63 days between July 2016 and October 2016.

Panels a and b of Fig. 9.4 indicate unidirectional Granger causality from Ethereum returns to volume at the 5% level. The estimated *MWald* statistics in Panel a of Fig. 9.4 are always above the 5% critical value sequence, indicating the episode of Granger causality from Ethereum returns to volume persists over the sample period. However, the results reported in Panels c to f of Fig. 9.4 suggest bi-directional Granger causality between Ethereum returns and volume at the 5%

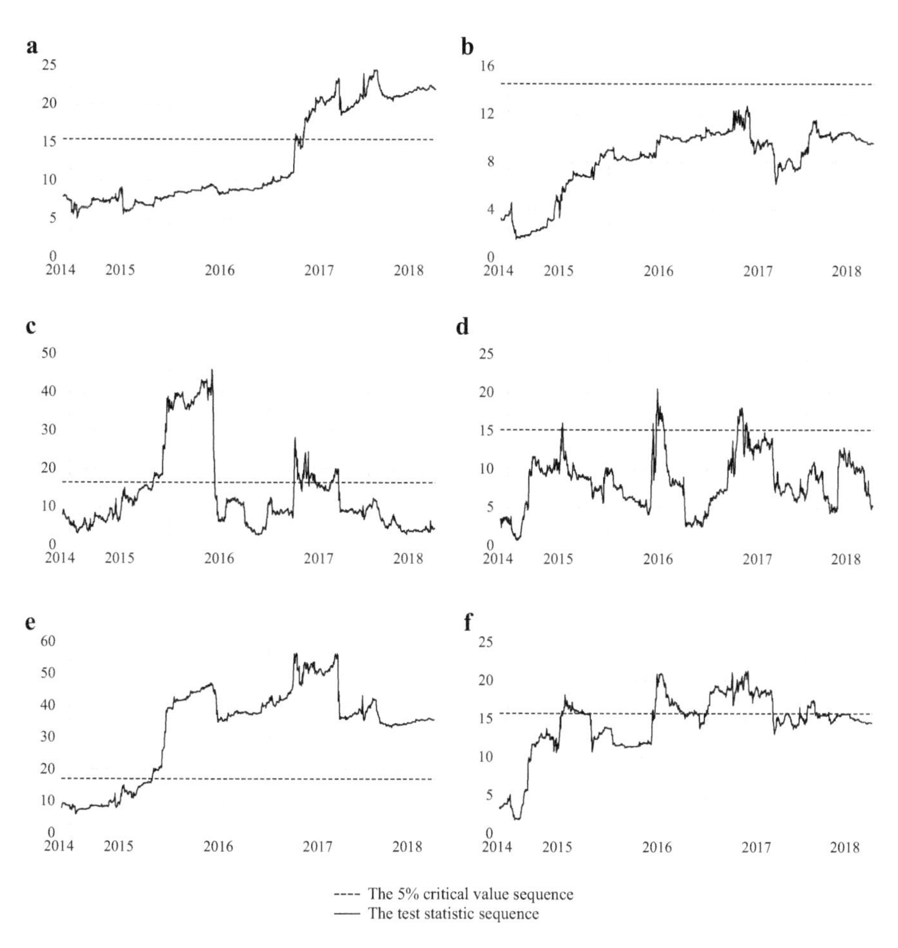

Fig. 9.5 Granger causality tests for Litecoin

level, or better. In Panel c of Fig. 9.4, the rolling window procedure detects two episodes of Granger causality from Ethereum returns to volume; the first lasts 39 days between the beginning of the sample period and April 2016; and the second lasting 24 days between May 2016 and June 2016. Panel d of Fig. 9.4 illustrates one episode of Granger causality from volume to Ethereum returns at the 5% level, lasting 30 days between January 2017 and March 2017. In Panel e of Fig. 9.4, the recursive evolving window procedure detects the episode of Granger causality from Ethereum returns to volume at the 5% level, persisting over the sample period. Panel f of Fig. 9.4 illustrates one significant episode of Granger causality from volume to Ethereum returns, lasting 49 days between January 2017 and March 2017.

In Panels a and b of Fig. 9.5, the forward expanding window procedure suggests unidirectional Granger causality from Litecoin returns to volume at the 5% level; the procedure detects the Granger causality episode, which starts in April 2017 and lasts 498 days, until the end of the sample period. However, the results based on the rolling and recursive evolving window procedures, reported in Panels c to f of Fig. 9.5, suggest bi-directional Granger causality between Litecoin returns and volume at the 5% level. The rolling window procedure detects three episodes of Granger causality from Litecoin returns to volume at the 5% level; the first lasts 225 days between November 2015 and June 2016; the second lasts 72 days between April 2017 and June 2017; and the third lasts 29 days between August 2017 and September 2017. In Panel d of Fig. 9.5, the chapter evidence three episodes of Grange causality from volume to Litecoin returns at the 5% level; the first lasts only 4 days in July 2015; the second lasts 29 days between June 2016 and July 2016; and the third lasts 27 days between April 2017 and May 2017. Panel e of Fig. 9.5 shows that the recursive evolving window procedure detects the episode of Granger causality from Litecoin returns to volume at the 5% level, starting in November 2015 and lasting 1037 days, until the end of the sample period. In Panel f of Fig. 9.5, the recursive evolving window procedure detects four episodes of Granger causality from volume to Litecoin returns at the 5% level; the first lasts 81 days between July 2015 and October 2015; the second lasts 141 days between June 2016 and November 2016; the third lasts 246 days between January 2017 and August 2017; and the fourth lasts 34 days between January 2018 and February 2018.

The procedures unanimously suggest unidirectional Granger causality from Nem returns to volume at the 5% level. Panels b, d, and f of Fig. 9.6 illustrate that *MWald* statistics are always below their 5% critical value sequences, suggesting not rejecting the null hypothesis of that volume does not Granger cause Nem returns. The forward expanding and recursive evolving window procedures detect identical episodes of Granger causality from Nem returns to volume at the 5% level, which starts in starting in June 2016 and continues until the end of the sample period. The rolling window procedure detects multiple episodes of Granger causality from Nem returns to volume at the 5% level; however, they are shorter than those detected by the other procedures; the first lasts 94 days between July 2016 and October 2016; the second lasts 15 days between November 2016 and December 2016; and the third lasts 169 days between April 2017 and September 2017.

The results reported in the panels of Fig. 9.7 suggest unidirectional Granger causality from Stellar returns to volume at the 5% level. According to the results reported in Panels b, d, and f of Fig. 9.7, the study cannot reject the null hypothesis of that volume does not Granger cause Stellar returns as the *MWald* statistics are always below their 5% critical value sequences. The estimations obtained from both the forward expanding and recursive evolving window procedures, reported in Panels a and e of Fig. 9.7, detect the episodes of Granger causality from Stellar returns to volume at the 5% level, persisting over the sample period. Panel c of Fig. 9.7 shows that the rolling window procedure detects three episodes of Granger causality from Stellar returns to volume at the 5% level; the first lasts only 2 days at the beginning of the sample period; the second lasts 17 days between February 2016 and March 2016; and the third lasts 607 days between June 2016 and February 2018.

As reported in Panels a to d of Fig. 9.8, the results obtained from the forward expanding and rolling window procedures suggest unidirectional Granger causality from Monero returns to volume at the 5% level. Panel a of Fig. 9.8 shows that the

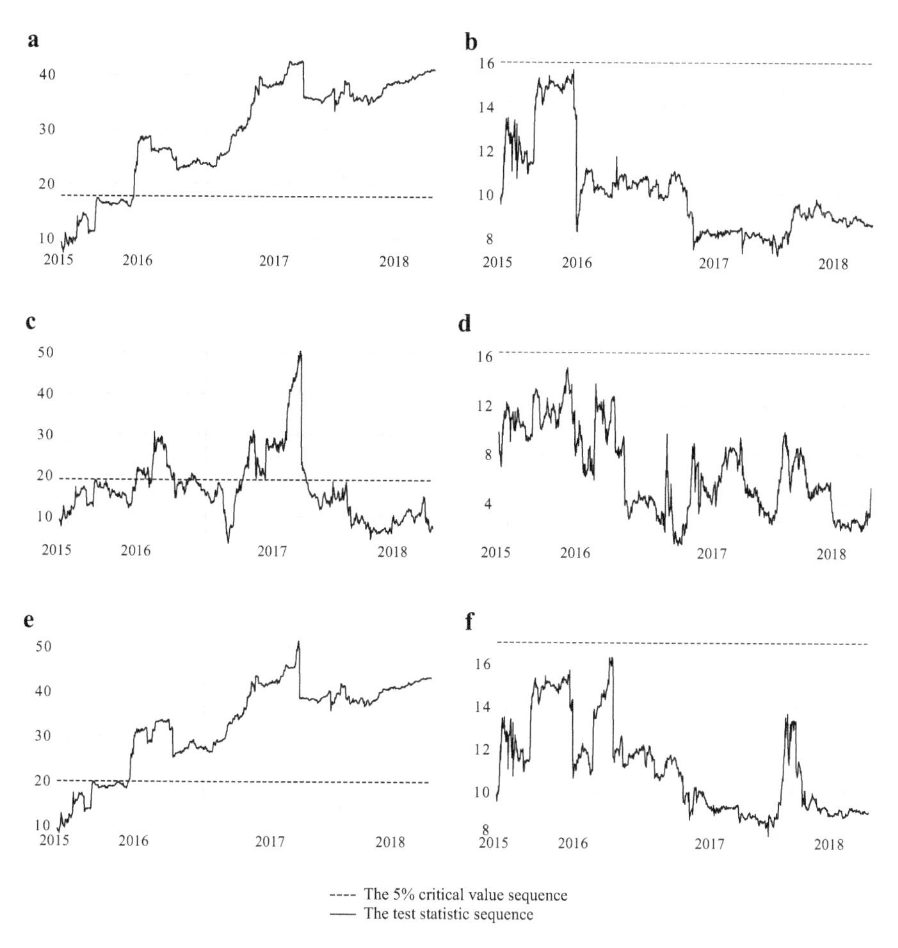

Fig. 9.6 Granger causality tests for Nem

forward expanding window procedure detects the episode of Granger causality from Monero returns to volume at the 5% level, persisting all over the sample period. In Panel c of Fig. 9.8, the rolling window procedure suggests five episodes of Granger causality from Monero returns to volume; the first lasts 235 days between March 2015 and December 2015; the second lasts 52 days between February 2016 and April 2016; the third lasts 388 days between August 2016 and September 2017; the fourth lasts 9 days in January 2018; and the fifth starts in March 2018 and lasts 173 days, until the end of the sample period. However, the study evidences bi-directional Granger causality between Monero returns and volume at the 5% level, based on the results obtained from the recursive evolving window procedure. In Panel e of Fig. 9.8, the estimated *MWald* statistics are always above the 5% critical value sequence, indicating Monero returns Granger cause volume over the whole sample period. Panel f of Fig. 9.8 illustrates that the recursive evolving window

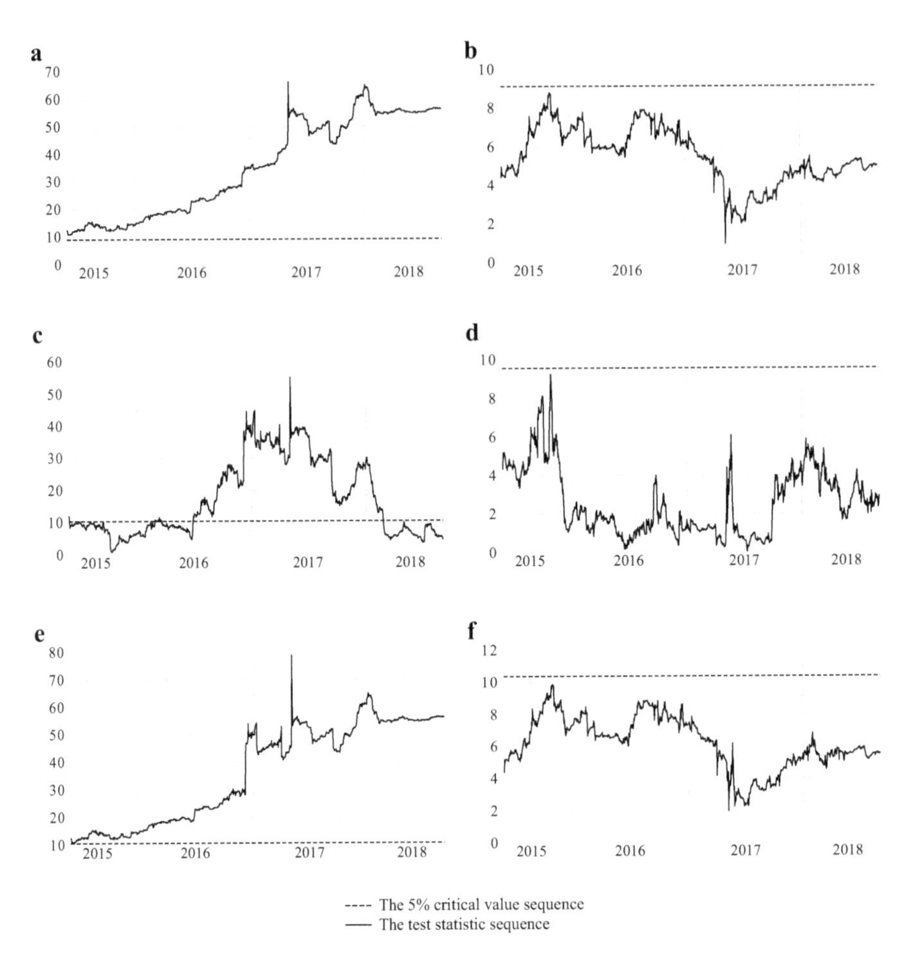

Fig. 9.7 Granger causality tests for Stellar

procedure detects two episodes of Granger causality from volume to Monero returns at the 5% level; the first is on March 11, 2016; and the second lasts 22 days between July 2016 and August 2016.

The procedures unanimously suggest bi-directional Granger causality from Ripple returns to volume at the 5% level. According to Panel a of Fig. 9.9, The forward expanding procedure detects one episode of Granger causality from Ripple returns to volume at the 5% level, starting in May 2017 and lasting 484 days until the end of the sample period. Panel b of Fig. 9.9 illustrates one episode of Granger causality from volume to Ripple returns at the 5% level, lasting 93 days between December 2014 and March 2015. In Panel c of Fig. 9.9, the rolling window procedure detects three episodes of Granger causality from Ripple returns to volume at the 5% level; the first lasts 62 days between February 2016 and April 2016; the second lasts 10 days in May 2016; and the third lasts 75 days between March 2017 and June 2017. Panel d

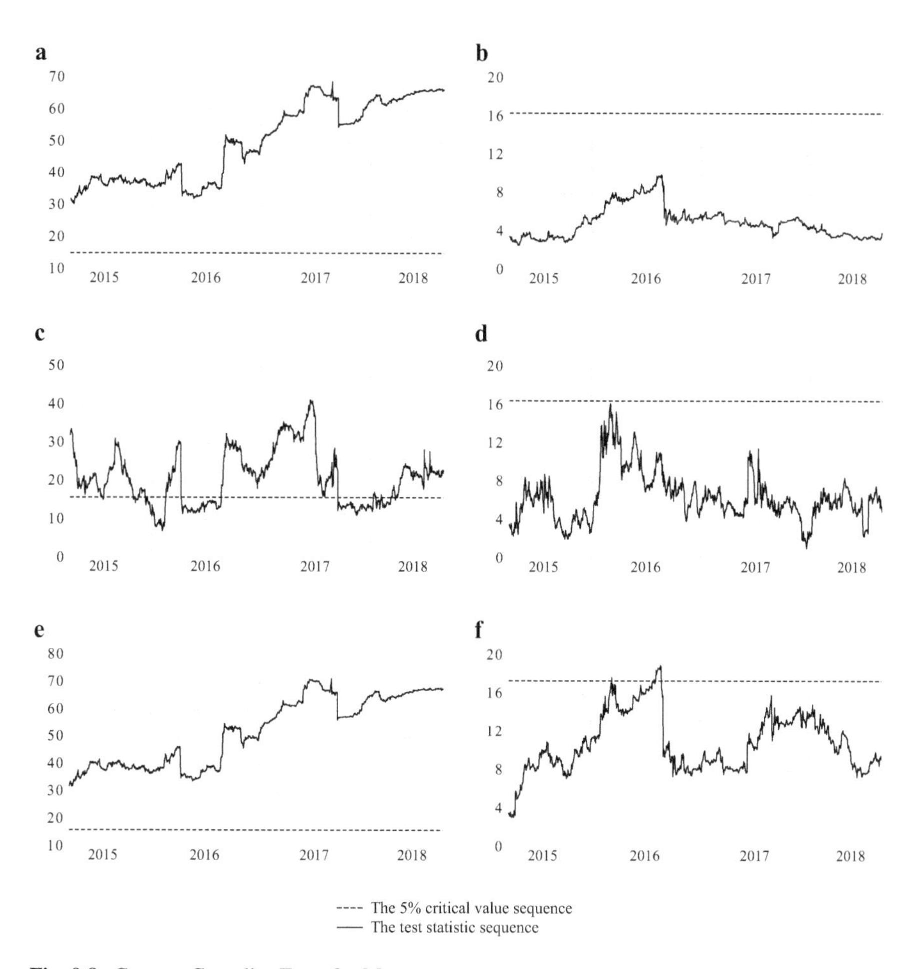

Fig. 9.8 Granger Causality Tests for Monero

of Fig. 9.9 illustrates that the rolling window detects two episodes of Granger causality from volume to Ripple returns at the 5% level; the first starts at the beginning of the sample period and ends in January 2015; and the second lasts 82 days between November 2015 and February 2016. In Panel e of Fig. 9.9, the recursive evolving window procedure detects four episodes of Granger causality from Ripple returns to volume at the 5% level; the first lasts 61 days between February 2016 and April 2016; the second lasts 7 days in May 2016; the third lasts 53 days between June 2016 and August 2016; and the fourth starts in March 2017 and lasts 523 days, until the end of the sample. Panel f of Fig. 9.9 illustrates four episodes of Granger causality from volume to Ripple returns; the first starts at the beginning of the sample period and terminates in February 2015; the second lasts 128 days between December 2015 and April 2016; the third lasts 9 days in June 2016; and the fourth lasts 209 days between August 2016 and March 2017.

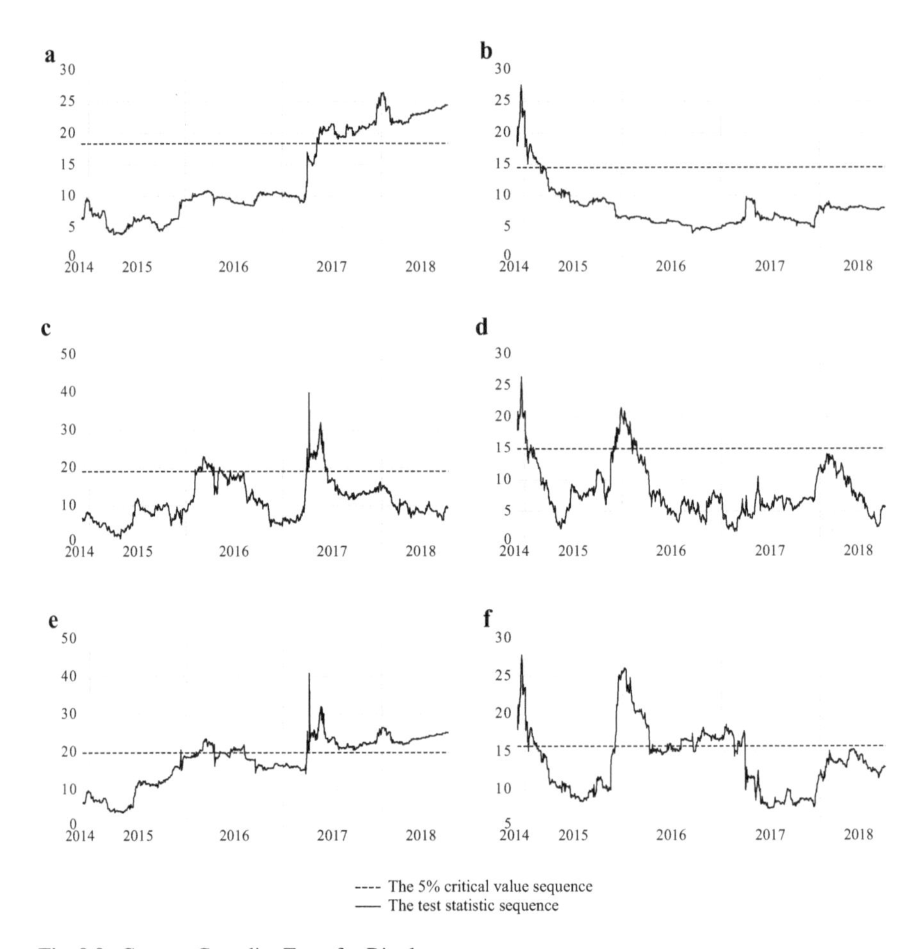

Fig. 9.9 Granger Causality Tests for Ripple

It is clear from the empirical results that the study may understate the predictability of volume series on the returns when testing simple Granger causality following forward expanding window procedure, which is based on the entire sample period. As reported in Table 9.2, the conclusion of no Granger causality from volume to returns changes depending on the procedures employed; the novel recursive evolving window procedure detects more episodes of Granger causality than both the forward expanding and rolling window procedures, consistent with the findings of Shi et al. (2018). The findings may relate any information about technological developments, regulations, or even the rumors in the cryptocurrency markets to the episodes of Granger causality between returns and volume since these markets are highly speculative, and, thus volatile. Moreover, trading volume does not cause the returns of the two least liquid cryptocurrencies in the sample, Nem and Stellar, implying that as average volume increases, it becomes a successful predictor of returns and the duration of the episodes of Granger causality from volume to returns tends to increase.

Table 9.2 Summary of Granger causality tests

	Forward		Rolling		Recursive	
	Vol > Ret	Ret > Vol	Vol > Ret	Ret > Vol	Vol > Ret	Ret > Vol
BTC	–	–	✓	✓	✓	✓
DSH	–	✓	–	✓	✓	✓
ETH	–	✓	✓	✓	✓	✓
LTC	–	✓	✓	✓	✓	✓
XEM	–	✓	–	✓	–	✓
XLM	–	✓	–	✓	–	✓
XMR	–	✓	–	✓	✓	✓
XRP	✓	✓	✓	✓	✓	✓

Note: > shows the Granger causality direction, for instance, Vol > Ret indicates testing for Granger causality running from volume to return. ✓ denotes the rejection of null hypothesis of no Granger causality at 5% significance level or better. – indicates not rejecting the null hypothesis of no Granger causality

9.5 Conclusion

The chapter investigates the causal relationship between returns and trading volume of Bitcoin and seven altcoins, employing the novel time-varying Granger causality framework of (Shi et al., 2018) that allows detecting and dating the causal changes. The Granger causality framework is employed for the first time in cryptocurrency markets.

Overall results suggest significant bi-directional causality between returns and volume for the considered cryptocurrencies except for Nem and Stellar, which are the two least liquid cryptocurrencies in the sample, on average. The findings show that return-volume relations are subject to change over time, highlighting the importance of testing Granger causality through employing the time-varying procedures, especially the recursive evolving window procedure of (Shi et al., 2018), which is most able to detect changes in the Granger causality relationships. It is also clear from the findings that the duration of the episodes of Granger causality from volume to returns tends to increase for a cryptocurrency as its trading volume increases, on average.

The evidence of significant bi-directional causality among the cryptocurrency returns and volume implies disagreement about the news as argued by the Mixed Distribution Hypothesis (Epps & Epps, 1976; Harris, 1986; Tauchen & Pitts, 1983) due to different frameworks (Harris & Raviv, 1993) and that trading strategies are built on new information (He & Wang, 1995). The evidence reveals the fact that information arrives sequentially as put forth by the Sequential Information Arrival Hypothesis (Copeland, 1976, 1977; Jennings et al., 1981) and that noise traders heavily employ feedback strategies in the market (De Long et al., 1990; Gebka, 2012). For Nem and Stellar, the procedures detect the unidirectional Granger causality running from returns to volume, suggesting that trading is driven by public information (Kyle, 1985) on which there exists a lack of consensus among equally informed investors (Gebka, 2012; Kandel & Pearson, 1995).

Trading volume has predictive power on returns in most of the cryptocurrency markets, implying potential benefits of constructing volume-based trading strategies for the investors and the necessity of considering trading volume information in developing pricing models to determine the fundamental value of the cryptocurrencies.

References

Aalborg, H. A., Molnár, P., & de Vries, J. E. (2018). What can explain the price, volatility and trading volume of bitcoin? *Finance Research Letters, 29*, 255–265. https://doi.org/10.1016/J.FRL.2018.08.010

Al-Yahyaee, K. H., Mensi, W., & Yoon, S.-M. (2018). Efficiency, multifractality, and the long-memory property of the bitcoin market: A comparative analysis with stock, currency, and gold markets. *Finance Research Letters, 27*, 228–234. https://doi.org/10.1016/J.FRL.2018.03.017

Balcilar, M., Bouri, E., Gupta, R., & Roubaud, D. (2017). Can volume predict bitcoin returns and volatility? A quantiles-based approach. *Economic Modelling, 64*, 74–81. https://doi.org/10.1016/J.ECONMOD.2017.03.019

Bariviera, A. F. (2017). The inefficiency of bitcoin revisited: A dynamic approach. *Economics Letters, 161*, 1–4. https://doi.org/10.1016/J.ECONLET.2017.09.013

Baur, D. G., & Dimpfl, T. (2018). *Price discovery in bitcoin spot or futures?* SSRN. https://doi.org/10.2139/ssrn.3171464

Becker, J., Breuker, D., Heide, T., Holler, J., Rauer, H. P., & Böhme, R. (2013). Can we afford integrity by proof-of-work? Scenarios inspired by the bitcoin currency. In *The economics of information security and privacy* (pp. 135–156). Berlin, Heidelberg: Springer. https://doi.org/10.1007/978-3-642-39498-0_7

Beneki, C., Koulis, A., Kyriazis, N. A., & Papadamou, S. (2019). Investigating volatility transmission and hedging properties between bitcoin and Ethereum. *Research in International Business and Finance, 48*, 219–227. https://doi.org/10.1016/J.RIBAF.2019.01.001

Böhme, R., Christin, N., Edelman, B., & Moore, T. (2015). Bitcoin: Economics, technology, and governance. *Journal of Economic Perspectives, 29*(2), 213–238. https://doi.org/10.1257/jep.29.2.213

Bouoiyour, J., & Refk, S. (2015). What does bitcoin look like? *Annals of Economics and Finance, 16*(2), 449–192. Retrieved from http://aeconf.com/Articles/Nov2015/aef160211.pdf

Bouri, E., Gupta, R., Tiwari, A. K., & Roubaud, D. (2017). Does bitcoin hedge global uncertainty? Evidence from wavelet-based quantile-in-quantile regressions. *Finance Research Letters, 23*, 87–95. https://doi.org/10.1016/J.FRL.2017.02.009

Bouri, E., Jalkh, N., Molnár, P., & Roubaud, D. (2017). Bitcoin for energy commodities before and after the December 2013 crash: Diversifier, hedge or safe haven? *Applied Economics, 49*(50), 1–11. https://doi.org/10.1080/00036846.2017.1299102

Bouri, E., Lau, C. K. M., Lucey, B., & Roubaud, D. (2018). Trading volume and the predictability of return and volatility in the cryptocurrency market. *Finance Research Letters., 29*, 340–346. https://doi.org/10.1016/j.frl.2018.08.015

Bouri, E., Molnár, P., Azzi, G., Roubaud, D., & Hagfors, L. I. (2017). On the hedge and safe haven properties of bitcoin: Is it really more than a diversifier? *Finance Research Letters, 20*, 192–198. https://doi.org/10.1016/J.FRL.2016.09.025

Brandvold, M., Molnár, P., Vagstad, K., & Andreas Valstad, O. C. (2015). Price discovery on bitcoin exchanges. *Journal of International Financial Markets, Institutions and Money, 36*, 18–35. https://doi.org/10.1016/J.INTFIN.2015.02.010

Brauneis, A., & Mestel, R. (2018). Price discovery of cryptocurrencies: Bitcoin and beyond. *Economics Letters, 165*, 58–61. https://doi.org/10.1016/J.ECONLET.2018.02.001

Brière, M., Oosterlinck, K., & Szafarz, A. (2015). Virtual currency, tangible return: Portfolio diversification with bitcoin. *Journal of Asset Management, 16*(6), 365–373. https://doi.org/10.1057/jam.2015.5

Cagli, E. C. (2018). Explosive behavior in the prices of bitcoin and altcoins. *Finance Research Letters., 29*, 398–403. https://doi.org/10.1016/j.frl.2018.09.007

Cheah, E.-T., & Fry, J. (2015). Speculative bubbles in bitcoin markets? An empirical investigation into the fundamental value of bitcoin. *Economics Letters, 130*, 32–36. https://doi.org/10.1016/J.ECONLET.2015.02.029

Cheung, A., Roca, E., & Su, J.-J. (2015). Crypto-currency bubbles: An application of the Phillips–Shi–Yu (2013) methodology on Mt. Gox bitcoin prices. *Applied Economics, 47*(23), 2348–2358. https://doi.org/10.1080/00036846.2015.1005827

Ciaian, P., Rajcaniova, M., & Kancs, D. (2016). The economics of BitCoin price formation. *Applied Economics, 48*(19), 1799–1815. https://doi.org/10.1080/00036846.2015.1109038

Clark, P. K. (1973). A subordinated stochastic process model with finite variance for speculative prices. *Econometrica, 41*(1), 135–155. https://doi.org/10.2307/1913889

Copeland, T. E. (1976). A model of asset trading under the assumption of sequential information arrival. *The Journal of Finance, 31*(4), 1149–1168. https://doi.org/10.1111/j.1540-6261.1976.tb01966.x

Copeland, T. E. (1977). A probability model of asset trading. *The Journal of Financial and Quantitative Analysis, 12*(4), 563–578. https://doi.org/10.2307/2330332

Corbet, S., Lucey, B., & Yarovaya, L. (2018). Datestamping the bitcoin and Ethereum bubbles. *Finance Research Letters, 26*, 81–88. https://doi.org/10.1016/j.frl.2017.12.006

Corbet, S., Meegan, A., Larkin, C., Lucey, B., & Yarovaya, L. (2018). Exploring the dynamic relationships between cryptocurrencies and other financial assets. *Economics Letters, 165*, 28–34. https://doi.org/10.1016/J.ECONLET.2018.01.004

Crouch, R. L. (1970). The volume of transactions and price changes on the New York stock exchange. *Financial Analysts Journal, 26*(4), 104–109.

De Long, J. B., Shleifer, A., Summers, L. H., & Waldmann, R. J. (1990). Noise trader risk in financial markets. *Journal of Political Economy, 98*(4), 703–738. https://doi.org/10.1086/261703

Demir, E., Gozgor, G., Lau, C. K. M., & Vigne, S. A. (2018). Does economic policy uncertainty predict the bitcoin returns? An empirical investigation. *Finance Research Letters, 26*, 145–149. https://doi.org/10.1016/j.frl.2018.01.005

Dickey, D. A., & Fuller, W. A. (1979). Distribution of the estimators for autoregressive time series with a unit root. *Journal of the American Statistical Association, 74*(366), 427–431. https://doi.org/10.2307/2286348

Dwyer, G. P. (2015). The economics of bitcoin and similar private digital currencies. *Journal of Financial Stability, 17*, 81–91. https://doi.org/10.1016/J.JFS.2014.11.006

Dyhrberg, A. H. (2016). Bitcoin, gold and the dollar – A GARCH volatility analysis. *Finance Research Letters, 16*, 85–92. https://doi.org/10.1016/J.FRL.2015.10.008

El Alaoui, M., Bouri, E., & Roubaud, D. (2018). Bitcoin price–volume: A multifractal cross-correlation approach. *Finance Research Letters.* https://doi.org/10.1016/J.FRL.2018.12.011

Epps, T. W., & Epps, M. L. (1976). The stochastic dependence of security price changes and transaction volumes: Implications for the mixture-of-distributions hypothesis. *Econometrica, 44*(2), 305–321. https://doi.org/10.2307/1912726

Feng, W., Wang, Y., & Zhang, Z. (2018). Can cryptocurrencies be a safe haven: A tail risk perspective analysis. *Applied Economics, 50*(44), 4745–4762. https://doi.org/10.1080/00036846.2018.1466993

Fry, J., & Cheah, E.-T. (2016). Negative bubbles and shocks in cryptocurrency markets. *International Review of Financial Analysis, 47*, 343–352. https://doi.org/10.1016/J.IRFA.2016.02.008

Gallant, A. R., Rossi, P. E., & Tauchen, G. (1992). Stock prices and volume. *Review of Financial Studies, 5*(2), 199–242. https://doi.org/10.1093/rfs/5.2.199

Garcia, D., Tessone, C. J., Mavrodiev, P., & Perony, N. (2014). The digital traces of bubbles: Feedback cycles between socio-economic signals in the bitcoin economy. *Journal of the Royal Society Interface, 11*(99), 20140623. https://doi.org/10.1098/rsif.2014.0623

Gebka, B. (2012). The dynamic relation between returns, trading volume, and volatility: Lessons from spillovers between Asia and the United States. *Bulletin of Economic Research, 64*(1), 65–90. https://doi.org/10.1111/j.1467-8586.2010.00371.x

Gebka, B., & Wohar, M. E. (2013). Causality between trading volume and returns: Evidence from quantile regressions. *International Review of Economics & Finance, 27*, 144–159. https://doi.org/10.1016/J.IREF.2012.09.009

Glaser, F., Zimmermann, K., Haferkorn, M., Weber, M. C., & Siering, M. (2014). Bitcoin - asset or currency? Revealing users' hidden intentions. In *Twenty second European conference on information systems* (pp. 1–14). Tel Aviv. Retrieved from https://papers.ssrn.com/sol3/papers.cfm?abstract_id=2425247

Harris, L. (1986). Cross-security tests of the mixture of distributions hypothesis. *The Journal of Financial and Quantitative Analysis, 21*(1), 39–46. https://doi.org/10.2307/2330989

Harris, M., & Raviv, A. (1993). Differences of opinion make a horse race. *Review of Financial Studies, 6*(3), 473–506. https://doi.org/10.1093/rfs/6.3.473

He, H., & Wang, J. (1995). Differential information and dynamic behavior of stock trading volume. *Review of Financial Studies, 8*(4), 919–972. https://doi.org/10.1093/rfs/8.4.919

Hiemstra, C., & Jones, J. D. (1994). Testing for linear and nonlinear granger causality in the stock price-volume relation testing for linear and nonlinear granger causality in the stock price-volume relation. *The Journal of Finance, 49*(5), 1639–1664. https://doi.org/10.1111/j.1540-6261.1994.tb04776.x

Jennings, R. H., Starks, L. T., & Fellingham, J. C. (1981). An equilibrium model of asset trading with sequential information arrival. *The Journal of Finance, 36*(1), 143–161. https://doi.org/10.1111/j.1540-6261.1981.tb03540.x

Jiang, Y., Nie, H., & Ruan, W. (2018). Time-varying long-term memory in bitcoin market. *Finance Research Letters, 25*, 280–284. https://doi.org/10.1016/j.frl.2017.12.009

Kandel, E., & Pearson, N. D. (1995). Differential interpretation of public signals and trade in speculative markets. *Journal of Political Economy, 103*(4), 831–872. https://doi.org/10.1086/262005

Kapar, B., & Olmo, J. (2019). An analysis of price discovery between bitcoin futures and spot markets. *Economics Letters, 174*, 62–64. https://doi.org/10.1016/J.ECONLET.2018.10.031

Karpoff, J. M. (1987). The relation between price changes and trading volume: A survey. *The Journal of Financial and Quantitative Analysis, 22*(1), 109–126. https://doi.org/10.2307/2330874

Kokkinaki, A., Sapuric, S., & Georgiou, I. (2019). The relationship between bitcoin trading volume, volatility, and returns: A study of four seasons. In M. Themistocleous & P. R. da Cunha (Eds.), *15th European, Mediterranean, and Middle Eastern Conference, EMCIS* (Vol. 2018, pp. 3–15). Limassol, Cyprus: Springer. https://doi.org/10.1007/978-3-030-11395-7_1

Koutmos, D. (2018). Bitcoin returns and transaction activity. *Economics Letters, 167*, 81–85. https://doi.org/10.1016/J.ECONLET.2018.03.021

Kristoufek, L. (2013). BitCoin meets Google trends and Wikipedia: Quantifying the relationship between phenomena of the Internet era. *Scientific Reports, 3*(3415), 1–7. Retrieved from https://www.nature.com/articles/srep03415%3FWT.ec_id%3DSREP-20131210

Kristoufek, L. (2015). What are the main drivers of the bitcoin price? Evidence from wavelet coherence analysis. *PLoS One, 10*(4), e0123923. https://doi.org/10.1371/journal.pone.0123923

Kyle, A. S. (1985). Continuous auctions and insider trading. *Econometrica, 53*(6), 1315–1335. https://doi.org/10.2307/1913210

Lakonishok, J., & Smidt, S. (1989). Past price changes and current trading volume. *The Journal of Portfolio Management, 15*(4), 18–24. https://doi.org/10.3905/jpm.1989.409223

Llorente, G., Michaely, R., Saar, G., & Wang, J. (2002). Dynamic volume-return relation of individual stocks. *Review of Financial Studies, 15*(4), 1005–1047. https://doi.org/10.1093/rfs/15.4.1005

Nadarajah, S., & Chu, J. (2017). On the inefficiency of bitcoin. *Economics Letters, 150*, 6–9. https://doi.org/10.1016/J.ECONLET.2016.10.033

Pagnottoni, P., Dimpfl, T., & Baur, D. (2018). Price discovery on bitcoin markets. *SSRN Electronic Journal.* https://doi.org/10.2139/ssrn.3280261

Panagiotidis, T., Stengos, T., & Vravosinos, O. (2018). On the determinants of bitcoin returns: A LASSO approach. *Finance Research Letters, 27*, 235–240. https://doi.org/10.1016/j.frl.2018.03.016

Phillips, P. C. B., Shi, S., & Yu, J. (2015). Testing for multiple bubbles: Historical episodes of exuberance and collapse in the S&P 500. *International Economic Review, 56*(4), 1043–1078. https://doi.org/10.1111/iere.12132

Plassaras, N. A. (2013). Regulating digital currencies: Bringing bitcoin within the reach of the IMF. *Chicago Journal of International Law, 14*(1), 377–407.

Rogojanu, A., & Badea, L. (2014). The issue of competing currencies. Case study-bitcoin. *Theoretical & Applied Economics, 21*(1), 103–114.

Sadeghi, A.-R. (Ed.). (2013). *Financial cryptography and data security* (Vol. 7859). Berlin, Heidelberg: Springer. https://doi.org/10.1007/978-3-642-39884-1

Schneider, J. (2009). A rational expectations equilibrium with informative trading volume. *The Journal of Finance, 64*(6), 2783–2805. https://doi.org/10.1111/j.1540-6261.2009.01517.x

Sensoy, A. (2018). The inefficiency of bitcoin revisited: A high-frequency analysis with alternative currencies. *Finance Research Letters, 28*, 68–73. https://doi.org/10.1016/J.FRL.2018.04.002

Shahzad, S. J. H., Bouri, E., Roubaud, D., Kristoufek, L., & Lucey, B. (2019). Is bitcoin a better safe-haven investment than gold and commodities? *International Review of Financial Analysis., 63*, 322–330. https://doi.org/10.1016/J.IRFA.2019.01.002

Shi, S., Phillips, P. C. B., & Hurn, S. (2018). Change detection and the causal impact of the yield curve. *Journal of Time Series Analysis, 39*(6), 966–987. https://doi.org/10.1111/jtsa.12427

Stosic, D., Stosic, D., Ludermir, T. B., & Stosic, T. (2019). Multifractal behavior of price and volume changes in the cryptocurrency market. *Physica A: Statistical Mechanics and Its Applications, 520*, 54–61. https://doi.org/10.1016/J.PHYSA.2018.12.038

Swanson, N. R. (1998). Money and output viewed through a rolling window. *Journal of Monetary Economics, 41*(3), 455–474. https://doi.org/10.1016/S0304-3932(98)00005-1

Tauchen, G. E., & Pitts, M. (1983). The price variability-volume relationship on speculative markets. *Econometrica, 51*(2), 485–505. https://doi.org/10.2307/1912002

Thoma, M. A. (1994). Subsample instability and asymmetries in money-income causality. *Journal of Econometrics, 64*(1–2), 279–306. https://doi.org/10.1016/0304-4076(94)90066-3

Tiwari, A. K., Jana, R. K., Das, D., & Roubaud, D. (2018). Informational efficiency of bitcoin—An extension. *Economics Letters, 163*, 106–109. https://doi.org/10.1016/J.ECONLET.2017.12.006

Urquhart, A. (2016). The inefficiency of bitcoin. *Economics Letters, 148*, 80–82. https://doi.org/10.1016/J.ECONLET.2016.09.019

van Wijk, D. (2013). *What can be expected from the BitCoin* (no. 345986). Rotterdam.

Vidal-Tomás, D., & Ibañez, A. (2018). Semi-strong efficiency of bitcoin. *Finance Research Letters, 27*, 259–265. https://doi.org/10.1016/J.FRL.2018.03.013

Yermack, D. (2015). Is bitcoin a real currency? An economic appraisal. In D. K. C. Lee (Ed.), *Handbook of digital currency* (pp. 31–43). London: Academic. https://doi.org/10.1016/B978-0-12-802117-0.00002-3

Ying, C. C. (1966). Stock market prices and volumes of sales. *Econometrica, 34*(3), 676–685. https://doi.org/10.2307/1909776

Efe Caglar Cagli is an Assistant Professor of Finance at Faculty of Business, Dokuz Eylul University, Turkey. He received his Ph.D. degree from Dokuz Eylul University. He was a Postdoctoral Visiting Scholar at the Department of Finance, Bentley University, Massachusetts, US, during the 2015/2016 academic term. His research interests include asset pricing, macro-finance, behavioral finance, applied time-series econometrics. He has published several articles in reputed journals including Energy Economics, Finance Research Letters, Expert Systems in Applications, Applied Financial Economics, and presented papers in national and international conferences. He has been teaching Managerial Finance, Investments and International Financial Management in both undergraduate and graduate levels.

Chapter 10
Assessment of the Crypto Market Efficiency: Empirical Evidence from Unit Root Tests with Different Approximations

Yuksel Iltas, Gulbahar Ucler, and Umit Bulut

Abstract The purpose of this study is to examine whether the weak form of the efficient market hypothesis (EMH) is valid for the Bitcoin market. To that end, we consider the recent developments in unit root analysis utilizing daily data from February 2, 2012 to November 23, 2018. More specifically, we employ unit root tests with and without sharp breaks and also a unit root test with gradual breaks in order to obtain efficient and unbiased output. Our findings show that the EMH appears to be valid for the Bitcoin market. We discuss theoretical and practical implications of these findings.

10.1 Introduction

Payment systems and payment instruments proceed to diversify and evolve to meet the new needs of financial markets. Globalization and high technological practices, shaping the modern world, lead to the emergence of new needs in the world. These needs sometimes cannot be met due to infrastructure deficiencies and/or the legal framework. The concept of virtual currency, showing up under these conditions, can be considered as a type of asset that meets new consumer demand which traditional payment instruments and financial markets have difficulty to fulfill (Uzer, 2017).

The European Central Bank (ECB) published a report on virtual currencies in October 2012 by making the first detailed investigation. In the report, the ECB defined the virtual currency as a digital money that is (1) issued and usually controlled by its developers, (2) not regulated by laws, (3) adopted and used by the members of a particular virtual community (ECB, 2012). In another report published by the ECB in 2015, the definition of the virtual currency was reviewed

Y. Iltas
Department of Business Administration, Kirsehir Ahi Evran University, Kirsehir, Turkey
e-mail: yiltas@ahievran.edu.tr

G. Ucler · U. Bulut (✉)
Department of Economics, Kirsehir Ahi Evran University, Kirsehir, Turkey
e-mail: gulbahar.ucler@ahievran.edu.tr; ubulut@ahievran.edu.tr

© Springer Nature Switzerland AG 2019

U. Hacioglu (ed.), *Blockchain Economics and Financial Market Innovation*, Contributions to Economics, https://doi.org/10.1007/978-3-030-25275-5_10

and virtual currency was defined as the following: virtual currency is a digital representation of value that (1) is not issued by any central banks, credit institutions or e-money institutions and (2) can be used as an alternative to money in some conditions (ECB, 2015). This new definition does not include the term "money" as virtual currencies are not widely accepted assets in the society. Besides, the term "not regulated by laws" is not included in this definition as legal regulations in some countries cover virtual currencies (Uzer, 2017).

Virtual currencies have attracted more attention and many studies have been conducted on virtual currencies since the emergence of Bitcoin in 2008. Bitcoin is the first virtual currency developed by Satoshi Nakamoto in 2008. Bitcoin is an encrypted computer record which has a monetary value (Guven & Sahinoz, 2018). Bitcoin is a crypto currency that enables a person to pay safely to another person without a financial instrument. The technology owned by Bitcoin (1) establishes a direct relationship between the buyer and the seller, (2) enables the actualization of transactions at a relatively low cost compared to the comparable alternatives, and (3) allows the transfer of wealth all over the world independently of time (Carpenter, 2016).

Bitcoin and similar digital currencies are called crypto currencies because they use crypto algorithms in their basic software and security. Crypto currencies are not centralized and they use encryption technology as the verification system. The system records every currency transaction and money transfer in a public database. This database is called "block chain" (Dwyer, 2015).

The common features of the virtual currencies (Bitcoin, Bitcoin Cash, Ethereum, Iota, Ripple, Monero, etc.) and the used infrastructures (Blockchain, tangle and its derivatives) are as follows (Guven & Sahinoz, 2018):

- They are not associated with a central authority.
- There is no central bank or state behind them.
- They have a distributed structure (no central server that will lead to be hacked or attacked).
- The amount of supply is limited.

The production of a crypto currency is called mining. In the Bitcoin network, currency is produced or mined using internet technology while it cannot be printed. Miners work independently to support and reinforce the first prepared algorithm. The feature of the system is that the mining network lets Bitcoins be exchanged between currencies while it is producing Bitcoins (Bradbury, 2013).

Bitcoin is the new digital currency based on a computer cryptology and a decentralized network (Li & Wang, 2017). According to Teknochain (2019) data, there are 894 crypto currencies today, whereas there are 2116 crypto currencies in the world with regard to Coinmarket (2019) data. Among these crypto currencies, top ten crypto currencies with the highest market capitalization are reported in Table 10.1. The most popular digital currency is Bitcoin in terms of market capitalization and trading volumes.

As Kocoglu, Çevik, and Tanrıöven (2016) point out, the first Bitcoin transaction was occurred in January 2009 by the founders of the system. Bitcoin met the real economy on May 22, 2010 when Laszlo Hanyecz purchased two pizzas in exchange

Table 10.1 Top ten crypto currencies with the highest market capitalization

Rank	Name	Symbol	Market capitalization (USD)	Circulating supply
1	Bitcoin	BTC	$62,691,192,330	17,492,175 BTC
2	Ripple	XRP	$13,130,888,481	41,040,405,095 XRP[a]
3	Ethereum	ETH	$ 12,449,943,817	104,480,979 ETH
4	Bitcoin cash	BCH	$ 2,160,366,300	17,577,163 BCH
5	EOS	EOS	$ 2,139,623,235	906,245,118 EOS[a]
6	Tether	USDT	$ 2,049,183,204	2,016,152,117 USDT[a]
7	Stellar	XLM	$ 1,995,763,889	19,129,175,185 XLM[a]
8	Litecoin	LTC	$ 1,867,310,350	60,106,150 LTC
9	TRON	TRX	$ 1,595,686,172	66,653,472,407 TRX
10	Bitcoin SV	BSV	$ 1,323,599,308	17,576,061 BSV

[a]Not Mineable
Source: Coinmarket (2019)

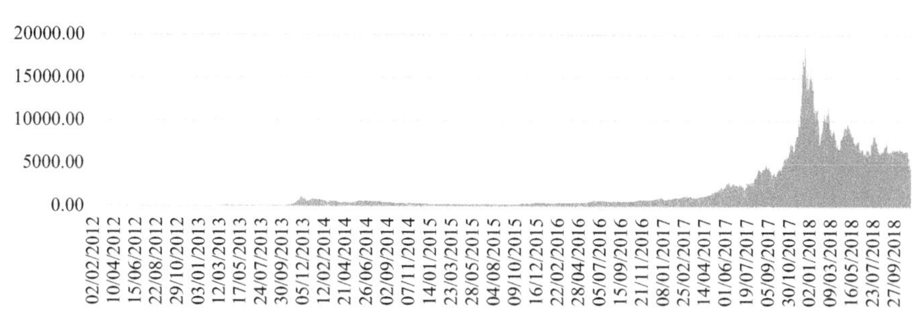

Fig. 10.1 Price of bitcoin (units of USD per unit of bitcoin price). Source: Blockchain (2019)

for 10,000 Bitcoins. After Bitcoin emerged, its price did not change too much for a long time. Figure 10.1 shows Bitcoin price from February 2, 2012 to October 23, 2018.

According to the data of Blockchain (2019), Bitcoin price followed a relatively horizontal path until the end of 2016 except for the break in 2013 and started to 2017 at 998 USD (Ozturk, Arslan, Kayhan, & Uysal, 2018). One can argue that 2017 is the year of Bitcoin in terms of price, trading volume, prevalence, and recognition. Bitcoin started to 2017 with a price of 998 USD. With the acquisition of a legal status in Japan, the price exceeded 2800 USD in June 2017. Although the price was 4800 USD in September 2017, it fell below 3000 USD as a result of the prohibitive attitude of China towards Bitcoin. In December 2017, Bitcoin futures began being traded in the world's largest options exchange, the CBOE (Chicago Board Options Exchange), and the world's largest derivatives exchange, the CME (Chicago Mercantile Exchange). When the CBOE and CME began to include Bitcoin futures, rapid price increases were observed and Bitcoin quickly approached 20,000 USD in the last month of the year. Bitcoin price closed the year at 15,000 USD. Accordingly, Bitcoin experienced a 20-fold price rise in 2017. In 2018, Bitcoin price did not exhibit rapid movements and had a tendency to decrease.

The rapid rise of Bitcoin price drew the attentions of investors and triggered the appetite of investors who wanted to make big gains. Bitcoin's trading volume increased rapidly in parallel with the fluctuations in prices. The easy use of crypto coins, the growing popularity, and the rapid earning opportunity have led to the emergence of crypto money exchanges. Bitcoin and other crypto currencies began being purchased through virtual markets in exchange for USD or other national currencies (Kilic & Cutcu, 2018). Table 10.2 exhibits the Bitcoin stock exchanges with the highest trading volumes.

As seen in Table 10.2, Bitfinex has 22.95% of the market share with 568,000 BTC trading volume. Bitfinex is followed by the Kraken with a market share of 16.55% and Coinbase with a share of 16.02%. The emergence of crypto currencies' own stock exchanges supports the idea that these currencies are investment instruments.

Some basic data about bitcoin are obtained from Blockchain (2019) and shown below. Figure 10.2–10.7 respectively show Bitcoin's price, Bitcoin's market capitalization, difficulty in mining, the number of Blockchain wallet users, daily miners' revenues, and total transaction fees.

This study examines whether investors forecast future Bitcoin prices using past values of Bitcoin prices. Put differently, we investigate whether the weak form of the efficient market hypothesis developed by Fama (1970) prevails in the Bitcoin market using daily data from February 2, 2012 to November 23, 2018. While doing that, to be able to produce efficient output, this study employs three types of unit root tests, namely Dickey and Fuller (1981, hereafter ADF) unit root test without breaks, Narayan and Popp (2010, hereafter N&P) unit root test with sharp breaks, and Enders and Lee (2012a, hereafter E&L) unit root test with gradual breaks. Therefore, we investigate whether regarding breaks have a considerable role and different approaches in modelling breaks result in different findings for the Bitcoin market.

The rest of the study is structured as follows: Section 10.2 gives brief literature. Methodology and findings are presented in Sect. 10.3. Section 10.4 concludes the study.

Table 10.2 Bitcoin stock exchanges with the highest trading volumes

Rank	Exchange	Volume (BTC)	Market share
1	Bitfinex	568,000	22.95%
2	Kraken	409,000	16.55%
3	Coinbase	396,000	16.02%
4	Bitstamp	296,000	11.97%
5	Bitflyer	206,000	8.34%
6	Bit-x	181,000	7.34%
7	Others	159,000	6.43%
8	Gemini	112,000	4.53%
9	Itbit	102,000	4.12%
10	Bitbay	43,200	1.75%

Source: Bitcoinity (2019)

Fig. 10.2–10.7 Some basic indicators for bitcoin. Source: Blockchain (2019)

10.2 Brief Literature

Whether or not prices and/or returns of financial assets can be forecasted is one of the most discussed topics in the finance literature. Efficient market hypothesis (EMH) indicates that future prices of financial assets cannot be forecasted. Put differently, the EMH prevails if investors cannot predict future movements in prices of financial assets (Hatemi-J, 2012). According to Fama (1970), there are three types of EMH, namely weak form, semi-strong form, and strong form. If prices reflect all available information and investors cannot predict future prices using past prices, then the weak form of the EMH dominates. In such a case, prices of financial assets follow a random walk (Mishkin, 2004). In other words, prices are not stationary as they have a unit root (Karadeniz, Ozturk, & Iskenderoglu, 2012). If investors cannot forecast future prices using past prices along with all publicly available information, one can determine that the semi-strong form of the EMH is valid. Finally, the strong form of the EMH prevails if some investors in the market cannot predict future prices although they (1) analyse past prices, (2) use all publicly available information, and (3) even obtain some publicly unavailable information due to their monopoly powers.

One can observe throughout the finance literature that there is an extending empirical literature on the efficiency of the Bitcoin market. One can also classify these papers under two groups. The first group of the papers yields the EMH is valid for the Bitcoin market. For instance, Bartos (2015) explores the EMH is valid in the Bitcoin market through the error correction model while Bariviera (2017) finds that the Bitcoin market is efficient by employing the Hurst exponent. Sensoy (2018) examines the time-varying weak form efficiency of the Bitcoin market by performing the permutation entropy in the high-frequency range and yields the EMH dominates. Tiwari, Jana, Das, and Roubaud (2018) discover the Bitcoin market is efficient in most of the observed period. Wei (2018), performing the wild-bootstrapped automatic variance test and the non-parametric BDS test, find evidence in favour of the Bitcoin market. The second group of the papers explores the EMH is not valid. For instance, Kurihara and Fukushima (2017), who yield Bitcoin market is inefficient, suggest that the Bitcoin market may be efficient in the future as a result of increases in trading volume in the Bitcoin market. Caporale, Gil-Alana, and Plastun (2018), employing R/S analysis and the fractional integration method, explore that the Bitcoin market is inefficient. Finally, Jiang, Nie, and Ruan (2018) discover the EMH is not valid in the Bitcoin market by employing the rolling window estimation method.

10.3 Data, Methodology, and Empirical Findings

In the study, we use daily Bitcoin price data from February 2, 2012 to November 23, 2018. Just like in Fig. 10.1, we consider USD units per unit of Bitcoin. BTC represents Bitcoin price in the study.

As was denoted in Sect. 10.1, the study employs ADF, N&P, and E&L unit root tests. N&P develop a unit root test with two sharp breaks. The data-generating process of a time series N&P determine has two components: a deterministic component (d_t) and a stochastic component (u_t). Therefore, the data-generating process is defined as $y_t = d_t + u_t$. The model allowing two sharp breaks in intercept and trend is demonstrated as the following:

$$d_t = \alpha + \beta t + \Psi^*(L)(\theta_1 DU_{1,t} + \theta_2 DU_{2,t} + \gamma_1 DT_{1,t} + \gamma_2 DT_{2,t}) \tag{10.1}$$

where $DU_{i,t} = 1(t > T_{B,i})$, $DT_{i,t} = 1(t > T_{B,i})(t\text{-}T_{B,i})$, $i = 1,2$ and $T_{B,i}$ indicates the break dates. The test regression is the reduced form of the corresponding structural model demonstrated as the following:

$$y_t = \rho y_{t-1} + \alpha^* + \beta^* t + \Omega_1 D(T_B)_{1,t} + \Omega_2 D(T_B)_{2,t} + \theta_1^* DU_{1,t-1} + \theta_2^* DU_{2,t-1} + \gamma_1^* DT_{1,t-1} + \gamma_2^* DT_{2,t-1} + \sum_{j=1}^{k} \beta_j \Delta y_{t-j} + e_t \tag{10.2}$$

The null hypothesis of a unit root of $\rho = 1$ is tested against the alternative hypothesis of $\rho < 1$. While t-statistic of $\hat{\rho}$ in Eq. (10.2) is utilized, critical values are generated via Monte Carlo simulations by N&P. To reject the null hypothesis of a unit root, test statistic has to be greater than the critical values.

Traditional unit root tests using dummy variables to captures changes in intercept and/or in intercept and trend assume that structural breaks in variables are presumed to occur instantaneously and they are considered as unit root tests with sharp breaks. Enders and Lee (2012b) remark that the break dates and the number of the breaks may be unknown. They also denote that breaks may be gradual. Hence, they develop a unit root test for the cases when breaks are gradual. E&L extend the work of Enders and Lee (2012b) and propound a Fourier unit root test in a Dickey-Fuller type regression equation.

E&L first use the following Dickey-Fuller test with the deterministic term which is a time dependent function stated by $\alpha(t)$:

$$y_t = \alpha(t) + \rho y_{t-1} + \gamma t + e_t \tag{10.3}$$

where e_t denotes the stationary error term and $\alpha(t)$ stands for the deterministic function of t. The null hypothesis of a unit root is defined as $\rho = 1$. When the form of $\alpha(t)$ is unknown, E&L consider the following Fourier expansion:

$$\alpha(t) = \alpha_0 + \Sigma_{k=1}^{n} \alpha_k \sin(2\pi kt/T) + \Sigma_{k=1}^{n} \beta_k \cos(2\pi kt/T), n \leq T/2 \tag{10.4}$$

where n is the number of frequencies included in the approximation, k denotes a particular frequency, and T stands for the number of observations.

E&L point out that at least one Fourier frequency must exist in the data generating process when there is a break or nonlinear trend. As the utilization of many frequency components decreases degrees for freedom and is likely to result in an overfitting problem, in their original work, E&L employ only one frequency k considering the following equation:

$$\Delta y_t = \rho y_{t-1} + c_1 + c_2 t + c_3 \sin(2\pi kt/T) + c_4 \cos(2\pi kt/T) + e_t \tag{10.5}$$

Table 10.3 Results of the unit root tests

Variable	No break ADF	Sharp breaks N&P	Gradual breaks E&L
BTC	−2.982	−4.182	−2.977
ΔBTC	−9.340[a]	−14.86[a]	−41.751[a]

Δ is the first difference operator
[a]indicates 1% statistical significance

In Eq. (10.5), to test the null hypothesis of a unit root defined as $\rho = 0$, E&L compare the statistic obtained from the test to the critical values that depend on the frequency k and the sample size T. When the test statistic is greater than the critical values generated by E&L, the null hypothesis of a unit root is rejected.

The results of the unit root tests are reported in Table 10.3. Accordingly, the null hypothesis of a unit root cannot be rejected at level, whereas it can be rejected at first difference with regard to the unit root tests. Put differently, all unit root tests indicate that the Bitcoin price data are not stationary. Hence, we explore that investors in the Bitcoin market cannot predict future Bitcoin prices using past values of the prices and that the weak form of the EMH prevails for the Bitcoin market.

Therefore, the results from our study are in line with the results from the studies of Bartos (2015), Bariviera (2017), Sensoy (2018), Tiwari et al. (2018), and Wei (2018), while they contradict with those of Kurihara and Fukushima (2017), Caporale et al. (2018), and Jiang et al. (2018).

10.4 Conclusion

In this study, we investigated whether the weak form of the EMH was valid for the Bitcoin market using daily data from February 2, 2012 to November 23, 2018 by employing unit root tests with different approaches in modelling breaks. We first exploited the ADF unit root test without breaks and second utilized the N&P unit root test with sharp breaks. Finally, we performed the E&L unit root test with gradual breaks based on the Fourier approximation. All unit root tests implied that Bitcoin price data were not stationary, indicating considering breaks and different approaches in modelling breaks did not lead to different findings. Overall, the results showed that the weak form of the EMH dominated the Bitcoin market.

These findings have important implications for the Bitcoin market. First, the results of the unit root tests imply that shocks to Bitcoin price are permanent and that Bitcoin price does not return to its mean after it is affected by a shock. Put differently, shocks to Bitcoin price have abiding effects which prevent Bitcoin price from returning to its mean. Second, investors in financial markets cannot make predictions about future values of Bitcoin price by only utilizing past values of Bitcoin price. Third, future papers can examine the semi-strong form of the EMH for the Bitcoin market. If these papers find some variables can be used to predict future Bitcoin prices, their findings can guide financial investors to invest in Bitcoin.

References

Bariviera, A. F. (2017). The inefficiency of bitcoin revisited: A dynamic approach. *Economics Letters, 161*, 1–4.

Bartos, J. (2015). Does bitcoin follow the hypothesis of efficient market? *International Journal of Economic Sciences, 4*(2), 10–23.

Bitcoinity. (2019). Retrieved January 01, 2019, from https://bitcoinity.org/

Blockchain. (2019). Retrieved January 01, 2019, from https://www.blockchain.com/

Bradbury, D. (2013). The problem with bitcoin. *Computer Fraud & Security, 11*, 5–8.

Caporale, G. M., Gil-Alana, L., & Plastun, A. (2018). Persistence in the cryptocurrency market. *Research in International Business and Finance, 46*, 141–148.

Carpenter, A. (2016). Portfolio diversification with bitcoin. *Journal of Undergraduate in France, 6* (1), 1–27.

Coinmarket. (2019). Retrieved January 03, 2019, from https://coinmarketcap.com/

Dickey, D. A., & Fuller, W. A. (1981). Likelihood ratio statistics for autoregressive time series with a unit root. *Econometrica, 49*(4), 1057–1072.

Dwyer, G. P. (2015). The economics of bitcoin and similar private digital currencies. *Journal of Financial Stability, 17*, 81–91.

ECB. (2012). Virtual currency schemes. Retrieved January 01, 2019, from https://www.ecb.europa.eu/pub/pdf/other/virtualcurrencyschemes201210en.pdf

ECB. (2015). Virtual currency schemes–a further analysis. Retrieved January 01, 2019, from https://www.ecb.europa.eu/pub/pdf/other/virtualcurrencyschemesen.pdf

Enders, W., & Lee, J. (2012a). The flexible Fourier form and Dickey–Fuller type unit root tests. *Economics Letters, 117*(1), 196–199.

Enders, W., & Lee, J. (2012b). A unit root test using a Fourier series to approximate smooth breaks. *Oxford Bulletin of Economics and Statistics, 74*(4), 574–599.

Fama, E. F. (1970). Efficient capital markets: A review of theory and empirical work. *The Journal of Finance, 25*(2), 383–417.

Guven, V., & Sahinoz, E. (2018). *Blokzincir, kripto paralar, bitcoin*. İstanbul: Kronik Kitabevi.

Hatemi-J, A. (2012). Asymmetric causality tests with an application. *Empirical Economics, 43*(1), 447–456.

Jiang, Y., Nie, H., & Ruan, W. (2018). Time-varying long-term memory in bitcoin market. *Finance Research Letters, 25*, 280–284.

Karadeniz, E., Ozturk, I., & Iskenderoglu, O. (2012). An investigation of efficient market hypothesis in OECD countries. *Actual Problems of Economics, 129*(3), 398–405.

Kilic, Y., & Cutcu, İ. (2018). Bitcoin Fiyatları ile Borsa İstanbul Endeksi Arasındaki Eşbütünleşme ve Nedensellik İlişkisi. *Eskişehir Osmangazi Üniversitesi İktisadi ve İdari Bilimler Dergisi, 13* (3), 235–250.

Kocoglu, S., Çevik, Y. E., & Tanrıöven, C. (2016). Bitcoin piyasalarının etkinliği, likiditesi ve oynaklığı. *Journal of Business Research Turk-Türk İşletme Araştırmaları Dergisi, 8*(2), 77–97.

Kurihara, Y., & Fukushima, A. (2017). The market efficiency of bitcoin: A weekly anomaly perspective. *Journal of Applied Finance and Banking, 7*(3), 57.

Li, X., & Wang, C. A. (2017). The technology and economic determinants of cryptocurrency exchange rates: The case of bitcoin. *Decision Support Systems, 95*, 49–60.

Mishkin, F. S. (2004). *The economics of money, banking, and financial markets* (7th ed.). New York: Pearson Education.

Narayan, P. K., & Popp, S. (2010). A new unit root test with two structural breaks in level and slope at unknown time. *Journal of Applied Statistics, 37*(9), 1425–1438.

Ozturk, B. M., Arslan, H., Kayhan, T., & Uysal, M. (2018). Yeni bir hedge enstrümanı olarak Bitcoin: Bitconomi. *Ömer Halisdemir Üniversitesi İktisadi ve İdari Bilimler Fakültesi Dergisi, 11*(2), 217–232.

Sensoy, A. (2018). The inefficiency of bitcoin revisited: A high-frequency analysis with alternative currencies. *Finance Research Letters, 28*, 68–73. https://doi.org/10.1016/j.frl.2018.04.002

Teknochain. (2019). Retrieved January 04, 2019, from https://teknochain.com/
Tiwari, A. K., Jana, R. K., Das, D., & Roubaud, D. (2018). Informational efficiency of bitcoin—An extension. *Economics Letters, 163*, 106–109.
Uzer, B. (2017). Sanal Para Birimleri. Retrieved January 02, 2019, from https://www.tcmb.gov.tr/wps/wcm/connect/f4b2db90-7729-4d94-8202-031e98972d0f/Sanal+Para+Birimleri.pdf?MOD=AJPERES&CACHEID=ROOTWORKSPACE-f4b2db90-7729-4d94-8202-031e98972d0f-m3fBagn
Wei, W. C. (2018). Liquidity and market efficiency in cryptocurrencies. *Economics Letters, 168*, 21–24.

Yuksel Iltas is an assistant professor in the Business Administration Department at Kirsehir Ahi Evran University in Turkey. He holds a Ph.D. degree from Erciyes University in Turkey. His research interests focus on Financial Markets and Institutions, Financial Risk Management, and Investments and Portfolio Management. He has been teaching fundamentals of financial management, money and banking, and evaluation of investment projects at Kirsehir Ahi Evran University since 2007.

Gulbahar Ucler is an associate professor in the Economics Department at Kirsehir Ahi Evran University in Turkey. She has a Ph.D. Degree from Selcuk University in Turkey. Her research interests cover topics in the fields of microeconomics, financial economics, and institutional economics. She has been teaching microeconomics and econometrics at Kirsehir Ahi Evran University since 2012.

Umit Bulut is an associate professor in the Economics Department at Kirsehir Ahi Evran University in Turkey. He received his Ph.D. degree from Gazi University in Turkey. His research interests include monetary economics, energy economics, and applied econometrics. He has been teaching monetary theory and policy and theories of business cycles at Kirsehir Ahi Evran University since 2018.

Chapter 11
Forecasting the Prices of Cryptocurrencies Using GM(1,1) Rolling Model

Cem Kartal and Mehmet Fatih Bayramoglu

Abstract Although cryptocurrencies initially emerged as a transnational payment instrument, it has become an investment tool by attracting the attention of investors within the functioning of the capitalist system. In this chapter, the use of cryptocurrencies as an investment tool rather than in commercial transactions is discussed. As is known, most investors remain in a dilemma between the risk of risk aversion and the maximization of returns. Investors in this dilemma try to predict the future price or returns of the financial instruments through various analyzes and thus make an effort to give direction to their investments. These analyses are generally carried out by analyzing the past values of prices or returns by adopting a technical analysis approach. However, since the cryptocurrencies are a relatively new investment tool, it is not possible to reach the previous period price and yield information for an extended period.

For this reason, the scope of the chapter is to explain the functioning of the cryptocurrencies as an investment tool in the market and to share information about the types of investors who have transferred their funds to cryptocurrencies by providing statistical information. Then, it is aimed to share the theoretical knowledge about GM(1,1) Rolling Model which has been proved by the literature in which it produces successful results especially in forecasting problems in uncertainty environment. Finally, the price forecasting of popular cryptocurrencies which are Bitcoin, Ethereum, Litecoin, and Ripple was made using the GM(1,1) Rolling Model, and it was tested whether this model is advisable for price forecasting of cryptocurrencies. Results of the Model show that the forecasting errors ranged from 1.35% to 7.76% for 10-days period. Also, direction forecasting results are between 40% and 50% in the same period. Also, returns of the bitcoin investment which

C. Kartal
Department of International Trade and Business, Zonguldak Bulent Ecevit University, Zonguldak, Turkey
e-mail: cem.kartal@beun.edu.tr

M. F. Bayramoglu (✉)
Department of Accounting and Finance, Zonguldak Bulent Ecevit University, Zonguldak, Turkey
e-mail: fatih.bayramoglu@beun.edu.tr

© Springer Nature Switzerland AG 2019

U. Hacioglu (ed.), *Blockchain Economics and Financial Market Innovation*, Contributions to Economics, https://doi.org/10.1007/978-3-030-25275-5_11

made by trusting the results are ranged from −0.60% to −8.18. The results may be considered that the model was successful in forecasting the prices but unsuccessful in the direction forecasting. Even though the estimates are made with low percentages, the time series analyzes made with the lagged data of Bitcoin prices are not successful. Therefore, the technical analysis approach can be interpreted as not sufficient for modeling Bitcoin prices. So, these results show that defining bitcoin price movements is not only a forecasting problem but also a classification problem.

11.1 Introduction

The blockchain is a technology used to read, store and verify transactions in a distributed database system. The first form of Blockchain began with a mixed tree, also known as the Merkle Tree. This data structure, patented by Ralph Merkle in 1979, was used to validate and use data from computer systems. In 1991, the Merkle tree was used to form a secure blockchain, each with a series of data records connected to the former. The newest record in this chain contains the history of the entire chain. Verification of data in a peer-to-peer computer network is essential to ensure that nothing changes during the transfer. This data structure also prevented the sending of incorrect data. This data structure is used to preserve the integrity of the shared data and to verify the accuracy of the data (Zheng, Xie, Dai, Chen, & Wang, 2017: 558).

Blockchain can be defined as a technology designed to securely store and manage data (such as currency, identity, valuable papers) with value. In the blockchain approach, the structures in which data is stored are called blocks. These block structures are arranged in the form of a chain (in the form of a linear array in time) called the "Blockchain" (FinTech Istanbul, 2019).

When a new block with operations is created, the miners process the information in this block. The mathematical formulas applied to this information, and the long list of transactions in the block is compiled with a *hash* of the figures. This hush is added to the end of the blockchain with the newly added block. *Hushes* help to minimize the information in the long list of transactions. Every hash produced is unique. If any character changes within a Bitcoin block, this will cause the entire *hash* to change Blockchain Bitcoin is the technology selected to run the process data structure. Some countries are even considering harmonization of such technologies with public infrastructures. With the success of Bitcoin, interest in the blockchain has increased in order to be used in different processing systems for use in multiple potential application areas. Recently, especially in the field of a communication network, many studies have been carried out in this direction (NRI, 2016: 3–4).

Global integration, deregulation, developments in Internet technologies significantly change the nature of financial services. Internet and related technologies enable new financial service providers to compete more effectively for investors. Technological changes accelerate the development of the financial sector by reducing costs, increasing width and quality, and expanding access to financial services (Shahrokhi, 2008: 366.) Cryptocurrencies are one of the most significant financial technologies emerging in recent years. In 2009, a Japanese programmer named Nakamoto entered the world of finance with the creation of Bitcoin. Although they emerged as a means of payment, some of the investors saw cryptocurrencies as an investment tool.

The Bitcoin system has never been suspended (called no/zero downtime), and Bitcoin users are increasing not only in the United States but also across the globe. In January 2009, the first block (Genesis block) was created by Nakamoto and mining, and transfers started. What makes Bitcoin different from the virtual coins before it is that it can be transferred from person to person (P2P) directly and no tool is needed, and it is based on blockchain technology. In Japan, the development of money change in early 2014 attracted the attention of people, and in 2015, people began to be interested in blockchains because of the growing momentum in the FinTech area (NRI, 2016: 5–10).

The organization of the rest of the chapter is as follows. Second section provides a general, albeit, a brief, theoretical background of the cryptocurrencies and also provides a brief literature review. Third section provides definition and description of the concept of the GM(1,1) Rolling Model. Fourth section consists of the motivation of the application, the data, modeling, and application of the Model. Fifth section presents the empirical findings and discussion. Lastly, Sixth section provides conclusions.

11.2 Theoretical Background and Literature Review

Brassard (1988) described cryptology as the art and science of safe communication over unsafe channels. Cryptology consists of the process of data re-emergence as a result of decoding the data (number, text or an encrypted message) in a system frame, sending the encrypted data through a security-based environment and decoding the sent passwords. Cryptology is defined as a science of cryptography. In the Blockchain approach, the data is kept in the structures called blocks. The concept of security for the block structure is used not to hide the information it contains from the outside world, but it cannot be changed without knowing the information it contains after being created. Cryptographic hashing and time information are used to provide this.

The cryptocurrency is the currency that uses cryptography in its structure. Cryptocurrencies can be evaluated in the virtual currencies category, which cannot be edited by the currency matrix laid down by the European Central Bank and in digital format. In other words, it is not structured by any central bank, government or similar official institution. The cryptocurrency has a digital format that is theoretically not represented by any physical material. Although there is no need for any central authority to be printed, an electronic money transfer company is not required for the existence and transfer of a commercial bank to be stored (Gültekin & Bulut, 2016: 83).

11.2.1 Blockchain

The block body consists of a process counter and operations. The maximum number of operations a block can contain depends on the size of the block and its size. Blockchain uses an asymmetric encryption path to verify that transactions are validated. A digital signature based on asymmetric encryption is used in an undependable environment (NRI, 2016: 8). Each user has a public key and a pair of private keys. A private key is used to keep transactions secret. Digitally signed transactions are broadcast across the whole network. The typical digital signature is in two stages: the verification phase and the signing phase (Zheng et al., 2017: 555–556).

In Blockchain, a blockchain network node with the role of approving new transactions is called a miner. After a miner verifies a transaction, it places it in a new block that he publishes to other nodes in the network. The blockchain is a block array that contains a complete list of transaction records, such as a traditional public book. The blockchain must be reliable and should not be damaged by any power. At this stage, miners are activated to ensure the reliability of the blockchain (Swanson, 2015: 5).

Figure 11.1 shows an example of a blockchain. The block consists of two main sections as a block header for controlling the data integrity within the block and data within the block. A block comprises the block body and block header as shown in Fig. 11.2. A block header contains the following information (Zheng et al., 2017: 558):

1. Block version: Indicates which block validation rules are followed.
2. Merkle tree root hash: Indicates the hash value of all transactions in the block.
3. Timestamp: The current time in seconds in the universal time since January 1, 1970.
4. nBits: The target threshold value of a valid block hash.
5. Nonce: An area of four bytes usually starting with 0 and increasing for each hash calculation.
6. Parent block hash: the 256-bit hash value that points to the previous block.

Due to the nature of its design, there are some disadvantages as well as the advantages of adopting a blockchain solution. The blockchain has strong and weak

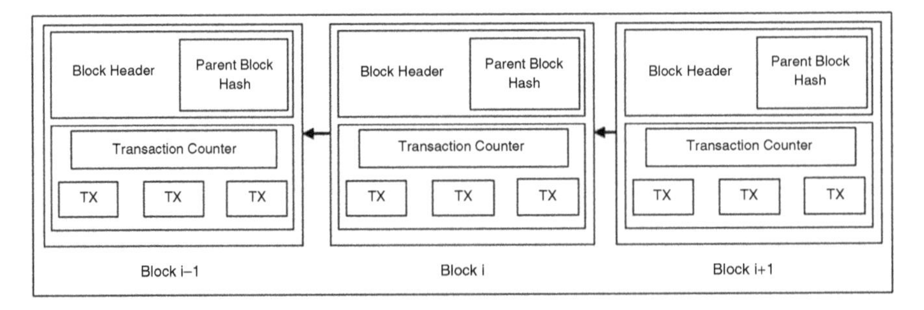

Fig. 11.1 Blockchain which involves a continuous sequence of blocks (Source: Zheng et al., 2017)

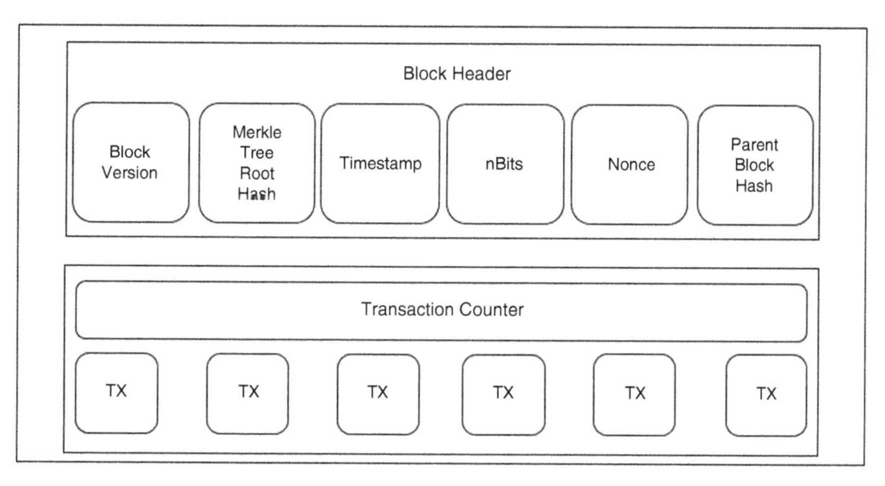

Fig. 11.2 Block structure (Source: Zheng et al., 2017)

Table 11.1 Blockchain vs. distributed databases and legacy centralized

Features	Blockchain	Distributed databases	Legacy centralized
Accessibility	High	Medium	Low
Records integrity	High	Medium	Medium
Fault tolerance	High	High	Low
Privacy	Low	Medium	High
Trustless nodes collaboration	High	Low	Low
Computing time	Low	Medium	High

Source: Bozic, Pujolle, and Secci (2016)

points compared to distributed and legacy database systems. Table 11.1 contains the key differences between blockchain, distributed databases, and legacy centralized.

The main differences between blockchain, legacy databases and distributed databases are summarized in Table 11.1. Positioning concerning legacy database systems are based on centralized databases, relying on central integrity, all nodes in a database scenario deployed in the blockchain are reliable. It is important to note that data storage in legacy database systems can depend on a trusted third-party stakeholder, that they can access and corrupt or destroy the data. In Blockchain, high availability and integrity have its unique guarantee: the blockchain consensus mechanism, which forces all records to be processed separately, to be validated by additional nodes and keep these nodes in sync. Compared to distributed database systems, it can be said that a blockchain can be as error-free as the old distributed database system. All computers involved in the system must approve any changes to Blockchain technology. However, the data transfer requires the approval of the whole system, and this situation improves more slowly than expected (Bozic et al., 2016: 3–4).

The basic value of a blockchain allows a database to be shared directly across trust boundaries without requiring a centralized administrator. This is likely because the blockchain operations do not require some centralized application logic to implement these constraints, rather than their proof of authorization and the proof of their validity. Thus, the processes can be independently verified and processed by multiple nodes that serve as a consensus mechanism to ensure that these nodes remain synchronized. Privacy is much lower than older systems, and especially those that are central to all database content requests being forwarded by database administrators. Public blockages benefiting from a public book are potentially accessible to internet users, unlike distributed/central databases, where the central authority determines access. However, there are recommendations on how to mitigate this problem; when processing under multiple blockchain addresses, zero information evidence is recommended as a way to mitigate privacy issues. These alternatives have additional disadvantages by making the blockchain more complex or less scalable. The contents of the database are stored in the memory of a specific computer or the memory of the computer system. Anyone with access to this system may corrupt the data within it. In summary, when data are entrusted to a regular database, a particular human organization is depended on it. If an organization checks a primary database, it needs a large number of people and processes to prevent interference with this database. Recruitment and design processes are time-consuming and costly. Therefore, block chains offer a way to replace these organizations with a distributed database that is locked by smart cryptography. When performing operations, a blockchain has to perform the same operations as a regular database but has three problems (Multichain, 2019).

- Signature verification; each block chain process since the processes are spread across peers; it must be digitally signed using a private-public encryption scheme, such as the ECDSA (Elliptic Curved Digital Signature Algorithm). This is necessary, so resources cannot be proven otherwise. Creating and verifying these signatures is difficult and complex for calculators and creates primary congestion in products like ours. However, once a connection is established in central databases, it is not necessary to verify each incoming request individually.
- Consensus mechanisms; efforts should be made to ensure that the nodes in the network reach consensus in a distributed database, such as a blockchain. According to the consensus mechanism used, this may include dealing with important forward-looking communications and/or forks and their consequent rollbacks. Though it is true that central databases should also manage conflicting and revoked transactions, they are less likely to occur where operations are queued and processed in one place.
- Redundancy; this is related to the total amount of calculation required by a blockchain rather than the performance of a single node. When central databases perform operations once (or twice), a blockchain must be processed freely by each node on the network. This requires the blockchain to perform much more for the same result.

11.2.2 Cryptocurrencies

In 2008, Satoshi Nakamoto conceptualized the distributed blockchain. This blockchain would include a secure data exchange history, benefit from a peer-to-peer network to stamp and validate the change of time and be autonomously managed without a central authority. It became the backbone of the Bitcoin, and it was born in the world of the blockchain, cryptocurrencies that we know today. At the end of November 2008, Nakamoto published a thesis with the title "Bitcoin: A Peer-to-Peer Electronic Cash System." In his thesis, Bitcoin's features are (NRI, 2016: 4–5):

- Enable direct transactions without the need for reliable third parties;
- Activate non-return operations;
- Reduce the cost of credit in small temporary transactions;
- Reduce transaction fees and
- Avoid double spending.

Bitcoin was first introduced to a closed e-mail group from Japan in 2008 by Satoshi Nakamoto. However, there is no information about who is Satoshi Nakamoto. Therefore, it is unclear what bitcoin is done by (Aslantaş, 2016: 354–355).

Bitcoin has been used to enable a payment system released as an open source project for the cryptocurrency. Bitcoin has been rapidly spreading and developing worldwide. After world economic crises, Bitcoin had the goal of running an independent payment system capable of storing and monitoring all transactions using distributed nodes (or notebooks) performed by participants of such a geographically distributed system. The purpose of Bitcoin is to ensure that transactions related to money transfers are legible, secure and transparent. Security is the ability to prevent fraudulent payments or at least reduce the likelihood of unauthorized and fraudulent payments. The entire process is managed between the nodes of the Bitcoin network, no need for the third trusted parties, such as banks or other financial institutions. Accomplishing this, it was necessary to found a system where records could be securely locked, stored and verified (Bozic et al., 2016: 2–3).

Bitcoin is a decentralized open-source electronic currency and monetary system that cannot be controlled by any state, company or authority that allows any person from anywhere in the world to make online payments (Atik, Köse, Yılmaz, & Sağlam, 2015: 248). The first Bitcoin production was started in 2009. Bitcoin supply is not carried out from a center. It is carried out by the processing powers of the volunteer computers in the global network. An open-source miner software is run to include the Bitcoin network, and anyone involved in the network can generate Bitcoin as a miner. Those involved in the Bitcoin production process are called *"Bitcoin Miner."* Bitcoins are supplied by Bitcoin Mining Software, a process called mining based on the solution of a complex mathematical problem. This production information is transmitted to all individuals in the P2P (Peer to Peer to Peer to Peer) network. Bitcoin miners try to solve this problem by competing with each other.

Bitcoin miners who produce the first solution are given as a reward in the amount of Bitcoin produced automatically (Çarkacıoğlu, 2016: 13). After the solution of the problem, a more difficult problem than the previous software is presented to the miners to be solved.

On the other hand, the prize is reduced by about half every 4 years. This production acceleration following a logarithmic course is defined in such a way that a total of 21,000,000 Bitcoins can be produced. No one and no authority can supply money to the Bitcoin system. Bitcoins are also called crypto-money because of the crypto technique used in Bitcoin production and circulation (Atik et al., 2015: 249–250).

Bitcoin is one of the examples of the degree of abstraction that money has reached today. Bitcoin expresses all of the concepts, definitions, and issues that make up the digital money market. It is digital and not physically represented (Antonopoulos, 2014: 3–4). The costs of Bitcoin operations are meager and have a global usage network. For these reasons, its use is rapidly increasing. To be able to trade with Bitcoin, one of the wallet programs must be installed. A Bitcoin wallet is a virtual wallet, and this application enables Bitcoin trading, transfer, and storage (Hileman & Rauchs, 2017: 50–51). Bitcoin can be converted to USD, Euro, and all other regular coins when requested (Weusecoins, 2019). There is no leverage effect in Bitcoin transactions, and the purchase is a full ballot, which is not the balloon. The price is determined in the market conditions, and there are unpredictable profits and losses due to the volatility in the price. The existence of Bitcoin is entirely virtual because it is not a person or institution, it has no representation, and it does not belong to the central bank of any country. For these reasons, it is not influenced by any regional situation (Velde, 2013: 4–5).

The Bitcoin system is generally classified as six sub-systems:

(a) Mining companies: These companies provide computational power for the mathematical operations required to verify the security of the transactions and also serve as a mint for adding Bitcoins as a reward to the system. An important point to consider here is that at any given moment theoretically it is known how much Bitcoin is or will be in the market. Accordingly, the new Bitcoins are driven as a result of the mining process at a decreasing rate.
(b) Companies providing e-wallet services: E-wallet is the application that allows the person to store the personal keys required for his/her Bitcoins. The e-wallet can be found in many different formats. The critical point is that the data stored here is not the money itself but the data that allows validation of transactions and access to publicly available Bitcoin addresses. Wallets can be found in desktop, mobile, online, paper, and hardware.
(c) Financial service providers: The financial services provided in the classical sense are also provided as a result of the transactions made through Bitcoin. The sub-group of companies that provide services such as buying and selling of financial assets, forex trading, buying and selling of securities, buying and selling of stock, stock exchanges, buying and selling of options, and giving interest to Bitcoin for their investment.

(d) Exchange markets: Only Bitcoin, or, in some cases, all predetermined cryptocurrencies in the classical sense of all other currencies that have the task to exchange markets. These companies receive a commission as a result of the change process, and users have the opportunity to change the crypto money and the classic coins with each other at any time.
(e) Payment processors: these companies are companies that can make payments and receive payments to Bitcoin or other parties who want to trade with other cryptocurrencies. The companies provide their customers with an exchange of goods and services by using Bitcoin while minimizing the possible risk of transactions by offering online exchange points to corporate customers and by offering instant exchange services in the currency they want as a result of the sales transaction.
(f) Universal companies: these companies offer more than one of the aforementioned services in different variations. For example, a multi-purpose company also serves as an e-wallet service and as a payment processor (Gültekin & Bulut, 2016: 87–88).

The Bitcoin Wallet is defined by users in the Bitcoin ecosystem, expressing several tremendous and complex characters, such as an account number. When a new Bitcoin account is requested, a password is requested from the user by generating credentials with another open source software called Bitcoin Wallet. After this process, the Bitcoin account is ready for use. The fact that Bitcoin is spreading and all account holders can track transfer transactions and that it offers a secure and global payment network has led to the emergence of Bitcoin payment service companies. Although Bitcoin transfer does not require intermediary institutions, these companies provide services in order to enable safe trade. Nowadays, only the equipment developed for Bitcoin mining has been produced. Since the first production in 2009, the total amount had exceeded 13,750,000 Bitcoin as of January 2015. However, in the Mt.Gox (Mount Gox) 750,000 Bitcoins (about $500 million) were considered to be very difficult to break with the stealing of crypto money, and there is no longer a future for Bitcoin, as it explains the bankruptcy of Mt.Gox. (Atik et al., 2015: 250).

Today, there are hundreds of cryptocurrencies with traded market values and thousands of cryptocurrencies that exist at some point. The common element of these different encryption parasitic systems is the shared public book (blockchain) between the uses of natural markers as a way of encouraging collaborators to run the network in the absence of a central authority with network participants.

Cryptocurrencies, especially Bitcoin, are one of the examples of the degree of abstraction that money has reached today. Cryptocurrencies represent the full range of concepts, definitions, and issues that make up the digital money market. It is digital and not physically represented (Antonopoulos, 2014). The cost of crypto money transactions is meager and has a global network of uses. For these reasons, its use is rapidly increasing; many investors are trying to define how to invest in this new asset class. There are many things to consider when investing in cryptocurrencies with an increase in the popularity of cryptocurrencies. The steady increase in Bitcoin's trading volume and the volatility of its value against the USD have attracted attention in financial markets and enabled investment as an investment tool. Although there are many types of cryptocurrencies today, Bitcoin has been the most well-known and widely used cryptocurrency since its inception.

Cryptocurrencies developed after Bitcoin, which are alternative to Bitcoin, are called Altcoin, which means Alternative Coin. Hundreds of sub-circulates such as Anoncoin, Bitshares Ripple, Counterparty, Darkcoin, Dogecoin, Ethereum, Litecoin, Namecoin, and Nextcoin are circulated, and new subcoins are circulated every day. The world's open source subcoins, such as Bitcoin, are made, and profit and loss are generated according to demand and demand. In addition to the exchanges that enable Bitcoin swap with traditional currencies, there are only exchanges (Bitcoin Exchange) that mediate an exchange between Bitcoin and Altcoins. The value of the altcoins is based on Bitcoin. Therefore, the changes in the market value of Bitcoin directly affect the market value of Altcoins (Aslantaş, 2016: 360).

However, there are significant differences between some cryptocurrencies regarding the level of innovation displayed (Fig. 11.3). The vast majority of cryptocurrencies are mainly clones of Bitcoin or other cryptocurrencies and have different parameter values (e.g., different block time foreign exchange supply and regulatory scheme). These cryptocurrencies do not contain any novelty and are called "altcoins." Examples include Dogecoin and Ethereum Classic. Blockchain and cryptocurrency innovations' can be classified two categories: new (public) blockchain systems that feature their blockchain (e.g., Zcash, Peercoin, Ethereum), and dApps (decentralized applications) /other that existing additional layers built on top of existing blockchain systems (Fig. 11.4).

The first Bitcoin operation took place on 12 January 2009 between Nakamoto and Hal Finney, who contributed to the project. As a test, Nakamoto sent 10 BTC to Finney. Ten months later, on October 5, 2009, New Liberty Standard set the first

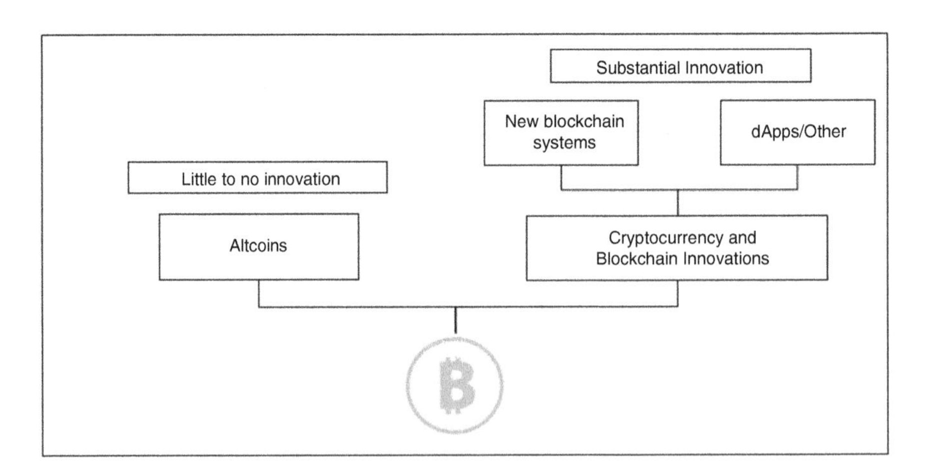

Fig. 11.3 The world of cryptocurrencies beyond Bitcoin (Source: Hileman & Rauchs, 2017)

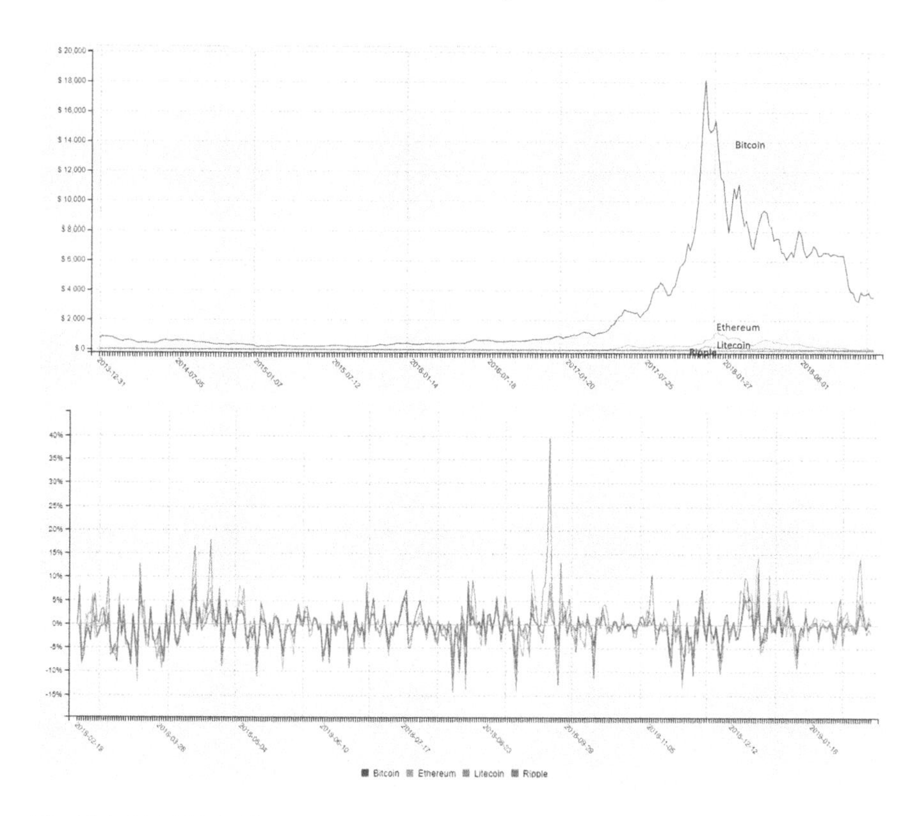

Fig. 11.4 31.12.2013–23.01.2019 date range Bitcoin, Ethereum, Litecoin and Ripple price movements (Source: Cryptocurrency Chart, 2019)

Bitcoin exchange rate as $12,300.03 BTC. Bitcoin's first transaction for physical goods took place on May 22, 2010. The process is a turning point for Bitcoin, but what is amazing today is that it can buy with the same amount of BTC and is often used as a reference point for the increase in the value of the cryptocurrency. On February 9, 2011, Bitcoin reached 1: 1 with the US dollar. After only 4 months, Bitcoin jumped from $1 to $31.91. In 2012, Bitcoin increased its value by 161.15% against the Dollar and the BTC price of 2012 ended at $13.58. In 2013, the value increased by 5290.86% against the Dollar and the BTC price of the year 2013 reached $731. In 2015, Bitcoin increased its value by 36% against the USD and by 122.03% against the US Dollar in 2016. In 2016, the average BTC price was $567.27. At the beginning of 2016, the price of 1 BTC was $432.32, and at the end of 2016, 1 BTC price was $959.87. Bitcoin gained 1319.79% in 2017 against the Dollar. In 2017, the average BTC price was $4001,16. At the beginning of 2017, the price of 1 BTC was $997.72, while at the end of 2017 1 BTC price was $14,165.57. Bitcoin depreciated by −72.55% against the US dollar in 2018. The value at the beginning of 2018 was $3791.5458, while the year-end value was $3791.5458. Bitcoin depreciated by about −6.75% against the Dollar in 2019 compared to the previous year. The value at the beginning of 2018 was $3752.2717 (Coin Market Cap, 2019).

Ethereum (ETH) is a smart contract platform that renders possible developers to build decentralized applications (dApps) conceptualized by Vitalik Buterin in 2013, a programmer working on Bitcoin in late 2013 to generate decentralized applications. Buterin argued that a different software language was needed for the development of Bitcoin. When the agreement could not be reached, he proposed to develop a platform with a more general software language. In January 2014, the announcement was made public by the core Ethereum team, Vitalik Buterin, Charles Hoskinson, Anthony Di Iorio, and Mihai Alisie. The official development of the Ethereum software project began in early 2014 through a Swiss company, Ethereum Switzerland GmbH (EthSuisse). The Ethereum Foundation (Stiftung Ethereum), a Swiss non-profit organization, was founded. An online public subsidized development during the period July–August 2014; participants purchased the Ethereum value coin (ether) with another digital currency, Bitcoin. Although Ethereum's early technical praise was praised, there were problems with security and scalability. ETH is the local currency for the Ethereum platform and also works as a processing fee for miners in the Ethereum network. Ethereum is a pioneer in smart contracts based on the blockchain. When working on the Blockchain, a smart contract is like a computer program that runs automatically after certain conditions are encountered. In the blockchain, smart contracts enable the code to be executed precisely as programmed, deprived of any interruption, censorship, fraud or third-party intervention. Money can facilitate the exchange of stocks, bonds, assets, content or anything of value (Coin Market Cap, 2019).

The price of Ethereum fluctuates wildly in short history. In July 2015, the price of an Ethereum token (Ether) was only $0.43. In the following years, Ethereum's price would see an increase of $1.422.47 in January 2018, before it fell below 80% after 9 months. This striking volatility has attracted global attention with its mainstream media, which publishes daily reports on Ether's price. The introduction was a great blessing for thousands of new developers and commercial enterprises. In 2018, the amount collected through Ethereum ICOs increased by more than $90 million in 2016 to $8 billion. Ethereum's price has increased over the years as it faces extreme volatility. After each explosion and bust cycle, Ethereum is on the other side with a stronger platform and a broader developer community that supports it. These key developments will bring a long-term positive outlook to the Ethereum price (Coin Market Cap, 2019).

Litecoin (LTC) is a peer-to-peer encrypted synchronization and open source software project that is available under the MIT/X11 license. The creation and transmission of coins are based on the open-source encryption protocol and is not governed by any central authority. Although Litecoin is technically the same as Bitcoin (BTC), it differs from Bitcoin and other cryptocurrencies by technical features such as Acceptance of Discrete Witnesses ile and ile Lightning Network Lit. These can effectively reduce the potential bottlenecks in Bitcoin, allowing the network to process more transactions within a certain period. The payment cost of Litecoin is almost zero and is paid about four times faster than Bitcoin. Litecoin was released on October 7, 2011, by former Google employee Charlie Lee through an open source client at GitHub. As a branch of the Bitcoin Core client; first, the block generation time was reduced, and then the maximum number of coins, the different hash algorithm (instead of scrypt, SHA-256) and a slightly modified GUI. In November 2013, the total value of Litecoin experienced massive growth of 100% in 24 h. Litecoin's value reached $1 billion in November 2013. As of May 9, 2017, the market value was approximately US $1,542,657,077, which was approximately US $30 per crypto money. In May 2017, Litecoin was the first crypto money to adopt the Detained Witness from the top five market-valued currencies. The first Lightning Network transaction was accomplished by transferring 0.00000001 LTC from Zurich to San Francisco in under a second in the same year (Litecoin, 2019).

Ripple XRP was launched in 2012 as a payment network (RippleNet) and also as crypto money. Unlike many public crypto money, Ripple is a coin that is entirely managed by a company whose blockchain-based notebooks are called mining. With the difference of finding a center, especially the big companies and banks that provide payment systems, they can find a contact for a common choice is a coin and network. Ripple has offices in San Francisco, New York, London, Singapore, Luxembourg Sydney, and India. American Express, Santander, UBS and Turkey's Akbank as a strong partner to more than 100 corporate and asset manages money transfers Ripple, blockchain technology offers the ability to process high-speed and at low cost on a global scale thanks. With Ripple, it is possible to send 4 s from one end to the other. The Ripple network can now perform up to 1500 transactions per second, and the company claims that the Ripple Network can handle the same number of transactions as Visa (50,000 per second). Ripple is the 4th largest digital currency after Bitcoin, Ethereum and Bitcoin Cash among the cryptocurrencies with a market share of $35 Billion. Ripple, which is traded on more than 50 crypto money exchanges globally, stands out as a highly accessible coin. It is not valid for Ripple, which is the point where centralized cryptocurrencies like Bitcoin are criticized most. Ripple differentiates itself from other cryptocurrencies by basing its value on a profit-making company. Currently working on 55 different processor networks, Ripple is aiming to develop its centerless structure by opening blockchain to third-party processors in the long term. In this way, Ripple itself is intended to adopt a structure that can not be manipulated on genius blocks and to increase confidence in its customers. The total number of Ripple produced is 99,993,093,880 XRP. However, for the moment, only 38,739,144,847 XRPs are in circulation in the market. The remaining coins are offered to the market at regular intervals by the company. This provides a different production strategy from mining, allowing the blockchain to run without the need for Bitcoin, Ethereum or Litecoin processing power (Ripple, 2019).

11.2.3 Cryptocurrencies as an Investment Tool

Investment means a permanent use of savings in order to generate income. The difference in consumption is that the source or value is not exhausted at the end of the process. Individuals consume a particular portion of their income in order to survive after their income. The remaining part of the consumed part is called savings. Many individuals want to evaluate their savings with investment. For individuals who want to invest their savings by investing, there are two options, capital investment, and financial investment. Capital Investments are generally investments in fixed assets. Financial Investment is to invest in financial instruments with specific maturities in order to provide a certain return. The investor should make some investment decisions before investing in financial investment instruments. First of all, it is necessary to answer questions such as which financial instruments to invest in, and how much risk they will take. With the globalization, the need for new financial instruments and financial transfer methods has emerged in order to make capital transfer faster and in different ways. Advances in information technology and widespread use of computers have led to the emergence of new financial instruments. Financial instruments contain a variety of risks, and therefore it is crucial to examine the nature of the financial instrument before deciding to invest. In the capital market, investors can invest directly in financial instruments such as government bonds, private sector bonds, and bonds, stock certificates, stock exchange traded funds, warrants, certificates, repo treasury bills, asset-based or asset-secured securities, lease certificates, futures, and options. They can also invest indirectly in capital market instruments by acquiring investment funds or in the private pension system (Investing, 2019).

Investors who do not like risk generally prefer to invest in fixed-income securities, such as government bonds and treasury bills, which are safer but lower. Liquid funds, repo and short-term investment instruments such as deposits are among the most preferred investment instruments in order to maintain the value of savings against inflation. Risk acceptant investors can achieve high rates of leverage by leveraging Forex, which is a popular type of investment in recent years. Investors with a high level of knowledge can use derivative products to hedge or increase their yields by taking a certain level of risk.

After the 2008 financial crisis, many financial firms and their customers agreed to the importance of asset allocation and the need to diversify their customer portfolios. This led portfolio managers to add alternative investments to customer asset allocation models. The most recent alternative investment on the stage is the cryptocurrency, and it seems to be as easy for US investors to buy a cryptocurrency exchange-traded fund (ETF) or some other stocks, traded in cryptocurrency before entering this alternative investment. For those who are not familiar with them, alternative investments are defined as non-correlated assets, which means that their performance does not match those of more traditional asset classes, such as stocks and bonds. As these assets move in the opposite direction of traditional investments, they can provide adequate protection against market declines. Even if the portfolio is looked at and anything is not seen directly that is known as an alternative investment, they can be many large corporate funds, such as ETFs or funds, as well as pensions and even pension fund proposals. Retail companies such as Morgan Stanley and Merrill Lynch have proposed allocation models for customers with alternatives that are close to or above 20% of the portfolio. Each customer is different, and the allocations will vary depending on their needs, but a current discussion with a financial advisor will probably include the subject of alternative investments in a portfolio. It is true that many people often associate with a hedge fund as the most common alternative investment and for many investors. However, most hedge funds can only be used by large investors and require considerable amounts of paperwork, high fees, and tax shortages. Many investors are achieving exposed to alternative investments through liquid alternatives such as mutual funds, ETFs and closed-end funds that provide daily liquidity, but have sophisticated investment strategies that seek to maintain their non-correlated status. Some financial advisors may consider that the inclusion of alternative investments is a prudent aspect of asset allocation for retirement accounts. A consultant may allocate five to 10% of the pension portfolio to this non-related investment class. If the alternative investment is wanted to be in a kind of cryptocurrency or related asset, this cryptocurrency market can be a risky investment until it matures. Investment performance will also vary depending on how cryptocurrency is invested in. An investment can be done directly in Bitcoin, Ethereum, or more than 2000 cryptocurrencies or companies with specialized equipment that specializes in the development of existing blockchain technology or specialized in mining cryptography (The Balance, 2019).

As an alternative investment option, it will not take long to see an ETF of companies that follow Blockchain technology. Some of the hedge funds' portfolios include Bitcoin and other crypto coins. If hedge funds are invested in crypto coins and considered as an alternative investment, then firms and media will start to publicize crypto coins as an alternative investment instrument. A risk-loving investor may want to invest in other cryptocurrencies other than Bitcoins. Bitcoin is not only a digital currency but also an investment tool. Some exchanges buy and sell most of these different cryptocurrencies, including the XRP used in blockchain projects such as Ripple Labs and ETH (Ethereum), which cause price movement every day. Due to the complex nature of the Blockchain technology (the underlying infrastructure of Bitcoin), many do not yet understand it and feel it is not very valuable, but financial companies such as Credit Suisse, Citi, Merrill Lynch, Bank of America and JPMorgan are conducting tests to improve their existing processes. Investments and financial firms' investments in Bitcoin are classified as alternative investments in cryptocurrencies and blockchain-based technologies, so it is time to take part in an appropriately assigned investment portfolio. In a few years, it is clear that there will be more opportunities to invest in them. As these investment opportunities are opened, they must be classified appropriately to be placed in the investor portfolios using appropriate asset allocation models (The Balance, 2019).

There are many studies about Bitcoin and cryptocurrencies in Turkish and foreign literature. Bitcoin, a cryptocurrency, is considered as a financial investment tool and some of the following are:

Sönmez (2014), the rise of Bitcoin, featured its place in the development of the economy in the world and Turkey and examined, researched the method of operation, has analyzed the current situation related to the new virtual currency. An innovative and successful technological design, Bitcoin, which is defined as a technological design, examines its weaknesses and strengths by examining it from different perspectives and discusses its opportunities and threats.

Pirinççi (2018) evaluated the historical development of digital money, the concept of virtual money and crypto money by reconsidering the definition, features, and functions of money. Bitcoin's historical background, its positive and negative aspects, and its market have been handled, and the particular server of Bitcoin cannot be found and therefore it cannot be intervened by the authorities and can provide investors with high earnings in the short term. It has even identified the risk of becoming a global crisis in the long run.

In a study conducted by Çalışır and Şanver (2018), they looked for answers to the question of how central banks will play a role in the implementation of monetary policy in the medium and long term. In the study, debates were discussed whether Bitcoin was a balloon or a commodity, currency or financial investment instrument in national and international platforms. Although the use of it as an alternative currency has been discussed, it is stated that cryptocurrencies are a challenge to the monopoly of money, which is one of the most influential forces of the nation-state. However, it is stated that the inability to fully penetrate payment systems and, more specifically, the appearance of the speculative asset tool in the short term will have an impact on the monetary policies and the weakening effect will be weak.

Elendner, Trimborn, Ong, and Lee (2017) investigated the value generating approaches of subcoins and their trade and information platforms. By investigating cryptocurrencies as alternative investment assets, they examined their returns and joint actions of the subcoin prices against Bitcoin and each other and evaluated their contributions to the investor portfolio. They evaluated the portfolios as one based on an equal weight, a value-weighted one and one based on the Cryptocurrency Index (CRIX). They determined that the CRIX portfolio has a lower risk than all liquid cryptocurrencies.

Liu (2019) examined the investment feasibility and diversity of the cryptographic currency as an alternative class of assets in the article. For this purpose, the cryptocurrencies, traded on the market, including Bitcoin, Dash, Ethereum, Litecoin, Monero, NEM, Ripple, Stellar, Tether, Verge and whose market value is more than 1 billion, were examined. The data set covers the periods from 07 August to 15 and 09-Apr-18 with 977 trading days in total. The study also shows whether the portfolio choice theory can benefit the cryptocurrency market. It has been found that the diversity between the cryptocurrencies can significantly increase the Sharpe rate and utility.

Troster, Tiwari, Shahbaz, and Macedo (2018) conducted a general GARCH and GAS analysis to model and estimate Bitcoin returns and risk. According to the results obtained from the econometric analysis carried out within the scope of the study, the strongly biased GAS models have the best fit for Bitcoin returns. According to the results of the study, GAS models provide the best protection for Bitcoin risk.

Symitsi and Chalvatzis (2019) using traditional performance measures to estimate the value-added of Bitcoin, it explored as a mainstream financial asset in the risky portfolios of various assets and, as a result of the econometric analysis, the Bitcoin portfolios, in most cases, did not require a statistically significant increase in variance or they have acquired.

Salisu, Isah, and Akanni (2019) examined the role of Bitcoin prices in G7 countries in estimating stock returns. For this purpose, the current forecasting models for stock returns taking country-specific and common factors into account are compared singly and commonly according to the Bitcoin-based forecasting model. According to the results of the econometric analysis conducted in the study, the predictive power of Bitcoin can be used during the modeling of stock returns especially in coincident periods with high volume Bitcoin operations.

Branvold, Molnar, Vagstad, and Valstad (2015) investigated the role of different stock exchanges in Bitcoin's price discovery process. The exchange of information on the stock exchange measures the actual price discovery rate. As the rate of price discovery is higher in the stock exchanges with high trading volume, the ratio between information sharing and share of activity is also examined. According to the results obtained from the econometric analysis carried out within the scope of the study, the market leaders with the highest share of knowledge during the sample period are Mt.Gox and BTC-e. The rest of the exchanges are relatively less informative.

Briere, Oosterlinck, and Szafarz (2015) examined the Bitcoin investment from a US investor perspective with a diversified portfolio, including both traditional assets and alternative investments. According to the results of the statistical analysis, it is obtained that Bitcoin investment provides significant diversification benefits. The inclusion of a small Bitcoin ratio can significantly increase the risk-return variation of well-diversified portfolios.

Wu and Pandey (2014) included Bitcoin in the portfolio of financial instruments such as major currencies, US equities, US bonds, and commodities in the world and examined the value of Bitcoin as an investment asset. According to the results of the statistical analysis carried out within the scope of the study, Bitcoins have the potential to increase the performance of an investor's portfolio. It can, therefore, be useful to keep Bitcoins as a component in a diversified investment portfolio.

Moore and Stephan (2016) aimed to provide an assessment of the potential benefits and costs of keeping Bitcoin as part of the international reserves portfolio using the Barbados example. According to the findings of the empirical analysis, as the ratio of the reserves held in Bitcoin increases, the variability of the reserves also increases. The transactions carried out by Barbados in the digital currency do not exceed 10% of all transactions in the short term. It is therefore recommended that Bitcoin should be relatively small if the Central Bank of Barbados is included in the foreign currency balances portfolio.

Lo and Wang (2014), in their work, analyzed how the various intermediary organizations emerged and developed in the Bitcoin network and the impact on the blockchain and the long run if the regulatory system did not improve the deficiencies of the digital system network due to the severe design errors and the mining cost of the Bitcoin network. They cannot be a permanent system.

Parker (2014) compared Bitcoin to electronic money and found that Bitcoin was an asset away from the banking system. He believed that Bitcoin would only create a generation of his financial services over time. Bitcoin's lack of legal clarity, such as electronic coins, was highly risky and therefore argued that cryptocurrencies should be included in the formal financial system.

Velde (2013) stated that people are betting with Bitcoin and that the reason is that they want to convert Bitcoin into a fully equipped money. He stated that Bitcoin, as such, is limited in its use as an exchange tool.

Gandal and Halaburda (2014) investigated the competition between cryptocurrencies and showed that the prices of some other cryptocurrencies increased more than the US dollar compared to Bitcoin. This leads to an increase in the demand for cryptocurrencies and the expansion of the market, rather than the popularity of Bitcoin.

Jonker (2018) stated in his study in the Netherlands that only 2% of the online payment method is in Bitcoin. He stated that currently those who have a cryptocurrencies account have the expectation of increasing the future value of money rather than using it for shopping and they have the purpose of investing.

Briere et al. (2015) stated that Bitcoin's return and volatility are high and that it is low about other known investment tools when it is considered as an investment tool, and therefore it would benefit in portfolio diversification to minimize risk.

Kristoufek (2015) examined the factors affecting the prices of Bitcoin and argued that the volume of use as a means of change is useful in the price of real money as the means of exchange is useful in price formation. Contrary to the general view, he stated that it was not a speculative formation.

Cheah and Fry (2015) pointed out that the extreme fluctuations in Bitcoin prices were remarkable because they were unstable due to their instability and that the price of balloons was in question.

Kanat and Öget (2018), the relationship between one of the cryptocurrency in working with Bitcoin prices, which stock indexes belonging to Turkey and the G7 countries examined using VECM Granger Causality/WALD testing, with long-term Bitcoin cannot be mentioned any relationship between other countries stock indexes in the short term with Bitcoin prices on the UK Stock Exchange (FTSE) index, S&P 500 and the Canadian Stock Exchange (CSE) index has been found to have a relationship between the results.

Baur, Hong, and Lee (2018) investigated whether Bitcoin was used as an alternative currency in the payment of goods and services or as an investment. According to the results of the statistical analysis done in the study, Bitcoin; it is not related to traditional asset classes such as stocks, bonds, and commodities. Analysis of the transaction data of Bitcoin accounts shows that Bitcoins are mainly used as a speculative investment and not used as an alternative currency and exchange tool.

Chu, Nadarajah, and Chan (2015) conducted a statistical analysis of the Bitcoin exchange rate against the US dollar using various parametric distributions known in the field of finance. According to the results of the statistical analysis conducted in the 2011–2014 period, Bitcoin investment yields very high returns with high fluctuations.

Karaağaç and Altınırmak (2018) investigated the effect of the prices of the crypto coins traded in many and various markets. In the study, total market values were taken into consideration in the selection of Bitcoin, Bitcoin Cash, Cardano, Ethereum, Litecoin, NEM, NEO, Ripple, Stellar and IOTA crypto coins and ten cryptocurrencies with the highest total market value were included in the analysis. Between December 15, 2017, and January 17, 2018, Johansen Cointegration Test and Granger Causality Test were applied to the series in order to examine the relationship between the daily price movements of the crypto coins. As a result of the study, Cardano is the reason for NEO Granger; Bitcoin is the reason for Bitcoin Cash's Granger, Litecoin is the Granger cause of Bitcoin Cash, NEM is the Bitcoin Cash Granger cause. It was revealed that NEO and Ethereum are Granger causes of each other, NEO and Litecoin are the Granger causes of each other and NEM is the Granger cause of Stellar and the price movements of these variables affect each other in the short term.

Sovbetov (2018), using the weekly data in his paper, examined the factors affecting the prices of the five most popular cryptocurrencies, such as Bitcoin, Dash, Ethereum, Litecoin and Monero during 2010–2018. ARDL technique was used in the study, and it was seen that cryptomarket-related factors such as market beta, transaction volume, and volatility were essential determinants for all five cryptocurrencies in both short and long term.

Dyhrberg (2016) analyzed the relationship between Bitcoin, gold and US dollar, and states that Bitcoin can be classified as something between gold and US dollars. Findings based on the original sample and extended sample time. It also shows that Bitcoin is very different from gold and other currencies. Bitcoin shows significantly different returns, volatility and correlation characteristics compared to other assets, including gold and US dollars. They found that Bitcoin had unique risk-return characteristics, followed a different volatility process compared to other assets, and had no relationship with other assets.

11.3 GM(1,1) Rolling Model

Gray System Theory was developed by Deng in 1982. By Deng, systems about which there is deficient information are defined as gray systems. In another saying, gray refers to cases characterized by uncertainty. Deng specifies that there must be one of four states with incomplete information to identify a system as gray (Lin, Chen, & Liu, 2004:197):

1. incompleteness of information about the parameters of a system,
2. incompleteness of information about the structure of a system,
3. incompleteness of information about the boundaries of a system, and/or
4. incompleteness of information about the behavior of variances in a system.

In Gray System Theory, GM (h,N) refers to a gray model. In a GM(h,N) model, the "GM" refers to Gray Model, the "h" in parentheses is the degree of the model and "N" is the number of variables in the model. Although there are several gray models used in the context of Gray System Theory, most of the empirical studies using the gray model prefer to use GM(1,1) Rolling Model because the model is simple to program and provides practical outputs.

GM(1,1) refers to a first-order gray model with one variable. The Model is used to explore relationships within time series, to model according to these relationships and to forecast using this model. The GM(1,1) Rolling Model can make effective forecasting by adapting new relevant data to the model. In order to use the GM(1,1) Rolling Model, the data to be used in the model must have a positive value and the same frequency. Also the forecasting with the GM(1,1) Rolling Model follows three main steps. These are, (i) Accumulated Generating Operation (AGO), (ii) Gray Modeling (GM) and (iii) Inverse Accumulative Generating Operation (IAGO). The GM(1,1) Rolling Model applies these steps in order to model and to forecast a system. These three steps can be shown with the following equations (Zhou & He, 2013:6235–6239):

Assume that time series of the price of Cryptocurrency denoted as P.

Step 1: AGO Process

Assuming $P^{(0)}$ is an original time series of the cryptocurrency;

$$P^{(0)} = \{P(1), P(2), \ldots, P(n)\} \tag{11.1}$$

where $P(k) \geq 0$, $k = 1, 2, \ldots, n$ and $n > 4$.

AGO process is done to reduce the randomness of the original series. The AGO process is applied to the original series to obtain the accumulated series. The GM (1,1) Rolling Model converts the original series to a series with an even increase. Thanks to this transformation created by AGO, the randomness in the original series is effectively reduced. This process can be shown as in Eq. (11.2):

$$P^{(0)} = \left[P^{(1)}(1), P^{(1)}(2), \ldots, P^{(1)}(n)\right] \tag{11.2}$$

where $P^{(1)}(k) = \sum_{i=1}^{k} P(i)$, $\widehat{P}(k) \geq 0$, $k = 1, 2, \ldots, n$ and $n > 4$.

Step 2: Gray Modeling

As the solution of first-order differential equations is in exponential form, new time series obtained from the AGO process are used to generate the first-order differential equation. The resulting equation is used to forecast the future behavior of the system. The first-order gray differential equation can be shown as in Eq. (11.3):

$$\frac{dP^{(1)}}{dt} + aP^{(1)} = b \tag{11.3}$$

Where b is the gray action quantity and a is the development coefficient.

The forecasting model for the gray system can be obtained by solving Eq. (11.4):

$$\widehat{P}^{(1)}(k+1) = \left[P(1) - \frac{b}{a}\right] \exp(-ak) + \frac{b}{a}, \quad k = 1, 2, \ldots, n \tag{11.4}$$

Where $\widehat{P}^{(1)}(k+1)$ represents an estimation of $P^{(1)}$.

The primary form of GM(1,1) Rolling Model corresponding to the unit t can be expressed as Eq. (11.5):

$$P(k) + \frac{a}{2}\left(P^{(1)}(k) + P^{(1)}(k-1)\right) = b \tag{11.5}$$

Optimal a and b are calculated by Least Squares Method.

$$\begin{bmatrix} \hat{a} \\ \hat{b} \end{bmatrix} = \left(Z^T Z\right)^{-1} Z^T Y \tag{11.6}$$

where

$$Y = \begin{bmatrix} P(2) \\ P(3) \\ \cdots \\ P(n) \end{bmatrix}, \quad Z = \begin{bmatrix} -\frac{1}{2}\left(P^{(1)}(2) + P^{(1)}(1)\right) & 1 \\ -\frac{1}{2}\left(P^{(1)}(3) + P^{(1)}(2)\right) & 1 \\ \cdots & \cdots \\ -\frac{1}{2}\left(P^{(1)}(n) + P^{(1)}(n-1)\right) & 1 \end{bmatrix} \tag{11.7}$$

Step 3: IAGO Process
Forecasted values for the original series are obtained by IAGO process. The forecasting of P can be calculated by using Eq. (11.8):

$$\widehat{P}^{(1)}(k+1) = \widehat{P}^{(1)}(k+1) - \widehat{P}^{(1)}(k) \quad k = 1, 2, \ldots, n \tag{11.8}$$

After the IAGO process, error analysis for gray predictions is performed using the Mean Absolute Percentage Error (MAPE) is shown in Eq. (11.9).

$$MAPE = e(k) = \left| \frac{P^{(0)}(k) - \widehat{P}^{(0)}(k)}{P^{(0)}(k)} \right| \times 100 \tag{11.9}$$

11.4 Application

11.4.1 The motivation of the Application

Although cryptocurrencies initially emerged as a transnational payment instrument, it has become an investment tool by attracting the attention of investors within the functioning of the capitalist system. As is known, most investors remain in a dilemma between the risk of risk aversion and the maximization of returns. Investors in this dilemma try to predict the future price or returns of the financial instruments through various analyzes and thus make an effort to give a path to their investments. These analyses are generally carried out by analyzing the past values of prices or returns by adopting a technical analysis approach. However, since the cryptocurrencies are a relatively new investment tool, it is not possible to reach the previous period price and yield information for an extended period. For this reason, in this application, GM(1,1) Rolling Model developed under the Gray System Theory which is recommended to be used in the short term forecasting is used. The price forecasting of popular cryptocurrencies which are Bitcoin (BTC), Ethereum (ETH), Litecoin (LTC), and Ripple (XRP) was calculated using the GM (1,1) Rolling Model, and it was tested whether this model is advisable for price forecasting of cryptocurrencies.

11.4.2 Data, Parameters, and Modeling

The data employed in this chapter includes only one variable which is the price of the cryptocurrency for each on a daily basis, taken from the Coin Desk (2019). The data are sectionalized into two parts named training dataset and forecasting dataset. The training dataset is used to determine the best GM(1,1) Rolling Model and to generalize k and alfa predictions.

The horizontal adjustment coefficient (α) and the length of subsequences (k) which are relevant factors in developing a successful model have been determined. The forecasting performance of the GM(1,1) Rolling Model is affected by α and k parameters. According to the theory of Gray Systems, these parameters are used as $\alpha = 0.5$ and $k = 4$. Besides, different values of these parameters may increase the forecasting performance of GM(1,1) Rolling Model. At this moment, the best value of these parameters is studied by the writers. This process is named as optimization of the Grey Rolling Model. α and k parameters are optimized by using Matlab R2018b. Table 11.2 shows the parameters of each optimized model.

In light of the parameters in Table 11.2, time series forecasting was made based on the technical analysis approach for each cryptocurrency.

11.5 Empirical Findings and Evaluation

Since the analysis by GM(1,1) Rolling Model is performed for four cryptocurrencies, the findings of the application should be evaluated separately for each. In Table 11.3, the results of the 10-day forecasts made with the GM(1,1) Rolling Models are shown according to Mean Absolute Percentage Error (MAPE) criteria.

Table 11.2 The parameters of the optimized GM(1,1) Rolling Models

		Cryptocurrencies			
		BTC	ETH	LTC	XRP
Parameters	Best α (The horizontal adjustment coefficient)	0.4	0.4	0.4	0.4
	Best k (The length of subsequences)	5	5	5	4
	Length of training data	749 days	1258 days	231 days	231 days
	Beginning of the training set	31.12.2016	9.8.2015	31.5.2018	31.5.2018
	Ending of the training set	25.1.2019	25.1.2019	25.1.2019	25.1.2019
	MAPE of the training process (%)	0.0252	0.0561	0.0724	0.0064
	Length of the forecasting period	10 days	10 days	10 days	10 days
	Beginning of forecasting period	26.1.2019	26.1.2019	26.1.2019	26.1.2019
	Ending of the forecasting period	4.2.2019	4.2.2019	4.2.2019	4.2.2019

Table 11.3 shows that the error percentages of forecasting of cryptocurrencies ranged from 1.35% to 7.76%. According to these results, it can be said that the model gives a high performance in terms of forecasted percentages for short-term forecasting.

Graph 11.1 shows daily and 10-days average errors. In Graph 11.1, daily errors are represented by APE (Absolute Percentage Error) and mean errors are represented by MAPE (Mean Absolute Percentage Error). As can be seen from Graph 11.1, APE values often have an error value equal to or lower than the MAPE values except for the LTC. This means that APE values are low, except for the large deviations over a few trading days.

However, it is not possible to make the same interpretation for 10-days direction forecasting, which is between 40% and 50% accurate. Although low deviations have obtained the results, the direction signals of the forecasting show that the signals cannot reach an acceptable success. In other words, despite the forecasting with low error percentages, if investors invest in these forecasting reliably, there is a high probability of losses in their portfolios. Therefore, investors may not have the opportunity to obtain a return above the average.

Also, Graph 11.2 support the comments about the findings of direction forecasting. When Graph 11.2 is analyzed, it can be seen that the actual directions (real values) and the forecasted directions (forecasted values) usually show different direction movements daily. Therefore, it can be stated that the forecasting results made by GM(1,1) Rolling Model is not reliable enough for investors.

Table 11.4 provides information about the yields that can be obtained in the 10-days investment period. When the data about the returns in the relevant investment period is analyzed from Table 11.4, it is seen that all of the investments made

Table 11.3 Forecasting result of the GM(1,1) Rolling Models

BTC				ETH			
Date	Real values	Forecasted values	MAPE (%)	Date	Real values	Forecasted values	MAPE (%)
26.1.2019	3563.28	3579.85	0.0047	26.1.2019	115.24	118.61	0.0292
27.1.2019	3539.62	3523.66	0.0045	27.1.2019	112.63	113.57	0.0083
28.1.2019	3425.26	3555.81	0.0381	28.1.2019	105.01	117.02	0.1144
29.1.2019	3395.02	3361.80	0.0098	29.1.2019	104.14	105.62	0.0143
30.1.2019	3441.03	3433.79	0.0021	30.1.2019	107.57	101.43	0.0571
31.1.2019	3420.63	3456.45	0.0105	31.1.2019	106.17	108.40	0.0209
1.2.2019	3439.81	3388.40	0.0149	1.2.2019	106.12	111.02	0.0461
2.2.2019	3433.04	3480.00	0.0137	2.2.2019	106.73	117.12	0.0974
3.2.2019	3404.50	3436.89	0.0095	3.2.2019	106.09	123.73	0.1662
4.2.2019	3412.23	3505.86	0.0274	4.2.2019	106.24	129.88	0.2225
Mean			0.0135	Mean			0.0776
Accuracy of direction forecasting			0.5000	Accuracy of direction forecasting			0.4000
LTC				**XRP**			
26.1.2019	32.96	31.27	0.0515	26.1.2019	0.31	0.32	0.0319
27.1.2019	32.34	30.30	0.0630	27.1.2019	0.31	0.32	0.0308
28.1.2019	30.58	30.69	0.0038	28.1.2019	0.29	0.31	0.0781
29.1.2019	30.60	27.28	0.1085	29.1.2019	0.29	0.32	0.1018
30.1.2019	31.58	28.90	0.0849	30.1.2019	0.31	0.31	0.0040
31.1.2019	31.26	30.54	0.0228	31.1.2019	0.31	0.30	0.0407
1.2.2019	32.70	30.15	0.0779	1.2.2019	0.30	0.29	0.0391
2.2.2019	32.82	31.27	0.0472	2.2.2019	0.30	0.30	0.0088
3.2.2019	33.05	31.01	0.0618	3.2.2019	0.30	0.29	0.0272
4.2.2019	33.62	32.71	0.0269	4.2.2019	0.30	0.30	0.0259
Mean			0.0548	Mean			0.0388
Accuracy of direction forecasting			0.4000	Accuracy of direction forecasting			0.5000

by relying on the forecasting results of GM(1,1) Rolling Models have a negative return despite being above the average.

As a result, the findings are important in terms of showing that not only price forecasting with low error percentages but also high accuracy direction forecasting should be made.

11.6 Conclusion

Although cryptocurrencies have been developed primarily as a means of payment, it has also become an investment instrument for portfolio investors and speculators to be interested in because of its rapid price increases and high return potential. In this

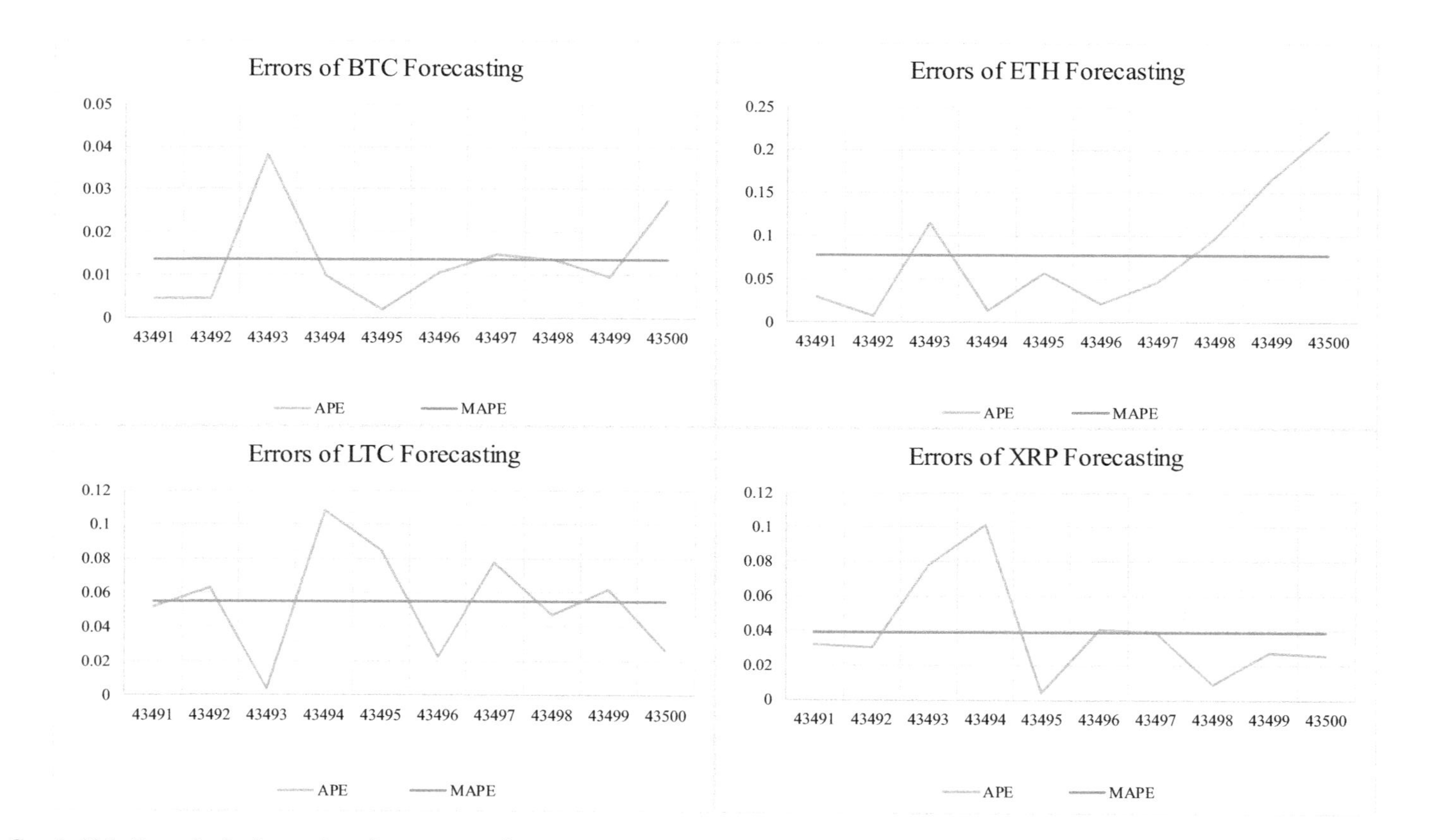

Graph 11.1 Errors in the forecasting of cryptocurrencies

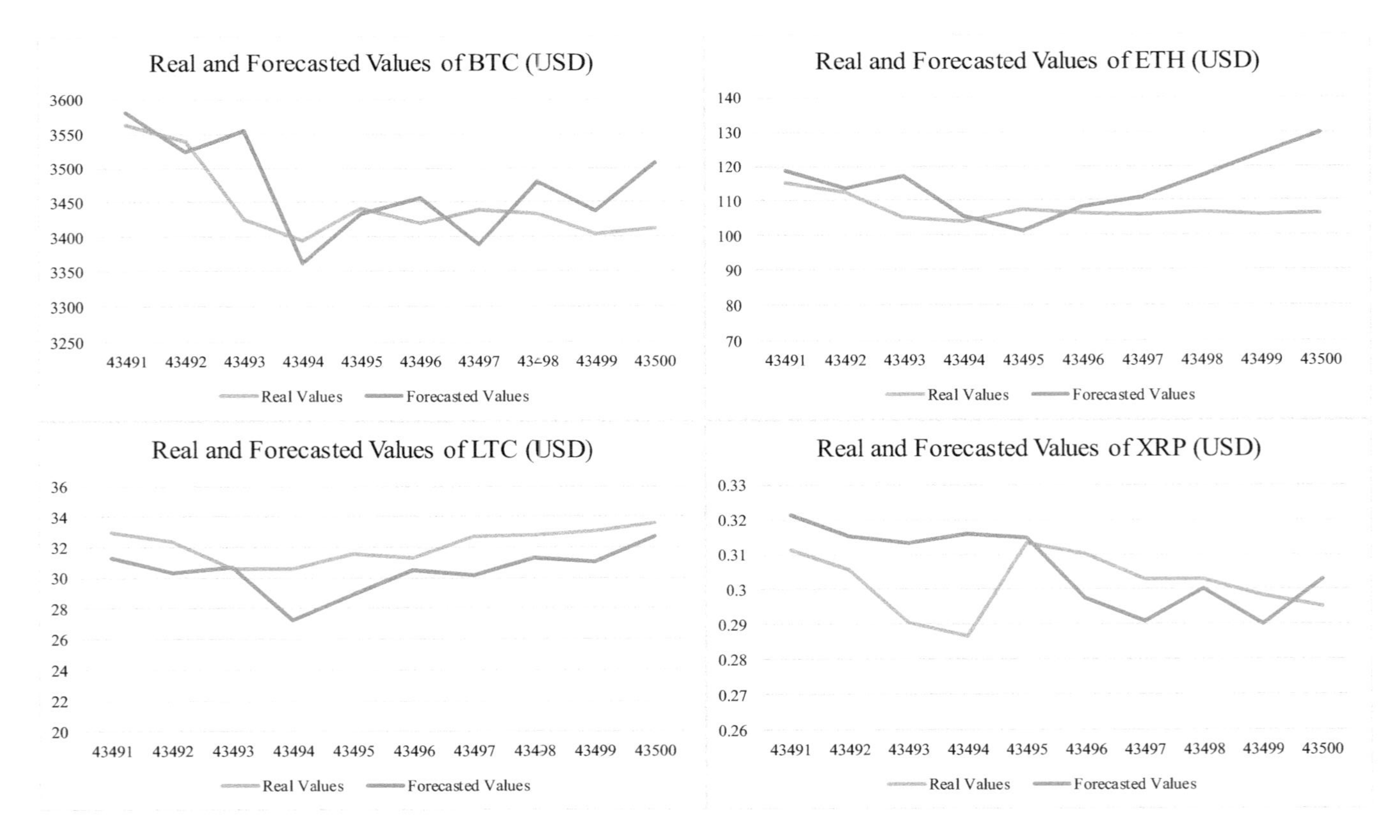

Graph 11.2 Real and forecasted values of cryptocurrencies

Table 11.4 Yields of the investment period

%	BTC	ETH	LTC	XRP
Potential maximum positive return of the investment period	2.14	3.98	12.15	8.93
Potential maximum negative return of the investment period	−6.41	−11.66	−8.35	−13.96
Average return of the investment period	−4.40	−8.15	2.78	−6.28
Average return of following the signals of the GM(1,1) Rolling Model	−2.63	−8.18	−0.60	−2.97

chapter, time series forecasting based on the technical analysis approach was made to forecast the prices of cryptocurrencies in the light of the Gray System Theory.

In the application, GM(1,1) Rolling Model developed under the Gray System Theory which is recommended to be used in the short-term time series forecasting is used. The price forecasting of popular cryptocurrencies which are Bitcoin (BTC), Ethereum (ETH), Litecoin (LTC), and Ripple (XRP) was made using the GM(1,1) Rolling Model, and it was tested whether this model is advisable for modeling of price movements cryptocurrencies.

Results of the Model show that the forecasting errors ranged from 1.35% to 7.76% for 10-days period. Also, direction forecasting results are between 40% and 50% in the same period. Also, returns of the bitcoin investment which made by trusting the results are ranged from −0.60% to −8.18.

The results may be considered that the model was successful in forecasting the prices but unsuccessful in the direction forecasting. Even though the estimates are made with low percentages, the time series analyzes made with the lagged data of Bitcoin prices are not successful. Therefore, the technical analysis approach can be interpreted as not sufficient for modeling Bitcoin prices. So, these results show that defining bitcoin price movements is not only a forecasting problem but also a classification problem.

Glossary

Bitcoin Bitcoin is a digital or virtual currency that uses peer-to-peer technology to facilitate instant payments.

Blockchain The blockchain is a technology used to read, store and verify transactions in a distributed database system.

Cryptocurrency A cryptocurrency (or crypto currency) is a digital asset designed to work as a medium of exchange that uses strong cryptography to secure financial transactions, control the creation of additional units, and verify the transfer of assets.

GM(1,1) Model GM(1,1) refers to a first-order gray model with one variable. The Model is used to explore relationships within time series, to model according to these relationships and to forecast using this model.

Gray System Theory A method which measures the degree of similarity between two systems.

Technical Analysis Technical analysis is a trading discipline employed to evaluate investments and identify trading opportunities by analyzing demographic trends gathered from trading activity, such as price movement and volume.

Time Series Analysis Time series analysis comprises methods for analyzing time series data in order to extract meaningful statistics and other characteristics of the data.

References

Antonopoulos, A. M. (2014). *Mastering bitcoin: Unlocking digital cryptocurrencies* (1st ed.). California: O'Reilly Media.

Aslantaş, A. B. (2016). Kripto Para Birimleri, Bitcoin ve Muhasebesi. *Çankırı Karatekin Üniversitesi Sosyal Bilimler Enstitüsü Dergisi, 7*(1), 349–366.

Atik, M., Köse, Y., Yılmaz, B., & Sağlam, F. (2015). Kripto Para: Bitcoin ve Döviz Kurları Üzerine Etkileri. *Bartın Üniversitesi İ.İ.B.F. Dergisi, 6*(11), 247–261.

Baur, D. G., Hong, K., & Lee, A. D. (2018). Bitcoin: Medium of exchange or speculative assets? *Journal of International Financial Markets, Institutions and Money, 54*, 177–189.

Bozic, N., Pujolle, G., & Secci, S. (2016). *A tutorial on blockchain and applications to secure network control-planes* (pp. 1–8). In: Proceedings of the IEEE 3rd smart cloud networks & systems.

Branvold, M., Molnar, P., Vagstad, K., & Valstad, O. (2015). Price discovery on bitcoin exchanges. *Journal of International Financial Markets, Institutions and Money, 36*, 18–35.

Brassard, G. (1988). *Modern cryptology: A tutorial*. New York: Springer.

Briere, M., Oosterlinck, K., & Szafarz, A. (2015). Virtual currency, tangible return: Portfolio diversification with bitcoin. *Journal of Asset Management, 16*(6), 365–373.

Çalışır, M., & Şanver, C. (2018). *Kripto Paralar ve Para & Maliye Politikalarına Muhtemel Yansımaları* (pp. 157–163). In: Proceedings of VII. IBANESS Conference.

Çarkacıoğlu, A. (2016). *Kripto-para Bitcoin*. Ankara: Sermaye Piyasası Kurulu.

Cheah, E. T., & Fry, J. (2015). Speculative bubbles in bitcoin markets? An empirical investigation into the fundamental value of bitcoin. *Economics Letters, 130*, 32–36.

Chu, J., Nadarajah, S., & Chan, S. (2015). Statistical analysis of the exchange rate of bitcoin. *PLoS One, 10*(7), 1–27.

Coin Desk. (2019). Accessed January 18, 2019, from https://www.coindesk.com/

Coin Market Cap. (2019). Accessed January 8, 2019, from https://coinmarketcap.com

Cryptocurrency Chart. (2019). Accessed January 28, 2019, from https://www.cryptocurrencychart.com/chart/BTC,ETH,XRP,LTC/priceChange/USD/linear/2018-02-19/2019-02-14

Dyhrberg, A. H. (2016). Bitcoin, gold and the dollar–A GARCH volatility analysis. *Finance Research Letters, 16*, 85–92.

Elendner, H., Trimborn, S., Ong, B., & Lee, T. M. (2017). The cross-section of crypto-currencies as financial assets: Investing in crypto-currencies beyond bitcoin. In D. L. K. Chen & R. Deng (Eds.), *Handbook of blockchain, digital finance, and inclusion* (Vol. 1, pp. 145–173). London: Academic.

FinTech Istanbul. (2019). Accessed January 4, 2019, from http://fintechistanbul.org

Gandal, N., & Halaburda, H. (2014). *Competition in the cryptocurrency market* (pp. 1–34). Bank of Canada.
Gültekin, Y., & Bulut, Y. (2016). Bitcoin Ekonomisi: Bitcoin Eko-Sisteminden Doğan Yeni Sektörler ve Analizi. *Adnan Menderes Üniversitesi, Sosyal Bilimler Enstitüsü Dergisi, 3*(3), 82–92.
Hileman, G., & Rauchs, M. (2017). *Global cryptocurrency benchmarking study*. Cambridge: The University of Cambridge.
Investing. (2019). Accessed January 5, 2019, from https://www.investing.com
Jonker, N. (2018). *What drives bitcoin adoption by retailers?* (Working Paper No. 585, pp. 1–35). De Nederlandsche Bank NV.
Kanat, E., & Öget, E. (2018). Bitcoin ile Türkiye ve G7 Ülke Borsaları Arasındaki Uzun ve Kısa Dönemli İlişkilerin İncelenmesi. *Finans Ekonomi ve Sosyal Araştırmalar Dergisi, 3*(3), 601–614.
Karaağaç, G. A., & Altınırmak, S. (2018). En Yüksek Piyasa Değerine Sahip On Kripto Paranın Birbirleriyle Etkileşimi. *Muhasebe ve Finansman Dergisi, 79*, 123–138.
Kristoufek, L. (2015). What are the main drivers of the bitcoin price? Evidence from wavelet coherence analysis. *PLoS One, 10*(4), 1–15.
Lin, Y., Chen, M., & Liu, S. (2004). Theory of grey systems: Capturing uncertainties of grey information. *The International Journal of Systems and Cybernetics, 33*(2), 196–218.
Litecoin. (2019). Accessed January 16, 2019, from https://litecoin.com
Liu, W. (2019). Portfolio diversification across cryptocurrencies. *Finance Research Letters, 29*, 200–205.
Lo, S., & Wang, J. C. (2014). Bitcoin as money? *Federal Reserve Bank of Boston, 14*(4), 1–28.
Moore, W., & Stephan, J. (2016). Should cryptocurrencies be included in the portfolio of international reserves held by the central banks? *Cogent Economics & Finance, 4*(1), 1–12.
Multichain. (2019). Accessed January 5, 2019, from http://www.multichain.com/blog/2016/03/blockchains-vs-centralized-databases/
NRI. (2016). Survey on blockchain technologies and related services FY2015 report (Technical Report 03 2016, pp. 1–78). Nomura Research Institute.
Parker, S. R. (2014). Bitcoin versus electronic money. *CGAP*, Accessed December 24, 2018, from https://www.cgap.org/research/publication/bitcoin-versus-electronic-money
Pirinççi, E. (2018). Yeni Dünya Düzeninde Sanal Para Bitcoin'in Değerlendirilmesi. *Batman Üniversitesi Uluslararası Ekonomi Politikaları Beşeri ve Sosyal Bilimler Dergisi, 1*(1), 45–52.
Ripple. (2019). Accessed January 16, 2019, from https://ripple.com/xrp/
Salisu, A. A., Isah, K., & Akanni, L. O. (2019). Improving the predictability of stock returns with bitcoin prices. *The North American Journal of Economics and Finance, 48*, 857–867.
Shahrokhi, M. (2008). E-finance: Status, innovations, resources and future challenges. *Managerial Finance, 34*(6), 365–398.
Sönmez, A. (2014). Sanal para bitcoin. *Türk Online Tasarım, Sanat ve İletişim Dergisi, 4*(2), 1–14.
Sovbetov, Y. (2018). Factors influencing cryptocurrency prices: Evidence from bitcoin, ethereum, dash, litcoin, and monero. *Journal of Economics and Financial Analysis, 2*(2), 1–27.
Swanson, T. (2015). *Consensus-as-a-service: A brief report on the emergence of permissioned, distributed ledger systems*. Accessed December 8, 2018, from https://www.ofnumbers.com/2015/04/06/consensus-as-a-service-a-brief-report-on-the-emergence-of-permissioned-distributed-ledger-systems/
Symitsi, E., & Chalvatzis, K. J. (2019). The economic value of bitcoin: A portfolio analysis of currencies, gold, oil and stocks. *Research in International Business and Finance, 48*, 97–110.
The Balance. (2019). Accessed January 10, 2019, from https://www.thebalance.com
Troster, V., Tiwari, A. K., Shahbaz, M., & Macedo, D. N. (2018, in press). Bitcoin returns and risk: A general GARCH and GAS analysis. Finance Research Letters, 1–13.
Velde, F. R. (2013). Bitcoin: A primer. *Chicago Fed Letter*. Accessed December 8, 2018, from file:///C:/Users/iibf021/Downloads/cfldecember2013–317-pdf.pdf

Weusecoins. (2019). Accessed January 5, 2019, from https://www.weusecoins.com/what-is-cryptocurrency/

Wu, C. Y., & Pandey, V. K. (2014). The value of bitcoin in enhancing the efficiency of an investor's portfolio. *Journal of Financial Planning, 27*(9), 44–52.

Zheng, Z., Xie, S., Dai, H., Chen, X., & Wang, H. (2017). *An overview of blockchain technology: Architecture, consensus, and future trends* (pp. 557–564). In: Proceedings of 6th IEEE International Congress on Big Data.

Zhou, W., & He, J. M. (2013). Generalized GM (1, 1) model and its application in forecasting of fuel production. *Applied Mathematical Modelling, 37*(9), 6234–6243.

Cem Kartal is an Assistant Professor of International Trade and Business at Zonguldak Bulent Ecevit University, Zonguldak, Turkey. Dr. Kartal has a BS in Civil Engineer from Kocaeli University (2006), an MBA from Marmara University (2008) and a Ph.D. in Accounting and Finance from Marmara University (2013). He has taught Corporate Finance, Business Finance, Capital Markets, Financial Statement Analysis, Working Capital Management, Derivatives Markets, Accounting Standards, Financial Risk Management, International Banking and Finance, Real Estate Taxation and Finance, Solvency.

Mehmet Fatih Bayramoglu is an Associate Professor of Finance at Zonguldak Bulent Ecevit University, Zonguldak, Turkey. Dr. Bayramoglu has a BS in Business Administration from Pamukkale University (2002), an MBA from Zonguldak Karaelmas University (2007) and a Ph. D. in Accounting and Finance from Marmara University (2012). His research interests lie in Capital Markets such as forecasting of stock prices, portfolio management, and investment strategies in financial crisis periods and data mining & decision-making methodologies such as ANNs, Grey Systems, VIKOR, TOPSIS, PROMETHEE, Decision Trees, OneR, and KStar. He has taught Capital Markets, Portfolio Management, Investment, Financial Modeling, and Financial Statement Analysis, among others, at both graduate and undergraduate levels. Dr. Bayramoglu completed his post-doctoral research at Lamar University College of Business, TX, USA.

Part III
Economic and Financial Assessment of Crypto-Currencies

Chapter 12
Is It Possible to Understand the Dynamics of Cryptocurrency Markets Using Econophysics? Crypto-Econophysics

Tolga Ulusoy and Mehmet Yunus Çelik

Abstract Closely related to the entire humanity, Finance, as a scientific field, seeks to meet humanity's endless needs and to continue its race against time. While doing so, it also benefits from other branches of science. Since speed, reliability, accessibility are at the forefront of model structures, finance continuously improves itself and tries to achieve the best interaction with other disciplines. Financial physics, also known as Econophysics, has brought new statistical methods and insights into the studies. Since thermodynamic laws, one of the most frequently used simulation systems, can explain the basics of all physical movements, the crypto money market, the stock market, and the dynamics of the foreign exchange market have been introduced. Thermodynamics describes heat movements; explain internal energy of economic systems, heat and jobs created (also called wealth or profits), and open a new page in quantitative/qualitative Economic Research. In this study, following the second law of thermodynamics, the Carnot cycle was written with a new point of view from the question of whether the amount of work given to the system in the crypto currency reserve can explain the possible trading (exchange) prices that occur or are likely to occur with the exchange of money.

This research was funded by Kastamonu University under the Scientific Research Project KU- BAP 01/2015-10.

T. Ulusoy (✉)
Faculty of Economics and Administrative Sciences, Department of Banking and Finance, Kastamonu University, Kuzeykent Campus, Kastamonu, Turkey
e-mail: tulusoy@kastamonu.edu.tr

M. Y. Çelik
Faculty of Economics and Administrative Sciences, Department of Economics, Kastamonu University, Kuzeykent Campus, Kastamonu, Turkey
e-mail: mycelik@kastamonu.edu.tr

© Springer Nature Switzerland AG 2019

U. Hacioglu (ed.), *Blockchain Economics and Financial Market Innovation*, Contributions to Economics, https://doi.org/10.1007/978-3-030-25275-5_12

12.1 Introduction

New theoretical approaches to predict prices may be proposed, by captivating formulation of the financial market in terms of statistical correlation to be given, where some simple (non-differential, non-fractal) expressions are also suggested as general price formula in a closed form that are able to generate a variety of possible price movements in time (Donmez, 2018). A given attributes of mechanics may be submitted as a plausible option to cover the price movements in terms of physical concepts and realization of growing of crypto portfolios is an asset.

One of the interesting field is econophysics combining Economics and Physics. There are many interesting and valuable open questions about market dynamics that interest researchers' econophysicists. Raising questions about how to measure and explain the important properties of crypto-market dynamics properly, about the stability of crypto-markets, and about what the differences are in the behavior of variants of markets (Donmez, 2018). Unsatisfied with the traditional explanations of economists, econophysics applied tools and methods from physics - first to try to match financial data sets, and then to explain more general economic phenomena (Sharma, Agrawal, Sharma, Bisen, & Sharma, 2011). For example players as agents interaction, econophysics often used Lagrange Equations and formations to upgrade theoretical framework. Some subsystems of huge-economic markets behave like rigid-body (Heukelom, Dopfer, Frantz, Mousavi, & Chen, 2016: 57). In that of systems mentioned as economic box is a closed system. These analyzable subsystems' interact with outer environment (thought like rigid-body properties) maybe mentioned derivative of kinetic energies written in classical Lagrange equations or Consensed Lagrange Equations in the future (Ivanov, 2018). The study of the dynamical behavior of crypto- markets using econophysics will be applied in latter researches.

Crypto money; the first application of crypto markets, is a kind of secure payment tool that is implemented by blockchain technology. Cryptography works with infrastructure, data management, network management and secure incentive elements, and all transactions on the internet to audit, execute and record the combined digital technology to support. This is a list of the files that are stored in the block, as well as a list of the process blocks. The parties who propose mutual action should register each information in the pool; open, but encrypted information is possible by verifying the posting. After the processing mechanism verifies the integrity and accuracy of the information, it saves the nodes in the blockchain collection by opening new blocks in the post. The replicated information obtained from geographical regions is stored without a single reliable third-party center with a security mechanism established in itself. The blockchain system provides accurate and quick information on the integrity of the blockchain notebook and its shared content at the end of all nodes.

12.2 Econophysics Literature

The term "Econophysics" was first proposed by Stanley (Mantegna & Stanley, 1999). Researches describe articles written by physicists in the context of capital and money markets. Then in Calcutta-India (1999) the first conference was held by the authors (Chakrabarti, 2005; Derman, 2004; Ghosh & Kozarević, 2018; Ghosh, Krishna, Shrikanth, Kozarević, & Pandey, 2018; Mandelbrot, 1963; Merton & Scholes, 1972; Oliveira & Stauffer, 1999; Silva, 2005; Levy, Levy, & Solomon, 1994; Sornette, 2004; Stanley et al., 1999; ul Haq, Usman, Bursa, & Özel, 2018; Zhou & Sornette, 2004). Easwaran et al. worked on a project of Bitcoin fluctuations. The exponent of the tail implies that Bitcoin fluctuations follow an inverse square law, in contrast to the inverse cubic law exhibited by most financial and commodities markets (Easwaran, Dixit, & Sinha, 2015).

Li et al. researched an article and it can be inferred that Bitcoin can be used as a hedge against market specific risk. Finally, Bitcoin bubbles would collapse due to the administrative intervention by economic authorities (Li, Tao, Su, & Lobonţ, 2019). Cocco et al. proposed a model which is able to reproduce some of the real statistical properties of the price absolute returns observed in the Bitcoin real market (Coco, Concas, & Marchesi, 2014; Sharma et al., 2011). Venegas wrote a investor guide for Etherium price prediction and used Econophysics dynamics (Venegas, 2017) (Table 12.1).

12.3 Blockchain and Crypto Money Market

What distinguishes the crypto money which was revealed by an unknown person or a group in 2009 from other financial instruments is that the crypto money has been on the internet since the first day of the system. When we look at the origins of the financial system, the opinion that money circulating in the financial system is based on money laundering is often expressed by critics. Crypto is a simple software code called money, or in other words a shopping protocol. The protocol has been translated into code and is designed to obtain approval from more than 15,000 computers around the world. In the case of a financial institution, the investor shall sell the financial instruments issued by a financial institution. Registers are kept by computers only. It is the state that determines the rules we call the "arbiter". Disputes related to the purchase are shaped around the rules set by the referee. In the Crypto world, all kinds of positive or negative situations starting from the sale are recorded on the system. In the system, hand-change of the asset that is sold on all computers is done in confidence by updating the records on the whole system and comparing the information on the existing system. It is not possible for the investor to remain passive before and after trading. The only situation here is that all unsolved problems and solved problems are resulted on a proactive system and can be done in a comparative way easily. When the swap is made (when any crypto is bought and

Table 12.1 Major studies on econophysics

Author (s) (Date)	Article	Research	Results
Bachelier (1900) and Boness (1964)	Theorie de la speculation. Annales Scientifiques de l'Ecole Normale Superieure, III-17. 21–86. Elements of a Theory of Stock-Option Value. Journal of Political Economy	Price fluctuations in Paris course	Brownian motion
Mandelbrot (1963) and Fama(1998)	Mandelbrot and the stable Paretian hypothesis. The Journal of Business. Market efficiency, long-term returns, and behavioral finance. Journal of financial economics	Fluctuation of prices and universalities in the context of scaling theories, etc	Open the way for the use of a physics approach in Finance. complementary to the widespread mathematical approach
Black and Scholes (1973)	The pricing of options and corporate liabilities. The Journal of Political Economy, 637–654.	Theory of option pricing	Black-sholes option pricing formula with a Nobel Prize. Uses in derivative markets
Bouchaud and Pottera (2000)	Theory of financial risks: From statistical physics to risk management, Cambridge University Press: Cambridge.	Determining financial risks by statistical physics	Theories for a belter overall control of financial risks
Mantegna and Stanley (1999)	Introduction to econophysics: correlations and complexity in finance. Cambridge University Press.	First uses of concepts from statistical physics in the description of financial systems	The term econophysics, concepts and the details of econophysics was met

Source: Ulusoy (2017)

sold) the system is kept and monitored in the books instantly as a result of which purchase is made. The approval phase is renewed by the entire system. This has made crypto money the fastest, safest, immediate, most cost-effective means of transferring transactions. It has created economic value for the communities.

Bitcoin is the most widely known currency in the world and the most speculative currency in the world. So much so that almost all crypto money transactions made on the block chain, which is identified with the bitcoin brand, are called "buy or sell Bitcoin". Then, brands such as Ripple and Etherium have the same popularity as Bitcoin.

The codes written in the initial stage of this market can not be interfered by anyone after they become financial markets. Looking at financial exchanges, the Real manager is accompanied by the algorithms written. Purchases are processed on a separate card (called a wallet), such as credit card systems. The records are created on the transaction books as mentioned previously. The advantage of this self-governing system against other systems is the inability to manipulate information, and the ability to verify personal data.

12.4 Why Econophysics Explaining the Crypto Markets?

All systems on earth can be explained by the physics infrastructure. When we look at system behavior, it is obvious that stochastic processes are exhibited, especially in atomic-sub-particle statistics. Econophysical studies related to the world capital market and money market tried to prove this. The attempt to explain the econophysical structure of money and capital markets shows that the crypto money markets that meet a current portion of these markets can also be explained by these methods. The world's crypto money portfolio behavior is very similar to the sub-atomic behavior in quantum physics. When collective decision is evaluated according to the degree of being affected by the decisions of the persons holding the similar portfolio, it is possible to say that the investor's portfolio behavior in the physical container shows physical-based behaviors.

Gas particles are not capable of remembering. It is a fact that investors have the ability to remember enough to change their behavior. The result is a fact that all physics theories can not be applied to the crypto money market. However, with a theoretical journey to the quantum world starting with Newton's mechanics and fundamentals, it is more meaningful to make predictions about what to trade with the prediction of the current crypto money trend direction. Thus, this kind of work is an opportunity for researchers to capture different perspectives that will not be found in other traditional studies written on crypto money pricing and trend analysis. Crypto will be able to assess the outlook on the money markets in the following way:

1. Starting from the Great Depression, all market players have a wish to predict the decline and rise.
2. Although the crypto currency markets are basically developed with similar trading techniques, there are situations that cannot be explained with traditional understanding.
3. Since modern financial instruments are emerging, traditional accounting techniques need to be reviewed.
4. In fact, in order to better assess the risk, new methods have to be tried to make sure that the trend in the crypto currency markets can be predicted which direction the trend may lead to.

It is possible to conclude the result proposal as follows:

> As with all other financial instruments, the possibility of making price forecasts in the next time t in the crypto currency markets is so low. The real answer is to find ways to predict how many percent of the crypto currency market will likely go up and how many percent will go down if there will be a fall and rise.

It is clear that prices cannot be fully predicted as many of the studies on the prediction of the crypto money markets are argued. The fact that the ups and downs in the crypto money markets follow a certain order should not be ignored in the studies to be carried out. Crypto money market with a holistic understanding of the "physical system" will be carried out with the research, the future of these markets, the evaluation of possible future situations will be easier.

The use of classical forecasting methods may be healthy for future movements after the event or effect that is required for boom and crash in the crypto currency markets is realized. On the other hand, since the crypto money market is a financial instrument that the entire financial system is looking at, crypto can create difficulties for estimating the behavior of those who hold high amounts of money in their hands. Classical methods are also made by looking at technical analysis and basic economic indicators and interpretations of the relevant analysis charts can be interpreted in an individual perspective. The problem starts here. When each individual analyses the future of the crypto money markets with the same methods, it increases the risk that each player will enter the same expectation. Investors using block chain technology are aware of each other and can see what are the good and bad decisions regarding investments. As with other markets, although the forecasts of prices are not clear in crypto currency markets, it is not possible for the traders to predict the future trend of fluctuations with a modern and accepted understanding.

It seems like a good way to try to explain the fluctuations in the crypto money markets based on the concept of Efficient Market Hypothesis (EMH) with a normal distribution. It is known that the oscillations in the crypto money markets show exponential distribution. At this point the situation; with the law Zipf's law and Pareto at one end on one end approaches demonstrate approaches to the concepts of statistical physics exponential probability be closer in nature to bring even closer.

The view that emerged in the stock market and the subsequent market studies, as stated below, appears to have existed in the crypto money markets.

> ...Although prices seem to be influenced by external influences, they are actually determined by internal dynamics of the market. Internal market events determine prices and external dynamics are shown only after events have occurred. The impression that prices are reacting to external influences affects all investors at the same time. What is important in a neutral environment is the willingness of investors not to be in the same energy situation as how they affect each other. This is a situation that can be evaluated within the statistical infrastructure of econophysics... (Ulusoy, 2008)

12.5 Newtonian Crypto Coins and Financial Entropy

As is known, there are four laws of thermodynamics, namely zero, first, second and third. It is known that the first two laws can only be applied in closed systems. The first law states that the energy in the whole universe is constant and that the energy cannot be destroyed. This is called the conservation of energy law. According to this law, energy can be obtained in more than one form. Mechanical energy, chemical energy, electrical energy and many other types of energy are covered. The conversion of energy types here to each other is a result of the first law of thermodynamics. Some types of energy can be converted directly to one another, while some types of energy cannot be converted to one, and they are lost during the transformation of energy. As can be explained in the next section, the lossless transformation of energy is called reversible process. Kelvin or Clausius the Carnot cycle is given here as an example with; its energy from heat energy in buildings of mathematical principles to what could be exchange related studies were found. In their study, they found that there is a sequence of different forms of energy and that there are some imbalances between energy transformations. These hierarchies and imbalances have laid the foundation for the next law of thermodynamics. The second law is based on the principle of irreversibility. In all energy systems, it has been proven that processes tend to shift from a low probability of realization to a high probability of realization with irreversible energy transformations. It is stated that the situation where energy differences are reduced and eliminated is the most appropriate situation. It can be turned into mechanical energy in the form of energy of heat energy. At this point, it should not always be reached that it is destroyed by the loss of energy in the system. If the first law of thermodynamics is true, it must be considered that the energy cannot be destroyed in the universe under this assumption.

Below is an example to explain this:

Let's assume that after the crypto money mining, the price difference in trading is a fortune. It is obvious that an investment in this way, which is thought to be physically, will generate energy. According to the first law of thermodynamics, energy cannot be destroyed. This means that some of the heat energy is lacking the ability to produce work. A specific crypto money price can jump at a higher price (energy level). This work (in physics $W = m.g.H$ is shown with the formula.) the energy required for the crypto currency is provided from the expectation of the rise in the price movement. This is called the kinetic energy (Finish). With this bounce, the expectation that it will rise further decreases. So the crypto portfolio is going to cool down. It is impossible to observe such an event in practice. In contrast to this process, it is possible that the price will move down from a certain level of energy (such as falling of a substance thrown into the air). When the price falls, the kinetic energy it gains can raise up to the same price again after it hits the price of its expected base. However, the price of the falling crypto money, after falling, may make a few small jumps and remains still. Agitation most of the energy is transformed into heat, and this energy is absorbed as its internal. In other words, like a physical being, the heat that the crypto absorbs during the price oscillation of money does not return to kinetic energy and it does not return to kinetic energy. The temperature of a substance increases as the temperature of the substance increases. Finally, a thermal equilibrium is established between crypto currency and other market instruments.

The above example can be found in the example of the oscillation of the Crypto money price, but it can only be simulated to show a similar behavior to the thermodynamic processes. On the other hand, there is an energy conversion in all natural processes. The direction of natural processes is determined by the direction of transformation of energy. The kinetic energy of an object is the sum of its kinetic energy and its potential energy. Other forms of energy such as mechanical energy in closed systems doesn't turn into heat energy. Heat energy flows from high temperature objects to low ones. This process is irreversible. So, without help from outside, low temperature liquid is heated to high temperatures would not be able to transfer it. Here, the measure used for energy falling quality "entropy" is called. For example, during the conversion of heat energy to mechanical energy in an internal combustion engine, involves some of this heat energy needed to take. The increase in energy that is not capable of producing work is measured by entropy in a way that cannot be recovered in the universe. The first to use this concept is Clausius (1850). In such a case, the concept of entropy in financial systems would be more appropriate to explain the event (Külahoglu, 2001).

As explained above, the concept of financial entropy is a concept of physics, which can be easily transferred to other disciplines and is widely accepted and used in many disciplines, from economics, philosophy, sociology, and business. Briefly, entropy and the second law of thermodynamics is accepted as a measure of the disorder is derived from. The entropy base entropy ceiling points calculated using statistical physics infrastructure provide additional information for the purchase and sale of online Analyses. In this study entropy is again an application of the second rule of thermodynamics. This rule, as stated in the previous sections, shows that all events will eventually come to a stable level. They have a lifetime of assets that are traded in financial markets, as in the case of people and other items. These lifetimes can be explained by the movements of the assets forward in the positive direction, or backward in the negative direction. When the concept of entropy is applied to the stock market, it is an indication that the movement of the financial instrument in one direction is over and that it can move the other direction. As in the case of physical systems, two acceptances are required to mention the existence of entropy in the crypto.

Axiom

1. *The individual trader trading in the currency market tries to be united with other investors around him and behaves in the direction that most of the market goes.*
2. *The individual investor tries to maximize his profit, not to minimize his losses at all costs.*

Two main systems can be mentioned in the systems where the crypto currency is traded. Microscopy crypto system (MiCS) and Macroscopic crypto system (MaCS). External events and effects can be explained under MaCS and internal events and effects can be explained under mics when the Crypto money market is considered as a physical container. If entropy is taken as a measure of irregularity in the monetary system, entropy can be seen as the basic and most important measure with temperature in mics. Entropy is such a physical concept in the internal system that separate entropy definitions are made in the disciplines formed by separate financial vessels. For example, thermodynamic entropy, financial entropy. Any function that increases as the irregularity of the crypto currency system increases can be an entropy function. For example, let's imagine that in the "Crypto cap" (traded market) there is any amount of investor/crypto currency/miner in the market and that we are eyeing a drop of new crypto currency and trying to imagine what is going on inside. The new crypto currency molecules will initially begin to stick together for a short period of time and then begin to spread into the existing mics. The investor/investment/ portfolio molecules that hit them (the investor who wants to trade) are scattering in different directions.

Let it be considered that all possible situations can be counted. When a system is said to have a state, we need to understand that a molecule has a specific coordinate and a specific speed, and another molecule has a specific coordinate and a specific speed. In the case of the supply molecules in the container (crypto money), it is obvious that the number of such situations is very high, but a large part of them corresponds to irregular, high entropical situations in which the requested crypto monetary molecules randomly disperse in all directions in the container. All of them are homogeneous. When we look at the mixture, it can be said that the money offered is distributed in the most probable form within the homogenous structure of the market, regardless of where the molecules are. In other words, an extraordinary number of different microscopic situations correspond to a single macroscopic state, a homogeneous state.

That's why every time we drop crypto money in a rough new supply, that's why it's falling apart. The excess number of microscopic States corresponding to a homogeneous macroscopic state increases its probability. Statistical physics laws say that the probability of a macroscopic state is proportional to the microscopic states that correspond to it. However, even though the possibility of existing monetary molecules re-forming a drop of supply, or perhaps a small drop of collection, is very close to zero, it is not zero. This is only possible in situations where molecules have very special speed and coordinates, and the number of these situations is almost negligible and unlikely to occur under efficient market conditions.

12.6 Carnot Cycle of Heat Machine in Physics

First of all, it is necessary to give information about what the Carnot cycle is. The most important feature of Carnot cycle is reversible. Carnot cycles are the highest productivity machines known in the literature. This is the most important feature of the system with the acceptance of reversible work or energy that is constantly in the system is not an escape from the system is accepted. Especially PV diagrams are used in Carnot cycle. Here, the specific volume specified by v, that is, in cubic meters can be an equivalent unit of kilograms and the diagram consists of isotherm lines. Th is generally named as high temperature line, TL is used as low temperature line. At high temperature, specific volume (v) increases, lowers the temperature of the system itself, and the specific volume decreases at low temperatures as observed from the Graph 12.1 that the temperature increases again. When isotherm lines are combined with expansion lines (T_h to T_L), circular lines compression line (T_L to T_h) emerges. At this point, it should be noted that the system is reversible. Going on an isothermal situation takes place between 1 and 2 for a single isotherm. Internal energy is the change of energy when isothermal $\delta u = 0$. Specific volume decreases with increasing pressure in between 1 and 2. There is expansion here. After this isothermal process, expansion between 2 and 3 continues and the so-called adiabatic process takes place. ($Q = 0$) in this section, pressure tends to fall in the specific

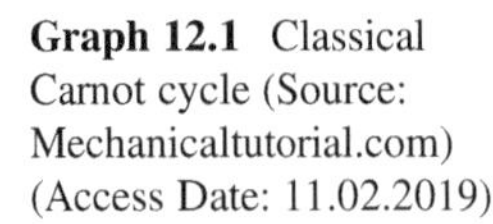
Graph 12.1 Classical Carnot cycle (Source: Mechanicaltutorial.com) (Access Date: 11.02.2019)

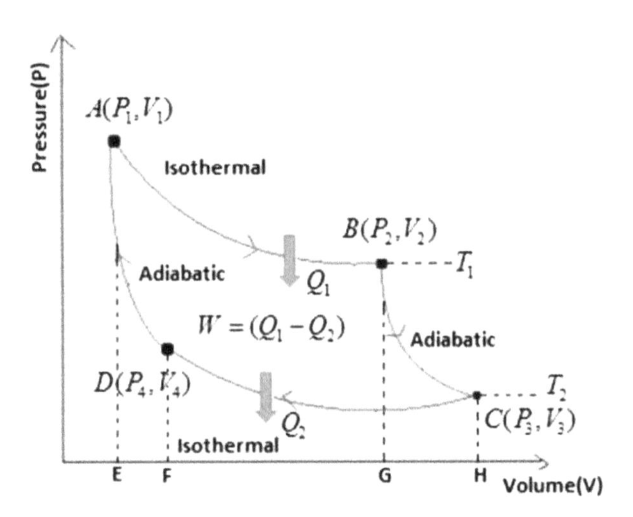

volume of the decrease is limited. Three to four isothermal single temperature on a single isotherm to the range compression is performed on a value increasingly our work is so ΔU = 0). Between 4 and 1, the cycle is completed and the adiabatic process takes place again and the specific volume reaches the lowest point and the pressure reaches the maximum point. As it is said at the beginning, Carnot cycles are reversed, so in all cases they have to return to their original state. At this point, the ideal gas formula PV = RT, which is the first law of thermodynamics Q − W = ΔU, so our work needs to think through.

When the efficiency of the closed system is considered, $\frac{W}{Q_h} = \frac{(Q_h - Q_L)}{Q_h} = \frac{Q_h}{Q_h} - \frac{Q_L}{Q_h}$ equality can be written. From here the yield is $1 - \frac{Q_L}{Q_h}$ Because the process is reversible and the system is closed, the proportional statements related to the business are equal to the proportional statement at the temperature. In this way, the efficiency (yield) can be written as 1 − (TL/Th). This makes it easier to find efficiency over temperature values if heat values are not known during the transition. The ΔU = 0 and q − w = 0 is the $q_h = w_{1-2}$ at this point. It can be said simply that there is a work arising from expansion here.

$\int_1^2 Pdv$ is due to volume change from the first state to the second state. The ideal gas law PV = RT is used to calculate the molar mass of a gas. If P = RT/V transformation is done $\int_1^2 \frac{RT_h}{v} dv$ is obtained. When basic simplifications formula witten as $RT_h . \ln\left(\frac{v_2}{v_1}\right)$.

12.7 Carnot Cycle and Crypto-Portfolios Interaction

The idea that the economic and thermodynamic processes of physical systems, such as markets and motors, may be identical, was first proposed by Mimkes (2004). In his research on thermodynamics formulation of the economy, the theoretical

Graph 12.2 Carnot cycle of basic *financial* heat machine. Mimkes (2004) described as:

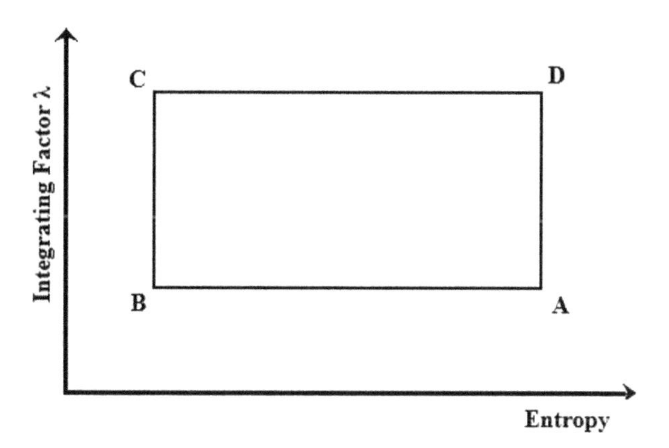

approach developed between the differential forms of Carnot cycle has shed light on the studies that came after him. Under the assumption that there are two hot and cold reservoirs, the amount of heat in the Carnot cycle goes to the low temperature reservoir, while the rest turns into work. From this point of view, it can be seen that the amount of work produced by the long and short positions in markets can be easily found by the difference between the heat provided and heat lost during the heat transfer.

The model is set up as described below with; v_i represents the trading volume in position i, m_i represents average market return of the currency in position i, δ represents the volatility of the market. Integrating factor $\lambda = \frac{m_i}{v_i}$ represents the market rate. Entropy $\frac{\delta_i}{v_i}$ gives with a standard deviation represents rf of risk-free rate or treasury bonds of market (Graph 12.2).

As seen in Graph 12.2 from point A → B; represents the cost of buying cryptocurrencies and point A → B with same mi vi values represents ***buying (long position)*** stage and points are represented as follows.

$$\left(\frac{\delta_1}{\sqrt{v_1}}, \frac{m_1}{v_1}\right) \rightarrow \left(\frac{\delta_2}{\sqrt{v_2}}, \frac{m_1}{v_1}\right)$$

From point B → C; at the same entropy level, this integrating factor increases $\frac{m1}{v1} \rightarrow \frac{m2}{v2}$ and in this ***holding*** stage she holds the cryptocurrency portfolio from time level t_1 to t_2. From point C → D; stage is the ***selling (short position)*** stage. Selling Income (SI) comes to portfolio and subtracting BC − SI = P is maximizing wealth. Buying and selling transactions, entropy increases from $\frac{\delta_2}{\sqrt{v_2}}$ to $\frac{\delta_1}{\sqrt{v_1}}$ with a high entropy rf of risk free rate also increases.

Crypto currency markets investments are started from $đϱ = (r_f, r_m)\, d_{rf} + b\, (r_f, r_m)\, d_{rm}$. It then goes to a general area formula of Carnot cycle

$$\int_{c}^{\Delta} dg_1 - \int_{A}^{B} dg_2 = \Delta P$$

It is expanded with an expressions to r_m and if as follows

$$\int_{C}^{D} f\left[\left(\frac{\delta_2}{\sqrt{v_2}}, \frac{m_2}{v_2}\right), \left(\frac{\delta_1}{\sqrt{v_1}}, \frac{m_2}{v_2}\right)\right] - \int_{A}^{B} f\left[\left(\frac{\delta_2}{\sqrt{v_2}}, \frac{m_1}{v_1}\right), \left(\frac{\delta_1}{\sqrt{v_1}}, \frac{m_1}{v_1}\right)\right] = \Delta P$$

With another point of view $đ_p = đ_w - đ_{BC}$ where w is total wealth and BC is the input or gain of portfolio given buying costs. Carnot cycling process of investment of cryptos written as;

$$-\int dBC = \int dP$$

$$-\int dBC = \int dP = \int_{C}^{D} dP_1 - \int_{A}^{B} dP_2 = SI\_BC = \Delta P$$

The profit change of ΔP is increasing when the investor increases the income from her wealth if we change the dP with an reverse integrating factor of 1/λ

$$dS = \frac{1}{\lambda} = dP \Rightarrow dP = \lambda dS$$

where d_S represents the entropy of a buy/sell crypto currencies at time t. Then the Carnot-cycle area of profit goes in terms of λ as follows

$$\int_{S}^{D} \lambda_2 dS - \int_{A}^{B} \lambda_1 dS = \int_{C}^{D} \frac{m_2}{v_2} dS - \int_{A}^{B} \frac{m_1}{v_1} dS = SI\_BC = \Delta P$$

We again explain the notation SI that calculated # of crypto currencies sold (n_2) with a selling price $\varepsilon_2 \Rightarrow SI = \varepsilon_2.n_2$

Buying cost-BC equals to ε_1 multiplied by n_1 where n_1 is # of crypto currencies bought with a price level ε_1. It then goes to an expansion of formula the rate or gain the entropy of buying/selling action (pressure of the market)

$$\begin{aligned} r_s &= \frac{SI - BC}{BC} = \frac{\varepsilon_2 n_2 - \varepsilon_1 n_1}{\varepsilon_1 n_1} = \frac{\lambda_2 - \lambda_1}{\lambda_1} \\ &= \frac{\frac{m_2}{v_2} - \frac{m_1}{v_1}}{\frac{m_1}{v_1}} = \frac{\frac{m_2 \cdot v_1 - m_1 \cdot v_2}{v_1 v_2}}{\frac{m_1}{v_1}} = \frac{m_2 \cdot v_1 - m_1 \cdot v_2}{v_1 v_2} \cdot \frac{v_1}{m_1} = \frac{m_2 \cdot v_1 - m_1 \cdot v_2}{v_1} \end{aligned}$$

An integrating factor 1/λ of market rate of return (r_m) and with a (temperature of the market) it then goes to;

$$r_\lambda = \frac{\frac{\delta_1}{\sqrt{v_1}} - \frac{\delta_2}{\sqrt{v_2}}}{\frac{\delta_2}{\sqrt{v_2}}} = \frac{\frac{\delta_1\sqrt{v_2} - \delta_2\sqrt{v_1}}{\sqrt{v_1}\cdot\sqrt{v_2}}}{\frac{\delta_2}{\sqrt{v_2}}} = \frac{\delta_1\sqrt{v_2} - \delta_2\sqrt{v_1}}{\sqrt{v_1}\cdot\delta_2}$$

12.8 Conclusion

Through the calculations above, we can give some information of adoption. From A → B investor collects the crypto-currencies from market and during this phase, a cost of investment is obtained from a buying price of ε_1 of n_1 amount of crypto money. The integrating factor 1/λ goes $\frac{m_1}{v_1}$. It may represent the market rate of return of cryptos.

$$\frac{m_1}{v_1} \rightarrow \frac{m_2}{v_2}$$

From B → C investors hold this crypto portfolio the integrating factor of 1/λ (heat) goes $\frac{m1}{v1} \rightarrow \frac{m2}{v2}$ at the same entropy. (Entropy changes are small enough to be neglected. It is about why the basis of graph to be given as smooth regtangle not a trapezoid) There is no change of distribution of crypto portfolio.

From C → D the crypto portfolio n_2 sold at price ε_2 and it can get $n_2.\ \varepsilon_2 = SI$.

From D → A explains the short position of crypto portfolio does not distribute again at time level until next A → B phase (buy) occurs. The area of representation of carnot cycle gives if the crypto currency investment is ***risky*** or ***unrisky***. Suppose there are two cycles with same area of P_1 and P_2. The small changes of entropy, if higher temperature occurs, it will be risky assets with a risky investment levels. If cycle is risky but profitable it is defined:

$$\frac{\delta_1}{\sqrt{v_1}} - \frac{\delta_2}{\sqrt{v_2}} < \frac{m_2}{v_2} - \frac{m_1}{v_1}$$

$$\frac{\delta_1\sqrt{v_2} - \delta_2\sqrt{v_1}}{\sqrt{v_1}\sqrt{v_2}} < \frac{m_2 v_1 - m_1 v_2}{v_1 v_2}$$

$$r\lambda < rs$$

$$r_f < r_m$$

$$r_m - r_f > 0 \Rightarrow \textit{Risk premium is an asset.}$$

The high temperature differences in low entropy changes and the risk parameters of crypto portfolios are increasing. High risk premium cryptography portfolios have high expectations of return.

References

Boness, A. J. (1964). Elements of a theory of stock-option value. *Journal of Political Economy, 72* (2), 163–175.

Chakrabarti, B. K. (2005). Econophys-Kolkata: A short story. In A. Chatterjee, S. Yarlagadda, & B. K. Chakrabarti (Eds.), *Econophysics of wealth distributions* (pp. 225–228). Milan: Springer.

Coco, L., Concas, G., & Marchesi, M. (2014). Using an artificial financial market for studying a cryptocurrency market. University of Cagliari, Italy Dipartimento Ingegneria Elettrica ed Elettronicaar Xiv:1406.6496v1 [q-fin.TR].

Derman, E. (2004). *My life as a quant – reflections on physics and finance*. Hoboken, NJ: Wiley.

Donmez, C. C. (2018). An econophysics approach to introduction uncertainty in dynamics of complex market structural models. In I. Nekrasova, O. Karnaukhova, & B. Christiansen (Eds.), *Fractal approaches for modeling financial assets and predicting crises* (pp. 1–22). Hershey, PA: IGI Global. https://doi.org/10.4018/978-1-5225-3767-0.ch001

Easwaran, S., Dixit, M., & Sinha, S. (2015). Bitcoin dynamics: The inverse square law of price fluctuations and other stylized facts. In F. Abergel, H. Aoyama, B. Chakrabarti, A. Chakraborti, & A. Ghosh (Eds.), *Econophysics and data driven modelling of market dynamics* (pp. 121–128). Cham: Springer. https://doi.org/10.1007/978-3-319-08473-2_4

Fama, E. F. (1998). Market efficiency, long-term returns, and behavioral finance. *Journal of Financial Economics, 49*(3), 283–306.

Ghosh, B., & Kozarević, E. (2018). Identifying explosive behavioral trace in the CNX Nifty Index: A quantum finance approach. *Investment Management and Financial Innovations, 15*(1), 208–223. https://doi.org/10.21511/imfi.15(1).2018.18

Ghosh, B., Krishna, M. C., Shrikanth, R., Kozarević, E., & Pandey, R. K. (2018). Predictability and herding of bourse volatility: An econophysics analogue. *Investment Management and Financial Innovations, 15*(2), 317–326. https://doi.org/10.21511/imfi.15(2).2018.28

Heukelom, F., Dopfer, K., Frantz, R., Mousavi, S., & Chen, S. H. (2016). *Routledge handbook of behavioral economics*. London: Taylor and Francis.

Ivanov, A. I. (2018). *Condensed lagrange equations*. XVIII International Scientific Conference, VSU'2018, Sofia, Bulgaria.

Külahoglu, T. (2001). *Termodinamik entropi ve iletişim teorisi*. Ankara: TMMOB Makine Mühendisleri Odası Yayınları.

Levy, M., Levy, H., & Solomon, S. (1994, May). A microscopic model of the stock market: Cycles, booms, and crashes. *Economics Letters, Elsevier, 45*(1), 103–111.

Li, Z. Z., Tao, R., Su, C. W., & Lobonţ, O. R. (2019). Does Bitcoin bubble burst? *Quality & Quantity, 53*(1), 91–105.

Mandelbrot, B. (1963). The variation of certain speculative prices. *Journal of Business, 36*(4), 394–419.

Mantegna, R. N., & Stanley, H. E. (1999). *Introduction to econophysics: Correlations and complexity in finance*. New York: Cambridge University Press.

Merton, R., & Scholes, B. (1972). The valuation of options contracts and a test of market efficiency. *Journal of Finance, 27*(2), 399–417.

Mimkes, J. (2004). A thermodynamic formulation of econophysics. In B. K. Chakrabarti, A. Chakraborti, & A. Chatterjee (Eds.), *Econophysics and sociophysics: Trends and perspectives*. Weinheim: Wiley.

Oliveira, S. M., & Stauffer, D. (1999). *Evolution, money, war, and computers – non-traditional applications of computational statistical physics*. Stuttgart-Leipzig: Teubner.

Sharma, B. G., Agrawal, S., Sharma, M., Bisen, D. P., & Sharma, R. (2011). Econophysics: A brief review of historical development, present status and future trends. arXiv preprint arXiv:1108.0977.

Silva, A. C. (2005). Applications of physics to finance and economics: Returns, trading activity and income. arXiv:physics/0507022v1 [physics.soc-ph]

Sornette, D. (2004). A complex system view of why stock markets crash. *New Thesis, 1*(1), 5–18.

Stanley, H. E., Amaral, L. A. N., Canning, D., Gopikrishnan, P., Lee, Y., & Liu, Y. (1999). Econophysics: Can physicists contribute to the science of economics? *Physica A: Statistical Mechanics and its Applications, 269*(1), 156–169.

ul Haq, M. A., Usman, R. M., Bursa, N., & Özel, G. (2018). McDonald power function distribution with theory and applications. *International Journal of Statistics and Economics, 19*(2).

Ulusoy, T. (2008). Ekonofizik ve finans: İMKB üzerine görgül bir çalışma/Econophysics and finance: An empirical study on ISE. Yayınlanmamış Doktora Tezi, Ankara Universitesi, PhD. Thesis, Ankara University.

Ulusoy, T. (2017). Price fluctuations in econophysics. In Ü. Hacioğlu & H. Dinçer (Eds.), *Global financial crisis and its ramifications on capital markets* (pp. 459–474). Cham: Springer.

Venegas, P. (2017). Ethereum price prediction: The value investor's guide initial coin offering (ICOS). In: *Blockchain trustless crypto markets*.

Zhou, W. X., & Sornette, D. (2004). Antibubble and prediction of China's stock market and real-estate. *Physica A, 337*(1–2), 243–268.

Tolga Ulusoy is an Associate Professor of Financial Management at Kastamonu University Department of Banking and Finance, Kastamonu, Turkey. Dr. Ulusoy has a BS in Mathematics from Hacettepe University (1998), an MBA from Baskent University (2001) and a Ph.D. in Management from Ankara University Faculty of Political Science (2008). His research interests lie in the methodological development of interdisciplinary techniques such as econophysics and financial simulation applications into fields including financial management, corporate finance, and capital markets. He has taught Financial Management, Corporate Finance, Future Market courses, among others, at both graduate and undergraduate levels. He has been an ad hoc reviewer for many journals. He is a member of the Finance Association.

Mehmet Yunus Çelik is an Assistant Professor in the Faculty of Economics and Administrative Sciences, Department of Economics at Kastamonu University in Kastamonu (Turkey). He received a Ph.D. in Economics at Dokuz Eylül University, Turkey. His research interests cover topics in the field of development, information society, and globalization, such as social and economic transformation. He is currently focusing on the new economy and new development approaches. He performed research projects at the national level. He took part in different international conferences and in several workshops.

Chapter 13
The Linkage Between Cryptocurrencies and Macro-Financial Parameters: A Data Mining Approach

Arzu Tay Bayramoğlu and Çağatay Başarır

Abstract Digital currencies have increased their effectiveness in recent years and have started to see significant demand in international markets. Bitcoin stands out from the other cryptocurrencies in considering the transaction volume and the rate of return. In this study, Bitcoin is estimated by using a decision tree method which is among the data mining methodology. The variables used in the decision tree created in the estimation of Bitcoin are the S&P 500 stock index, gold prices, oil prices, Euro/Dollar exchange rate, and FED Treasury bill interest rate. When the experimental results were examined, it was observed that the decision tree C4.5 algorithm was an appropriate method with the correct classification percentage of 73% in estimating the direction of Bitcoin. Also, the results obtained from the decision tree show that Bitcoin is related to S&P 500 index among macro-financial indicators similar to the results of econometric models used in the literature.

13.1 Introduction

Cryptocurrencies are an innovation which emerged with financial transactions becoming online in the last decade. Money, in its traditional definition, is required to serve functions of a medium of exchange, a unit of account which facilitates calculations, and a store of value. In order for money to serve these functions, it must be readily available to everyone, low cost, durable, exchangeable, movable, and reliable. Since precious metals such as gold, silver, and bronze satisfy the criteria mentioned above as well, they had been used instead of money for a certain period (Ferguson, 2008: 25–26). While the assets mentioned above or systems based on

A. T. Bayramoğlu (✉)
Faculty of Economics and Administrative Sciences, Department of Economics, Bulent Ecevit University, Zonguldak, Turkey
e-mail: arzutb@beun.edu.tr

Ç. Başarır
Faculty of Applied Sciences, Department of International Trade and Logistics, Bandırma Onyedi Eylul University, Bandırma, Turkey
e-mail: cbasarir@bandirma.edu.tr

© Springer Nature Switzerland AG 2019

U. Hacioglu (ed.), *Blockchain Economics and Financial Market Innovation*, Contributions to Economics, https://doi.org/10.1007/978-3-030-25275-5_13

these assets with the intrinsic value used to be used as a payment tool, replaced by paper banknotes and coins. Together with the rapid change in technology, money or payment systems have responded to this change as well and today, money has become completely digital (Koçoğlu, Çevik, & Tanrıöven, 2016: 79). These currencies referred to as virtual money, electronic money, digital money, etc. are usually used as cryptocurrencies. Cryptocurrencies are a subset of digital currencies (Antonopoulos, 2014).

Developments between 1998 and 2009 are considered as a milestone for cryptocurrencies. Although Bitcoin was the first encrypted currency introduced to the market, attempts at creating online currencies with encryption-protected books had started in 1999. B-Money and Bit Gold are examples of cryptocurrencies which were formulated, yet were never completely developed (Marr, 2017).

The most popular and successful cryptocurrency, Bitcoin has been receiving more recognition throughout the world (Chiu & Koeppl, 2017: 1). Cryptocurrencies such as Bitcoin are explained as a new digital currency system based on computer cryptography and built upon decentralized (peer-to-peer) network architecture (Li & Wang, 2017, 50). Although there are cryptocurrencies other than Bitcoin as well, it is the first, most widely known, and most preferred cryptocurrency. One of the most significant reasons behind this is suggested to be the financial crisis of 2008. One of the essential elements of money, trust, was lost during the financial crisis of 2008 due to problems experienced in the banking sector, which is in the center of the entire monetary system. Banks had created a centralized trust system and placed themselves at the heart of this system. Thus, no one, not even countries, could do business with each other without banks and countries had become economically dependent on each other. The collapse of this system with the financial crisis of 2008 affected the trust that people had in this system. Seizing this opportunity, Nakamoto (2008) introduced the concept of Bitcoin to international markets for the first time with a whitepaper released in 2008. Bitcoin is described as "a peer-to-peer version of electronic cash would allow online payments to be sent directly from one party to another financial institution." It is a decentralized monetary system and currency which is not controlled by any state, company, or authority (Bilir & Çay, 2016: 24). The emergence of Bitcoin as a response to financial policies attracted considerable attention. Widely used in online games in the initial stages, Bitcoin was later accepted by giants such as PayPal, Microsoft, Dell, and Expedia, which played a significant role in increasing recognition of Bitcoin (Dulupçu, Yiyit, & Genç, 2017: 2241).

The increasing use of cryptocurrencies and their recognition as investment tools show the importance of the relationship between virtual currencies and various financial variables. After providing information about widely-used cryptocurrencies' general properties and status in financial markets, the study examines the relationship between Bitcoin and certain macro-financial variables using the data mining approach. In this way, the study aims to reveal the interactions between macro and financial variables and cryptocurrencies, which have rapidly attained a place in the market and are used as both an investment tool and a speculation tool.

The organization of the rest of the chapter is as follows. The second section of the study describes crypto-money technology; the third section describes the interaction between crypto coins and financial markets. In the fourth section, brief information about general principles and algorithms of decision tree method is given. In the fifth section, the data set and the findings, and in the conclusion part, the overall evaluation of the study is given.

13.2 Cryptocurrency Technology

Although cryptocurrencies are referred to as money, they do not possess any of the properties attributed to money by society. Cryptocurrencies are intangible, cannot be printed by any government, and are not supported by any official institution, which is very important in terms of the trust. Finally, they are not made from a precious metal such as gold, which is accepted by everyone. The European Central Bank defines virtual money as "a digital representation of value that is neither issued by a central bank or public authority nor necessarily attached to a fiat currency, but is used by natural or legal persons as a means of exchange and can be transferred, stored or traded electronically" (EBA, 2014: 11). As is evident from the definition, cryptocurrencies do not reflect the properties of fiat currencies.

However, cryptocurrencies will change the banking and commerce methods entirely, and many people will take their place in the modern, integrated, digital, and global economy with this digital transformation. The concept of an intermediary will disappear thanks to the cryptocurrency technology; however, people who do not know each other will be able to maintain their business using the network infrastructure. This system, referred to as blockchain in the literature, can determine whether or not the target account is eligible in a money transfer. Thus, the task of mediating between people is assumed by computer technology. Cryptocurrency technology eliminates intermediaries and fees charged by intermediaries, which allows for reduced transaction costs (Vigna & Casey, 2017: 15–17).

The concept of blockchain consisting of data blocks is generally known as an open ledger where all transactions are recorded. Two elements stand out in the structure of these blocks. Firstly, peer-to-peer technology is consolidated by conventional key encryption, which makes the concept of blockchain a chain of blocks that allows for performing and recording Bitcoin transactions. Secondly, it is not possible to remove or change these blocks one they are formed, a rule protected by some strict codes. The algorithms and calculation infrastructure which allow for forming, joining, and using these blocks are referred to as blockchain technology (Zhao, Fan, & Yan, 2016: 1). The blockchain technology involves a decentralized network. As a measure against problem caused by any central error, each transaction is stored independently in end computers in this network structure. Since all data is stored in all blocks, deciphering a block does not mean that it is possible to reach other blocks as well. Therefore, it is described as a very safe system. Today, the essential infrastructure behind the use of cryptocurrencies is the blockchain technology (Antonopoulos, 2014: 330–331).

Cryptocurrencies are still new and continually fluctuating. Thus, they are not accepted by many businesses, and their use is mostly for experimental purposes. Some reasons why people might prefer cryptocurrencies instead of traditional currencies, particularly in payment systems, or in other words, advantages of using cryptocurrencies are as follows (Kılıç & Çütcü, 2018; Rogojanu & Badea, 2014):

- Since transactions take place in a digital environment, there is no need for a physical entity, which saves time.
- Transactions can be performed at any time and in any place, which provides flexibility.
- Transportation, storage, and security costs associated with transactions using traditional currencies do not apply to cryptocurrencies. Also, there are no bureaucratic procedures when issuing money.
- Since cryptocurrencies are valid anywhere in the world, they allow for eliminating costs incurred from currency trade. There is no need for currency exchange. Possessing a certain amount of cryptocurrency (Bitcoin) should be sufficient to carry out operations anywhere in the world.
- Since generating Bitcoin by mining requires much power, time, and hardware and the total supply is limited to 21 million, it can be said that Bitcoin has the same properties with gold, which is one of the most valuable precious metals.
- Commission fees charged by banks are eliminated since banks are not used as intermediaries for money transfer. Transfers can be performed directly between blocks. Transactions are usually free; however, there may be a small fee if a transaction is to be prioritized.
- The use of money may cause inflation, while the use of Bitcoin does not cause inflation. Because the total supply of Bitcoin is limited and limiting the money supply is regarded as an essential tool to fight against headline inflation.
- Cryptocurrencies allow people to make transactions rapidly.
- Since payment systems use virtual environments, advancements in computer technology lead to financial innovations.
- The system has an encryption mechanism, which provides transaction privacy. Users have an infinite number of digital identities that they can generate and use. Different digital identities can be used in different transactions. This enables users to protect their privacy. Money transfers can be performed thanks to these features safely.

Similar to all traditional currencies, cryptocurrencies have certain disadvantages as well. For example, in terms of system security, it is believed that the system may be hacked at any time. There are also regulatory issues. Since a central bank or a government do not back cryptocurrencies, there is no regulation to protect the users of the system in case of a problem. Also, Bitcoin was used for unethical purposes in its initial stages and is therefore considered politically, socially, and ethically problematic. Finally, since cryptocurrencies are not controlled by a central bank, it is not possible to intervene in these currencies via monetary policy, which is considered an economic risk (İçellioğlu & Öztürk, 2018; Polasik, Piotrowska, Wisniewski, Kotkowski, & Lightfoot, 2015).

Despite all these advantages and disadvantages, almost all cryptocurrencies, Bitcoin, in particular, are used for both speculation and investment purposes in financial markets. Although some economists suggest that cryptocurrencies cannot be used for hedging purposes due to their higher volatility compared to both precious metals and other currencies (Cheah & Fry, 2015; Tschorsch & Scheuermann, 2016; Yermack, 2013), some authors advocate that cryptocurrencies can be used for both speculation and hedging purposes in financial markets (Baek & Elbeck, 2014).

Despite the abovementioned disadvantages, cryptocurrencies are widely used by both retail and institutional investors. The remaining sections of the study briefly discuss the relationship between prominent cryptocurrencies and financial markets.

13.3 Cryptocurrencies and Financial Markets

About 10 years after Nobel laureate economist Milton Friedman's claim that an electronic currency which is not controlled by any state and allows for transfers between parties would undoubtedly emerge as a result of the proliferation of the internet and advancement of technology, Bitcoin, the first cryptocurrency, was introduced to the world with an e-mail sent in 2008 by an individual or a group identifying as Nakamoto (2008) (Koçoğlu et al., 2016: 78). After the introduction of Bitcoin, the cryptocurrency market started to improve very rapidly and was followed by almost all sections of society with great interest. In this stage, cryptocurrencies did not only play the role of a currency, but they also created a different framework for businesses. Cryptocurrency markets went beyond the role of digital money and started to be used as a speculative investment tool. Today, cryptocurrencies are regarded as financial assets for investment (Corbet, Meegan, Larkin, Lucey, & Yarovaya, 2018: 1). However, the high volatility of cryptocurrencies indicates that any investment in cryptocurrencies would be risky (Howell, Niessner, & Yermack, 2018; Katsiampa, 2017).

Figure 13.1 shows the data of cryptocurrencies with the highest market capitalization rate in the cryptocurrency market. As of 10 January 2019, there are 2096 cryptocurrencies which participants can invest in the cryptocurrency market. The market capitalization of these currencies is about USD 123 billion. As seen in Fig. 13.1, Bitcoin accounts for approximately 57% of the total market capitalization, followed by Ripple with 13% and Ethereum with 12%. Prominent cryptocurrencies, namely Bitcoin, Ripple, and Ethereum, seem to account for approximately 82% of the total market capitalization. The remaining 2093 cryptocurrencies account for about 18% of the market capitalization. As evident from the chart, Bitcoin has an undeniable weight among cryptocurrencies.

As can be seen in Fig. 13.1, there is a significant capital influx to cryptocurrencies. One of the most significant factors behind this capital influx is the expectation of high returns. When we examine the price change of the

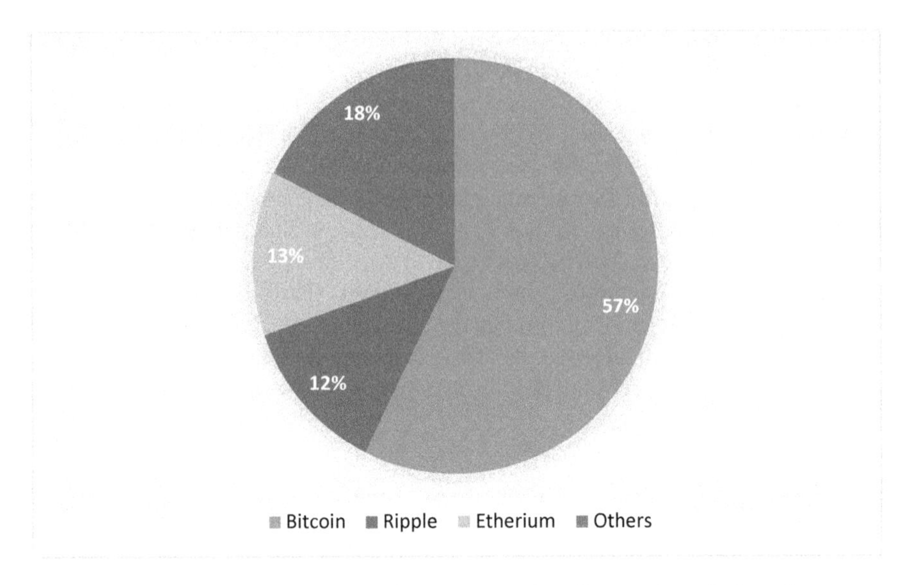

Fig. 13.1 Market capitalization rate of cryptocurrencies (Source: https://coinmarketcap.com [10.01.2019])

cryptocurrencies mentioned above, it is seen that the price of Bitcoin increased from USD 430 in January 2016 to 4030 in January 2019, the price of Ripple climbed from USD 0.006 to USD 0.33, and finally the price of Ethereum went from USD 0.93 to about USD 151 (Coin Market Cap, 2019).

This high return rate has attracted the attention of investors significantly. Therefore, these currencies have become an investment tool with the introduction of exchanges where investors can trade cryptocurrencies with the expectation of high returns. One of the critical points here is whether or not these currencies with high volatility could be used together with other investment tools for portfolio diversification. What needs to be done in this stage is to determine the relationship of cryptocurrencies with other financial investment tools.

13.4 Literature Review

Cryptocurrencies have seen significant demand from investors due to advancements in technology and expectations of high return and have been a subject of research for both investors and scholars. When studies on cryptocurrencies are examined, it is seen that these studies can be divided into three groups. A group of researchers deal with and examine the technological construct of cryptocurrencies (Li & Wang, 2017; Radziwill, 2018), while another group of investors primarily investigate the market structure, existence of a bubble in the market, and whether or not cryptocurrencies can be used for speculation and hedging purposes (Baek & Elbeck, 2014;

Blau, 2017; Bouoiyour & Selmi, 2014; Bouri, Molnár, Azzi, Roubaud, & Hagfors, 2017; Cheah & Fry, 2015; Cheung, Roca, & Su, 2015; Dulupçu et al., 2017; Dyhrberg, 2015; Glaser, Zimmermann, Haferkorn, Weber, & Siering, 2015; Katsiampa, 2017; Koçoğlu et al., 2016; Malhotra & Maloo, 2014; Urquhart, 2016; Yermack, 2013). Some other studies investigate relationships between cryptocurrencies, Bitcoin in particular, and other currencies (USD, EUR, JPY, etc.) as well as certain financial and macroeconomic variables (Atik, Köse, Yılmaz, & Sağlam, 2015; Baek & Elbeck, 2014; Baur et al., 2018; Carrick, 2016; Corbet et al., 2018; Çütcü & Kılıç, 2018; Dirican & Canoz, 2017; Georgoula, Pournarakis, Bilanakos, Sotiropoulos, & Giaglis, 2015; İçellioğlu & Öztürk, 2018; Kılıç & Çütcü, 2018; Vockathaler, 2015).

Widespread usage of cryptocurrencies led the questioning whether there is a bubbling movement in the cryptocurrency market. Many studies find bubble movement (Bouoiyour & Selmi, 2014; Bouri et al., 2017; Cheung et al., 2015; Glaser et al., 2015; Malhotra & Maloo, 2014). Furthermore, Baek and Elbeck (2014) state that the volatility of bitcoin is higher than the S&P 500 index and thus the market is highly speculative. Koçoğlu et al. (2016) also show speculative movements in the market. Similiarly, Cheah and Fry (2015) found that the fundamental value of Bitcoin is zero. Dyhrberg (2015) adds that Bitcoin can be used as a financial asset for hedging or minimize specific risk against the FTSE Index. The market structure of the cryptocurrencies is analyzed by Urquhart (2016). The study shows that the Bitcoin market is inefficient over the full sample but efficient over the subsample periods. The market is less liquid than other assets (Bouri et al., 2017).

Conversely, daily return for Bitcoin shows that speculative trading is not associated with the volatility of the bitcoin market (Blau, 2017). Also, Katsiampa (2017) concluded that volatility of the market is best explained by AR-CGARCH model. Table 13.1 shows a summary of studies in the literature which examine specific properties of cryptocurrency markets.

In Table 13.2, studies investigating the relationship between cryptocurrencies and macroeconomic variables and the usage of cryptocurrencies as an investment tool are summarized. Some studies emphasize that the value of the cryptocurrencies are related to the USD and the Euro, and the stock exchange indices of different countries (Dirican & Canoz, 2017; Georgoula et al., 2015) and some studies conclude that the value of the cryptocurrencies are independent of these factors (İçellioğlu and Öztürk, 2018). The volatility of the cryptocurrencies can result from shocks (Vockathaler, 2015) or market-related factors (Sovbetov, 2018). While some data shows that the value of the cryptocurrencies is connected (Corbet et al., 2018) other studies show that cryptocurrencies are a complement to fiat currencies (Carrick, 2016). The relationship between cryptocurrencies and domestic currencies are revealed in some studies (Atik et al., 2015), but other data concludes that cryptocurrencies act independently from exchange rates in short and long term (İçellioğlu and Öztürk, 2018).

Table 13.1 Cryptocurrencies market structure, bubbles, hedging or speculative

Author(s)	Data set	Method	Result
Baek and Elbeck (2014)	CPI, Indus. Prod., R.Per. Cons.Exp., 10 years treasury Euro Exc., U.R., S&P 500, Bitcoin prices daily return data (06.2010 to 02.2014)	Regression analysis and detrended ratios for volatility	–Bitcoin had more volatility than the S&P 500 –Bitcoin market is highly speculative –Bitcoin prices are affected only by itself
Malhotra and Maloo (2014)	Daily data of Bitcoin prices (13.09.2011 to 28.02.2013)	Sequential unit root tests	–Bubble movement in the market is found
Bouoiyour and Selmi (2014)	Daily data of the number of variables (economic, technical and financial) (05.12.2010 to 04.06.2014)	ARDL bounds testing	–Bitcoin market is highly speculative. Bitcoin market is not a safe haven
Glaser et al. (2015)	Daily data of Open, High, Low and Close prices as well as exchange volumes in BTC and Bitcoin network volume (01.01.2011 to 08.10.2013)	ARCH and GARCH regression	Keep Bitcoins for speculation, Bitcoin is used as an asset Bitcoin returns react on news
Cheah and Fry (2015)	Daily closing prices Bitcoin coindesk index (18.07.2010 to 17.07.2014)	Stochastic bubble model from January 1st, 2013–November 30th, 2013 Likelihood ratio tests over a moving time window	Bitcoin prices are prone to speculative bubbles The fundamental value of Bitcoin is zero
Dyhrberg (2015)	Bitcoin price index, FTSE index, dollar-euro exchange rate and the dollar-sterling exchange rate (19.07.2010 to 22.05.2015)	Asymmetric GARCH methodology	Bitcoin is used for hedging or minimizes specific risk against the FTSE Index like gold. It can be used as a financial asset with gold to reduce portfolio and market risks
Cheung et al. (2015)	Daily Bitcoin prices (17.07.2010 to 18.02.2014)	The bubble detecting method of Phillips, Shi, and Yu (2013)	Some short-lived bubbles and three huge bubbles found
Urquhart (2016)	Bitcoin price index (01.08.2010 to 3.08.2016)	Autocorrelation analysis, run tests, and variance ratio test	Bitcoin market is inefficient over the full sample but efficient over the subsample periods
Koçoğlu et al. (2016)	Bitfinex (USD), Bitstamp (USD), Mt.Gox (USD), Btce (USD), Okcoin (CNY), Kraken (EUR), Anx (JPY), Coinfloor (GBP) (02.06.2014 to 02.06.2015)	ILLIQ method, standard error analysis	Bitcoin market is highly volatile and Speculative

(continued)

Table 13.1 (continued)

Author(s)	Data set	Method	Result
Blau (2017)	Daily return for Bitcoin (17.07.2010 to 01.06.2014)	GARCH model	Speculative trading is not associated with the volatility of the bitcoin market
Katsiampa (2017)	Bitcoin coindesk index (18.07. 2010 to 01.02.2016)	AR-CGARCH	The best model explaining the conditional variance of the market is AR-CGARCH
Bouri et al. (2017)	Price index values for Bitcoin, the stock market indices for the US, the UK, Germany, Japan, and China respectively are the S&P 500, FTSE 100, DAX 30, Nikkei 225 and Shanghai A-share. (18.06.2011 to 22.12.2015)	Bivariate DCC model of Engle (2002), regression analysis	Bitcoin investments are far less liquid than conventional assets; Bitcoin prices showed high volatility

13.5 Data Mining Approach

Data mining is an interdisciplinary field of study, combining methods and algorithms from many different fields such as statistics, mathematics, and computer science, which aim to obtain unknown and useful information from large data sets. Data mining methods find applications in many different areas. Business and industrial areas are the leading areas. The Data mining analysis consists of several steps as follows: Business Understanding, Data Understanding, Data Preparation, Modelling, Evaluation, and Deployment.

Data mining functions are divided into predictive and descriptive data mining. In the prediction models, it is aimed to develop a model based on the known data and to estimate the result values for the unknown datasets by using this model. In descriptive data mining models, in contrast to the predictive model, patterns in the existing data used to guide decision-makers are defined. With one data mining model, one or more of the following operations can be performed: Classification and Regression models, Clustering models, Association Rules and Sequential Patterns models. Classification and regression models are predictive, clustering, association rules, and sequential pattern models are descriptive. In Classification models, the process involves two steps. In the second step, the class is estimated by applying this class of data on unspecified data. Primary classification techniques are Artificial Neural Networks, Genetic Algorithms, K–Nearest Neighbour, BMemory Based Reasoning, Naive—Bayes, Logistic Regression and Decision Trees (Albayrak & Koltan Yılmaz, 2009: 36).

Table 13.2 Relationship between the cryptocurrencies and financial, macroeconomic variables and exchange rates

Author (s)	Data set	Method	Result
Vockathaler (2015)	Value of Bitcoin, The financial stress index, S&P 500 index, Gold Price (XAU) and SSE (Shanghai Stock Exchange) index (19.08.2010 to 27.05.2015)	GARCH model	The majority of the volatility of Bitcoins came from unexpected shocks. These unexpected shocks are by far the largest contributor to the price fluctuations of Bitcoins
Georgoula et al. (2015)	Daily data between Bitcoin with 11 variables, including basic economic variables, technological factors, Twitter posts, Wikipedia and Google Bitcoin, search queries (27.10.2014 to 12.01.2015)	Time-series (vector error correction model) and sentiment analysis (Support vector machines)	Interest (Twitter posts, Wikipedia and Google Bitcoin search queries has a positive relationship) in Bitcoin was found to be effective in increasing market prices. While there is a negative relationship between the value of Bitcoin between the USD and the Euro and the S&P 500 index
Atik et al. (2015)	Daily price of Bitcoin EUR, GBP, JPY, CAD, AUD, CHF in dollars (01.06.2009 to 01.02.2015)	Johansen cointegration test and Granger causality analysis	One-way causality relationship was found between Bitcoin and Japanese Yen. Moreover, there is no causal relationship with other currencies could be determined
Carrick (2016)	Bitcoin and six major market currencies and 23 emerging market currencies prices (01.01.2011 to 31.12.2015)	Correlation, Sharpe and Sortino ratios	Bitcoin is in harmony with all currencies, especially in developing country currencies. Bitcoin can act as a complement to fiat currencies
Dirican and Canoz (2017)	Bitcoin prices and DOW30, NASDAQ100, FTSE100, NIKKEI225, and CHINAA50 index (24.05.2013 to 05.11.2017)	ARDL bound test	There is a cointegration relationship between bitcoin and the stock exchange indices
Corbet et al. (2018)	Bitcoin Ripple Lite VIX Bond Gold FX SP500 GSCI (Daily data between 2013 and 2016)	GARCH	Value of the cryptocurrencies is connected
Kılıç and Çütcü (2018)	Daily data between 02.02.2012 and 06.03.2018	Engle-Granger and Gregory-Hansen cointegration test	There is no cointegration relationship between data in the middle and

(continued)

Table 13.2 (continued)

Author (s)	Data set	Method	Result
	Bitcoin and BIST 100 logarithmic Prices	Toda-Yamamoto and Hacker-Hatemi casualty test	long term. According to the causality analysis, only the Toda-Yamamoto causality test fixed a one-way causality from the Borsa Istanbul to Bitcoin prices
İçellioğlu and Öztürk (2018)	Bitcoin, USD, Euro, Yen, Pound, and Yuan (29.04.2013 to 22.09.2017)	Engel-Granger and Johansen cointegration analysis	Bitcoin acts independently from exchange rates in the short and long term. There is no short and long term relationship between Bitcoin, USD, Euro, Yen, and Yuan
Çütcü and Kılıç (2018)	Weekly data between 24.11.2013 and 04.03. 2018 period, Bitcoin Price and USD Prices logarithmic	Maki cointegration test Hacker-Hatemi-J Bootstrap causality test	There is a long term relationship with the structural breaks between the variables. In the results of Hacker-Hatemi-J Bootstrap Causality test, causality relation was determined at 1% significance level from Bitcoin prices to dollar exchange rates
Sovbetov (2018)	Five cryptocurrencies (Bitcoin, Ethereum, Dash, Litecoin, and Monero) and S&P500 Index (Daily data between 2010 and2018)	ARDL method	Cryptocurrencies are affected by market-related factors (market beta, trading volume, and volatility). Attractiveness of Cryptocurrencies are subjected to the time factor. SP500 index have weak positive long-run impact on Bitcoin, Ethereum, and Litecoin, and negative and losing the effect in short-run
Baur et al. (2018)	Prices of Bitcoin, Gold, and USD (19.07.2010 to 14.07.2017)	GARCH model	Bitcoin is in a different state of volatility compared to other assets. It also has different risk-return characteristics. There is no relationship between Bitcoin with gold and USD

13.5.1 Decision Trees

The decision tree method is used as a convenient method for classification and estimation problem due to it is simple and easy to understand the structure. The decision tree method is a classification method created by conditional probabilities and can be presented as a tree in the form of outputs (results, costs, events, etc.). When the output variable used in decision trees is categorical, it is called the classification tree, and when the output variable is a continuous variable, it is called the regression tree. The estimation analysis performed with decision trees is visualized richly. This allows decision trees to produce outputs that are easy to evaluate, even by non-experts.

The decision tree chart contains the root node, branches, and leaves. The leaves are places where the classification occurs; the branches contain the results of each event. In the formation of the classification rules, the paths from the root node to the leaf nodes are considered (Geetha & Nasira, 2014).

The decision tree is a directional tree consisting of a root node with no input and internal nodes, each of which takes a single input. The nodes received as input by another node are called internal, or test nodes and nodes that do not output to another node are called leaf nodes. Each internal node in the decision tree divides the sample space into two or more parts based on subjecting the input attribute values to a specific function (Aytuğ, 2015: 11). The internal nodes of the decision tree represent the tests performed on the attributes; the branches test results and each leaf node class label.

It consists of decision trees, building the tree and cutting the tree processes. In the process of creating the tree, the root node is decided first. All the observations in the training set are assigned to this root node. Then, with the help of the iterative process, new nodes are created, and as a result, a fully developed tree emerges. When creating nodes, it is intended to maximize the calculated Gini index and thus reach the optimal solution (Ma, 2013: 160).

Various factors are competent in using decision trees in classification. The first of these is that the decision trees are in a simple structure. Thus, the classification model created can be easily understood. Other features, such as the fact that decision trees are not parametric, provide an appropriate structure for the discovery of information, and are partly faster than other classification methods, have also extended the use of decision trees (Aytuğ, 2015: 11). Also, obtaining rules from decision trees can also be carried out very quickly. Decision trees can be used to classify both categorical and numerical data. Despite the mentioned superior features, decision trees are faced with problems such that they do not enable output with multiple attributes, produce partially variable results, are sensitive to even small changes in test data, and form a complicated tree structure for numerical datasets (Zhao & Zhang, 2008). Since data mining methods face extreme harmony, the tree must be pruned. Complexity parameter is calculated for sub-trees, and high complexity sub-trees are truncated (Ma, 2013: 160).

The decision tree method creates a tree-like graph and provides a visual evaluation. This is a significant advantage of the decision tree method. It also has advantages such as being ideal for Business Intelligence, being used as a substitute for traditional statistical methods, and supporting a qualitative and quantitative data set (Geetha & Nasira, 2014). Some of the disadvantages of decision trees can be listed as follows (Dahan, Cohen, Rokach, & Maimon, 2014: 8). The healthy functioning of decision trees is compromised in cases where too many interrelated variables are used in the input set. When used as a large number of variable inputs, the size of the tree becomes more complex, making it difficult to examine. In many applications, less than one of the variables presented in the decision trees (and rules) is used.

The decision tree method for classification and estimation works as follows: first, a decision tree model is created from the training data. The model generated is then evaluated by appropriate test criteria using the test data, and finally, the future values of the model are estimated.

Among the approaches used to increase the generalization performance obtained from the decision tree, the pruning method is also included. The pruning method eliminates the trees of the tree with low statistical validity, resulting in a smaller size tree, thereby improving the generalization accuracy rate. Pruning methods are performed by scanning the nodes from top to bottom or from bottom to top. Some nodes improve a criterion by pruning (Kotsiantis, 2013).

13.5.2 Decision Tree Algorithms

There are many decision tree algorithms developed to create a decision tree structure from data sets automatically. Decision tree algorithms are usually aimed at creating the optimal decision tree structure that minimizes generalization error, but it is also possible to aim to minimize the number of nodes, mean depth or other objective functions (Maimon & Rokach, 2010). Decision tree algorithms are intended to form small-sized and low-depth trees. Large and complex decision trees created as a result of decision tree algorithms have low generalization performance. Therefore, many approaches have been developed to create small-sized decision trees. One of the approaches used to create decision trees is the use of criteria for node allocation. Criteria such as knowledge gain, chi-square statistics, and GINI index are among the most commonly used node allocation criteria (Kothari & Dong, 2001).

There are many applications of decision tree algorithms which are applied successfully in different fields. The C4.5 algorithm is one of the most known decision tree algorithms. In the C4.5 algorithm, the rate of knowledge gain is used as the test attribute selection criterion, and the attribute with the highest information gain ratio is selected for each set. The C4.5 algorithm is based on the ID3 algorithm and eliminates some of the limitations of this algorithm. The C4.5 algorithm can work with both continuous and discrete attributes. Also, it can work with training datasets that contain missing attribute values. Also, the decision tree eliminates the problem of excessive compliance with deletion of some nodes or sub-trees during or after the creation of the tree and allows the extraction of exceptional and noisy values in the training set (Niuniu & Yuxun, 2010).

The Decision Stump algorithm is a method that creates a single-level decision tree. The root node in the tree formed by this method is directly connected to the leaf nodes. Decision Stump directly performs the classification process based on a single input attribute value. The Decision Stump algorithm is usually used in conjunction with boosting methods (Witten & Frank, 2005).

The Hoeffding Tree algorithm is a decision tree classifier that works effectively in large datasets by reading each instance at most one time and processing it at an appropriate time interval. Also, the Hoeffding Tree algorithm eliminates the storage problems of traditional decision tree algorithms such as ID3, C4.5, and SLIQ, making it possible to create highly complex decision trees with an acceptable computational cost. In each node of the decision tree, the algorithm uses the statistical value, called the Hoeffding limit, in deciding how to break the node. One of the important features of the Hoeffding Tree algorithm is that the decision tree, which is the result of the algorithm, is almost identical to the classifiers that use all samples to test each node (Domingos & Hulten, 2000).

Logistic Model Trees is a method that combines decision tree induction and logistic regression models. In this algorithm, the tree structure is expanded similar to the C4.5 algorithm. In each fragmentation, logistic regressions of the parent node are passed to the lower nodes. This ensures that leaf nodes contain information about all the parent nodes and generate probability estimates for each class. The tree structure created as a result of the algorithm is pruned, and the model is simplified, and the generalization performance is increased (Doetsch et al., 2009).

The Random Forest algorithm is a model consisting of uncorrected classification and regression trees created by randomly selecting samples in the training data. In this model, the generalization error of classifiers is based on the individual power of all trees and the relationship between these trees. The random selection of the features to be used for the breakdown of each node causes the algorithm to give the results to compete with AdaBoost and to be more resistant to noisy values (Breiman, 2001).

The tree created as a result of the Random Tree algorithm is randomly selected from within the possible tree cluster. Here, each tree within the tree cluster has the chance to be tested as an equal example. The distribution of the trees is distributed uniformly. Random trees can be created effectively, and the models of many random trees usually have high accuracy (Fan, Wang, Yu, & Ma, 2003).

The REPTree algorithm is one of the fast decision tree classification algorithms. The algorithm uses the information gain criterion to construct a decision or regression tree and subjects the tree to pruning based on the method of reduced error pruning. In the REPTree algorithm, only the numerical attributes are sorted. For incomplete values, examples of the C4.5 algorithm are applied to the corresponding fragmentation approach (Zhao & Zhang, 2008).

13.6 Data Set and Findings

The data employed in the application consist of five independent and one dependent variable. The independent variables are numerical data, while the dependent variable is categorical data. Table 13.3 shows the descriptive information of independent and dependent variables.

The data employed in the study has a daily frequency. The values of the independent variables are obtained for the period between 3 January 2017 and 30 January 2019, and the values of the dependent variable are obtained for the period between 4 January 2017 and 31 January 2019. The length of the data set is determined as 510 days, and data in the 1–472 interval (92.5% of the database) are used as the test data set for model development. 473–510 (2 months) interval employed for classification. Decision trees in this study are from WEKA (The Waikato Environment for Knowledge Analysis). WEKA is a tool for data analysis and includes implementations of data preprocessing, classification, regression, clustering, association rules, and visualization by different algorithms. Implemented methods include instance-based learning algorithms, statistical learning like Bayes methods and tree-like algorithms like ID3 and J48 (slightly modified C4.5). An application has been made to classify the Bitcoin Prices as "will increase" or "will decrease" using the J48 decision tree.

A decision stump is a one-level decision tree where the split at the root level is based on a specific attribute/value pair. J48 is slightly modified C4.5 in WEKA. The C4.5 algorithm generates a classification-decision tree for the given data-set by recursive partitioning of data. The decision is grown using Depth first strategy.

Table 13.3 Definition of independent and dependent variables of the model

Independent variables				
Symbol	**Definition of the independent variables**	**Type of the variable**	**Frequency of data**	**Data source**
WTI	West Texas intermediate crude oil price (Closing price, USD, per barrel)	Numeric	Daily	Yahoo Finance
S&P 500	Standard and poor's 500 index (Closing price, USD)	Numeric	Daily	Yahoo Finance
GOLD	Gold price (Closing price, 1 Ounce, USD)	Numeric	Daily	Yahoo Finance
EUR/USD	EUR to USD exchange rate (Closing price)	Numeric	Daily	Yahoo Finance
TB	Thirteen week treasury bill interest rate (Closing price, Chicago options, %)	Numeric	Daily	Yahoo Finance
Dependent variable				
Symbol	**Definition of the dependent variable**	**Type of the variable**	**Frequency of data**	**Data source**
BTC	Bitcoin price (Closing price, USD)	Categorical	Daily	Yahoo Finance

Table 13.4 Classification results of Bitcoin prices

Confusion matrix			Correctly classified		Total correctly classified		Mean absolute percentage error (MAPE)	Kappa statistic (κ)
	Will increase	Will decrease						
Will increase	17	4	80.95%	17/21	28/38	73.68%	44.66%	46.18%
Will decrease	6	11	67.71%	11/17				

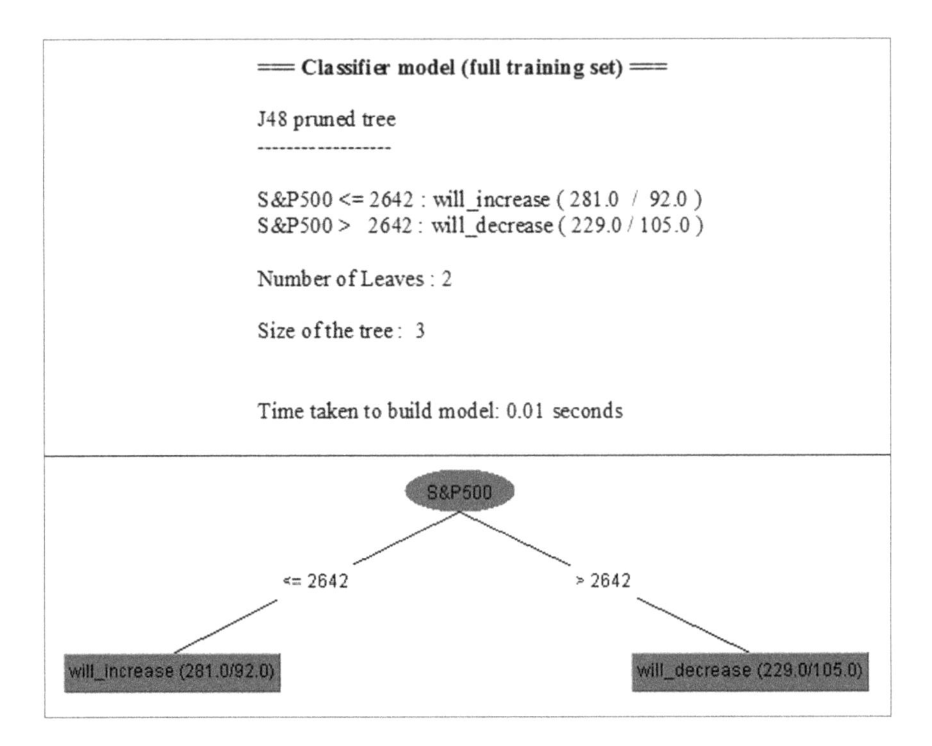

Graph 13.1 Summary of the J48 decision tree

The algorithm considers all the possible tests that can split the data set and selects a test that gives the best information gain (Malangsa & Bacalla, 2015).

When looking at the 38-days (2 months) classification results of the Bitcoin prices from Table 13.4, the J48 tree has classified the price directions of Bitcoin over the next 38 days with a precision of 73.68%. Also, Graph 13.1 shows the J48 decision tree for Bitcoin prices. As can be seen from Graph 13.1, the only variable determined by the J48 decision tree to be associated with the dependent variable is the S&P 500 Index. In other words, no relation was found between the other independent variables and Bitcoin prices by the J48 decision tree.

Table 13.4 also demonstrates the results for the change in the direction of Bitcoin prices; in other words, results show whether the price of bitcoin prices "will decrease" or "will increase" in comparison with rates in the previous day. If the model shows that the real value and estimated value change in the same direction for the related day, the classification is successful. However, if the model shows that the real value and estimated value change in different directions, the classification is unsuccessful (Basarir & Bayramoglu, 2018: 344). According to Table 13.4, the J48 decision tree was performed with 73.68% precision on the classification of Bitcoin prices. This accuracy means that the S&P500 Index is descriptive in the estimation of Bitcoin prices.

From Table 13.4, the Kappa statistic shows that there is a moderate agreement between the real direction of Bitcoin prices and the estimated Bitcoin prices (Viera & Garrett, 2005). In other words, the classification performance of the J48 decision tree is successful on a moderate level. Therefore, it can be stated that there is a moderate relationship between the S&P500 index and Bitcoin prices.

13.7 Conclusion

Bitcoin, which has the highest transaction volume among cryptocurrencies, was used as regular money in clearing and payment transactions and it is considered as an essential investment tool today.

Investors who want to get returns from rapid price changes have turned to a new asset, digital money. In this respect, the possibility of digital money as an alternative to traditional securities has been discussed. In this study, the relationship between Bitcoin, which is an international investment instrument and S&P 500 stock market index, gold price, the interest rate of treasury bills and oil price, which are considered to be affected by this investment tool and which have an important role in financial markets, have been examined. In this study, the relationship between Bitcoin and other variables was analyzed by decision trees method from data mining techniques.

In this study, the decision tree method, which is one of the important classification methods of machine learning and data mining fields, was used in Bitcoin estimation. In the study, C4.5 (J48) algorithm was used from the main decision tree algorithms. The tree structure and properties, working time, the percentage of correct classification, average absolute error and Kappa statistics are presented. It is seen that the decision tree algorithm obtain the correct classification percentage partially successful (average 73%) in estimating how the Bitcoin price will change.

When the findings of the decision tree analysis were evaluated, the following conclusions were reached: (1) The relationship between Bitcoin and S&P 500 Stock Exchange index indicates that Bitcoin could be an investment tool in portfolio diversification. (2) Bitcoin is not related to interest rate, gold, exchange rate, and oil price. Thus Bitcoin could be a safe haven for investors to hedge. (3) Finally, for investors who think that the international banking system is at risk, Bitcoin is an alternative financial instrument, to traditional investment instruments, like gold. It should also be noted that this result of the relationship between Bitcoin and macro-financial indicators is very similar to the results obtained by structural econometric models in the literature.

References

Albayrak, S., & Koltan Yılmaz, Ş. (2009). Data mining: Decision tree algorithms and an application on ISE data. *Suleyman Demirel University the Journal of Faculty of Economics and Administrative Sciences, 14*(1), 31–52.

Antonopoulos, A. M. (2014). *Mastering Bitcoin: Unlocking digital cryptocurrencies* (1st ed.). Sebastopol, CA: O'Reilly Media.

Atik, M., Köse, Y., Yılmaz, B., & Sağlam, F. (2015). Kripto para: Bitcoin ve Döviz Kurları Üzerine Etkileri. *Bartın Üniversitesi İİBF Dergisi, 6*(11), 247–261.

Aytuğ, O. (2015). Şirket İflaslarının Tahmin Edilmesinde Karar Ağacı Algoritmalarının Karşılaştırmalı Başarım Analizi. *Bilişim Teknolojileri Dergisi, 8*(1), 9–19.

Baek, C., & Elbeck, M. (2014). Bitcoins as an investment or speculative vehicle? A first look. *Applied Economics Letters, 22*(1), 30–34. https://doi.org/10.1080/13504851.2014.916379

Basarir, C., & Bayramoglu, M. F. (2018). Global macroeconomic determinants of the domestic commodity derivatives. In H. Dincer, Ü. Hacioglu, & S. Yüksel (Eds.), *Global approaches in financial economics, banking, and finance: Contributions to economics* (pp. 331–349). Cham: Springer.

Baur, D. G., Hong, K., & Lee, A. D. (2018). Bitcoin: medium of exchange or speculative assets? *Journal of International Financial Markets, Institutions and Money, 54*, 177–189. https://doi.org/10.1016/j.intfin.2017.12.004

Bilir, H., & Çay, Ş. (2016). Elektronik Para ve Finansal Piyasalar Arasındaki İlişki. *Niğde Üniversitesi İktisadi ve İdari Bilimler Fakültesi Dergisi, 9*(2), 21–31.

Blau, B. M. (2017). Price dynamics and speculative trading in bitcoin. *Research in International Business and Finance, 41*, 493–499.

Bouoiyour, J., & Selmi, R. (2014). *What does crypto-currency look like? Gaining insight into Bitcoin phenomenon* (MRPA Paper No. 57907, pp. 1–29). https://mpra.ub.uni-muenchen.de/id/eprint/57907

Bouri, E., Molnár, P., Azzi, G., Roubaud, D., & Hagfors, L. I. (2017). On the hedge and safe haven properties of Bitcoin: Is it really more than a diversifier? *Finance Research Letters, 20*, 192–198.

Breiman, L. (2001). Random forests. *Machine Learning, 45*(1), 5–32.

Carrick, J. (2016). Bitcoin as a complement to emerging market currencies. *Emerging Markets Finance and Trade, 52*(10), 2321–2334.

Cheah, E. T., & Fry, J. (2015). Speculative bubbles in Bitcoin markets? An empirical investigation into the fundamental value of Bitcoin. *Economics Letters, 130*, 32–36.

Cheung, A., Roca, E., & Su, J. J. (2015). Crypto-currency bubbles: An application of the Phillips–Shi–Yu (2013) methodology on Mt. Gox Bitcoin prices. *Applied Economics, 47*(23), 2348–2358.

Chiu, J., & Koeppl, T. V. (2017, September). The economics of cryptocurrencies – Bitcoin and beyond. *SSRN*. Available from https://doi.org/10.2139/ssrn.3048124

Coin Market Cap. (2019). Accessed January 8, 2019, from https://coinmarketcap.com

Corbet, S., Meegan, A., Larkin, C., Lucey, B., & Yarovaya, L. (2018). Exploring the dynamic relationships between cryptocurrencies and other financial assets. *Economics Letters, 165*, 28–34.

Çütcü, İ., & Kılıç, Y. (2018). Döviz Kurları İle Bitcoin Fiyatları Arasındaki İlişki: Yapısal Kırılmalı Zaman Serisi Analizi. *Yönetim ve Ekonomi Araştırmaları Dergisi, 16*(4), 349–366. https://doi.org/10.11611/yead.474993

Dahan, H., Cohen, S., Rokach, L., & Maimon, O. (2014). *Predictive data mining with decision trees*. New York: Springer.

Dirican, C., & Canoz, İ. (2017). Bitcoin Fiyatları ile Dünyadaki Başlıca Borsa Endeksleri Arasındaki Eşbütünleşme İlişkisi: ARDL Modeli Yaklaşımı ile Analiz. *Journal of Economics, Finance and Accounting, 4*(4), 377–392.

Doetsch, P., Buck C., Golik P., Hoppe N., Kramp M., Laudenberg J., et al., (2009). *Logistic model trees with AUCsplit criterion for the KDD Cup 2009 small challenge* (pp. 77–88). In: Proceedings of JMLR Workshop and Conference.

Domingos P., & Hulten P. G., (2000). *Mining high-speed data streams* (pp. 71–80). In: Proceedings of the Sixth ACM SIGKDD International Conference on Knowledge Discovery and Data Mining.

Dulupçu, M. A., Yiyit, M., & Genç, A. G. (2017). Dijital Ekonominin Yükselen Yüzü: Bitcoin'in Değeri ile Bilinirliği Arasındaki İlişkinin Analizi. *Süleyman Demirel Üniversitesi İktisadi ve İdari Bilimle Fakültesi Dergisi, 22.*, Kayfor15 Özel Sayısı, 2241–2258.

Dyhrberg, A. H. (2015). *Hedging capabilities of Bitcoin. Is it the virtual gold?* (Working Paper Series, UCD Centre for Economic Research, WP2015/21).

EBA. (2014). *EBA opinion on virtual currencies*. Available from https://www.eba.europa.eu/documents/10180/657547/EBA-Op-2014-08+Opinion+on+Virtual+Currencies.pdf

Fan, W., Wang, H., Yu, P. S., & Ma, S. (2003). *Is random model better? On its accuracy and efficiency* (pp. 51–88). In: Proceedings of the Third IEEE International Conference on Data Mining.

Ferguson, N. (2008). *The ascent of money: A financial history of the world*. London: Penguin Books.

Geetha, A., & Nasira, G. M. (2014). *Data mining for meteorological applications: Decision trees for modeling rainfall prediction*. In: Proceedings of IEEE International Conference on Computational Intelligence and Computing Research.

Georgoula, I., Pournarakis, D., Bilanakos, C., Sotiropoulos, D., & Giaglis, G. M. (2015). *Using time series and sentiment analysis to detect the determinants of Bitcoin prices* (pp. 1–12). In: Proceedings of Ninth Mediterranean Conference on Information Systems.

Glaser, F., Zimmermann, K., Haferkorn, M., Weber, M. C., & Siering, M. (2015). *Bitcoin-asset or currency? Revealing users' hidden intentions* (pp. 1–14). In: Proceedings of Twenty Second European Conference on Information Systems.

Howell, S. T., Niessner, M., & Yermack, D. (2018). *Initial coin offerings: Financing growth with cryptocurrency token sales* (National Bureau of Economic Research No. w24774).

İçellioğlu, C. Ş., & Öztürk, M. B. E. (2018). Bitcoin ile seçili döviz kurları arasındaki ilişkinin araştırılması: 2013-2017 Dönemi için Johansen testi ve Granger nedensellik testi. *Maliye ve Finans Yazıları, 1*(109), 51–70.

Katsiampa, P. (2017). Volatility estimation for Bitcoin: A comparison of GARCH models. *Economics Letters, 158*, 3–6.

Kılıç, Y., & Çütcü, İ. (2018). Bitcoin Fiyatları ile Borsa İstanbul Endeksi Arasındaki Eşbütünleşme ve Nedensellik İlişkisi. *Eskişehir Osmangazi Üniversitesi İktisadi ve İdari Bilimler Dergisi, 13* (3), 235–250. https://doi.org/10.17153/oguiibf.455083

Koçoğlu, Ş., Çevik, Y. E., & Tanrıöven, C. (2016). Efficiency, liquidity and volatility of Bitcoin markets. *Journal of Business Research, 8*(2), 77–97.

Kothari, R., & Dong, M. (2001). Decision trees for classification: A review and some new results pattern recognition: From classical to modern approaches. In S. K. Pal & A. Pal (Eds.), *Pattern recognition from classical to modern approaches* (pp. 169–184). River Edge, NJ: World Scientific Press.

Kotsiantis, S. B. (2013). Decision trees: A recent overview. *Artificial Intelligence Review, 39*(4), 261–283.

Li, X., & Wang, C. A. (2017). The technology and economic determinants of cryptocurrency exchange rates: The case of Bitcoin. *Decision Support Systems, 95*, 49–60.

Ma, Y. (2013). *The research of stock predictive model based on the combination of CART and DBSCAN*. In: Proceedings of Ninth International Conference on Computational Intelligence and Security.
Maimon, O., & Rokach, L. (2010). Classification trees. In O. Maimon & L. Rokach (Eds.), *Data mining and knowledge discovery handbook* (pp. 149–175). New York: Springer.
Malangsa, R. D., & Bacalla, A. C. (2015). Performance comparison of decision stump and J48 classification algorithm on the programming skill of IT students. *Journal of Science, Engineering and Technology, 3*, 199–212.
Malhotra, A., & Maloo, M. (2014). *Bitcoın-is it a bubble? Evidence from unit root tests*. Retrieved May, 28, 2018, from https://papers.ssrn.com/sol3/papers.cfm?abstract_id=2476378
Marr, B. (2017, December). A short history of Bitcoin and crypto currency everyone should read. *Forbes*. https://www.forbes.com/sites/bernardmarr/2017/12/06/a-short-history-of-bitcoin-and-crypto-currency-everyone-should-read/#36bf83003f27
Nakamoto, S. (2008). *Bitcoin: A peer-to-peer electronic cash system*. https://bitcoin.org/bitcoin.pdf
Niuniu, X. & Yuxun, L. (2010). *Review of decision trees* (pp. 105–109). In: Proceedings of the Third IEEE International Conference on Computer Science and Information Technology.
Polasik, M., Piotrowska, A. I., Wisniewski, T. P., Kotkowski, R., & Lightfoot, G. (2015). Price fluctuations and the use of Bitcoin: An empirical inquiry. *International Journal of Electronic Commerce, 20*(1), 9–49.
Radziwill, N. (2018). Blockchain revolution: How the technology behind Bitcoin is changing money, business, and the world. *Quality Management Journal, 25*(1), 64–65.
Rogojanu, A., & Badea, L. (2014). The issue of competing currencies: Case study-Bitcoin. *Theoretical & Applied Economics, 21*(1), 103–114.
Sovbetov, Y. (2018). Factors influencing cryptocurrency prices: evidence from bitcoin, ethereum, dash, litcoin, and monero. *Journal of Economics and Financial Analysis, 2*(2), 1–27.
Tschorsch, F., & Scheuermann, B. (2016). Bitcoin and beyond: A technical survey on decentralized digital currencies. *IEEE Communications Surveys & Tutorials, 18*(3), 2084–2123.
Urquhart, A. (2016). The inefficiency of Bitcoin. *Economics Letters, 148*, 80–82.
Viera, A. J., & Garrett, J. M. (2005). Understanding inter observer agreement: The kappa statistic. *Family Medicine, 37*(5), 360–363.
Vigna, P., & Casey, J. (2017). *Kripto Para Çağı* (2nd ed.). Ankara: Buzdağı Yayın Evi.
Vockathaler, B. (2015). *The Bitcoin boom: An in depth analysis of the price of Bitcoins* (Major Research Paper). University of Ottawa. Retrieved January 10, 2019, from https://ruor.uottawa.ca/bitstream/10393/32888/1/Vockathaler_Brian_2015_researchpaper.pdf
Witten, I. H., & Frank, E. (2005). *Data mining: Practical machine learning tools and techniques* (2nd ed.). San Francisco, CA: Elsevier.
Yermack, D. (2013). *Is Bitcoin a real currency?* (Working Paper Series, National Bureau of Economic Research No. 19747, pp. 1–14).
Zhao, Y., & Zhang, Y. (2008). Comparison of decision tree methods for finding active objects. *Advances in Space Research, 41*(12), 1955–1959.
Zhao, J. L., Fan, S., & Yan, J. (2016). Overview of business innovations and research opportunities in blockchain and introduction to the special issue. *Financial Innovation, 2*(28), 1–7. https://doi.org/10.1186/s40854-016-0049-2

Arzu Tay Bayramoğlu is an Associate Professor of Economics at Bulent Ecevit University, Zonguldak, Turkey. Dr. Tay Bayramoglu has a BS in Economics from Marmara University (2004), a Master in Economics from Zonguldak Karaelmas University (2007) and a Ph.D. in Economics from Istanbul University (2012). Her research interests lie in Macroeconomics. She has taught Macroeconomics, at both graduate and undergraduate levels. Dr. Tay Bayramoglu has her post-doctoral research at Lamar University Department of Economics and Finance, Beaumont, TX, USA.

Çağatay Başarır is an Associate Professor of Accounting and Finance at Bandirma Onyedi Eylul University, Bandirma-Balikesir-Turkey. Dr. Basarir has a BS in Business Administration from Ege University (2003), an MBA from Balikesir University (2006) and a Ph.D. in Business Administration-Finance from Balikesir University (2013). His research interests lie in Risk Management and Financial and Capital Markets such as forecasting of stock prices, commodity prices, and investment strategies in financial crisis periods and in forecasting methodologies such as time series analysis and panel data analysis. He has taught Capital Markets, Portfolio Management, International Finance, Accounting, Financial Management, and Financial Statement Analysis, among others, at both graduate and undergraduate levels.

Chapter 14
Impact of Digital Technology and the Use of Blockchain Technology from the Consumer Perspective

Hande Begüm Bumin Doyduk

Abstract The Blockchain concept, promising breakthrough innovations has captured interest in the business world and also among potential investors and consumers. Block chain technology can be used for various purposes in the finance industry, public sector, smart contracts and internet of things, shared economy and financial transformation. However, from the individual consumer side, it is mainly linked to the cryptocurrency. Cryptocurrency is seen as a subset of alternative currencies from a general perspective and as a subset of digital currencies in a narrow sense. According to Gartner Hype Cycle for Emerging Technologies 2017, block chain technology is in the peak of inflated expectations stage. There are 5–10 years for maturity and widespread usage. Many consulting firms share the same insight, stating that block chain technology will be a main technological platform by 2025. Blockchain technology is a relatively new area of study mainly analyzed by engineering and finance scholars. There are very few academic studies including the consumers' standpoint. Analyzing a new business idea or a technological novelty without considering the consumer side will be ungrounded.

In this chapter, crypto currencies, specifically Bitcoin and the underlying technology Blockchain, will be studied from the consumer point of view. Awareness of crypto currencies, attitudes toward it, purchase intentions, user profiles, usage motivation around the world and in Turkey will be discussed. Following this section, the adoption of Blockchain technology and Bitcoin adoption will be analyzed by different technology acceptance and adoption models.

H. B. Bumin Doyduk (✉)
Department of International Trade, Faculty of Business, Altınbaş University, Istanbul, Turkey
e-mail: hande.doyduk@altinbas.edu.tr

© Springer Nature Switzerland AG 2019

U. Hacioglu (ed.), *Blockchain Economics and Financial Market Innovation*, Contributions to Economics, https://doi.org/10.1007/978-3-030-25275-5_14

14.1 Introduction

The concept of block chain was first mentioned in October 2008, in the Bitcoin article (Nakamoto, 2008). The author of the article, whose pseudonym is Satoshi Nakamoto developed the Bitcoin system. In the article Bitcoin is construed as "a system for electronic transactions without relying on trust". Bitcoin, as the first and the biggest one in terms of market size, is the most substantial crypto currency (Karaoglan, Arar, & Bilgin, 2018). In February 2019, the number of Bitcoins was around 17.5 million with a market value of US$63 billion. The Cryptocurrency market has not yet reached the maturity stage. Notwithstanding the adoption has been increasing (Gazali, Ismail, & Amboala, 2018; Grant & Hogan, 2015).

Cryptocurrency can be characterized as a currency, technology platform, protocol, and a payment system. Cryptocurrency is seen as a subset of alternative currencies from a general perspective and as a subset of digital currencies in a narrow sense (Gültekin, 2017). Cryptocurrency can be summarized as an open source software, providing public ledger transaction and security through protocols (Athey, Parashkevov, Sarukkai, & Xia, 2017). Indeed it can be said that there is no consensus about how it can be categorized. A very popular cryptocurrency, Bitcoin, is accepted as accepted as a commodity from a tax perspective in United States. On the other hand, the European Central Bank (ECB) accepted Bitcoin as a virtual currency (ECB, 2012).

The basis of Bitcoin rests on Blockchain technology. Blockchain is a series of blocks keeping records of transactions similar to a public ledger (Chuen, 2015).

14.2 Blockchain Mechanism

The first block in the Blockchain is named as genesis block. The following blocks indicate the former block namely the parent block with a hash value. The difference between classical link and Blockchain is that any block in the chain cannot be changed, by means of the hash value (Narayanan, Bonneau, Felten, Miller, & Goldfeder, 2016). The Blockchain architecture is shown in Fig. 14.1 (Zheng et al, 2018).

Each block has a block header consisting of block version, parent block hash, Merkle tree root hash, timestamp, nBits and Nonce and a block body comprises transaction counter and transactions as seen in Fig. 14.2.

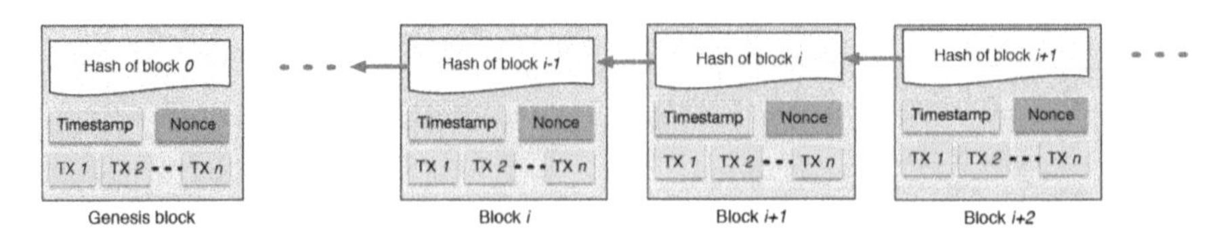

Fig. 14.1 An example of blockchain which consists of a continuous sequence of blocks (see online version for colors)

Block version	02000000
Parent Block Hash	b6ff0b1b1680a2862a30ca44d346d9e8 910d334beb48ca0c0000000000000000
Merkle Tree Root	9d10aa52ee949386ca9385695f04ede2 70dda20810decd12bc9b048aaab31471
Timestamp	24d95a54
nBits	30c31b18
Nonce	fe9f0864

Transaction Counter

TX 1 TX 2 ... TX n

Fig. 14.2 Structure of a block

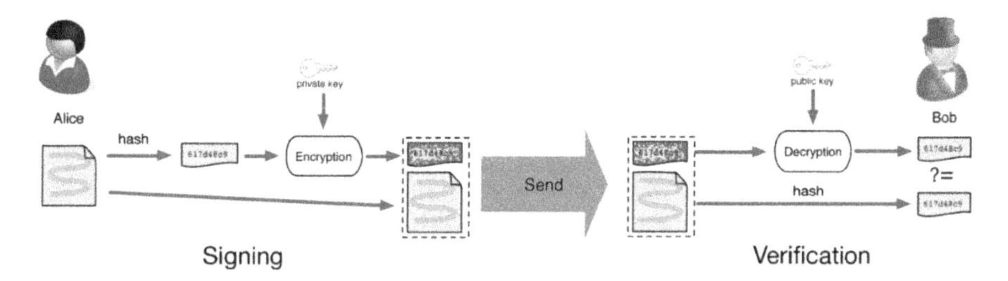

Fig. 14.3 Digital signature

There are private and a public keys in the digital signature procedure of Blockchain. Through the private key, transactions are signed and circulated by the network. These signed transactions are obtained by public keys, which can be seen by everyone. The process has two phases. Firstly during the signing phase, a hash value is achieved from the transaction. Then through the private key, this value is encrypted and sent to the receiver together with the original data. Secondly, the verification phase starts. The receiver verifies the transaction through comparing decrypted hash and the hash value from the data as shown in Fig. 14.3. The digital signature procedure provides not only transaction security but also anonymity of the sender and receiver, since the procedure is completed through digital signatures only known and confirmed by the sender himself or herself (Ünsal & Kocaoglu, 2018).

Each transfer added to the Blockchain should be verified by cryptography. This verification is done to confirm that the sender actually owns the Bitcoins that s/he is sending. In peer to peer network, the transactions are verified by using cryptographic proof. This process is known as mining or Proof of Work (PoW). Each miner, node, of the network calculates hash value of block header continuously. When one of the miners acquire the appropriate value, the other nodes (miners) should confirm. After the confirmation the transaction is agreed to be the authenticated result and it is shown by a new block in the chain (Aksoy, 2018; Bohr & Bashir, 2014; Zheng et al., 2018). The miner which solves the crypto is incentivized by new bitcoins. This role of miners in the Blockchain system is crucial.

The Bitcoins on Blockchain can be transferred through the use of digital signatures. In order to access Bitcoins and use them, Bitcoin owners should manage the keys used during the digital signature formation. Key management is a critical subject on its own in the information security area. There are many risks associated with the capture of the keys by others. In the Bitcoin world, an individual can create as many different keys as s/he pleases. The effective management of different keys can be done through digital wallets. Bitcoin wallet is a software that enable digital keys and consequently Bitcoins to be managed through simple interfaces (Ünsal & Kocaoglu, 2018). According to Eshkandri et al., there are six different ways of storing the Bitcoin. The first method is using keys in the user's home or application directory. The keys are stored on the device's local storage. Some mobile wallets like Android Bitcoin Wallet use this approach. The second method is encrypted wallets in which a locally stored wallet is encrypted with a key derived from a password or passphrase. This method only protects against the physical theft of the storage device. The third method is offline wallets in which, wallets are stored offline like a USB drive. The fourth method is air-gapped key storage. In this method wallets are stored on a device that is never connected to a network. The fifth way is to store through password derived keys. In this method cryptographic keys are derived from a password. The last method is hosted wallets in which the user accounts are hosted on a web service (Eskandari, Clark, Barrera, & Stobert, 2018).

14.3 The Characteristics of Blockchain

Bitcoin as a distributed ledger technology, has certain characteristics that endow the substantial place it has in the new technological era. Blockchain, as the main technology behind Bitcoin, provides these characteristics. Blockchain is irreversible, transparent, secure, fast, and trustworthy (Underwood, 2016). Some of the basic characteristics of blockchain are decentralization, persistency, anonymity, and auditability.

14.3.1 Decentralization

Unlike traditional transaction systems in which validation by a central agency like the central bank is needed; Blockchain is decentralized. Blockchain transactions are operated between two peers, without the verification of a central agency. This character of Blockchain does not only annihilate trust issues but also reduces server costs and performance inefficiencies at the central server (Zheng et al., 2018).

14.3.2 Persistency

It is almost impossible to interfere with the Blockchain system as the transactions are validated and recorded in blocks which are shared in the whole network. Along with that, falsification is easy to discover thanks to the validation of each single block by other nodes.

14.3.3 Anonymity

Blockchain technology eliminates central agencies through which the user's information is saved. Along with that, as users operate in Blockchain with generated addresses, they can prevent identity disclosure by using many different generated addresses.

14.3.4 Controllability

The validation and recording of transactions enable tracing of previous records by reaching any node of the network.

14.4 Challenges

Blockchain offers many benefits with the above mentioned characteristics. Nonetheless there are still some unsolved issues which require amelioration. Current challenges of the technology are as follows: data privacy, scalability, energy and technical requirements and illegal use options.

14.4.1 Scalability

The size limit of Bitcoin blocks is 1 MB. A block is created on average in ten minutes and seven transactions can be completed in one second. This brings the inability of handling high frequency trading. As of January 2018, Bitcoin Blockchain storage is more than 100 GB. Due to the small capacity of the blocks, miners would choose large transactions with high transaction fee over smaller transactions. This would cause long delays for small transactions.

14.4.2 *Privacy & Security*

The public visibility of all public key transactions creates the possibility of transactional privacy errors. Using generated addresses instead of identity information cannot guarantee privacy in all cases (Biryukov, Khovratovich, & Pustogarov, 2014; Kosba, Miller, Shi, Wen, & Papamanthou, 2016; Meiklejohn et al., 2013).

Blockchain technology has been seen as a trust machine. However the Bitcoin hacks have reduced the general trust in this technology. Mt Gox was hacked two times. In the first incidence, in 2011, 2609 BTC were hacked. After that in 2014, 750,000 BTCs were lost. Bitfloor lost 24,000 BTC in 2012. Poloniex was hacked in 2014. In 2015, Bitstamp lost 19,000 BTC. And in 2016, Bitfinex lost 120,000 BTC to hackers (Khatwani, 2018).

14.4.3 *Technical Requirements*

Proof of Work (PoW) requires big amount of electric energy consumption. Apart from high electricity requirement, mining also necessitates graphic processing units instead of central processing units as hardware. The personalized computers are not sufficient to solve difficult crypto puzzles anymore.

14.4.4 *Illegal Use*

The decentralized, peer to peer trust based system of Bitcoin created anonymity. This anonymity has allowed some users to purchase illicit goods and services such as drugs, illegal pornography, and contract killing (Brezo & Bringas, 2012; Trautman, 2014). Along with that, the other illegal use areas are money laundering, capital control avoiding and terrorism funding (Bryans, 2014; Dostov & Shust, 2014; Pagliery, 2014). It is suggested that bitcoin have promoted the growth of illegal online markets (Foley, Karlsen, & Putnins, 2018). Dark online markets such as Silk Road and Sheep marketplace encouraged crypto currency usage.

14.5 Bitcoin Around the World

Around the world the interpretation and the legal status of Bitcoin are not settled yet. There is still controversy about the main purpose of Bitcoin. Both in the academic and business world, a consensus has not been reached about whether it is a payment or an investment tool. (Glaser, Zimmermann, Haferkorn, Weber, & Siering, 2014; Hur, Jeon, & Yoo, 2015). Bitcoin is regarded as an investment tool by spectators, which caused a dilution of the image of Bitcoin as a currency and a fluctuation of its value (Bohr & Bashir, 2014). For tax functions, US Internal Revenue Service regards Bitcoin as an asset instead of a currency. From an investment tool point of view, the benefits of Bitcoin are high return rates, low cost of entry, and high growth anticipation (Brimble, Vyvyan, & Ng, 2013). US is one of the markets which has a positive attitude towards Bitcoin along with Canada and Australia (Bajpai, 2010). As of December 2018, almost 5% of Americans have Bitcoins. The European Union has not issued any common official decision. Thus each European country decides independently. According to a report by ING, 66% of Europeans are aware of the Bitcoin. In Europe males outnumber females in terms of Bitcoin awareness. Approximately one third of Europeans believe that Bitcoin is the future of investing and online transactions (Exton & Doidge, 2018).

A website listing the firms accepting Bitcoin, 'useBitcoins', reports 5071 companies as of February 2019. Some of the big firms accepting Bitcoin are Microsoft, Dell, Expedia, Save the Children, Shopify, Overstock, Subway, PayPal, DISH Network and Stream (Ayhan, 2018; Sloan, 2018).

It is suggested that in order to overcome the risks impeding the Bitcoin adoption, it is essential to form a legal and regulatory framework (Descoteaux, 2014).

In September 2017, China outlawed Initial Coin Offerings and cryptocurrency exchanges. Following that action, access to internal and domestic cryptocurrency exchanges were forbidden for people in January 2018. Countries such as Vietnam, Bolivia, and Columbia have a negative stance towards Bitcoin.

In Turkey, Bitcoin is primarily operated in the market through BTCTurk.com and Travelers Box firms. These firms operate under a Northern Cyprus originated company (Can, 2013). The Central Bank of Turkey does not have a direct control or execution over digital currencies. The Banking Regulation and Supervision Agency (BDDK), has not accepted bitcoin as a digital currency (Khalilov, Gündebahar, & Kurtulmuşlar, 2017). Along with that there is no legal regulation of cryptocurrencies in terms of personal income tax. The ways of taxation of income gained from cryptocurrencies are not identified yet (Kaplanhan, 2013).

After 2017, Blockchain focused workshops have been conducted at universities. The Blockchain Research Laboratory was founded by The Scientific and Technological Research Council of Turkey (Tubitak, 2018). Around the world, after Canada and Cyprus, Turkey is among the first countries which has a Bitcoin ATM. It is placed at Ataturk Airport and operations are conducted through an e-wallet named "Traveler's Box" (Sonmez, 2014). According to ING's study, 18% of Turkish people own cryptocurrency. The results might be biased towards younger tech-savvy professionals (Exton & Doidge, 2018). Recently more and more technological entrepreneurial firms, which provide new business solutions with Blockchain technology like 'Kimlik.io' and 'Colendi' are founded (Ünsal & Kocaoglu, 2018). A study conducted in 2018 analyzed the motivation of firms accepting cryptocurrencies. According to the study the universe of crypto currency accepting businesses were 122 firms. These firms were mostly located in Istanbul and Ankara. The age range of the firm executives was between 27 and 40. The main reasons of

Bitcoin acceptance were stated by the respondents as; ubiquitous and fast transactions, low service fees and future expectations (Karaoğlan, 2018).

The penetration of Bitcoin accepting firms affects the level of consumer adoption, since the absence of places to spend or use Bitcoin will trigger consumer hesitation (Connolly & Kick, 2015).

14.6 Applications

Blockchain technology serves many areas with various different applications. Finance, IoT, public and social services, reputation system, security and privacy are stated in many studies as the main Blockchain application areas. These applications serve both the business and consumer market.

14.6.1 Finance

Blockchain technology and specifically Bitcoin has changed the traditional viewpoint in the financial and business world. Blockchain is predicted to be a disruptive technology in the finance sector (Peters, Panayi, & Chapelle, 2015). Some of the main applications in finance industry are payment procedures, fund transfer, exchange platforms, barters, verification, digital identity management, document management, Islamic banking, clearing and settlement of financial assets, enterprise formation, P2P Transfers and risk management (Cognizant, 2016; Deloitte, 2015; Evans, 2015; Jaag & Bach, 2016; Noyes, 2016).

14.6.2 Public and Social Services

Some of the applications areas are stated to be energy distribution, smart contracts, digital passport, digital id, social security system, taxation system, polls, and land registration (Atkins, Chapman, & Gordon, 2014; Gogerty & Ziloti, 2011; NRI, 2015). Some countries such as Dubai, Switzerland, UK, and Singapore have already started investing in Blockchain technology for public sector applications. Also the American Ministry of Defense and NATO have been considering the intense use of this technology (Kar, 2016).

In emerging countries, the trust attribute of Blockchain is very valuable. It has been extensively used for land registration purposes. For instance it was used by the National Agency of Registration of Georgia in order to operate land titles and to reduce registration fees. Also in Honduras and Ukraine similar projects have been implemented (Underwood, 2016). It is believed that Blockchain technology will create added value for achieving '2030 Sustainable Development Goals' through Blockchain based land registration and digital identity applications.

14.6.3 Internet of Things

The term "internet of things" was first used by Kevin Ashton in 1999 (Ashton, 2009). The Internet of Things (IoT) is the network of physical objects that contain embedded technology to communicate and sense or interact with their internal states or the external environment (Gartner, 2014). IoT enables analog world objects to be connected with other objects, communicate and operate ubiquitously on their own, without human interaction. Apart from B2B applications, through this technology various important and useful applications such as smart homes and e-health have been served/developed for consumers. It is expected that the IoT applications will make life easier for consumers. With all the promised benefits, there are some risks concerning privacy and security. Without human interaction, machines, things, will have access to consumers' data and communicate with others in the network.

Blockchain technology can help to solve security and privacy issues in IoT systems. These two challenges impede the adaptation of IoT in business and on the consumer side. Blockchain technology is supporting the mainstream, widespread usage of IoT applications (Hardjono & Smith, 2016).

14.7 Consumer Applications

The opportunities, developments and benefits Blockchain has provided and will continue providing in the future to the business world and public are only one side of the mirror. The Blockchain technology has current and potential applications that can embellish the consumers', the end users', lives.

14.7.1 Consumer Electronics

The consumer electronics sector is an important sector with a very high growth expectation of 5% from 2016 to 2023. It is expected to reach US$1.8 trillion by 2023 (Research, 2017). With new technological improvements such as internet of things, many advancements in consumer electronics applications are also expected. These novelties bring some challenges such as data security. Blockchain is regarded as a platform which can provide security. From the consumer point of view, protecting their data is of vital importance for technology acceptance.

Technological innovations, consumer involved open innovation through social media, sustainability concerns, increasing middle class in developing countries with augmenting purchasing power, shorter product life cycles, planned obsolescence, cost cutting strategies of consumer electronics firms are the main factors requiring supply chain management advancements (Du, Yalcinkaya, & Bstieler, 2016).

Blockchain with its decentralized structure is useful for centralized architectures through supplying unified, verified information for all members of the network, such as supply chain. Blockchain technology enables firms to satisfy consumer demand through increased responsiveness, lower costs, and increased flexibility.

Due to the current status of inadequacy of transparency in supply chains, consumers cannot reach and validate the value of products and services. The prices paid in most cases are based on unreliable cost of production. Apart from the price concern, lately consumers are more and more conscious and demanding in environmental protection, sustainability, transparency and social responsibility issues. In order to confirm product authenticity, sustainability and ethical standards of production and consequently achieve consumer participation, supply chain management needs to reconsider the transparency concept (Pilkington, 2016). Blockchain technology enables transparency and traceability throughout the network. For instance, gemstones can be traced from mines to the consumers (Boucher, 2017; Iansiti & Lakhani, 2017). Ethical issues are more serious in some industries such as the diamond industry. Thanks to Blockchain technology, consumers can make sure they are not supporting unethical firms by their purchases. Some establishments such as Everledger and Provenance are specialized in ethical standards tracking in supply chains through Blockchain (Mendez, 2012). Apart from control of ethicality of the supply chain, Blockchain solutions are also used to track counterfeit. For example, BlockVerify is an anti-counterfeit solution recording ownership by consumer-level authentication (Hulseapple, 2015).

Blockchain also enables cooperation and collaboration among producers and consumers, thus consumers become prosumers, interacting with the producers and takeing active role in the process (Lee & Pilkington, 2017; Migirov, 2016).

14.7.2 Blockchain Crowding and Art Consumption

Artists and authors had to negotiate with intermediaries for the sake of receiving prevailing global coverage, distribution and advance payments. In return, a high profit margin of the art sale was given to the intermediaries. Due to the high profit margins shared with intermediaries, artists started looking for alternative methods of funding and disseminating their work (Geith, 2008; Lessig, 2004).

The technological advancements, especially in the area of information and communication, change the traditional way of obtaining global coverage and distribution (Manovich, 2009). The digitalization enabled low cost digital format production, thus minimized the initial investment for production cost. Apart from production, digitalization also provides fast and no cost global distribution of art material. For instance, social network sites, online music stores and platforms became the main media for consumer reach. The developments in information and communications technology started the process of intermediary disappearance, the disintermediation (De Filippi, 2016; Gellman, 1996). Blockchain is based on the idea of decentralization and peer to peer network without a central authority or intermediary. Artists can make their work available to the public through smart contracts. Smart contracts work with the principle of integrating small codes into the transactions. Thus through smart contracts, consumers can have access to the work of artists under certain constraints which are dismissed through fee payments. Another way of funding the art projects is crowdfunding in which a large group of people independent from each other contribute to a funding. Through "crypto-equity", new types of securities with cryptographic tokens can be created by using Blockchain technology. These tokens act like shares of the funded project. Art projects can be financed by crypto-equities (De Filippi, 2016). All of these above mentioned novel financing methods through Blockchain technology enable artists to have access to finance from their audience and eliminate intermediaries.

From the consumer point of view, this situation enables direct contact with their favorite artists. Consumers can also donate money to the projects of artists through the use of Blockchain. Through crowdfunding and crypto-equity shares, consumers evolve from the passive consuming part of the art projects into the active stakeholder.

14.7.3 Blockchain and Education

Another application of Blockchain technology can emerge in the education sector. The instructors could pack learning blocks and plant them in the Blockchain. The achievements in learning would be the coins in such a system (Alexander & Camilleri, 2017; Devine, 2015).

14.7.4 Reputation System

Reputation is an essential criterion of level of trust by the community. An individuals' reputation is defined by how reliable h/she is seen by other people. It is possible to falsify the reputation records of individuals and firms. The examples are numerous in e-trade as firms engage numerous fake customers to their membership system in order to have better reputation results. Blockchain technology can provide a solution for this problem. In the academic world, reputation has a great significance. Thus a Blockchain based system was proposed by Sharples and Domingue (2016) for education and reputation recording of academicians.

14.7.5 Security & Privacy

Bitcoin technology can provide advancements in distributed network security. Along with security, data privacy is also a very important issue for consumers. From the initiation until 2014, Facebook had accumulated more than 300 petabytes of data (Vagata & Wilfong, 2014). In September 2018, the Facebook data breach scandal broke. It was stated that more than 50 million peoples' accounts were affected (O'Flaherty, 2018). Personal data of users are collected through network and mobile sites and stored on central servers every day. The servers are exposed to attacks by hackers. Rather than trusting third parties with the security of personal data privacy, users should control their personal data. Blockchain technology can be used to assure sensitive data privacy. Zyskind, Nathan, and Pentland (2015) introduced a blockchain technology based decentralized data management system. A protocol turning a blockchain into an automated access-control manager was implemented. In this decentralized system, Blockchain was remodeled and utilized as an access control moderator. Thus users can realize when and how their personal data are collected and used. Consequently it is not a must to solely trust a third party with data privacy (Zyskind et al., 2015).

14.7.6 Smart Contracts

Smart contracts will empower consumers as the terms accepted are protected by the smart contracts. Even without an intermediate to facilitate the process for the consumer, they are protected from fraud attempts from the vendor side. It will be burdensome for the seller to evade the previously accepted terms since the payment will not be completed unless the contract terms are satisfied (Fairfield, 2014).

14.7.7 Digital Identity

Blockchain technology can assure potent, autonomous digital identity around personal data. Decentralization provided by Blockchain technology will prevent central control over digital entities (Pilkington, 2016; Underwood, 2016). It is suggested that Blockchain based digital identities will supply trusted measures of reputation to people in areas where documentation, registration and regulations are insufficient.

14.8 Consumers' Point of View

On the consumer side also there is not a general consensus about what Bitcoin is. It is perceived as both a payment method and an investment method. The hindrances of credit cards such as high fees, fraud risk and probability of charge backs necessitated merchants and consumers to look for an alternative payment method. PayPal and Bitcoin are among the many online payment methods created to overcome the credit card challenges (Baur, Büler, Bick, & Bonorden, 2015).

According to Glaser et al.'s study in 2014, the new Bitcoin users' interest has an impact on the volume of Bitcoin at exchange but no impact within the Bitcoin system. The scholars read this as the new Bitcoin buyers purchasing it for speculative purposes and not for payment of goods and services. Also the news related to Bitcoin affect the returns of it, as it is seen more like an asset than a currency (Glaser et al., 2014).

General public awareness about bitcoin is not very high although Bitcoin was very popular in the media due to its fast increase of value especially during the period of 2017–2018 (Coinbase, 2019). The cryptocurrency investments' returns were extraordinary, thus created fear of missing out (FOMO) in the society (Popper, 2013). The investment behavior became irrational, as people felt that they needed to take urgent action. Along with the fear of missing out, the subjective norm also played an important role in consumer purchase behavior of Bitcoin. Subjective norm can be summarized as the perceived social pressure to display a behavior. As people perceived that the important others in their lives thought that they should buy Bitcoin, they started to do so (Gao, Clark, & Lindqvist, 2015).

It is suggested that there are five adoption stages of bitcoin. The first stage is named experimentation stage. Between 2009 and 2010, only developers and technical people knew this new concept, exploring the source code and trying the exchange platform. The second stage, the early adopters phase, was between 2011 and 2013. At this stage, considerable media coverage attracted investors' and entrepreneurs' attention. Merchant processors, wallet providers and exchange firms were founded. The third stage started in 2013 with the substantial investments of distinguishing venture capitals. In 2013, US$90 million was invested in Bitcoin, the following year the investments rose up to US$300 million. The fourth phase was the Wall Street Phase. As the name implies at this stage banks, brokers, dealers started to take part in the game. The growing volume and price increases pulled people and started mass adoption. The last phase was suggested to be the Global Consumer Adoption Phase. The timing of this stage is unknown. This stage is expected to start only if companies continue to make it user-friendly for consumers; merchants accept Bitcoin as a payment method and general public awareness increases (Mauldin, 2014). A similar adaptation phase evolution was proposed by Accenture as seen in Fig. 14.4.

Risks associated with Bitcoin are very important for adoption. Consumers' perceptions about the risks of Bitcoin can be grouped as market risks, counterparty risks, transaction risks, operational risks, privacy risks, and legal risks (Böhme,

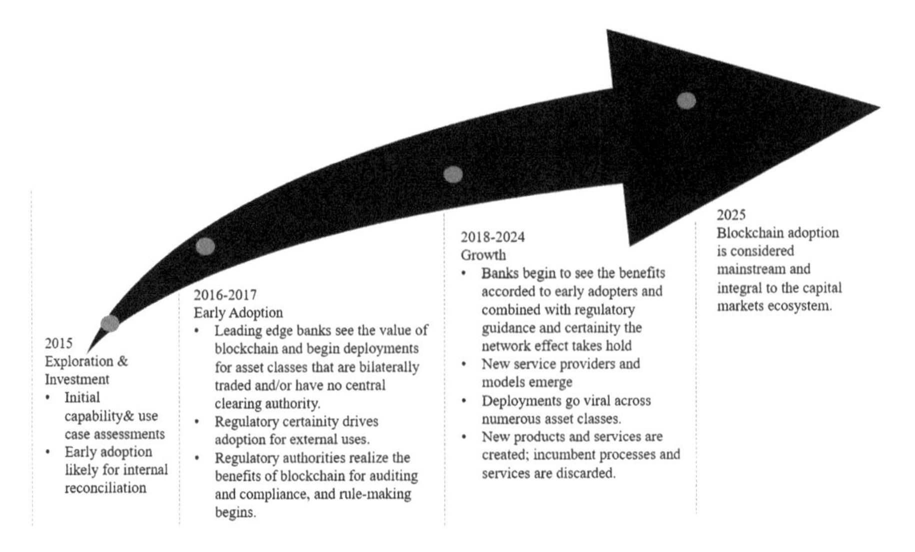

Fig. 14.4 Adoptation of Blockchain (Source: Accenture, 2015)

2013). The fluctuating prices and exchange rates are considered as market risks. Security infringements, the failure of exchange and wallet firms are grouped under the counterparty risks. The irreversibility of transactions is seen both as a risk factor and a benefit of the Bitcoin. Irreversibility means that it is not possible to retake a transaction, for example transfer of money, without the approval of the receiver. As there is no central arbitrator to validate the retake and the Blockchain system does not allow reversibility (Meiklejohn et al., 2013; Zohar, 2015). Security defects, incidences like forgetting passwords, exposure to bugs, adoption inefficiencies are placed under operational risk. Revealing real identities is the privacy risk of Bitcoin. Lastly as the legal status of Bitcoin is not clear in many countries, it constitutes an important perceived risk for consumers.

Despite the numerous risks associated with Bitcoin, there are many perceived benefits. Firstly, sending and receiving Bitcoins ubiquitously to anyone is one of the most important benefits of Bitcoin from the consumers' point of view (Nian & Chuen, 2015). Secondly, decentralized, faster and cheaper transactions are perceived as the technology benefits of Bitcoin (Ali, Barrdear, Clews, & Southgate, 2014; Kronbholz, Judmayer, Gusenbauer, & Weippl, 2016).

It is hard to talk about a global adaptation state currently. The backbones of adopting are user interfaces, user adoption and price stability. In order to achieve a global consumer adoption state, the interfaces must be easy to use for the general public. Along with that, there should be many businesses accepting Bitcoin as a payment method, so that it will be considered as a currency. Last but not least, a relatively stable price range would bring more users to the network (Torpey, 2014; Walch, 2015).

14.9 Bitcoin User Profile

There were approximately 32 million bitcoin wallets and 7.1 million active users globally by December 2018 (Lielacher, 2019). The seven million active users have different user motivations and different agendas which require further investigation. For this purpose, there are studies analyzing the bitcoin users and consumers in different countries. The awareness levels, adaptation stages, demographics of the users and the main motivations to acquire Bitcoin differ from one country to another. According to a study in the US conducted in 2013, the average Bitcoin user in US is male, in his 30'ies, libertarian and non-religious (Simulacrum, 2014). 13% of Americans would prefer investing in Bitcoin rather than gold (Wile, 2014). Bohr and Bashir (2014) suggested that Bitcoin accumulation magnitude can be predicted by age, the state of early adoption, illicit good purchase and participation in Bitcoin forums. According to this study, during the beginning stages of Bitcoin, miners were acquiring BTC twice as much as the users (non-miners). As time progressed, no significant difference between the accumulations of these two groups remained.

Greek consumers were stated to be unaware of Bitcoin. Although it is not possible to mention a general negative attitude towards Bitcoin, there is a certain level of skepticism (Tsanidis, Nerantzaki, Karavasilis, Vrana, & Paschaloudis, 2015).

In Karaoglan et al.'s study Turkish crypto currency awareness was examined. According to the study, the most well-known crypto currencies were; Bitcoin, Ethereum, Ripple, Cardano, Stellar, Litecoin, Bitcoin cash and IOTA (Karaoglan, Arar, & Bilgin, 2018).

14.10 Adoption Studies

The widespread usage of Blockchain technology and specifically Bitcoin, requires the adoption of this new technology. In this part, the technology diffusion and adoption models will be shortly summarized. Following that, studies analyzing the adoption of Bitcoin will be examined.

Blockchain technology and Bitcoin specifically can be regarded as an innovation. Diffusion of Innovation Theory is the most prevalent theory explaining how an idea or a new product diffuses (or spreads) through a community, social system. As a result of this diffusion people adopt a new idea, behavior, or product. The most important point for adoption is the perception of innovation. The adopted thing should be perceived as an innovation, a novelty. The diffusion of innovation theory by Rogers in 1962 utilizes a bell shaped curve for explaining the adoption diffusion over time (Rogers, 2010). According to the theory, there are five groups based on the time of adoption. The first group is innovators. Innovators are venturesome and they want to be the first to try the innovation. After innovators comes early adopters. They are usually the opinion leaders and have the potential to affect the rest of the population. Early majority adopt the innovation before an average person does.

The late majority is skeptical of innovations and the last group, the laggards, are very conservative and skeptical. It is suggested that Bitcoin is in the early adopters stage (McDougall, 2014).

The Technology Acceptance Model (TAM) is an information systems theory which is developed to explain technology acceptance (Davis, 1986). The model has a strong background from prior theories, such as the Theory of Reasoned Action (Fishbein & Ajzen, 1975). According to the Theory of Reasoned Action, the best way to forecast the behavior is the individual's intention to complete the behavior. Among the many models suggested to explain innovation acceptance and usage, TAM has been the mostly accepted model. (Chau & Hu, 2001; Svendsen, Johnsen, Almas-Sorensen, & Vitterso, 2013; Venkatesh & Ramesh, 2006). According to TAM, perceived ease of use and perceived usefulness determine the behavioral intention of using a technology. Perceived ease of use can be defined as the degree of belief that the technology usage will be effortless (Featherman, Miyazaki, & Sprott, 2010). On the other hand perceived usefulness refers to the belief of degree that using the technology will enhance the users' performance. In the model it is also proposed that perceived ease of use can explain the variance in perceived usefulness.

For understanding the Bitcoin adoption, studies were conducted using the Technology Acceptance Model in different countries. A study by Folkinshteyn and Lennon (2016) tried to analyze the Technology Acceptance Model's dimensions from both the end user side and the developers' (technical) side. According to the study, one of the perceived risks of the end users is the changing, evolving nature of the Bitcoin system. The operating mechanism is not very easy to comprehend even for technical people. During the early days of Bitcoin as the value of BTC was not high, people started to experience it without fully figuring out the whole system. At the early stages it was perceived that the potential losses were negligible. As the value of BTC has risen, potential losses due to user errors are not acceptable. Security risk is also another important perceived risk. The personal computer security risk is very high considering that 30–50% of all computers are under risk of crime ware attack in the US (Danchev, 2010). Consequently end users do not perceive personal computers to be a safe medium for Bitcoin storage. End users perceive Bitcoin to be easy to use because of certain attributes of it. Firstly Bitcoin has an open source software and there is no cost of getting started to use it. Secondly the Bitcoin client has a simple, clear cut interface. Thirdly there is sufficient national currency and Bitcoin convertibility. For an innovation to be adopted it should be perceived as useful. Excluding the traditional intermediaries, being accessible 24/7, fast and low cost and transnational are some of the characteristics of Bitcoin which are perceived as useful by the end user (Folkinshteyn & Lennon, 2016).

The Technology Acceptance Model is the simplest model for explaining the technology acceptance process. The model has been expanded by different scholars as time changes and technological environment necessitates change. Venkatesh, Morris, Davis, and Davis (2003) proposed another model, the Unified Theory of Acceptance and Use of Technology Model, which accommodates the findings of eight previous acceptance theories (Venkatesh et al., 2003). The core dimensions of the model are performance expectancy, effort expectancy, social influence, and facilitating conditions which affect both behavioral intention and actual behavior. According to the model gender, age, experience, and voluntariness moderate the relationships stated above. Jung et al. analyzed the intention towards cryptocurrency in three different countries (Korea, China and Vietnam) through the application of the Unified Theory of Acceptance and Use of Technology Model. The results of the study suggested that performance expectancy, effort expectancy, social influence and facilitating factors positively influence intention to use (Jung, Park, Phan, Bo, & Gim, 2018). A similar study was conducted in Indonesia, suggesting that performance expectancy and facilitating conditions have positive effect on the use behavior (Gunawan & Novendra, 2017).

Apart from general benefits, Bitcoin also provides specific benefits to developing country citizens. For countries with unstable monetary systems and high inflation rates, Bitcoin can be a solution for citizens to protect their savings against depreciation of national currency.

14.11 Conclusion

Technological advancements in industry 4.0 era started to change the way people do business, daily activities and pretty much in every aspect of life. The Blockchain technology is no exception. There are many application areas which have the potential to create substantial improvements in life. The business applications are already prevailing. The most well-known use of this technology, Bitcoin, has been seen like a breakthrough development and has changed the rule of the game in the recent years. There is no consensus about how to treat Bitcoin globally. It is regarded as a currency, an asset, and a novelty. The governments have not yet come to a conclusion about the legitimacy of it. In such a scene the adoption process of Bitcoin and the underlying technology, Blockchain, is not easy to forecast. With the advancement promises and the sui generis characteristics of the technology, the mainstream adoption in the future is obvious. The pace and the process of the adoption may differ from the business world to the individual user. Nevertheless, the adoption of each nourishes the other. The technology acceptance theories and models used in different academic studies exhibit the benefits that are provided by the technology, and the ease of use leads to usage behavior. As stated in the chapter, the agendas and motivations of developed and developing country citizens may differ, but in general the Blockchain technology and the applications of it will globally dominate the near future.

References

Accenture. (2015). Blockchain technology: preparing for change. https://www.accenture.com/pl-en/~/media/Accenture/next-gen/top-ten-challenges/challenge4/pdfs/Accenture-2016-Top-10-Challenges-04-Blockchain-Technology.pdf

Aksoy, E. (2018). *Bitcoin*. Istanbul: Abakus.

Alexander, G., & Camilleri, A. F. (2017). *Blockchain in education*. Luxemburg: JRC Science for Policy Report.

Ali, R., Barrdear, J., Clews, R., & Southgate, J. (2014). Innovations in payment technologies and the emergence of digital currencies. *Bank of England Quarterly Bullettin, 54*(3), 262–275.

Ashton, K. (2009). That 'internet of things' thing. *RFID Journal*. http://www.rfidjournal.com/articles/pdf?4986

Athey, S., Parashkevov, I., Sarukkai, V., & Xia, J. (2017). Bitcoin pricing, adoption, and usage: Theory and evidence. *SIEPR Working Paper*.

Atkins, B., Chapman, J., & Gordon, M. (2014). A whole new world: Income tax considerations of Bitcoin Economy. *Pitt Tax Review, 12*(1), 25–56.

Ayhan, G. (2018). *Ödemeleri Bitcoin'le Kabul Eden 8 Büyük Firma*. Banka Kredileri. https://www.banka-kredileri.com/odemeleri-bitcoinle-kabul-eden-8-buyuk-firma

Bajpai, P. (2010). *Countries where Bitcoin is legal & illegal (DISH, OTSK)*. Investopedia, https://www.investopedia.com/articles/forex/041515/countries-where-bitcoin-legal-illegal.asp

Baur, A. W., Büler, J., Bick, M., & Bonorden, C. S. (2015). Cryptocurrencies as a disruption? Empirical findings on user adoption and future potential of bitcoin and co. In *Conference on e-business, e-services and e-society* (pp. 63–80). Springer

Biryukov, A., Khovratovich, D., & Pustogarov, I. (2014). Deanonymisation of clients in Bitcoin P2P network. In *ACM SIGSAC conference on computer and communications security* (pp. 15–29). ACM.

Böhme, R. (2013). Internet protocol adoption: Learning from Bitcoin. *IAB workshop on internet technology adoption and transition* (pp. 319–327). ITAT.

Bohr, J., & Bashir, M. (2014). Who uses bitcoin? An exploration of the bitcoin community. In *Twelfth annual conference on privacy, security and trust (PST)* (pp. 94–101). IEEE.

Boucher, P. (2017). *How blockchain technology could change our lives: In-depth analysis*. Brussels: European Parliament.

Brezo, F., & Bringas, P. G. (2012). Issues and risks associated with cryptocurrencies such as bitcoin. In *The second international conference on social eco-informatics* (pp. 20–26). Venice: IARIA.

Brimble, M., Vyvyan, V., & Ng, C. (2013). Belief and investing: Preferences and attitudes of the faithful. *Australasian Accounting, Business and Finance Journal, 7*(1), 23–41.

Bryans, D. (2014). Bitcoin and money laundering: Mining for an effective solution. *Indiana Law Journal, 89*, 441–472.

Can, A. (2013). *Bitcoin, KKTC üzerinden geldi*. Hurriyet, http://www.hurriyet.com.tr/ekonomi/bitcoin-kktc-uzerinden-geldi.

Chau, P. Y., & Hu, P. J. (2001). Information technology acceptance by individual professionals: A model comparison approach. *Decision Sciences, 32*, 699–719.

Chuen, L. K. (2015). *Handbook of digital currency: Bitcoin, innovation, financial instruments, and big data*. London: Academic Press.

Cognizant. (2016). *Blockchain in banking: A measured approach*. Cognizant, https://www.cognizant.com/whitepapers/Blockchain-in-Banking-A-Measured-Approach-codex1809.pdf.

Coinbase. (2019). *Price bitcoin*. https://www.coinbase.com/price/bitcoin

Connolly, A. J., & Kick, A. (2015). What differentiates early organization adopters of bitcoin from non-adopters? In *Twenty-first Americas conference on information systems*. Puerto Rico.

Danchev, D. (2010). *Report: 48% of 22 million scanned computers infected with malware*. ZDNet, https://www.zdnet.com/article/report-48-of-22-million-scanned-computers-infected-with-malware

Davis, F. D. (1986). A technology acceptance model for empirically testing new end-user acceptance of information technology. *Doctoral dissertation.* Boston: Massachusetts Institute of Technology.

De Filippi, P. (2016). *Blockchain-based crowdfunding: What impact on artistic production and art consumption?.* HAL, http://hal.archieves-ouverts.fr

Deloitte. (2015). *Blockchain lab.* Deloitte, https://www2.deloitte.com/ie/en/pages/technology/topics/blockchain-lab.html

Descoteaux, D. (2014). Bitcoin: More than a currency, potential for innovation. Montreal Economic Institute.

Devine, P. (2015, December 7–10). Blockchain learning: Can crypto-currency methods be appropriated to enhance online learning? In *ALT online winter conference 2015.*

Dostov, V., & Shust, P. (2014). Cryptocurrencies: An unconventional challenge to the AML/CFT regulators? *Journal of Financial Crime, 21*(3), 249–263.

Du, S., Yalcinkaya, G., & Bstieler, L. (2016). Sustainability, social media driven open innovation, and new product development performance. *Journal of Product Innovation Management, 33*, 55–71.

ECB. (2012). *Virtual currency schemes.* https://www.ecb.europa.eu/pub/pdf/other/virtualcurrencyschemes201210en.pdf

Eskandari, S., Clark, J., Barrera, D., & Stobert, E. (2018). *A first look at the usability of bitcoin key management.* Arxiv, https://arxiv.org/pdf/1802.04351.pdf

Evans, C. W. (2015). Bitcoin in Islamic banking and finance. *Journal of Islamic Banking and Finance, 3*(1), 1–11.

Exton, J., & Doidge, F. (2018). *Cracking the code on cryptocurrency..* ING.

Fairfield, J. A. (2014). Smart contracts, bitcoin bots, and consumer protection. *Washington and Lee Law Review Online, 71*, 35–50.

Featherman, M. S., Miyazaki, A. D., & Sprott, D. E. (2010). Reducing online privacy risk to facilitate e-service adoption: The influence of perceived ease of use and corporate credibility. *Journal of Services Marketing, 24*, 219–229.

Fishbein, M., & Ajzen, I. (1975). *Belief, attitude, intention and behavior: An introduction to theory and research.* Reading, MA: Addison-Wesley.

Foley, S., Karlsen, J. R., & Putnins, T. (2018). *Sex, drugs, and bitcoin: How much illegal activity is financed through cryptocurrencies?* Oxford Law, https://www.law.ox.ac.uk/business-law-blog/blog/2018/02/sex-drugs-and-bitcoin-how-much-illegal-activity-financed-through

Folkinshteyn, D., & Lennon, M. (2016). Braving bitcoin: A technology acceptance model (TAM) analysis. *Journal of Information Technology Case and Application Research, 18*(4), 220–249.

Gao, X., Clark, G. D., & Lindqvist, J. (2015). *Of two minds, multiple addresses, and one history: Characterizing opinions, knowledge, and perceptions of bitcoin across groups.* Arxiv, https://arxiv.org/abs/1503.02377

Gazali, H. M., Ismail, C. H., & Amboala, T. (2018). Exploring the intention to invest in cryptocurrency: The case of bitcoin. In *International conference on information and communication technology for the Muslim world (ICT4M)* (pp. 64–68). CPS.

Geith, C. (2008). Unleashed with web 2.0 and open educational resources. In *The tower and the cloud.*

Gellman, R. (1996). Disintermediation and the Internet. *Government Information Quarterly, 13*(1), 1–8.

Glaser, F., Zimmermann, K., Haferkorn, M., Weber, M., & Siering, M. (2014). Bitcoin-asset or currency? revealing users' hidden intentions. In *Twenty second European conference on information systems.* Tel Aviv: ECIS.

Gogerty, N., & Ziloti, J. (2011). *DeKo—Currency proposal using a portfolio of electricity linked assets.* SSRN, https://papers.ssrn.com/soL3/papers.cfm?abstract_id=1802166

Grant, G., & Hogan, R. (2015). Bitcoin: Risks and controls. *Journal of Corporate Accounting & Finance, 26*(5), 29–35.

Gartner (2014, March 19). *Gartner says the internet of things will transform the data center – IoT – internet of things*. Retrieved from https://iot.do/gartner-says-internet-things-will-transform-data-center-2014-03

Gültekin, Y. (2017). Turizm Endüstrisinde Alternatif Bir Ödeme Aracı Olarak Kripto Para Birimleri: Bitcoin. *Güncel Turizm Araştırmaları Dergisi, 1*(2), 96–113.

Gunawan, F. E., & Novendra, R. (2017). An analysis of bitcoin acceptance in Indonesia. *ComTech: Computer, Mathematics and Engineering Applications, 8*(4), 241–247.

Hardjono, T., & Smith, N. (2016). Cloud-based commissioning of constrained devices using permissioned blockchains. In *Proceedings of ACM IoT privacy, trust & security*. Xian: IoTPTS.

Hulseapple, C. (2015). *Block verify uses blockchains to end counterfeiting and 'make world more honest'*. Cointelegraph, https://cointelegraph.com/news/block-verify-uses-blockchains-to-end-counterfeiting-and-make-world-more-honest

Hur, Y., Jeon, S., & Yoo, B. (2015). Is bitcoin a viable E-business?: Empirical analysis of the digital currency's speculative nature. In *International conference on information systems*. Texas.

Iansiti, M., & Lakhani, K. R. (2017). The truth about blockchain. *Harvard Business Review, 95*, 118–127.

Jaag, C., & Bach, C. (2016). *Blockchain technology and cryptocurrencies :Opportunities for postal financial services*. Technical report.

Jung, K. J., Park, J. B., Phan, N. Q., Bo, C., & Gim, G. Y. (2018). An international comparative study on the intension to using crypto-currency. In *International conference on applied computing and information technology* (pp. 104–123). Springer.

Kaplanhan, F. (2013). Kripto Paranın Türk Mevzuatı Açısından Değerlendirilmesi: Bitcoin Örneği. Vergi Mevzuatı ve Maliye Teorisi, 105–123.

Kar, I. (2016). *The latest customers for the technology behind bitcoin are NATO and the US military*. Quartz, https://qz.com/681580/the-latest-customers-for-the-technology-behind-bitcoin-are-nato-and-the-us-military/

Karaoğlan, S. A. (2018). Türkiye'de Kripto Para Farkındalığı ve Kripto Para Kabul Eden İşletmelerin Motivasyonları. In *Isletme ve Iktisat Calismalari Dergisi* (pp. 15–28).

Karaoglan, S., Arar, T., & Bilgin, O. (2018). Türkiye'de Kripto Para Farkındalığı ve Kripto Para Kabul Eden İşletmelerin Motivasyonları. *Isletme ve Iktisat Calismalari Dergisi, 6*(2), 15–28.

Khalilov, M. C., Gündebahar, M., & Kurtulmuşlar, I. (2017). Bitcoin ile Dünya ve Türkiye'deki Dijital Para Çalışmaları Üzerine Bir İnceleme. In *Proceedings of 19. Akademik Bilişim Konferansı*.

Khatwani, S. (2018). *Top 5 biggest bitcoin hacks ever*. Coinsutra, https://coinsutra.com/biggest-bitcoin-hacks/

Kosba, A., Miller, A., Shi, E., Wen, Z., & Papamanthou, C. (2016). Hawk: The blockchain model of cryptography and privacy-preserving smart contracts. In *IEEE symposium on security and privacy (SP)* (pp. 839–858). IEEE.

Kronbholz, K., Judmayer, A., Gusenbauer, M., & Weippl, E. (2016). The other side of the coin: User experiences with bitcoin security and privacy. In *International conference on financial cryptography and data security* (pp. 555–580). Berlin: Springer.

Lee, J. H., & Pilkington, M. (2017). How the blockchain revolution will reshape the consumer electronics industry [future directions]. *IEEE Consumer Electronics Magazine, 6*(3), 19–23.

Lessig, L. (2004). The creative commons. *Montana Law Review, 65*(1), 1–14.

Lielacher, A. (2019). How many people use bitcoin in 2019? *Bitcoin Market Journal*, https://www.bitcoinmarketjournal.com/how-many-people-use-bitcoin/

Manovich, L. (2009). The practice of everyday (media) life: From mass consumption to mass cultural production? *Critical Inquiry, 35*(2), 319–331.

Mauldin, J. (2014). *The 5 phases of bitcoin adoption*. Forbes, https://www.forbes.com/sites/johnmauldin/2014/12/23/the-5-phases-of-bitcoin-adoption/#5ccb50351764

McDougall, M. (2014). An investigation of the theory of disruptive innovation: Does the cryptocurrency bitcoin have the potential to be a disruptive innovation relative to an existing market? *Master's Thesis Edinburgh Napier University*. Edinburg: Edinburgh Napier University.

Meiklejohn, S., Pomarole, M., Jordan, G., Levchenko, K., McCoy, D., Voleker, G. M., et al. (2013). A fistful of bitcoins: Characterizing payments among men with no names. In *Proceedings of the 2013 conference on internet measurement* (pp. 127–140). New York: ACM.

Mendez, C. (2012). *BBVA*. Diamonds, blockchain and banks: The story of Everledger, https://www.bbva.com/en/diamonds-blockchain-and-banks-the-story-of-everledger/

Migirov, R. (2016). The supply circle: How blockchain technology disintermediates the supply chain, https://media.consensys.net/the-supply-circle-how-blockchain-technology-disintermediates-the-supply-chain-6a19f61f8f35

Nakamoto, S. (2008). *Bitcoin: A peer-to-peer electronic cash system*.

Narayanan, A., Bonneau, J., Felten, E., Miller, A., & Goldfeder, S. (2016). *Bitcoin and cryptocurrency technologies: A comprehensive introduction*. Princeton, NJ: Princeton University Press.

Nian, L. P., & Chuen, D. L. (2015). Introduction to bitcoin. In D. L. Chuen (Ed.), *Handbook of digital currency* (pp. 6–29). Elsevier.

Noyes, C. (2016). *Efficient blockchain-driven multiparty computation markets at scale*. Technical Report.

NRI. (2015). *Survey on blockchain technologies and related services*. Technical Report.

O'Flaherty, K. (2018). *Facebook data breach—What to do next*. Forbes., https://www.forbes.com/sites/kateoflahertyuk/2018/09/29/facebook-data-breach-what-to-do-next/#7a1fecfc2de3

Pagliery, J. (2014). *Bitcoin and the future of money*. Chicago: Triumph Books.

Peters, G., Panayi, E., & Chapelle, A. (2015). *Trends in cryptocurrencies and blockchain technologies: A monetary theory and regulation perspective*. Arxiv, https://arxiv.org/pdf/1508.04364.pdf

Pilkington, M. (2016). 11 Blockchain technology: Principles and applications. In F. X. Olleros & M. Zhegu (Eds.), *Research handbook on digital transformations* (pp. 225–253). Glos: Edward Elgar.

Popper, N. (2013). *The rush to coin virtual money with real value*. DealBook, https://dealbook.nytimes.com/2013/11/11/the-rush-to-coin-virtual-money-with-real-value/

Research, C. (2017). *Consumer electronics market*. Credence Research, https://www.credenceresearch.com/report/consumer-electronics-market

Rogers, E. M. (2010). *Diffusion of innovations*. New York: Free Press.

Sharples, M., & Domingue, J. (2016). The blockchain and kudos: A distributed system for educational record, reputation and reward. In *European conference on technology enhanced learning* (pp. 490–496). Cham: Springer.

Simulacrum. (2014). *The demographics of bitcoin*. Simulacrum, https://spacedruiddotcom.wordpress.com/2013/03/04/the-demographics-of-bitcoin-part-1-updated/

Sloan, K. (2018). *7 major companies that accept cryptocurrency*. Due, https://due.com/blog/7-companies-accept-cryptocurrency/

Sonmez, A. (2014). Sanal Para Bitcoin. *The Turkish Online Journal of Design, Art and Communication, 4*(3), 1–14.

Svendsen, G. B., Johnsen, J. A., Almas-Sorensen, L., & Vitterso, J. (2013). Personality and technology acceptance: The influence of personality factors on the core constructs of the technology acceptance model. *Behaviour & Information Technology, 13*(4), 323–334.

Torpey, K. (2014). *Andreas Antonopoulos lays out the 3 pillars of bitcoin mass adoption*. Inside Bitcoins, https://insidebitcoins.com/news/andreas-antonopoulos-lays-out-the-3-pillars-of-bitcoin-mass-adoption/26925

Trautman, L. J. (2014). Virtual currencies; Bitcoin & what now after Liberty Reserve, Silk Road, and Mt. Gox? *Richmond Journal of Law and Technology, 20*(4).

Tsanidis, C., Nerantzaki, D. M., Karavasilis, G., Vrana, V., & Paschaloudis, D. (2015). Greek consumers and the use of Bitcoin. *The Business & Management Review, 6*, 295–301.

Tubitak. (2018). *Blockchain research laboratory*. Tubitak, http://blokzincir.tubitak.gov.tr/

Underwood, S. (2016). Blockchain beyond Bitcoin. *Communications of the ACM, 59*, 15–17.

Ünsal, E., & Kocaoglu, O. (2018). Blok Zinciri Teknolojisi: Kullanım Alanları, Açık Noktaları ve Gelecek Beklentileri. *Avrupa Bilim ve Teknoloji Dergisi, 13*, 54–64.

Vagata, P., & Wilfong, K. (2014). *Scaling the Facebook Data Warehouse to 300PB*. Technical report.

Venkatesh, V., & Ramesh, V. (2006). Web and wireless site usability: Understanding differences and modeling use. *MIS Quarterly, 30*, 181–206.

Venkatesh, V., Morris, M. G., Davis, G. B., & Davis, F. D. (2003). User acceptance of information technology: Toward a unified view. *MIS Quarterly, 27*(3), 425–478.

Walch, A. (2015). The bitcoin blockchain as financial market infrastructure: A consideration of operational risk. *Legislation and Public Policy, 18*(4), 837–893.

Wile, R. (2014). *13% of Americans would choose bitcoin over gold*. Business Insider, https://www.businessinsider.com/bitcoin-awareness-survey-2014-3

Zheng, Z., Xie, S., Dai, H. N., Chen, X., & Wang, H. (2018). Blockchain challenges and opportunities: A survey. *International Journal of Web and Grid Services, 14*(4), 352–375.

Zohar, A. (2015). Bitcoin: Under the hood. *Communications of the ACM, 58*(9), 104–113.

Zyskind, G., & Nathan, O., & Pentland, A. (2015). Decentralizing privacy: Using blockchain to protect personal data. In *Security and privacy workshops (SPW)* (pp. 180–184). IEEE.

Hande Begüm Bumin Doyduk completed her BA in Business Administrations at İhsan Doğramacı Bilkent University, MBA at HEC Paris and Ph.D. in Marketing at Yeditepe University. She worked as a brand manager for seven years at Burger King, Alcatel Lucent Paris, and Colgate Palmolive Turkey. She taught various marketing courses at Yeditepe University, Okan Universityersity, Bahçeşehir University and Medipol University as part-time faculty. She is a full-time faculty member in Business Administration Faculty at Altınbaş University. Her research interests are in the areas of corporate social responsibility, sustainability, consumer behavior, and digital marketing.

Chapter 15
Empirical Evidence of the Relationships Between Bitcoin and Stock Exchanges: Case of Return and Volatility Spillover

M. Kamisli, S. Kamisli, and F. Temizel

Abstract Especially with the sharp increase in the trading volume of Bitcoin, researchers have focused on the topic of cryptocurrencies. Besides the high risks they carry, these vehicles give investors the opportunity of gaining high returns. For this reason, many investors consider cryptocurrencies as an investment vehicle and include them into their portfolios, notably Bitcoin. Bitcoin is a new alternative for investors who desire to invest in different assets besides traditional ones. This new investment vehicle is also used for portfolio diversification. But, to provide the desired benefits, the relationships between the bitcoin and asset or assets that will be included in the portfolio. Therefore, the purpose of this study is to analyze the return and volatility relationships between Bitcoin and stock markets from different regions. For this purpose, Diebold and Yilmaz spillover test are applied to the return series. The empirical results indicate both return and volatility spillovers between the Bitcoin and the selected stock markets that should be considered in portfolio and risk management processes.

15.1 Introduction

In today's financial world economic units desire to make their money transfers by online payment systems. Online payment systems provide saving of time, and transaction costs are generally lower in these systems because there is no intermediary in these transactions. In this context, the cryptocurrencies present users the opportunity of lower transaction cost and faster transaction. These tools also provide decentralized transaction process, in other words, they are free from the government

M. Kamisli (✉) · S. Kamisli
Department of Banking and Finance, Faculty of Applied Sciences, Bilecik Seyh Edebali University, Bozüyük, Bilecik, Turkey
e-mail: melik.kamisli@bilecik.edu.tr; serap.kamisli@bilecik.edu.tr

F. Temizel
Department of Business Administration, Faculty of Economics and Administrative Sciences, Anadolu University, Eskişehir, Turkey
e-mail: ftemizel@anadolu.edu.tr

© Springer Nature Switzerland AG 2019

U. Hacioglu (ed.), *Blockchain Economics and Financial Market Innovation*, Contributions to Economics, https://doi.org/10.1007/978-3-030-25275-5_15

control and regulations, and they provide privacy to the users (Hameed & Farooq, 2016). Although they are named as currency, there are some properties that the currency should carry; firstly they are used as a medium of exchange, secondly they have store value, and finally, they are used as a unit of account (Bariviera, Basgall, Hasperué, & Naiouf 2017). The cryptocurrencies may have these properties but, there are some features that distinguish cryptocurrencies from traditional currencies. They do not represent any underlying asset or tangible wealth source and not issued by governments or central banks (Teichmann, 2018).

The first decentralized cryptocurrency Bitcoin which is introduced by Nakamoto in 2008 is a peer-to-peer electronic payment system that allows payments from one party to another without an intermediary (Urquhart and Zhang, 2019). Although it is initially developed as a payment system, cryptocurrencies are now considered as an investment tool. Notably Bitcoin, cryptocurrencies are now added to portfolios for speculative purpose or used with the other assets for diversification. According to Modern Portfolio Theory (MPT) to construct a portfolio of different assets, on average provide a higher return and less risk in comparison to the portfolio of constructed any individual asset. In other words, diversification gives the opportunity of reducing risk of the portfolio by keeping or increasing the return of the portfolio. Therefore, it may improve the portfolio performance to add Bitcoin into the portfolios. On the other hand, in some of the studies it is revealed that there are bubbles in cryptocurrency markets. (Cheah & Fry, 2015; Cheung, Roca & Su, 2015; Corbet, Lucey, & Yarovaya, 2018; Fry & Cheah, 2016). In the occurrence phase of the bubbles, investors are motivated to undertake more risk, and risky investments are realized more in this phase. But if the bubble burst, the return of the investor is lower than the expected. Thus, in order to obtain the expected benefits from cryptocurrencies, the drivers of these assets should be determined, and the relationships with the other assets should be investigated appropriately.

To investigate the relationships of an asset with the other assets is essential for both determining the drivers of the asset and determining the portfolio and risk management strategies in case of constructing a portfolio with the analyzed assets. In case of both it is invested in only the asset or portfolio of the asset with the other assets, identifying the factors that affect the price and hence the return of the asset helps investors to make decision by following the changes in the variables that affect the asset. It is also important to determine the relationships between the assets for deciding the assets that will be included to the portfolio and determining the weights of the assets in the portfolio, if the investor desires to invest in the portfolio constructed of different assets. On the other hand, in case of analyzing the relationships between the assets before the investment, investors will have the chance of revising the portfolio if there occur changes in the assets.

In the light of the explanations above, the purpose of this study is to determine the relationships between Bitcoin and the stock markets from different regions of the world and present information to the investors that they may use in portfolio and risk management processes. For this purpose, the return and volatility spillovers between Bitcoin and the selected stock markets are investigated by Diebold and Yilmaz spillover test. Before the analyses, the literature on Bitcoin and the stock markets are given and the data and methodology part is presented. After the empirical tests, the results are interpreted and given at the conclusion part.

15.2 Literature Review

Due to the sharp increases in the trading volume of Bitcoin, the cryptocurrencies notably bitcoin, has aroused the interest of academic, investors and regulators. For this reason, the number of studies on the Bitcoin has increased day by day. There are studies in the finance literature that examine the market efficiency of Bitcoin market (Bariviera, 2017; Kristoufek, 2018; Nadarajah and Chu, 2017; Tiwari, Jana, Das, & Roubaud, 2018; Urquhart, 2016), drivers of bitcoin prices (Baek & Elbeck, 2015; Bouoiyour, Selmi, Tiwari, & Olayeni 2016; Ciaian, Rajcaniova, & Kancs, 2016; Kristoufek, 2015; Phillips & Gorse, 2018) and volatility of Bitcoin (Bouoiyour & Selmi, 2016; Katsiampa, 2017; Pichl & Kaizoji, 2017). However, most of the existing literature on Bitcoin focus on the hedging capability and safe-haven properties of Bitcoin. In most of these studies it is indicated that adding Bitcoin to the portfolio reduce the total risk and improve the performance of the portfolio. In other words, Bitcoin may be used as a financial asset and as a diversifier. To determine the relationships of Bitcoin with the other investment vehicles also will also present valuable information to the investors to be used in portfolio and risk management processes.

Öztürk, Arslan, Kayhan, and Uysal (2018) examined the relationships between Bitcoin and the traditional assets including stocks. The cointegration and causality results showed that there is no relationship between Bitcoin and the stock market indicating that Bitcoin may be used for portfolio diversification. In order to evaluate the diversification benefit of Bitcoin, Carpenter (2016) used a modified mean-variance framework and created a portfolio of Bitcoin and indices of equity, real estate, commodity and bond. The results revealed that Bitcoin is an attractive investment vehicle and increases the return/risk ratio of the portfolio. Similarly, Kajtazi and Moro (2018) analyzed the effect of adding Bitcoin to the portfolio on the performance of the portfolio. They constructed two portfolios by U.S., European and Chinese assets with and without Bitcoin, and compared the performances of the portfolios to determine the effect of Bitcoin. They concluded that Bitcoin improves portfolio performance and may be used for diversification. However, the empirical results indicated that the positive effect of Bitcoin on portfolio performance resulted from the increase in Bitcoin returns rather than the reduction of volatility.

Dyhrberg (2016) investigated the hedging capability of Bitcoin. According to empirical results, Bitcoin may be used as a hedge against Financial Times Stock Exchange Index and minimize market risk. Similarly, Bouri, Gupta, Tiwari, and Roubaud (2017) aimed to analyze that if Bitcoin may be used as a hedge or not. For this purpose wavelet-based quantile-in-quantile regression method was employed to Bitcoin and volatility index of selected stock markets from both developed and emerging countries. The results showed that Bitcoin may be used as a hedge against global uncertainty at short investment horizons. Bouri, Azzi, and Dyhrberg (2017) constructed a portfolio of Bitcoin and US equities in order to determine the effectiveness of Bitcoin in risk reduction. Empirical results proved that Bitcoin is was a safe-haven before the price crash of 2013. Findings also showed that adding Bitcoin to the portfolio reduces the portfolio risk and may be used as a financial asset supporting the other analyses of the study.

Including stocks, Andrianto and Diputra (2017) created portfolios in order to examine the effectiveness of cryptocurrencies; Bitcoin, Ripple and Litecoin on portfolio performance. They concluded that cryptocurrencies improve the performance of the portfolio by minimizing the standard deviation of the portfolio and creating allocation options for the investors. They also stated that the optimal weight for cryptocurrencies changes between 5 and 20% depending on the risk perception of the investor. Chan, Le and Wu (2019) investigated hedge and diversification ability of Bitcoin against stock indices for different frequencies. Pairwise GARCH models and constant conditional correlation model were applied to the data of Bitcoin price and Euro STOXX, Nikkei, Shanghai A-Share, S&P500, the TSX Index indices. The empirical results showed that for monthly frequency Bitcoin is a strong hedge for all indices. But, for daily and weekly frequency it does not have strong hedge property.

When the studies in the literature are examined, it is seen that there are limited studies on the relationships between Bitcoin and stock markets. Therefore, it is thought that it will contribute to the literature to analyze the relationships between Bitcoin and the stock markets with a large dataset and a contemporary econometric approach. For this reason, it is aimed to investigate the return and volatility spillovers between Bitcoin and stock markets from different regions of the world. In line with this purpose, the spillover method developed by Diebold and Yilmaz is applied in the study.

15.3 Data and Methodology

The principal aim of the study is to analyze the relationships between bitcoin and stock markets. Therefore, the method developed by Diebold and Yilmaz (2009, 2012) applied to Bitcoin and the stock market returns from different regions. This method have been widely used in the literature to determine the spillovers between different financial variables (Antonakakis, 2012; Awartani, Maghyereh, & Al Shiab, 2013; Chevallier & Ielpo, 2013; Cronin, 2014; Lee & Chang, 2013; Lucey, Larkin, & O'Connor, 2014; Maghyereh & Awartani, 2016; Maghyereh, Awartani, & Bouri, 2016; Narayan, Narayan, & Prabheesh, 2014; Sehgal, Ahmad, & Deisting, 2015; Sugimoto, Matsuki & Yoshida, 2014; Tsai, 2014; Yarovaya, Brzeszczyński & Lau, 2016). Prior to volatility spillover analysis, in order to obtain volatility series GARCH (1,1) model was applied to the series taken into consideration in the study.

Autoregressive Conditional Heteroscedasticity test developed by Engle (1982), indicates that the conditional mean and variance of the series can be discretely modeled simultaneously. The ARCH model and its limits are;

$$y_t = \varepsilon_t \sqrt{h_t} \tag{15.1}$$

$$y_t | \psi_{t-1} \sim N(0, h_t) \tag{15.2}$$

$$h_t = \alpha_0 + \alpha_1 y_{t-1}^2 \tag{15.3}$$

$$\alpha_0 > 0, \qquad \alpha_i \geq 0, \tag{15.4}$$

$$0 < \alpha_0 < 1 \tag{15.5}$$

The Generalized Autoregressive Conditional Heteroscedasticity model which is an extension of ARCH model developed by Bollerslev (1986), is the volatility model illustrating that the conditional variance is related to its own lags in addition to lags of error terms. The GARCH model and its limits are;

$$\varepsilon_t | \psi_{t-1} \sim N(0, h_t) \tag{15.6}$$

$$h_t = \alpha_0 + \sum_{i=1}^{q} \alpha_i \varepsilon_{t-i}^2 + \sum_{i=1}^{p} \beta_i h_{t-i} \tag{15.7}$$

$$p \geq 0, \qquad q > 0 \tag{15.8}$$

$$\alpha_0 > 0, \qquad \alpha_i \geq 0, \quad i = 1, \ldots, q \tag{15.9}$$

$$\beta_i \geq 0, \quad i = 1, \ldots, p \tag{15.10}$$

$$\sum_{i=1}^{q} \alpha_i + \sum_{i=1}^{p} \beta_i < 1 \tag{15.11}$$

The return and volatility spillovers in the model developed by Diebold and Yilmaz (2009, 2012) are based on Vector Autoregressive (VAR) models. A two-variable VAR model with stationary covariance is shown in (12);

$$X_t = \Phi x_{t-1} + \mathcal{E}_t \tag{15.12}$$

Here, $X_t = X_{1t}, X_{2t}$ and Φ are 2×2 parameter matrices. Covariance stationarity and moving averages of the VAR model are;

$$X_t = \Theta(L)\mathcal{E}_t \tag{15.13}$$

Here, $\Theta(L) = (I - \Theta L)^{-1}$. Additionally, the model can be rewritten to show moving averages as;

$$X_t = A(L)u_t \tag{15.14}$$

Here, $A(L) = \Theta(L)Q_t^{-1}, u_t = Q_t\mathcal{E}_t, E(u_t u_t')$ and Q_t^{-1}, $\mathcal{E}_t$ are sub triangular Cholesky factors of the covariance matrix.

The spillover index based on two first order variables VAR model is;

$$S = \frac{a_{0,12}^2 + a_{0,21}^2}{trace\left(A_0 A_0'\right)} * 100 \tag{15.15}$$

Spillover index based on a VAR model with N p[th] order variables can be calculated, using 1 step pre-estimation by the following equation;

$$S = \frac{\sum_{i,j}^{N} = 1 a_{0,ij}^2}{trace\left(A_0 A_0'\right)} * 100 \tag{15.16}$$

And, generally a spillover index based on a VAR model of N p[th] order variables can be calculated using H step pre-estimation by the following equation;

$$S = \frac{\sum_{h=0}^{H-1} \sum_{i,j=1}^{N} a_{h,ij}^2}{\sum_{h=0}^{H-1} trace\left(A_h A_h'\right)} * 100 \tag{15.17}$$

Diebold and Yilmaz (2012) have defined net spillovers from i to j as;

$$S_i^g(H) = S_{.i}^g(H) - S_{i.}^g(H) \tag{15.18}$$

The net spillover in Eq. (15.18) presents a summary of the volatility contribution of one investment vehicle to another. Net pairwise spillover can be shown as;

$$\begin{aligned} S_{ij}^g(H) &= \left(\frac{\tilde{\theta}_{ji}^g(H)}{\sum_{i,k=1}^{N} \tilde{\theta}_{ik}^g(H)} - \frac{\tilde{\theta}_{ij}^g(H)}{\sum_{j,k=1}^{N} \tilde{\theta}_{jk}^g(H)} \right) \\ &= \left(\frac{\tilde{\theta}_{ji}^g(H) - \tilde{\theta}_{ij}^g(H)}{N} \right) * 100 \end{aligned} \tag{15.19}$$

The study analyzed weekly data on bitcoin and regional stock index returns from 8.26.2011 to 2.15.2019. Data was obtained from Thomson & Reuters DataStream. The regional stock indices included in the study are;

- Stock markets of Americas Region: Argentina (ARG), Brazil (BRA), Canada (CAN), Colombia (COL), Mexico (MEX), PERU (PER), United States (USA), Venezuela (VEN).
- Stock markets of Asia Region: Australia (AUS), China (BRA), India (IND), Japan (JPN), Sri Lanka (LKA), Philippines (PHL), Singapore (SGP), Thailand (THA), Taiwan (TWN), Vietnam (VNM).

- Stock markets of Europe Region: Germany (DEU), Spain (ESP), France (FRA), United Kingdom (GBR), Greece (GRC), Ireland (IRL), Italy (ITA), Portugal (PRT), Turkey (TUR),
- Stock markets of Islamic Cooperation (OIC) Member Countries: Egypt (EGY), Indonesia (IDN), Malaysia (MYS), Pakistan (PAK), Qatar (QAT), Saudi Arabia (SAU), United Arab Emirates (UAE).

15.4 Empirical Results

Determination of the relationship between financial instruments to be included in the portfolio is very important for investors. In addition to this, descriptive statistics of financial assets present investors important information that can be used in selecting investment alternatives too. Therefore, descriptive statistics and unconditional correlations of Bitcoin and regional stock indexes were calculated and presented at Appendix 1

As can be seen in the descriptive statistics table that the variables with highest average returns in Americas region were; Venezuela (4.55%) Bitcoin (1.47%) and Argentina (0.66%) respectively. On the other hand, the Colombian stock index is observed to have negative return (–0.03%). The stock markets with highest risk following Bitcoin (13.6%) are Venezuela (13.1%) and Argentina (4.7%). In the region, only Venezuelan, Brazilian and Peruvian stock markets have positive skewness values. Results show that all return series examined in the study have high kurtosis values. The calculated descriptive statistics explain the effects of Venezuelan crisis on the region. Unconditional correlation coefficients show a negative relationship between Bitcoin and Mexican (–2.31%) and Colombian (–0.80%) stock markets while a small positive relationship can be observed for the other stock markets.

As can be seen from Table 15.9, in the Asia region Vietnam (1.47%), Taiwan (0.22%), Thailand (0.21%) and Singapore (0.19%) stock markets have the highest rate of return. For the period being examined only Australian stock market (–0.04%) have negative returns. Also, the Vietnamese (13.6%) and Australian (1.50%) stock markets were observed to have the highest and lowest level of risk respectively, in the region. Results indicate that for the Asia region, stock market returns for countries other than Australia, China and Thailand, have negative skewness values. Also, similar to the Americas region, stock returns for Asian markets have high kurtosis values. Unconditional correlation coefficients show a negative relationship between Bitcoin returns and stock market returns of China (–4.21%), India (–4.20%) and Thailand (–1.24%), and low but positive relationship for other countries in the region.

In the European region, Ireland (0.22%), Germany (0.17%) and Turkey (0.17%) have the highest return following Bitcoin. On the other hand, Greek (–0.09) and Portuguese (–0.05%) stock markets have negative returns. Of the stock markets taken into consideration, Greek (3.81%) and Italian (2.56%) markets have the highest level of risk. Also, returns for all stock markets in the region have negative skewness and high kurtosis values. Unconditional correlation results show a low level of positive relationship between Bitcoin and stock market returns of all the markets in the region. The descriptive statistics calculated can be interpreted as illustrating the effects of Greek and European debt crises.

All OIC member countries considered during the timeframe of the study have positive stock market returns. Following Bitcoin, Pakistan (0.22%), Egypt (0.17%), and United Arab Emirates (0.17%) have the highest average rate of return. In the group of countries being examined, Egypt (2.75%), Saudi Arabia (2.07%) and Qatar (1.91%) have highest stock market volatility. Also, in a manner similar to the European region, stock market returns for all OIC countries in the study have negative skewness and high kurtosis values. The unconditional correlation coefficients show low positive relationship between Bitcoin returns and the returns of Malaysian (3.3%), Egyptian (2.5%) and Indonesian (1.1%) stock markets and a negative relationship for all the other countries.

Results of Jarque-Bera test show that returns of Bitcoin and the stock markets taken into consideration are not normally distributed. Additionally, stationarity of the return series was tested using Ng-Perron unit root test and these results are presented at Appendix 2. According to results of both models; the model with intercept and the model with intercept and trend, all of the returns examined are stationary. In the next step of the study, GARCH (1,1) model was used to generate volatility series in order to analyze volatility spillover, and results are presented in Appendix 3. In the GARCH model; c denotes the constant, while α is the ARCH parameter and β is the GARCH parameter, and the series included in the analysis satisfy the constraints of the GARCH model. After obtaining the volatility series with the GARCH (1,1) model, return and volatility spillovers were tested using the method developed by Diebold and Yilmaz.[1] The tables below show the return and volatility spillovers between Bitcoin and the selected regional stock indexes.

As can be seen from Table 15.1, in the Americas region the spillover from the Canadian stock market to the others in the region is the highest. Additionally, there are significant spillovers from US, Mexico and Brazil stock markets to Bitcoin and the other stock markets. The spillover from Venezuelan stock market to the others is the lowest in the region. Return spillover from Bitcoin and other stock markets to Canadian stock market is the highest in the region. Also, significant return spillovers from other markets to the stock markets of United States, Mexico and Brazil were observed. Results obtained from the analysis show that the highest level of spillover to Bitcoin is from the Peruvian stock market, and the lowest level is from Brazilian stock market. The highest return spillover from Bitcoin is to Argentinian stock market while the lowest is to the Venezuelan stock market. Significant return spillovers were determined between the stock markets of the region with the exception of Bitcoin and the Venezuelan stock market. On the other hand, Brazilian, Canadian, Mexican and the United States stock markets were found to cause net spillovers. The volatility spillovers for the Americas region is presented at Table 15.2.

[1]The optimal lag length for VAR model was determined based on AIC information criteria.

Table 15.1 Return spillovers in Americas region

	ARG	BRA	BTC	CAN	COL	MEX	PER	USA	VEN	From other markets
ARG	43.45	11.21	1.04	10.34	4.52	8.73	6.72	11.91	2.08	56.55
BRA	9.75	38.98	0.58	11.11	8.67	11.45	9.97	8.52	0.97	61.02
BTC	3.11	0.99	85.45	1.99	0.95	2.29	2.63	2.07	0.51	14.55
CAN	7.84	9.27	0.64	33.73	8.69	9.98	7.56	21.24	1.06	66.27
COL	5.37	10.86	0.59	11.62	45.26	11.32	7.01	7.43	0.55	54.74
MEX	7.58	11.58	0.84	11.35	9.62	38.88	5.74	13.72	0.71	61.12
PER	6.51	13.81	2.53	10.96	9.41	6.93	42.22	6.63	1.00	57.79
USA	8.70	8.11	0.99	22.11	5.48	12.51	5.30	35.71	1.10	64.29
VEN	1.40	3.95	0.94	2.13	1.74	1.46	1.01	2.98	84.40	15.61
Contribution to other markets	50.25	69.78	8.14	81.61	49.07	64.67	45.93	74.50	8.00	451.9
Contribution including own	93.69	108.8	93.59	115.3	94.34	103.5	88.15	110.2	92.39	Total Spillover Index
Net spillovers	–6.31	8.76	–6.41	15.34	–5.66	3.54	–11.86	10.21	–7.61	50.22

Table 15.2 Volatility spillovers in Americas region

	ARG	BRA	BTC	CAN	COL	MEX	PER	USA	VEN	From other markets
ARG	71.30	6.09	0.08	9.77	5.72	1.04	0.79	2.37	2.85	28.70
BRA	11.87	60.57	1.83	9.85	4.20	2.29	7.58	1.38	0.43	39.43
BTC	0.13	0.55	89.96	1.16	1.15	4.98	1.15	0.70	0.23	10.04
CAN	9.44	3.38	0.75	39.84	17.78	6.82	2.49	19.23	0.28	60.16
COL	8.60	7.26	0.64	17.72	51.87	7.40	0.31	5.92	0.28	48.13
MEX	9.28	1.15	1.82	17.36	10.37	44.93	2.11	11.66	1.32	55.07
PER	3.25	13.98	10.76	8.84	0.52	1.16	59.96	1.32	0.23	40.04
USA	4.18	0.44	2.28	27.44	9.21	8.68	2.18	45.14	0.46	54.86
VEN	4.11	0.19	2.10	0.63	3.62	2.22	0.09	9.31	77.74	22.26
Contribution to other markets	50.86	33.02	20.26	92.76	52.56	34.59	16.70	51.88	6.06	358.7
Contribution including own	122.2	93.59	110.2	132.6	104.4	79.52	76.65	97.02	83.81	Total Spillover Index 39.85
Net spillovers	22.16	−6.41	10.22	32.60	4.43	−20.48	−23.35	−2.98	−16.19	

In the Americas region, volatility spillovers from Canadian stock market to Bitcoin and the other stock markets is the highest. Similar to return spillover, the spillover from Venezuelan stock market to the others is the lowest. The volatility spillovers from other stock markets to Canadian stock market are highest. As can be seen from Table 15.2, the highest risk spillover to Bitcoin is from the Peruvian stock market while the lowest spillover originated from Argentinian stock market. On the other hand, the highest spillover from Bitcoin is to Mexican stock markets, and the lowest is to Argentinian stock market. Results indicate that the lowest volatility spillover is from other stock markets to Bitcoin. Also, in addition to Bitcoin Argentinian, Canadian and Colombian stock markets are the markets that cause net risk spillovers in the region. When return and volatility spillovers for Americas are taken into consideration together, it is seen that there is limited relationship between Bitcoin and other markets. Analysis shows that in general, portfolio risk can be reduced by using Bitcoin and the Argentinian stock markets as reference. The return spillovers for Asia region are presented at Table 15.3.

For Asia region, the highest return spillover to Bitcoin and the other stock markets is from Singapore. The lowest spillover to the other stock markets is from Bitcoin. Additionally, the highest spillover from other markets is to Singapore while the lowest is to Bitcoin. As can be seen from Table 15.3, spillover to Bitcoin is highest from Sri Lanka and lowest from Singapore stock market. From Bitcoin, the highest return spillover is to Chinese stock market while the lowest is to Sri Lanka. Results generally indicate that a high level of return spillover between markets exists with the exception of Bitcoin and Sri Lankan stock market. Also, Singapore and Taiwan stock markets were determined to cause net return spillovers to the other markets. The volatility spillover for Asian region are presented at Table 15.4.

Volatility spillover results show that the dynamics of the return and volatility spillovers between the investigated financial assets are different in Asia region. In the mentioned region, spillover from Australia to the others is the highest and spillover from Bitcoin to the others is the lowest. Also, the highest spillover in the region is from Bitcoin and the other stock markets to Singapore, and the lowest spillover is from Bitcoin and the other stock markets to China. Analyses indicate that the highest risk spillover to Bitcoin is from Sri Lanka stock market and the lowest risk spillover to Bitcoin is from Japanese stock market. On the other hand, the highest risk transmission from Bitcoin is to Thailand stock market, and the lowest risk transmission from Bitcoin is to China stock exchange. Also, China, Philippines, Taiwan, Australia and Thailand stock markets are the markets that cause risk in the region. Differently from Americas region, Bitcoin is one of the net risky assets in the region. When the results for Asia region are considered together, it is seen that there are limited return and volatility spillovers between Bitcoin and the stock exchanges. In terms of portfolio management, analysis results indicate that investors may diversify their portfolios by investing in Bitcoin with China and Sri Lanka stock markets. The results of the return spillovers for Europe region is given at Table 15.5.

As can be seen from Table 15.5, in Europe region, spillover from France to the other stock markets is the highest and spillover from Bitcoin to the other stock markets is the lowest. Similarly, spillover from other stock markets to France is the

Table 15.3 Return spillovers in Asia region

	AUS	BTC	CHN	İND	JPN	LKA	PHL	SGP	THA	TWN	VNM	From other markets
AUS	37.19	1.32	1.86	8.07	11.40	1.28	4.21	13.59	7.44	10.98	2.68	62.81
BTC	1.66	78.74	5.00	1.69	1.20	0.49	2.72	2.81	3.05	1.02	1.64	21.26
CHN	2.50	0.79	53.29	3.45	4.99	4.58	4.69	8.33	2.87	8.81	5.70	46.71
İND	9.03	1.04	1.89	36.76	8.50	2.09	7.24	11.06	9.79	11.37	1.25	63.24
JPN	11.76	1.53	2.65	8.29	38.74	1.15	4.29	14.55	5.69	8.96	2.41	61.26
LKA	2.33	4.69	4.64	4.33	1.97	64.41	3.31	2.69	3.87	3.62	4.15	35.60
PHL	5.58	0.37	3.28	9.62	4.75	1.64	42.50	10.67	10.49	8.11	2.99	57.50
SGP	11.06	0.33	4.23	9.46	11.90	1.47	7.80	29.81	8.01	12.77	3.17	70.19
THA	7.66	1.06	2.17	10.21	6.34	2.70	10.32	10.46	38.62	7.36	3.10	61.38
TWN	10.17	0.87	4.30	10.15	8.08	1.51	6.47	14.59	6.24	32.92	4.69	67.08
VNM	4.51	2.28	5.09	3.13	3.15	1.61	4.98	5.83	7.64	9.37	52.43	47.57
Contribution to othermarkets	66.24	14.26	35.11	68.39	62.29	18.50	56.01	94.58	65.08	82.36	31.78	594.6
Contribution including own	103.4	92.99	88.40	105.1	101.0	82.91	98.51	124.4	103.7	115.3	84.20	Total Spillover Index 54.05
Net spillovers	3.43	−7.01	−11.60	5.15	1.03	−17.09	−1.49	24.39	3.71	15.28	−15.80	

Table 15.4 Volatility spillovers in Asia region

	AUS	BTC	CHN	İND	JPN	LKA	PHL	SGP	THA	TWN	VNM	From other markets
AUS	36.25	0.53	7.25	12.43	11.64	0.52	3.58	11.05	1.02	14.62	1.13	63.75
BTC	2.00	69.67	0.34	0.84	0.95	1.54	5.45	1.24	12.80	2.60	2.59	30.34
CHN	13.12	0.88	73.02	1.05	1.09	0.70	0.53	2.84	0.74	3.78	2.25	26.98
İND	9.76	0.29	3.97	49.90	12.82	1.99	2.77	8.04	3.04	6.90	0.54	50.10
JPN	11.15	0.22	5.54	6.03	51.05	2.13	7.84	6.75	0.45	7.10	1.74	48.95
LKA	2.87	1.92	1.63	3.39	2.43	68.43	1.94	5.40	2.83	3.54	5.64	31.57
PHL	5.96	1.56	0.89	2.04	1.37	2.21	68.23	7.36	6.23	3.27	0.89	31.77
SGP	11.98	0.42	11.31	7.79	11.23	1.61	10.39	28.60	2.36	12.42	1.89	71.40
THA	1.16	0.77	0.92	2.05	0.20	1.87	17.71	2.57	71.79	0.52	0.43	28.21
TWN	11.41	0.90	13.37	7.47	4.10	0.97	1.31	9.73	0.51	46.14	4.09	53.86
VNM	2.21	1.61	11.28	0.23	2.79	2.62	1.38	2.26	0.52	9.45	65.66	34.34
Contribution to other markets	71.62	9.09	56.49	43.30	48.63	16.16	52.88	57.23	30.50	64.19	21.18	471.3
Contribution including own	107.9	78.75	129.5	93.20	99.68	84.59	121.1	85.83	102.3	110.3	86.83	Total Spillover Index 42.84
Net spillovers	7.86	–21.25	29.51	–6.80	–0.32	–15.41	21.12	–14.17	2.29	10.34	–13.17	

Table 15.5 Return spillovers in Europe region

	BTC	DEU	ESP	FRA	GBR	GRC	IRL	ITA	PRT	TUR	From other markets
BTC	80.89	2.13	2.09	1.94	2.40	1.82	3.45	1.31	2.16	1.84	19.11
DEU	0.51	18.71	11.87	16.19	11.97	4.80	11.62	12.15	8.90	3.28	81.29
ESP	0.27	12.05	19.22	14.56	9.40	5.82	9.43	15.19	11.10	2.96	80.78
FRA	0.27	15.04	13.07	17.58	12.35	4.85	11.04	13.11	9.37	3.33	82.42
GBR	0.42	14.15	10.26	15.17	21.54	4.72	10.44	10.21	9.10	4.00	78.46
GRC	1.15	8.21	10.19	9.25	7.26	30.40	8.04	11.01	10.80	3.69	69.60
IRL	0.42	13.75	10.55	14.01	10.65	5.76	21.94	11.54	8.60	2.78	78.06
ITA	0.29	12.38	14.70	14.49	9.35	6.48	10.11	18.64	10.78	2.79	81.36
PRT	0.47	10.39	12.78	12.14	9.64	7.58	8.64	12.94	22.39	3.05	77.61
TUR	1.52	7.81	6.47	7.97	8.17	5.30	5.29	5.93	5.82	45.71	54.29
Contribution to other markets	5.31	95.91	91.98	105.7	81.19	47.13	78.07	93.37	76.62	27.70	703.0
Contribution including own	86.20	114.6	111.2	123.3	102.7	77.53	100.0	112.0	99.01	73.41	Total Spillover Index
Net spillovers	–13.80	14.63	11.20	23.28	2.73	–22.47	0.01	12.01	–0.99	–26.59	70.30

highest and spillover from other stock markets to Bitcoin is the lowest. The results indicate that there are significant return spillovers between the markets except for Bitcoin. On the other hand, the bidirectional return spillovers between Bitcoin and the stock markets of the region are quite limited. There are spillovers from Bitcoin to Ireland at most and to Italy least, and there are spillovers to Bitcoin from Turkey at most and from France least. Based on the analyses, it is also concluded that France, Germany, Italy, Spain, United Kingdom and Ireland stock markets are the markets that cause net return spillovers in the European region.

In Europe region, volatility spillover from France stock exchange to Bitcoin and the other stock markets is at highest level. On the other hand, it is determined that there are risk spillovers from stock markets of Germany, Italy, Spain and United Kingdom to the other stock markets. Also, it is concluded that volatility spillover from Bitcoin and the other stock markets to Italian stock market is the highest and spillover from Bitcoin and the other stock markets to Turkish stock exchange is the lowest. As can be seen from Table 15.6 that there are high levels of risk transmissions between the variables except for Bitcoin and stock market of Turkey. Results indicate that risk spillover from Bitcoin to Greek stock exchange is the highest and risk spillover from Bitcoin to Turkish stock exchange is the lowest. In Europe region, the volatility spillovers from stock markets to Bitcoin are at quite low levels. It is also determined that the highest risk spillover to Bitcoin is from Turkey and the lowest risk spillover to Bitcoin is from Spain. Volatility analysis reveals that France, Germany, Italy and Portugal stock markets are the markets that cause net risk spillovers and in the European region. The stock markets of Spain, Greece, United Kingdom, Turkey and Ireland are the assets under the net risk with Bitcoin, in the region. When the return and volatility spillover results are considered together, it may be said that investors may provide benefit from diversification by investing in Bitcoin with Spain.

As can be seen from Table 15.7 that return spillover from United Arab Emirates to the others is the highest and return spillover from Bitcoin to the others is the lowest among Bitcoin and OIC stock markets. Similarly, there are spillovers from the others to United Arab Emirates stock market at most and to Bitcoin least. The results show that there are significant return spillovers between the markets except for Egypt stock exchange and Bitcoin. Spillover from Bitcoin to the stock market of Saudi Arabia is the highest and Spillover from Bitcoin to the stock market of Egypt is the lowest. It is also seen that the highest return spillover to Bitcoin is from Egypt and the lowest return spillover to Bitcoin is from Saudi Arabia. Bitcoin and the stock markets of countries except Indonesia, Pakistan and Egypt are the markets that cause net return spillovers in OIC.

Volatility spillover to Saudi Arabia stock exchange is the highest and volatility spillover to Pakistan stock exchange is the lowest among Bitcoin and OIC stock markets. Similar to results of return spillover, the risk spillover from other financial assets to the stock market of United Arab Emirates is the highest and the risk spillover from other financial assets to Bitcoin is the lowest. Results show that there are significant volatility spillovers among the markets except for Pakistan stock exchange and Bitcoin. As can be seen from Table 15.8 that spillover from Bitcoin to Saudi Arabia stock market is the highest and spillover from Bitcoin to

Table 15.6 Volatility spillovers in Europe region

	BTC	DEU	ESP	FRA	GBR	GRC	IRL	ITA	PRT	TUR	From other markets
BTC	88.83	0.81	0.55	0.95	1.06	2.77	0.93	0.98	2.76	0.37	11.17
DEU	0.35	25.29	8.73	20.31	15.38	5.67	6.37	9.41	8.25	0.26	74.71
ESP	0.05	11.32	25.60	14.32	6.33	7.92	9.22	18.00	7.11	0.13	74.40
FRA	0.28	18.66	11.44	23.03	15.15	5.63	6.87	12.07	6.56	0.32	76.98
GBR	0.57	17.39	7.75	18.90	24.61	6.32	7.68	9.15	6.70	0.93	75.39
GRC	0.32	7.97	4.43	6.74	6.32	47.59	6.89	7.65	11.96	0.14	52.41
IRL	0.06	12.37	12.33	13.94	7.32	6.75	25.43	15.16	6.58	0.05	74.57
ITA	0.06	11.76	17.19	14.34	7.33	7.04	10.82	22.00	9.37	0.10	78.00
PRT	0.07	9.58	8.62	9.11	7.04	6.10	3.89	10.55	44.90	0.15	55.10
TUR	4.39	0.65	0.48	0.65	2.31	1.28	0.64	0.40	0.22	88.98	11.02
Contribution to other markets	6.15	90.50	71.52	99.26	68.23	49.48	53.31	83.36	59.51	2.43	583.7
Contribution including own	94.97	115.8	97.11	122.3	92.84	97.08	78.74	105.4	104.4	91.41	Total Spillover Index
Net spillovers	–5.03	15.80	–2.89	22.29	–7.16	–2.92	–21.26	5.36	4.40	–8.59	58.37

Table 15.7 Return spillovers in OIC

	BTC	EGY	IDN	MYS	PAK	QAT	SAU	UAE	From other markets
BTC	90.46	0.61	0.97	1.30	0.78	1.74	2.41	1.75	9.54
EGY	3.20	68.73	3.61	6.83	0.67	2.61	10.11	4.25	31.27
IDN	1.63	2.56	60.39	14.18	5.74	5.45	4.36	5.69	39.61
MYS	1.53	2.88	13.79	54.12	3.82	7.03	7.51	9.33	45.88
PAK	2.00	0.87	5.35	5.24	65.59	7.05	6.58	7.33	34.41
QAT	1.22	2.22	4.20	6.86	3.41	45.62	13.90	22.58	54.38
SAU	0.65	5.91	3.27	6.39	2.57	14.79	51.35	15.07	48.65
UAE	3.02	2.18	5.44	8.52	4.07	19.97	13.76	43.03	56.97
Contribution to other markets	13.24	17.22	36.64	49.31	21.06	58.63	58.62	66.00	320.7
Contribution including own	103.7	85.94	97.03	103.4	86.65	104.3	110.0	109.0	Total Spillover Index
Net spillovers	3.70	−14.06	−2.97	3.43	−13.35	4.26	9.97	9.03	40.09

Table 15.8 Volatility spillovers in OIC

	BTC	EGY	IDN	MYS	PAK	QAT	SAU	UAE	From other markets
BTC	93.22	0.45	0.84	0.78	1.30	0.41	2.36	0.65	6.78
EGY	2.64	65.43	3.86	2.76	1.54	6.52	11.34	5.93	34.57
IDN	1.94	3.26	78.66	7.85	0.73	0.14	6.98	0.44	21.34
MYS	0.73	0.79	6.10	84.56	1.08	1.12	2.72	2.90	15.44
PAK	0.62	0.28	6.79	5.89	79.11	2.87	2.86	1.59	20.90
QAT	0.41	0.90	0.15	3.94	0.25	45.34	23.43	25.60	54.66
SAU	1.13	3.30	0.55	11.97	0.75	18.88	46.59	16.82	53.41
UAE	0.19	2.57	1.04	6.94	0.91	27.35	20.73	40.27	59.73
Contribution to other markets	7.67	11.53	19.32	40.12	6.56	57.29	70.42	53.92	266.8
Contribution including own	100.9	76.96	97.98	124.7	85.66	102.6	117.0	94.19	Total Spillover Index 33.35
Net spillovers	0.89	−23.04	−2.02	24.68	−14.34	2.63	17.02	−5.81	

Qatar stock market is the lowest. On the other hand, the risk spillover to Bitcoin from Egypt stock market is the highest and the risk spillover to Bitcoin from United Arab Emirates stock market is the lowest. The results also reveal that stock markets of Malaysia, Saudi Arabia and Qatar and Bitcoin are the markets that cause net return spillovers in OIC. Analysis results indicate that investors may decrease the total risk of their portfolios by investing in Bitcoin with United Arab Emirates stock market.

15.5 Conclusion

In investment decision process, investors make choice from different alternatives depending on their risk perception. In order to make this choice, besides the risk and the return of the alternative assets, the investor also should know how the performance of the portfolio will be if the portfolio is constructed by using these assets. In other words, an investor should analyze the relationships between the variables that will be included in the portfolio. Especially, depending on the financial liberalization the relationships between the assets and between the markets increase day by day. In this context, it is thought that to investigate the relationships between stock markets and Bitcoin, which draws considerable attention as a new financial asset, is important for portfolio decisions. Therefore, the main purpose of this study is to determine the return and volatility spillovers between Bitcoin and stock markets on regional basis.

In line with the purpose of the study, the return and volatility spillovers between Bitcoin and stock markets of the countries from Americas, Asia, Europe and OIC were analyzed by the method developed by Diebold and Yilmaz. The results showed that there are limited bidirectional return and volatility spillovers between Bitcoin and regional stock markets, and the spillovers between the Bitcoin and the selected countries differ across the regions. The results also revealed that there are high degrees of both return and volatility spillovers from Bitcoin to OIC countries compared to other countries. The degree of volatility spillovers to Bitcoin from Asia region are higher compared to other regions. However, there are more risk spillovers from Bitcoin to stock markets of America compared to others. The degree of the volatility spillovers from Bitcoin to stock markets is higher only in Asia region compared to return spillovers. On the other hand, the return spillovers to Bitcoin is higher only from Asia region compared to volatility spillovers.

The empirical results indicate that Bitcoin may be used for diversification based on the risk perception of the investor. Investors may diversify their portfolio by investing in Bitcoin with the stock markets of Argentina from Americas region, China and Sri Lanka from Asia region, Spain from European region and United Arab Emirates from OIC region. But, it will be suitable for the investors, who desire to include different investment vehicles to their portfolios, to analyze the relationships between Bitcoin and the determined vehicles too. In this context, in further studies to examine the relationships between Bitcoin and investment vehicles such as bond, real estate and derivatives will present valuable information to the investors and contribute to the finance literature. Also for healthier investment decisions, the weights of the Bitcoin and the other assets in the portfolio should be determined too by further studies.

Appendix 1

Table 15.9 Descriptive statistics

	Mean	Median	Max.	Min.	Std. dev.	Skewness	Kurtosis	Jarque-Bera	Unconditional correlation
BTC	0.0147	0.0161	0.540	–0.594	0.136	–0.190	6.104	158.96*	–
Americas region stock markets									
ARG	0.0066	0.0082	0.155	–0.149	0.047	–0.299	3.698	13.751*	0.100
BRA	0.0015	0.0019	0.166	–0.087	0.030	0.281	4.866	61.739*	0.008
CAN	0.0006	0.0018	0.052	–0.068	0.016	–0.469	4.646	58.318*	0.024
COL	–0.0003	0.0001	0.067	–0.105	0.021	–0.468	5.120	87.243*	–0.008
MEX	0.0006	0.0013	0.066	–0.077	0.020	–0.113	3.827	11.937*	–0.023
PER	0.00004	–0.00006	0.106	–0.089	0.023	0.072	4.754	50.336*	0.031
USA	0.0022	0.0031	0.071	–0.073	0.019	–0.589	5.262	105.64*	0.067
VEN	0.0455	0.0142	1.100	–0.259	0.131	3.116	19.72	5174.2*	0.009
Asia region stock markets									
AUS	0.0009	0.0020	0.073	–0.061	0.018	–0.262	4.382	35.501*	0.055
CHN	0.0004	0.0014	0.107	–0.140	0.031	–0.596	5.524	126.65*	–0.042
IND	0.0021	0.0027	0.071	–0.069	0.020	0.040	3.594	5.8468**	–0.042
JPN	0.0019	0.0041	0.085	–0.135	0.026	–0.563	4.932	81.220*	0.083
LKA	–0.0004	–0.0013	0.075	–0.056	0.015	0.367	6.324	188.26*	0.024
PHL	0.0016	0.0018	0.074	–0.099	0.021	–0.444	4.856	68.827*	0.067
SGP	0.0004	0.0013	0.071	–0.047	0.017	0.319	4.960	69.023*	0.008
THA	0.0012	0.0031	0.062	–0.077	0.020	–0.559	5.053	88.752*	–0.012
TWN	0.0008	0.0019	0.059	–0.073	0.020	–0.454	4.242	38.505*	0.035
VNM	0.0022	0.0037	0.082	–0.099	0.025	–0.512	4.625	59.953*	0.063

Europe region stock markets									
DEU	0.0017	0.0021	0.074	–0.072	0.021	–0.402	3.992	26.417*	0.050
ESP	0.0001	0.0010	0.078	–0.090	0.024	–0.268	3.916	18.261*	0.059
FRA	0.0012	0.0025	0.067	–0.078	0.020	–0.292	4.295	32.724*	0.038
GBR	0.0008	0.0014	0.053	–0.066	0.015	–0.322	4.582	47.271*	0.053
GRC	–0.0009	0.0008	0.161	–0.157	0.038	–0.346	4.900	66.300*	0.123
IRL	0.0022	0.0038	0.045	–0.113	0.019	–1.090	7.233	367.39*	0.071
ITA	0.0007	0.0023	0.081	–0.102	0.026	–0.308	4.222	30.348*	0.058
PRT	–0.0005	0.0008	0.076	–0.099	0.023	–0.458	4.429	46.706*	0.045
TUR	0.0017	0.0041	0.071	–0.110	0.025	–0.464	3.736	22.712*	0.018
Stock markets of OIC member countries									
EGY	0.0018	0.0042	0.108	–0.095	0.028	–0.187	5.074	72.008*	0.025
IDN	0.0013	0.0032	0.058	–0.077	0.016	–0.900	6.257	224.54*	0.011
MYS	0.0004	0.0011	0.033	–0.044	0.011	–0.479	4.621	57.437*	0.033
PAK	0.0034	0.0039	0.066	–0.052	0.019	–0.188	3.314	3.8795	–0.011
QAT	0.0006	0.0018	0.069	–0.095	0.019	–0.603	6.290	199.03*	–0.110
SAU	0.0009	0.0028	0.074	–0.100	0.021	–0.882	6.988	308.22*	–0.022
UAE	0.0017	0.0017	0.069	–0.095	0.017	–0.550	7.070	288.16*	–0.018

*1%, **10% significance level

Appendix 2

Table 15.10 Ng-Perron unit root test results of return series

	Intercept				Intercept & trend			
	MZa	MZt	MSB	MPT	MZa	MZt	MSB	MPT
BTC	−131.6	−8.097	0.062	0.212	−215.0	−10.37	0.048	0.424
ARG	−177.5	−9.414	0.053	0.146	−182.4	−9.549	0.052	0.502
BRA	−51.19	−5.049	0.099	0.504	−112.3	−7.485	0.067	0.844
CAN	−68.31	−5.832	0.085	0.386	−118.3	−7.687	0.065	0.780
COL	−120.3	−7.753	0.064	0.209	−164.5	−9.067	0.055	0.563
MEX	−65.25	−5.695	0.087	0.414	−129.5	−8.043	0.062	0.714
PER	−98.79	−7.010	0.071	0.284	−163.5	−9.040	0.055	0.558
USA	−145.2	−8.511	0.059	0.184	−141.9	−8.422	0.059	0.651
VEN	−150.4	−8.662	0.058	0.176	−141.7	−8.396	0.059	0.716
AUS	−119.3	−7.717	0.065	0.213	−135.6	−8.235	0.061	0.672
CHN	−110.5	−7.381	0.067	0.314	−169.0	−9.186	0.054	0.558
IND	−22.00	−3.261	0.148	1.308	−62.34	−5.582	0.090	1.468
JPN	−126.7	−7.955	0.063	0.200	−150.7	−8.661	0.057	0.664
LKA	−111.7	−7.460	0.067	0.244	−159.8	−8.940	0.056	0.570
PHL	−123.7	−7.833	0.063	0.251	−173.3	−9.306	0.054	0.532
SGP	−36.97	−4.293	0.116	0.683	−83.66	−6.448	0.077	1.170
THA	−85.47	−6.528	0.076	0.306	−148.0	−8.598	0.058	0.627
TWN	−41.21	−4.533	0.110	0.614	−95.58	−6.895	0.072	1.025
VNM	−23.98	−3.457	0.144	1.040	−65.37	−5.677	0.087	1.578
DEU	−24.29	−3.471	0.143	1.056	−24.29	−3.471	0.143	1.056
ESP	−51.41	−5.062	0.098	0.497	−51.41	−5.062	0.098	0.497
FRA	−42.27	−4.580	0.108	0.629	−42.27	−4.580	0.108	0.629
GBR	−123.3	−7.850	0.064	0.204	−123.3	−7.850	0.064	0.204
GRC	−68.80	−5.846	0.085	0.399	−68.80	−5.846	0.085	0.399
IRL	−147.6	−8.585	0.058	0.173	−147.6	−8.585	0.058	0.173
ITA	−50.63	−5.015	0.099	0.525	−50.63	−5.015	0.099	0.525
PRT	−167.9	−9.161	0.055	0.147	−167.9	−9.161	0.055	0.147
TUR	−96.73	−6.950	0.072	0.261	−96.73	−6.950	0.072	0.261
EGY	−102.7	−7.164	0.070	0.243	−106.5	−7.292	0.068	0.877
IDN	−107.9	−7.329	0.068	0.257	−142.0	−8.425	0.059	0.642
MYS	−72.69	−6.018	0.083	0.361	−136.2	−8.249	0.061	0.682
PAK	−100.4	−7.047	0.070	0.318	−124.9	−7.900	0.063	0.742
QAT	−124.1	−7.823	0.063	0.291	−126.2	−7.921	0.063	0.798
SAU	−112.9	−7.508	0.066	0.227	−155.2	−8.808	0.057	0.591
UAE	−138.1	−8.306	0.060	0.184	−136.2	−8.246	0.061	0.685

Appendix 3

Table 15.11 GARCH (1,1) model results

	c	α	β
BTC	0.0012*	0.3429*	0.6228*
ARG	0.0001*	0.0930*	0.8829*
BRA	0.0001*	0.0340*	0.8932*
CAN	< 0.0001*	0.1892*	0.7242*
COL	< 0.0001**	0.1071*	0.7927*
MEX	< 0.0001***	0.1146**	0.7696*
PER	< 0.0001***	0.0584*	0.8703*
USA	< 0.0001*	0.2322*	0.6749*
VEN	< 0.0001*	0.0406*	0.9712*
AUS	< 0.0001*	0.0662*	0.8918*
CHN	< 0.0001*	0.1362*	0.8402*
IND	< 0.0001***	0.0522***	0.8891*
JPN	0.0001**	0.1545*	0.7384*
LKA	0.0001*	0.4142*	0.2905*
PHL	< 0.0001***	0.1250*	0.8148*
SGP	< 0.0001**	0.0958*	0.8679*
THA	< 0.0001**	0.0720*	0.8804*
TWN	< 0.0001**	0.0490*	0.9067*
VNM	< 0.0001**	0.0662*	0.9145*
DEU	< 0.0001**	0.0782*	0.8791*
ESP	< 0.0001**	0.0829*	0.8878*
FRA	< 0.0001**	0.1153*	0.8278*
GBR	< 0.0001**	0.1312*	0.7940*
GRC	< 0.0001***	0.0440*	0.9484*
IRL	< 0.0001**	0.1733*	0.7104*
ITA	< 0.0001***	0.0733*	0.8932*
PRT	< 0.0001***	0.0433*	0.9450*
TUR	< 0.0001***	0.0485**	0.9166*
EGY	0.0003*	0.2996*	0.6302**
IDN	< 0.0001*	0.1685*	0.7544*
MYS	< 0.0001**	0.1972*	0.6694*
PAK	< 0.0001***	0.1319**	0.7439*
QAT	< 0.0001*	0.0608*	0.9334*
SAU	< 0.0001**	0.2265*	0.7428*
UAE	0.0001*	0.3278*	0.5575*

*1%, **5%, ***10% significance level

References

Andrianto, Y., & Diputra, Y. (2017). The effect of cryptocurrency on investment portfolio effectiveness. *Journal of Finance and Accounting, 5*(6), 229–238. https://doi.org/10.11648/j.jfa.20170506.14

Antonakakis, N. (2012). Exchange return co-movements and volatility spillovers before and after the introduction of euro. *Journal of International Financial Markets, Institutions and Money, 22* (5), 1091–1109. https://doi.org/10.1016/j.intfin.2012.05.009

Awartani, B., Maghyereh, A. I., & Al Shiab, M. (2013). Directional spillovers from the US and the Saudi market to equities in the Gulf Cooperation Council countries. *Journal of International Financial Markets, Institutions and Money, 27*, 224–242. https://doi.org/10.1016/j.intfin.2013.08.002

Baek, C., & Elbeck, M. (2015). Bitcoins as an investment or speculative vehicle? A first look. *Applied Economics Letters, 22*(1), 30–34. https://doi.org/10.1080/13504851.2014.916379

Bariviera, A. F. (2017). The inefficiency of Bitcoin revisited: A dynamic approach. *Economics Letters, 161*, 1–4. https://doi.org/10.1016/j.econlet.2017.09.013

Bariviera, A. F., Basgall, M. J., Hasperué, W., & Naiouf, M. (2017). Some stylized facts of the Bitcoin market. *Physica A: Statistical Mechanics and its Applications, 484*, 82–90. https://doi.org/10.1016/j.physa.2017.04.159

Bollerslev, T. (1986). Generalized autoregressive conditional heteroskedasticity. *Journal of Econometrics, 31*(3), 307–327. https://doi.org/10.1016/0304-4076(86)90063-1

Bouoiyour, J., & Selmi, R. (2016). Bitcoin: A beginning of a new phase. *Economics Bulletin, 36*(3), 1430–1440.

Bouoiyour, J., Selmi, R., Tiwari, A. K., & Olayeni, O. R. (2016). What drives Bitcoin price? *Economics Bulletin, 36*(2), 843–850.

Bouri, E., Azzi, G., & Dyhrberg, H. (2017). On the return-volatility relationship in the Bitcoin market around the price crash of 2013. *Economics-The Open-Access, Open-Assessment E-Journal, 11*, 1–16. https://doi.org/10.5018/economics-ejournal.ja.2017-2

Bouri, E., Gupta, R., Tiwari, A. K., & Roubaud, D. (2017). Does Bitcoin hedge global uncertainty? Evidence from wavelet-based quantile-in-quantile regressions. *Finance Research Letters, 23*, 87–95. https://doi.org/10.1016/j.frl.2017.02.009

Carpenter, A. (2016). Portfolio diversification with Bitcoin. *Journal of Undergraduate Research in Finance, 6*(1), 1–27.

Chan, W. H., Le, M., & Wu, Y. W. (2019). Holding Bitcoin longer: The dynamic hedging abilities of Bitcoin. *The Quarterly Review of Economics and Finance, 71*, 107–113. https://doi.org/10.1016/j.qref.2018.07.004

Cheah, E. T., & Fry, J. (2015). Speculative bubbles in Bitcoin markets? An empirical investigation into the fundamental value of Bitcoin. *Economics Letters, 130*, 32–36. https://doi.org/10.1016/j.econlet.2015.02.029

Cheung, A., Roca, E., & Su, J. J. (2015). Crypto-currency bubbles: An application of the Phillips–Shi–Yu (2013) methodology on Mt. Gox bitcoin prices. *Applied Economics, 47*(23), 2348–2358. https://doi.org/10.1080/00036846.2015.1005827

Chevallier, J., & Ielpo, F. (2013). Volatility spillovers in commodity markets. *Applied Economics Letters, 20*(13), 1211–1227. https://doi.org/10.1080/13504851.2013.799748

Ciaian, P., Rajcaniova, M., & Kancs, D. A. (2016). The economics of Bitcoin price formation. *Applied Economics, 48*(19), 1799–1815. https://doi.org/10.1080/00036846.2015.1109038

Corbet, S., Lucey, B., & Yarovaya, L. (2018). Datestamping the Bitcoin and Ethereum bubbles. *Finance Research Letters, 26*, 81–88. https://doi.org/10.1016/j.frl.2017.12.006

Cronin, D. (2014). The interaction between money and asset markets: A spillover index approach. *Journal of Macroeconomics, 39*, 185–202. https://doi.org/10.1016/j.jmacro.2013.09.006

Diebold, F. X., & Yilmaz, K. (2009). Measuring financial asset return and volatility spillovers, with application to global equity markets. *The Economic Journal, 119*(534), 158–171. https://doi.org/10.1111/j.1468-0297.2008.02208.x

Diebold, F. X., & Yilmaz, K. (2012). Better to give than to receive: Predictive directional measurement of volatility spillovers. *International Journal of Forecasting, 28*(1), 57–66. https://doi.org/10.1016/j.ijforecast.2011.02.006

Dyhrberg, A. H. (2016). Hedging capabilities of bitcoin. Is it the virtual gold? *Finance Research Letters, 16*, 139–144. https://doi.org/10.1016/j.frl.2015.10.025

Engle, R. (1982). Autoregressive conditional heteroscedasticity with estimates of the variance of United Kingdom inflation. *Econometrica, 50*(4), 391–407. https://doi.org/10.2307/1912773

Fry, J., & Cheah, E. T. (2016). Negative bubbles and shocks in cryptocurrency markets. *International Review of Financial Analysis, 47*, 343–352. https://doi.org/10.1016/j.irfa.2016.02.008

Hameed, S., & Farooq, S. (2016). The art of crypto currencies: A comprehensive analysis of popular crypto currencies. *International Journal of Advanced Computer Science and Applications, 7* (12), 426–435. https://doi.org/10.14569/IJACSA.2016.071255

Kajtazi, A., & Moro, A. (2018). The role of bitcoin in well diversified portfolios: A comparative global study. *International Review of Financial Analysis.* https://doi.org/10.1016/j.irfa.2018.10.003

Katsiampa, P. (2017). Volatility estimation for Bitcoin: A comparison of GARCH models. *Economics Letters, 158*, 3–6. https://doi.org/10.1016/j.econlet.2017.06.023

Kristoufek, L. (2015). What are the main drivers of the bitcoin price? Evidence from wavelet coherence analysis. *PLoS One, 10*(4), 1–15. https://doi.org/10.1371/journal.pone.0123923

Kristoufek, L. (2018). On Bitcoin markets (in)efficiency and its evolution. *Physica A: Statistical Mechanics and its Applications, 503*, 257–262. https://doi.org/10.1016/j.physa.2018.02.161

Lee, H. C., & Chang, S. L. (2013). Spillovers of currency carry trade returns, market risk sentiment, and US market returns. *The North American Journal of Economics and Finance, 26*, 197–216. https://doi.org/10.1016/j.najef.2013.10.001

Lucey, B. M., Larkin, C., & O'Connor, F. (2014). Gold markets around the world—who spills over what, to whom, when? *Applied Economics Letters, 21*(13), 887–892. https://doi.org/10.1080/13504851.2014.896974

Maghyereh, A. I., & Awartani, B. (2016). Dynamic transmissions between Sukuk and bond markets. *Research in International Business and Finance, 38*, 246–261. https://doi.org/10.1016/j.ribaf.2016.04.016

Maghyereh, A. I., Awartani, B., & Bouri, E. (2016). The directional volatility connectedness between crude oil and equity markets: New evidence from implied volatility indexes. *Energy Economics, 57*, 78–93. https://doi.org/10.1016/j.eneco.2016.04.010

Nadarajah, S., & Chu, J. (2017). On the inefficiency of Bitcoin. *Economics Letters, 150*, 6–9. https://doi.org/10.1016/j.econlet.2016.10.033

Narayan, P. K., Narayan, S., & Prabheesh, K. P. (2014). Stock returns, mutual fund flows and spillover shocks. *Pacific-Basin Finance Journal, 29*, 146–162. https://doi.org/10.1016/j.pacfin.2014.03.007

Öztürk, M. B., Arslan, H., Kayhan, T., & Uysal, M. (2018). Bitcoin as a new hedge instrument tool: Bitconomy. *Ömer Halisdemir Üniversitesi İktisadi ve İdari Bilimler Fakültesi Dergisi, 11*(2), 217–232. https://doi.org/10.25287/ohuiibf.415713

Phillips, R. C., & Gorse, D. (2018). Cryptocurrency price drivers: Wavelet coherence analysis revisited. *PLoS One, 13*(4), 1–21. https://doi.org/10.1371/journal.pone.0195200

Pichl, L., & Kaizoji, T. (2017). Volatility analysis of Bitcoin price time series. *Quantitative Finance and Economics, 1*(4), 474–485. https://doi.org/10.3934/QFE.2017.4.474

Sehgal, S., Ahmad, W., & Deisting, F. (2015). An investigation of price discovery and volatility spillovers in India's foreign exchange market. *Journal of Economic Studies, 42*(2), 261–284. https://doi.org/10.1108/JES-11-2012-0157

Sugimoto, K., Matsuki, T., & Yoshida, Y. (2014). The global financial crisis: An analysis of the spillover effects on African stock markets. *Emerging Markets Review, 21*, 201–233. https://doi.org/10.1016/j.ememar.2014.09.004

Tsai, I. C. (2014). Spillover of fear: Evidence from the stock markets of five developed countries. *International Review of Financial Analysis, 33*, 281–288. https://doi.org/10.1016/j.irfa.2014.03.007

Teichmann, F. M. J. (2018). Financing terrorism through cryptocurrencies—a danger for Europe? *Journal of Money Laundering Control, 21*(4), 513–519. https://doi.org/10.1108/JMLC-06-2017-0024

Tiwari, A. K., Jana, R. K., Das, D., & Roubaud, D. (2018). Informational efficiency of Bitcoin—An extension. *Economics Letters, 163*, 106–109. https://doi.org/10.1016/j.econlet.2017.12.006

Urquhart, A. (2016). The inefficiency of Bitcoin. *Economics Letters, 148*, 80–82. https://doi.org/10.1016/j.econlet.2016.09.019

Urquhart, A., & Zhang, H. (2019). Is Bitcoin a hedge or safe haven for currencies? An intraday analysis. *International Review of Financial Analysis.* https://doi.org/10.1016/j.irfa.2019.02.009

Yarovaya, L., Brzeszczyński, J., & Lau, C. K. M. (2016). Intra-and inter-regional return and volatility spillovers across emerging and developed markets: Evidence from stock indices and stock index futures. *International Review of Financial Analysis, 43*, 96–114. https://doi.org/10.1016/j.irfa.2015.09.004

M. Kamisli received his MS in Business Administration Master Program from Osmangazi University in 2008. He was hired as a lecturer by Bilecik University in 2009. He received his Ph.D. in Finance program from Anadolu University in 2015 and became an assistant professor at Bilecik University in 2015. He currently gives international finance, investment decisions, and portfolio analysis, financial crisis and risk management courses in Finance and Banking Programme.

S. Kamisli received her Ph.D. in Finance program from Anadolu University in 2015 and became an assistant professor at Bilecik University in 2015. She currently gives financial management, financial markets and institutions, financial mathematics and business finance courses in Finance and Banking Programme.

F. Temizel received his Ph.D. in Business Administration program from Anadolu University in 2005 and became an assistant professor at Anadolu University in 2015. He currently gives capital markets, securities analysis and financial management courses in the department of accounting and finance.

Chapter 16
Cryptocurrencies as an Investment Vehicle: The Asymmetric Relationships Between Bitcoin and Precious Metals

M. Kamisli

Abstract One of the main purposes of the investors is to reduce the total risk of the portfolio. In order to achieve this purpose, investors increase the variety of the assets that they include to the portfolio. In today's financial world the number of the alternatives for the investors increases day by day. One of these alternatives is cryptocurrencies which start with Bitcoin (BTC) in 2008 and having increasing number of variety and popularity. Cryptocurrencies have become an attractive investment alternative for many investors due to their high returns and low correlations with the other assets. But, since these assets are relatively new, it is necessary to determine the return dynamics and analyze their relations with other financial assets in order to achieve the desired benefit for portfolio or risk management. On the other hand, another alternative for the investors who want to invest in different financial instruments, differently from traditional assets such as stocks and bonds, is precious metals. Besides their usage for industrial purposes, precious metals are considered as financial assets to be included in the portfolio, and they are used for diversification by being invested with different financial assets. In this context, the purpose of this study is to analyze the relationships between cryptocurrencies and precious metals, and to investigate their usability in portfolio and risk management. In line with this purpose, causality relationships between the most popular cryptocurrency Bitcoin and gold, silver, platinum, palladium, ruthenium, rhodium, iridium, osmium, rhenium are analyzed by asymmetric causality in frequency domain approach.

16.1 Introduction

Cryptocurrency market maintains its development for volume and currency variety since 2008. Although it is relatively a new asset, the demand for these currencies increases with each passing day, and despite their risks, the opportunities they

M. Kamisli (✉)
Faculty of Applied Sciences, Department of Banking and Finance, Bilecik Seyh Edebali University, Bilecik, Turkey
e-mail: melik.kamisli@bilecik.edu.tr

© Springer Nature Switzerland AG 2019

U. Hacioglu (ed.), *Blockchain Economics and Financial Market Innovation*, Contributions to Economics, https://doi.org/10.1007/978-3-030-25275-5_16

provide arouse the interest of economists, investors and policy makers. There are three reasons for the increasing popularity of cryptocurrencies (Ali, Barrdear, Clews, & Southgate, 2014). First of these is ideological reasons. The main reason for the design of digital currencies is that they do not have any central control as a money supply or payment system and minimize the confidence that should be had to a third party. The second reason is that the digital currencies have been seen as a new asset class. The price of these assets mainly depends on the changes in demand, and there is no intrinsic demand for the assets. In other words, they are not used as a production factor or consumer good. The factor that determines the price changes in medium and long term is the expectations for increase in the operational usage of currency in the future. The last reason that increases the attractiveness of cryptocurrencies is that they present lower transaction costs in comparison with the existing electronic payment systems and international transfers. Because there is no intermediary between buyer and seller in digital payment systems, the transaction costs are lower and the transactions are executed faster (Brito & Castillo, 2013).

Economists have not yet reached a consensus on the definition of cryptocurrencies. Some of their features require them to be defined as currency while some features cause to be considered as commodity. However, the studies made in recent years reveals that the similarities between cryptocurrencies and especially gold are high, and they suggest that it is more appropriate to consider the cryptocurrencies as an asset rather than a currency (Glaser, Haferhorn, Weber, Zimmarmann, & Siering, 2014). In this context, the cryptocurrencies are accepted as a new investment vehicle besides being payment instrument, and they are considered as a new instrument that investors may include in their portfolios besides traditional assets such as stock, bond and gold. As Dyhrberg (2016) and Baur, Dimpf, and Kuck (2018) stated in their study, cryptocurrencies become suitable investment vehicles especially when negative shocks occurred in the markets, and they become appropriate for risk management.

The data constraints, since it is a relatively new asset class, and the absence of future market for comparison are the features that distinguish this instrument from the others. But, high return opportunity they provide besides the high risks they carry because of the high volatility makes these assets attractive. The significant increases in cryptocurrency markets in recent years also support this view. Especially in 2016 and 2017 price and total market capitalization of Bitcoin which is the most popular cryptocurrency increased significantly. According to many economists, these increases are speculative and indicate that balloons have occurred in these markets. They also state that these balloons may show contagion and weaken financial stability. Therefore, to make studies on cryptocurrency markets via traditional and new methods is important in order to understand the features and dynamics of these new assets and to determine the transmissions between cryptocurrency markets and other markets (Peng, Albuquerque, Camboim de Sá, Padula, & Montenegro, 2018).

Some investors may prefer to invest in alternative investment vehicles instead of traditional investment vehicles such as bond and stocks. Cryptocurrencies are a good option for these investors but investing only in this instrument will increase the risk that portfolio exposed. At this point, precious metals are another alternative to invest in for mentioned investors. Precious metals are elements that are naturally occurred and having high economic value. Besides their usage for industrial purposes, precious metals are included in portfolios as a financial asset, and they provide diversification opportunity to the investors since they have low correlation with stocks. Especially in times that the volatility is high in the markets, precious metals provide hedging to the investors (Hillier, Draper, & Faff, 2006). Another factor that makes the precious metals attractive is that these assets generally have positive correlation with inflation, in other words, protect the investors against inflation and maintain the value of portfolio against the changes in exchange rates (McCown & Shaw, 2017).

Investors may improve their portfolio by investing in the stocks of the precious metal firms as they may directly add precious metals to the portfolio. Conover, Jensen, Johnson, and Mercer (2009) stated in their study that the performance of the portfolio increase significantly if the 25% percentage of the portfolio consists of the stocks of the metal firms. But, in order to provide the desired benefit from adding the precious metals to the portfolio, it is necessary to determine the relationships of the metals with the other assets in the portfolio. In this context, it is aimed in the study to analyze the relationships between precious metals and Bitcoin and provide information to investors to be used in portfolio and risk management. In line with this purpose, in the next section the existing literature on cryptocurrencies and precious metals is given, then the relations between precious metals and Bitcoin are examined with econometric methods, and the findings are interpreted.

16.2 Literature Review

Cryptocurrency is a quite new topic for finance literature. But increasing interest worldwide and fluctuations in these currencies reveal the necessity of analyzing cryptocurrencies. The prior studies in the literature on cryptocurrencies mostly focused on defining these assets (Carrick, 2016; Dyhrberg, 2016; Klein, Thu, & Walther, 2018; Malović, 2014; Sontakke & Ghaisas, 2017; Yermack, 2014). Most of the findings gained from these studies indicated that cryptocurrencies are neither tangible asset like gold nor currency like dollar, they should be considered as a new financial asset. Another part of the studies on cryptocurrencies was to answer the question of what are the factors that determine the price of cryptocurrencies. However, in recent years the studies that analyze the relationships of the cryptocurrencies, which were accepted as new financial asset depending on the prior studies, with other financial assets started to be made. But the number of the studies made in this area quite limited since they are newly developed instruments. On the other hand, determining the relationships between cryptocurrencies and financial assets will present valuable information that investors may use in the determination of portfolio and risk management strategies.

Trabelsi (2018) investigated the connectedness between cryptocurrency markets and commodity, stock and currency markets. Analysis results indicated that there is no volatility spillover between cryptocurrency markets and traditional assets and cryptocurrencies are independent financial instruments. Bouri, Jalkh, Molnár, and Roubaud (2017) studied the relationships of Bitcoin with commodities in order to investigate usability in risk management. The results of dynamic conditional correlation (DCC) showed that Bitcoin is a strong hedge and safe haven in pre-crisis periods against the changes in commodity indices. Similarly, Baur, Hong, and Lee (2018) examined the correlations between Bitcoin and traditional assets such as stock, bond and commodity. Results revealed that there is no correlation between the Bitcoin and selected assets in both normal and financial stress periods. Rehman and Apergis (2018) analyzed the relationships between cryptocurrencies and gold, silver, copper, crude oil, Brent oil, natural gas and wheat futures. Causality on quantiles test results demonstrated that there is causality running from cryptocurrencies to commodity futures in both return and volatility.

In their study, İçellioğlu and Öztürk (2018) researched the relationships between Bitcoin and Dollar, Euro, Pound, Yen and Yuan. Causality analyses showed that there is no causal relationship between Bitcoin and selected currencies in both short and long term. Corbet, Meegan, Larkin, Lucey, and Yarovaya (2018) analyzed the relationships between popular cryptocurrencies and financial assets such as gold and stock in time and frequency dimension. The empirical results indicated that cryptocurrencies are isolated from other financial assets, and they may be used for diversification by the investors who have short investment horizon. Results also revealed that external economic and financial shocks have effects on the relationships. By using directed acyclic graph Ji, Bouri, Gupta, and Roubaud (2018) investigated the contemporaneous and lagged relations between Bitcoin and dollar, gold, selected indices for stocks, commodities and energy. The results from the contemporaneous analysis showed that Bitcoin is an isolated asset and has no relationship with the selected assets, in other words, there is no asset that directly affects the Bitcoin markets. However, when the lagged relations were analyzed it was seen that Bitcoin has time-varying relations with some of the assets especially in periods that the Bitcoin market declined.

Bouri, Molnár, Azzi, Roubaud, and Hagfors (2017) examined the relationships between Bitcoin and major stock indices, bond, oil, gold, commodity index and dollar index in the study, which they used dynamic conditional correlation model. The empirical results indicated that Bitcoin may be used for diversification, and also it may be used as safe haven in periods that Asian stock markets decline sharply. Similarly, in order to analyze the usability of cryptocurrencies in portfolio and risk management Guesmi, Saadi, Abid, and Ftiti (2018) tested the spillovers between Bitcoin and gold, stock and oil. Findings of GARCH model showed that there are significant return and volatility spillovers between the variables. Results of the study also proved that when Bitcoin is included in a portfolio that is constructed of stock, gold and oil, the portfolio risk decreases. Similarly, Hong (2017) stated in his study, which examines the usability of Bitcoin market for investment, that adding Bitcoin to a portfolio constructed of traditional assets such as stock, bond and commodity will provide benefit to the investors.

In order investigate the usability of Bitcoin for diversification, Brière, Oosterlinck, and Szafarz (2015) analyzed the performance of a portfolio constructed of Bitcoin, traditional assets such as bond, stock, currency and alternative investment vehicles such as commodity, hedge fund, real estate. Results of the study revealed that, despite high volatility Bitcoin provide high return, and it has low correlations with the other assets, indicating that it may be used for diversification. Chuen and Wang (2018) tested the usability of Bitcoin as an investment vehicle by constructing a portfolio. The findings of study prove that the correlations between Bitcoin and traditional assets are low, and the performance of portfolios constructed by adding Bitcoin is better than the performance of portfolios constructed of only traditional assets. In their study in which it is aimed to examine the effects of traditional assets on Bitcoin prices, Giudici and Hashish (2018) concluded that there are low correlations between Bitcoin prices and gold, oil, stock and exchange rates. Baek and Elbeck (2015) also showed that there is no effect of stock index, treasury note and Euro on Bitcoin returns. According to the results, the main factor that determines the volatility of Bitcoin is internal dynamics.

When the literature on the precious metals is investigated, it is seen that most of the studies focused on gold, silver and platinum. In these studies the hedging capability of precious metals, notably gold, was analyzed (Baur & McDermott, 2010; Beckmann, Berger, & Czudaj, 2015; Bredin, Conlon, & Potì, 2015; Hood & Malik, 2013; Iqbal, 2017; Lucey & Li, 2015; Mensi, Hammoudeh, & Kang, 2015; Taylor, 1998). The number of the studies that examine the relationships between cryptocurrencies and precious metals is quite limited. In this context, it is thought that it will fill a gap in the literature to study on the relationships between the mentioned variables by a contemporary method and large dataset.

16.3 Data and Methodology

The main purpose of this study is to determine the relationships between Bitcoin and precious metals in frequency and asymmetry dimension. In this context, the fundamental hypothesis of the study may be stated as follows;

H_0: There is no asymmetric causality relationship between Bitcoin and precious metals in any frequencies.

H_1: There is asymmetric causality relationship between Bitcoin and precious metals in any frequencies.

In order to test this hypothesis, asymmetric causality in frequency domain test developed by Ranjbar, Chang, Nel, and Gupta (2017) will be used in the study. This test approach is an extended version of asymmetric causality test of Hatemi-J (2012). Differently from the methodology of Hatemi-J, frequency dimension of the relationships between the variables is considered in this approach.

Based on the study of Granger and Yoon (2002) and Hatemi-J (2012) stated that causal effects of positive and negative shocks may be different. The positive and negative shocks of y_t and x_t variables, which have random walk process, may be written as follows in cumulative form;

$$y_{1i}^{+} = \sum_{i=1}^{t} \varepsilon_{1i}^{+}, y_{1i}^{-} = \sum_{i=1}^{t} \varepsilon_{1i}^{-}, y_{2i}^{+} = \sum_{i=1}^{t} \varepsilon_{2i}^{+}, y_{2i}^{-} = \sum_{i=1}^{t} \varepsilon_{2i}^{-} \tag{16.1}$$

In Hatemi-J (2012) test the causality between the positive and negative shocks are examined. The method is frequently used in the literature due to its mentioned advantage (Aloui, Hkiri, Lau, & Yarovaya, 2016; Arouri, Uddin, Kyophilavong, Teulon, & Tiwari, 2014; Faisal, Tursoy, & Berk, 2018; Gozgor, 2015; Halkos & Tzeremes, 2013; Nguyen, Sousa, & Uddin, 2015; Shahbaz, Van Hoang, Mahalik, & Roubaud, 2017). But, this approach produces only one test statistic, and it is assumed that the result will be valid for each frequency. However, the relationships between the variables may vary for different frequencies. On the other hand, Breitung and Candelon (2006) introduced frequency domain causality test that produces more than one test statistic for different frequencies.

Frequency domain causality test which is based on the study of Geweke (1982) allows the examination of the causality dynamics at different frequencies instead of depending on a single test statistic as in traditional causality analysis (Ciner, 2011; Hosoya, 1991). In order to evaluate both the asymmetric effects and time-varying relationships Ranjbar et al. (2017) extended the asymmetric causality test of Hatemi-J (2012) in frequency dimension. Asymmetric causality test of Ranjbar et al. (2017) contains both the asymmetric and frequency in domain causality tests.

Finite VAR model for x_t^+ and y_t^+;

$$\begin{pmatrix} \theta_{11}(L) & \theta_{12}(L) \\ \theta_{21}(L) & \theta_{22}(L) \end{pmatrix} \begin{pmatrix} y_t^+ \\ x_t^+ \end{pmatrix} = \begin{pmatrix} v_{1t} \\ v_{2t} \end{pmatrix} \tag{16.2}$$

Here, $\theta(L) = I - \sum_{i=1}^{p} \theta_i L^i$ autoregressive polynomials. Error vector v_t, is white noise. $E(v_t) = 0$ and $E(v_t v_t') = \Sigma$. Here Σ is positive definite and symmetric. When it is assumed that the system is stationary, the expression of moving average is as follows;

$$\begin{pmatrix} y_t^+ \\ x_t^+ \end{pmatrix} = \begin{bmatrix} \psi_{11}(L) & \psi_{12}(L) \\ \psi_{21}(L) & \psi_{22}(L) \end{bmatrix} \begin{bmatrix} \eta_{1t} \\ \eta_{2t} \end{bmatrix} \tag{16.3}$$

Here, $\psi(L)^{-1} = \Phi(L)^{-1} G^{-1}$ and by using this notation spectral density of y_t^+ may be shown as follows;

$$f_{y^+}(\omega) = \frac{1}{2\pi} \left\{ ||\psi_{11}(e^{-i\omega})||^2 + ||\psi_{12}(e^{-i\omega})||^2 \right\} \tag{16.4}$$

Frequency domain causality criterion developed by Geweke (1982) is defined as;

$$M_{X_t^+ \to Y_t^+}(\omega) = log\left[1 + \frac{||\psi_{12}(e^{-i\omega})||^2}{||\psi_{11}(e^{-i\omega})||^2}\right] \tag{16.5}$$

If $|\psi_{12}(e^{-i\omega})| = 0$, there is no Granger causality from X_t^+ to Y_t^+ at frequency ω. Based on the study of Breitung and Candelon (2006) and Ranjbar et al. (2017) expressed the VAR equation as follows;

$$Y_t^+ = \sum_{k=1}^{p} \theta_{11,k} y_{t-k}^+ + \sum_{k=1}^{p} \theta_{12,k} x_{t-k}^+ - \varpi_t \tag{16.6}$$

At frequency ω, the hypothesis $M_{X_t^+ \to Y_t^+}(\omega) = 0$ may be written as follows;

$$H_0 = R(\omega)\theta_{12} = 0 \tag{16.7}$$

Here, $\theta_{12} = [\theta_{12,\ 1}, \theta_{12,\ 2}, \ldots, \theta_{12,\ p}]$ and

$$R(\omega) = \begin{bmatrix} \cos(\omega) & \cos(2\omega)\ldots & \cos(p\omega) \\ \sin(\omega) & \sin(2\omega)\ldots & \sin(p\omega) \end{bmatrix} \tag{16.8}$$

Thereby, the null hypothesis that there is no Granger causality may be tested at frequency ω. However, as indicated in the study of Hatemi-J (2012) financial data generally do not have normal distribution, and the existence of ARCH effect affect the asymptotic distribution of Wald test distributed X^2 with degree of freedom as the number of restrictions. For this reason, in the study of Ranjbar et al. (2017) the critical values are obtained from bootstrap simulations. Therefore, the shortcomings of the causality tests in the literature were eliminated with the mentioned methodology.

In the study, for the purpose of analyzing the causal relations between Bitcoin (BTC) and precious metals in frequency and asymmetry dimension, gold (G), iridium (IR), osmium (OS), palladium (PD), platinum (PT), rhenium (RE), rhodium (RH), ruthenium (RU) and silver (S) prices were used. Troy ounce price data for gold, iridium, palladium, platinum and silver, ounce price data for rhodium and ruthenium and kg price data for osmium and rhenium were gained from "Thomson and Reuters Datastream". In the analyses, weekly logarithmic price data were used for the period of 8.23.2011–2.12.2019. Based on the purpose of the study, the asymmetric relationships between Bitcoin and precious metals are examined firstly by asymmetric causality test developed by Hatemi-J (2012) then asymmetric causality in frequency domain test introduced by Ranjbar et al. (2017). In line with this purpose, the analysis steps of the study are as follows;

Calculation of the descriptive statistics of Bitcoin and precious metals and determination of the correlations between them,

Application of traditional causality tests,

Table 16.1 Descriptive statistics

	Mean	Median	Max.	Min.	Std. Dev.	Skewness	Kurtosis	Jarque–Bera
BTC	5.7528	6.0637	9.8497	0.8197	2.3112	−0.4352	2.3626	18.962*
G	7.1949	7.1566	7.5430	6.9673	0.1335	0.8704	2.7220	50.633*
IR	6.6755	6.7855	7.2998	5.9914	0.3600	0.0162	1.7449	25.680*
OS	9.2696	9.2830	9.4247	9.0943	0.0857	−0.0704	1.6204	31.330*
PD	6.6281	6.6120	7.2442	6.1527	0.2049	0.4890	3.1277	15.852*
PT	7.0603	7.0317	7.5422	6.6489	0.2395	0.0791	1.5556	34.395*
RE	7.9619	7.7898	8.6074	7.4999	0.3505	0.7036	1.8124	55.240*
RH	7.0432	7.0030	7.8594	6.4216	0.3706	0.4980	2.6761	17.870*
RU	4.3599	4.1972	5.5909	3.6375	0.6082	0.7107	2.3797	39.186*
S	2.9833	2.8661	3.7546	2.6195	0.2771	0.9165	2.6154	57.150*

*%1 significance level

Application of asymmetric causality tests,
Application of asymmetric causality in frequency domain test.

16.4 Empirical Results

Based on the purpose of the study, the relationships between Bitcoin and precious metal will be analyzed by asymmetric causality in frequency domain test. But, before the causality analyses the descriptive statistics of the variables were calculated, and results were presented in Table 16.1.

As can be seen from Table 16.1 that gold, platinum, iridium and palladium have the highest average price per troy ounce respectively for the selected period. On the other hand, Bitcoin, ruthenium, rhodium and iridium are the variables which have the highest volatility for the same period. All of the variables have positive skewness values except Bitcoin and osmium. Also, high kurtosis values with high volatility indicate that large shocks may occur and series are away from normal distribution. According to Jarque–Bera test results all of the variables do not have normal distribution. When the descriptive statistics interpreted with the time period of the study, it may be said that increase in the demand for the raw materials and usage of the precious metal in high technology causes the changes in the price of precious metals.

After the calculation of the descriptive statistics, the stationarity of the series was tested by Ng-Perron (2001) unit root test, and results were given at Appendix 1.[1] According to the results of both intercept and trend and intercept model, price series of Bitcoin and precious metal are not stationary. Unit root test results also show that

[1]It is needed to determine the additional lag length to be added to the model for the later analyses. Also, in order to test the stationarity of the series intercept and tend model was chosen since the series of positive and negative shocks are cumulative.

negative and positive components of the series are not stationary too. After the stationarity testing, the unconditional correlations between Bitcoin and precious metals were calculated, and results were shown at Table 16.2.

According to the unconditional correlation results, there are negative correlations between the prices of Bitcoin and platinum, silver, rhenium, gold and iridium, and there are low correlations between Bitcoin and ruthenium and rhodium. Results also give us the correlations between precious metals. For example, osmium is negatively correlated with the all other precious metals. Rhodium and ruthenium are positively correlated with the all precious metals except osmium. Therefore, considering only unconditional correlation results it may be said that portfolio risk may be decreased by investing in Bitcoin and platinum, silver, rhenium, gold, iridium, ruthenium and rhodium. On the other hand, unconditional correlation coefficients do not consider the asymmetric relationships between the variables. In this context, in the next step of the study, the relationships between Bitcoin and precious metals will be analyzed by asymmetric causality test besides the traditional causality test. Firstly traditional causality test was applied to the series and results were given at Table 16.3.

Traditional Granger causality test results indicate that there is no bidirectional causality between Bitcoin and none of the precious metals. Also, there is no causality from Bitcoin to any of the precious metals. There are only unidirectional causalities from iridium and palladium to Bitcoin. In this context, according to the traditional causality test an investor who wants to provide benefit from diversification may decrease portfolio risk by investing in precious metals except for iridium and palladium.

Before the asymmetric causality test, Doornik and Hansen (2008) test for multivariate normality and Hacker and Hatemi-J (2005) test for multivariate ARCH effect were applied to the series. The results of the tests given at Appendix 2 indicate that all of the series have multivariate normality problem and most the series have ARCH effect. These results reveal the necessity of using bootstrap based tests instead of traditional methods in the analysis of the causal relations between Bitcoin and precious metals. For this reason, in the next step asymmetric causality test was applied to the series and results were presented at Table 16.4.

Asymmetric causality test results indicate the existence of asymmetric relationships between Bitcoin and precious metals. Similar to the results of traditional causality test, causal relations were determined between Bitcoin and iridium and palladium. There are significant causalities between the increases-decreases in Bitcoin and increases-decreases in iridium and between positive shocks in palladium prices and negative shocks in Bitcoin price. On the other hand, differently from the traditional causality test results it was found asymmetric relationships between Bitcoin and gold, platinum, rhenium and silver.

According to the results, there are no bidirectional relationships between Bitcoin and osmium, rhenium and ruthenium prices. Results show that causal relationships are generally unidirectional. There are unidirectional causalities from gold, iridium and silver to Bitcoin and these relationships are in different asymmetric structure. For example, increases in gold, iridium, palladium, rhenium and silver prices are cause of decreases in Bitcoin prices, and there are significant causalities between

Table 16.2 Unconditional correlations between BTC and precious metals

	BTC	G	IR	OS	PD	PT	RE	RH	RU	S
BTC		−0.437	−0.026	0.377	0.678	−0.782	−0.531	0.104	0.046	−0.63
G	−0.437		0.528	−0.640	−0.116	0.689	0.630	0.320	0.509	0.85
IR	−0.026	0.528		−0.192	0.316	0.035	0.528	0.754	0.835	0.25
OS	0.377	−0.640	−0.192		−0.051	−0.719	−0.604	−0.342	−0.481	−0.69
PD	0.678	−0.116	0.316	−0.051		−0.364	−0.214	0.588	0.454	−0.29
PT	−0.782	0.689	0.035	−0.719	−0.364		0.534	0.013	0.083	0.89
RE	−0.531	0.630	0.528	−0.604	−0.214	0.534		0.365	0.590	0.58
RH	0.104	0.320	0.754	−0.342	0.588	0.013	0.365		0.882	0.07
RU	0.046	0.509	0.835	−0.481	0.454	0.083	0.590	0.882		0.20
S	−0.631	0.859	0.251	−0.691	−0.291	0.891	0.586	0.077	0.2042	

Table 16.3 Results of traditional causality test between BTC and precious metals

	Prob. value		Prob. value
BTC ≠> G	0.2587	PT ≠> BTC	0.6251
G ≠> BTC	0.2230	BTC ≠> RE	0.6983
BTC ≠> IR	0.3459	RE ≠> BTC	0.5152
IR => BTC	0.0181	BTC ≠> RH	0.6944
BTC ≠> OS	0.9489	RH ≠> BTC	0.4273
OS ≠> BTC	0.9139	BTC ≠> RU	0.7426
BTC ≠> PD	0.9082	RU ≠> BTC	0.6073
PD => BTC	0.0713	BTC ≠> S	0.6199
BTC ≠> PT	0.9454	S ≠> BTC	0.5017

The symbol A ≠> B indicates that there is no causality from variable A to variable B. Optimal lag length in VAR models were determined based on the Akaike Information Criteria (AIC)

decreases in Iridium and Palladium prices and increases in Bitcoin prices. Also, it is determined that positive and negative shocks in silver prices are cause of positive and negative shocks in Bitcoin prices and increases in gold prices are cause of increases in Bitcoin price. On the other hand, there is unidirectional causality from Bitcoin to rhenium. Accordingly, increases in Bitcoin price are cause of decreases in rhenium price.

When the asymmetric causality test results evaluated together, it may be said that to construct a portfolio from Bitcoin, osmium, rhenium and ruthenium will provide benefit to the investors. However, investors may construct their portfolios depending on their attitude towards risk. For example, if an investor is a risk lover then, the portfolio may be constructed of Bitcoin and precious metal, such as gold and silver, that are expected increase for both. On the other hand, asymmetric causality analysis produces only one test statistic for the period of 2011–2019, and it is not realistic to accept the assumption that the relationships between the variables do not change over the time especially when the dynamic structure in the precious metal markets and difference in the responses of investors are considered. For this reason, in the final step of the study, Ranjbar et al. (2017) asymmetric causality in frequency dimension test which allows to analyze the asymmetric relationship in frequency dimension was applied to the series and results were given at Table 16.5.

The most important finding gained from the analysis is the existence of the causality from all precious metals to Bitcoin even if in different frequencies. Asymmetric causality in frequency domain test results show that there are causalities from osmium, rhenium, rhodium and ruthenium to Bitcoin differently from the other test results. It is also determined that there causalities from negative shocks of Bitcoin to negative shocks of iridium; from negative shocks of Bitcoin to positive shocks of osmium and rhenium; from positive shocks of Bitcoin to negative shocks of rhenium, rhodium and ruthenium; from positive shocks of Bitcoin to positive shocks of palladium, platinum, rhenium and ruthenium.

As can be seen from Table 16.5, another important finding of the analysis is that increases in the price of rhenium and ruthenium are cause of increases in Bitcoin for all frequencies. Similarly, positive shocks in gold price are cause of negative shocks

Table 16.4 Results of asymmetric causality test between BTC and precious metals

	Wald statistic	Critical bootstrap values				Wald statistic	Critical bootstrap values		
		%1	%5	%10			%1	%5	%10
$G^{-} \neq> BTC^{-}$	2.621	8.312	7.134	4.594	$BTC^{-} \neq> PT^{-}$	1.430	17.30	8.241	5.418
$G^{-} \neq> BTC^{+}$	2.309	15.36	6.104	4.517	$BTC^{-} \neq> PT^{+}$	2.217	6.262	3.775	2.335
$G^{+} => BTC^{-}$	16.78^{*}	8.696	5.342	4.700	$BTC^{+} \neq> PT^{-}$	0.056	9.767	5.722	5.142
$G^{+} => BTC^{+}$	5.813^{**}	6.952	4.884	3.569	$BTC^{+} \neq> PT^{+}$	1.584	11.04	5.498	4.463
$BTC^{-} \neq> G^{-}$	0.765	16.10	7.362	5.921	$RE^{-} \neq> BTC^{-}$	3.031	9.697	6.664	4.704
$BTC^{-} \neq >G^{+}$	0.112	12.98	4.352	2.603	$RE^{-} \neq> BTC^{+}$	0.480	8.488	5.300	4.174
$BTC^{+} \neq> G^{-}$	1.158	11.20	4.622	3.768	$RE^{+} \neq> BTC^{-}$	4.267	6.794	5.213	4.393
$BTC^{+} \neq> G^{+}$	0.733	10.30	6.654	5.137	$RE^{+} \neq> BTC^{+}$	2.016	11.15	5.595	4.198
$IR^{-} \neq> BTC^{-}$	7.277	28.72	17.55	13.81	$BTC^{-} \neq> RE^{-}$	0.622	15.82	6.324	5.100
$IR^{-} =>BTC^{+}$	14.38^{*}	14.24	11.72	9.738	$BTC^{-} \neq>RE^{+}$	3.339	9.643	6.084	5.133
$IR^{+} => BTC^{-}$	27.57^{*}	17.52	14.58	12.79	$BTC^{+} => RE^{-}$	5.884^{***}	19.06	10.17	5.013
$IR^{+} \neq> BTC^{+}$	6.289	17.92	12.12	10.43	$BTC^{+} \neq> RE^{+}$	2.817	14.64	7.775	4.088
$BTC^{-} \neq> IR^{-}$	2.328	10.46	6.931	5.729	$RH^{-} \neq> BTC^{-}$	0.134	7.035	3.150	2.082
$BTC^{-} \neq> IR^{+}$	0.609	11.97	4.486	3.555	$RH^{-} \neq> BTC^{+}$	0.122	8.826	3.884	3.105
$BTC^{+} \neq> IR^{-}$	0.014	10.06	3.051	2.434	$RH^{+} \neq> BTC^{-}$	0.003	7.177	3.565	1.747
$BTC^{+} \neq> IR^{+}$	0.095	6.310	4.051	2.773	$RH^{+} \neq> BTC^{+}$	0.113	22.901	3.569	2.132
$OS^{-} \neq> BTC^{-}$	0.146	13.93	6.722	5.607	$BTC^{-} \neq> RH^{-}$	1.575	14.07	8.582	7.262
$OS^{-} \neq> BTC^{+}$	0.264	9.366	3.640	2.564	$BTC^{-} \neq> RH^{+}$	0.160	7.248	1.917	1.548
$OS^{+} \neq> BTC^{-}$	2.222	12.22	6.195	4.775	$BTC^{+} \neq> RH^{-}$	0.204	14.07	10.11	7.553
$OS^{+} \neq> BTC^{+}$	0.010	7.622	5.600	3.809	$BTC^{+} \neq> RH^{+}$	1.090	9.757	5.655	3.801
$BTC^{-} \neq> OS^{-}$	0.645	10.79	5.893	5.607	$RU^{-} \neq> BTC^{-}$	0.419	58.69	7.008	4.584
$BTC^{-} \neq> OS^{+}$	0.479	9.663	6.450	2.564	$RU^{-} \neq> BTC^{+}$	0.008	6.071	3.369	2.517
$BTC^{+} \neq> OS^{-}$	0.188	7.780	5.920	4.775	$RU^{+} \neq> BTC^{-}$	0.021	17.79	7.197	4.470
$BTC^{+} \neq> OS^{+}$	1.223	7.661	5.555	3.809	$RU^{+} \neq> BTC^{+}$	0.087	10.30	6.164	4.735
$PD^{-} \neq> BTC^{-}$	3.331	8.883	5.183	4.554	$BTC^{-} \neq> RU^{-}$	0.049	15.88	4.096	2.464

$PD^{-} \neq> BTC^{+}$	0.189	9.536	7.150	4.763	$BTC^{-} \neq> RU^{+}$	0.358	14.53	3.999	2.880
$PD^{+} => BTC^{-}$	3.623***	5.620	3.824	2.934	$BTC^{+} \neq> RU^{-}$	1.034	6.141	3.783	2.422
$PD^{+} \neq> BTC^{+}$	0.320	8.405	5.741	4.457	$BTC^{+} \neq> RU^{+}$	0.017	7.870	3.017	1.589
$BTC^{-} \neq> PD^{-}$	2.819	8.352	5.480	4.543	$S^{-} => BTC^{-}$	4.984**	7.307	5.730	3.688
$BTC^{-} \neq> PD^{+}$	3.708	12.95	6.780	6.024	$S^{-} \neq> BTC^{+}$	0.003	8.164	4.024	3.033
$BTC^{+} \neq> PD^{-}$	4.204	10.91	7.817	5.255	$S^{+} => BTC^{-}$	9.420**	10.31	6.236	4.098
$BTC^{+} \neq> PD^{+}$	3.690	8.384	6.365	4.613	$S^{+} => BTC^{+}$	5.123***	13.48	6.828	5.064
$PT^{-} \neq> BTC^{-}$	3.331	8.883	5.183	4.554	$BTC^{-} \neq> S^{-}$	0.472	10.68	8.314	4.320
$PT^{-} => BTC^{+}$	3.623***	5.620	3.824	2.934	$BTC^{-} \neq> S^{+}$	0.101	11.55	3.005	1.890
$PT^{+} \neq> BTC^{-}$	0.189	9.536	7.150	4.763	$BTC^{+} \neq> S^{-}$	1.142	7.870	6.634	3.842
$PT^{+} \neq> BTC^{+}$	0.320	8.405	5.741	4.457	$BTC^{+} \neq> S^{+}$	0.629	13.04	7.251	4.820

The symbol A ≠> B indicates that there is no causality from variable A to variable B. Optimal lag length in VAR models were determined based on the HJC information criteria. *%1, **%5, ***%10 indicate significance level

Table 16.5 Results of asymmetric causality in frequency domain test from precious metals to BTC

	Frequency							
	0.02 (314 week)	0.13 (48 week)	0.26 (24 week)	0.52 (12 week)	0.78 (8 week)	1.57 (4 week)	2.09 (3 week)	3.13 (2 week)
G^{-} ≠> BTC^{-}	3.515	3.793	2.645	1.155	1.172	3.564	3.523	2.519
G^{-} ≠> BTC^{+}	2.315	1.652	0.388	0.731	1.13	1.552	1.36	1.234
G^{+} => BTC^{-}	**8.824***	**8.770***	**8.512***	**7.762****	**6.393****	0.019	1.572	**3.686*****
G^{+} = > BTC^{+}	3.338	3.325	3.609	**4.563*****	**5.266*****	**7.291****	**4.555*****	2.364
IR^{-} => BTC^{-}	0.013	0.351	0.417	0.304	0.126	2.248	**4.405*****	**4.506*****
IR^{-} => BTC^{+}	**11.01***	**10.44***	**11.12***	**9.388***	4.074	1.153	**4.531*****	**7.152***
IR^{+} => BTC^{-}	**6.206****	**6.267****	2.847	0.878	0.643	0.043	0.023	0.094
IR^{+} ≠> BTC^{+}	2.304	2.149	1.899	1.875	1.756	0.264	0.059	0.180
OS^{-} ≠> BTC^{-}	2.695	2.660	1.394	0.804	0.673	0.040	0.064	0.334
OS^{-} => BTC^{+}	**7.017****	**7.703****	**5.307*****	1.785	1.301	2.381	2.520	1.900
OS^{+} ≠> BTC^{-}	0.306	0.339	0.558	0.481	0.505	0.970	0.867	0.607
OS^{+} ≠> BTC^{+}	2.692	2.689	1.329	0.792	0.686	0.083	0.081	0.339
PD^{-} ≠> BTC^{-}	2.250	2.189	2.058	2.285	1.968	1.002	3.075	4.237
PD^{-} ≠> BTC^{+}	3.075	3.257	2.642	1.040	0.669	0.831	1.474	1.626
PD^{+} => BTC^{-}	**5.510*****	**5.346*****	3.418	1.865	1.382	0.080	1.620	2.746
PD^{+} = > BTC^{+}	**8.357****	**6.922****	3.318	2.898	3.693	**6.423****	4.051	2.079
PT^{-} ≠> BTC^{-}	2.710	1.714	0.763	1.857	2.238	1.929	1.837	1.967
PT^{-} = > BTC^{+}	4.353	**4.678*****	2.903	0.976	0.659	0.308	0.277	0.322
PT^{+} ≠> BTC^{-}	2.494	2.901	2.471	1.204	0.915	0.502	0.423	0.505
PT^{+} => BTC^{+}	**8.113****	**8.388****	**6.190*****	3.255	3.087	3.603	1.644	0.545
RE^{-} ≠> BTC^{-}	0.677	0.836	0.916	0.470	0.370	1.298	1.876	1.749
RE^{-} = > BTC^{+}	1.259	1.293	2.472	4.137	4.489	**4.807*****	3.879	3.305
RE^{+} => BTC^{-}	3.468	3.208	2.070	1.568	2.100	**5.461****	**4.501*****	2.645
RE^{+} = > BTC^{+}	**6.971****	**10.08***	**13.82***	**11.36***	**9.880****	**6.606****	**9.076****	**11.31***
RH^{-} ≠> BTC^{-}	2.829	2.796	2.977	3.381	3.340	1.419	0.762	1.257

$RH^{-} \not\Rightarrow BTC^{+}$	3.921	3.966	3.426	2.593	2.597	1.967	0.440	0.030
$RH^{+} \Rightarrow BTC^{-}$	**13.19****	**12.87****	**7.700*****	1.057	0.529	2.192	2.404	1.816
$RH^{+} \not\Rightarrow BTC^{+}$	2.899	2.993	2.140	0.785	0.411	0.660	2.176	2.716
$RU^{-} \not\Rightarrow BTC^{-}$	2.342	0.682	0.151	0.241	0.223	0.561	0.935	1.060
$RU^{-} \not\Rightarrow BTC^{+}$	0.211	1.260	2.039	1.947	1.873	2.027	2.235	2.306
$RU^{+} \Rightarrow BTC^{-}$	**11.03***	**7.843****	0.880	0.254	0.626	1.008	1.089	1.119
$RU^{+} \Rightarrow BTC^{+}$	**5.134*****	**5.316*****	**5.570*****	**6.926*****	**9.803****	**16.45***	**11.68***	**8.307****
$S^{-} \not\Rightarrow BTC^{-}$	**7.952****	**6.913****	2.495	1.922	1.689	0.133	0.284	0.934
$S^{-} \not\Rightarrow BTC^{+}$	2.681	2.023	0.418	0.235	0.551	0.799	0.871	0.907
$S^{+} \Rightarrow BTC^{-}$	1.799	1.399	0.526	0.367	0.310	1.953	**6.208****	**5.984****
$S^{+} \Rightarrow BTC^{+}$	3.982	**4.617*****	3.738	1.491	1.179	2.922	3.239	2.671

The symbol $A \not\Rightarrow B$ indicates that there is no causality from variable A to variable B. Optimal lag length in VAR models were determined based on the HJC information criteria. *%1, **%5, ***%10 indicate significance level

Wald values written in bold indicate significant causality between the variables at the stated frequency of Bitcoin for almost all frequencies. Results show that there are causalities from precious metals to Bitcoin in mid and long term in general. For example, in mid and long term positive shocks in platinum and silver price are cause of positive shocks in Bitcoin price, and similarly positive shocks in ıridium, palladium, rhenium and ruthenium price are cause of negative shocks in Bitcoin price. On the other hand, in short term there are causalities from decreases in iridium price and increases in rhenium and silver price to decreases in Bitcoin price.

The results also indicate that there is no causality between precious metals and Bitcoin in some asymmetric dimensions. For example, there is no significant causality from decreases in gold, osmium, palladium, platinum, rhenium, rhodium, ruthenium price and increases in osmium, platinum price to decreases in Bitcoin price for all frequencies. After determination of the causalities from precious metals to Bitcoin, the causalities from Bitcoin to precious metals were determined and results were presented at Table 16.6.

Table 16.6 shows that Bitcoin is cause of limited number of precious metals. There are causalities from Bitcoin to only iridium, osmium and rhenium in different frequencies. According to the results, there is no causality from Bitcoin prices to gold, palladium, platinum, rhodium, ruthenium and silver prices. The most important finding is that increases in Bitcoin price are cause of decreases in iridium price almost in all frequencies. Results also indicate that there are causalities from increases in Bitcoin price to decreases in iridium and rhenium price and from increases in Bitcoin price to increases in iridium price. On the other hand, decreases in Bitcoin price are cause of increases in iridium and osmium price. All the significant causalities from Bitcoin price to precious metal prices are mid and long term except iridium. For example, decreases in Bitcoin price are cause of increases in iridium and osmium price at 24–314 week cycle. Similarly, increases in Bitcoin price are cause of increases in iridium price at 48 week cycle.

When the results of asymmetric causality in frequency domain test are evaluated together, it is seen that there are causalities, that the other methods did not catch, between Bitcoin and precious metals in different frequencies. Results show that the direction of the relationships is from precious metal to Bitcoin except for iridium, osmium and rhenium. Findings also indicate that there are generally mid and long term causality relationships between the variables. In this context, it may be advised to the investors to construct their portfolios based on gold, palladium, platinum, rhenium, ruthenium and Bitcoin for short investment horizon.

16.5 Conclusion

Investors desire to decrease total risk of their portfolios. For this reason, they diversify their portfolios by investing in different asset classes such as bonds, stocks and currencies. However, some of the investors look for alternative investment vehicles instead of traditional ones. Precious metals are good alternatives for these investors, and these assets have been used widely for investment purposes besides

Table 16.6 Results of asymmetric causality in frequency domain test from BTC to precious metals

	Frequency							
	0.02 (314 week)	0.13 (48 week)	0.26 (24 week)	0.52 (12 week)	0.78 (8 week)	1.57 (4 week)	2.09 (3 week)	3.13 (2 week)
$BTC^{-} \nRightarrow G^{-}$	0.968	0.855	0.452	0.209	0.162	0.153	0.641	0.855
$BTC^{-} \nRightarrow G^{+}$	1.377	1.409	0.972	0.322	0.153	0.568	1.500	1.719
$BTC^{+} \Rightarrow G^{-}$	1.007	0.887	0.805	0.872	0.771	0.216	0.422	0.647
$BTC^{+} \Rightarrow G^{+}$	1.922	1.873	1.081	0.666	0.646	0.268	0.885	1.408
$BTC^{-} \Rightarrow IR^{-}$	2.310	2.787	2.263	1.795	1.421	3.677	2.216	0.735
$BTC^{-} \Rightarrow IR^{+}$	**6.463****	**7.646****	**6.219****	2.932	2.239	1.511	1.210	1.235
$BTC^{+} \Rightarrow IR^{-}$	**8.063****	**6.722****	**11.29***	**14.98***	**15.24***	**12.32***	**4.632*****	2.461
$BTC^{+} \Rightarrow IR^{+}$	**9.263***	**7.362***	1.161	3.169	3.726	3.707	3.434	3.363
$BTC^{-} \nRightarrow OS^{-}$	0.306	0.339	0.558	0.481	0.505	0.970	0.867	0.607
$BTC^{-} \nRightarrow OS^{+}$	**5.535*****	**5.709*****	**5.475*****	1.950	1.065	0.514	0.442	0.471
$BTC^{+} \nRightarrow OS^{-}$	1.765	1.643	0.169	0.220	0.402	0.924	0.890	0.780
$BTC^{+} \nRightarrow OS^{+}$	1.822	1.780	1.726	1.540	1.067	0.681	2.222	3.112
$BTC^{-} \nRightarrow PD^{-}$	0.300	0.276	0.059	0.113	0.146	1.148	2.317	2.445
$BTC^{-} \nRightarrow PD^{+}$	0.588	0.598	0.636	0.795	1.109	3.616	3.542	2.377
$BTC^{+} \Rightarrow PD^{-}$	0.214	0.631	1.143	1.009	0.950	1.687	2.060	2.097
$BTC^{+} \nRightarrow PD^{+}$	2.456	2.372	2.104	2.354	1.844	1.192	2.873	3.986
$BTC^{-} \nRightarrow PT^{-}$	0.153	0.158	0.173	0.234	0.369	1.738	1.982	1.487
$BTC^{-} \nRightarrow PT^{+}$	0.018	0.011	0.323	1.038	1.163	1.369	1.454	1.416
$BTC^{+} \nRightarrow PT^{-}$	0.839	0.944	0.771	0.429	0.256	0.388	0.898	1.132
$BTC^{+} \nRightarrow PT^{+}$	1.785	1.655	1.188	1.120	1.255	0.965	0.383	0.143
$BTC^{-} \Rightarrow RE^{-}$	0.065	0.059	0.041	0.013	0.003	1.231	2.498	2.565
$BTC^{-} \nRightarrow RE^{+}$	1.304	1.350	1.434	1.358	1.168	0.183	0.356	0.652
$BTC^{+} \Rightarrow RE^{-}$	**5.476*****	**4.683*****	1.253	0.044	0.003	0.012	0.016	0.019
$BTC^{+} \nRightarrow RE^{+}$	1.446	1.606	1.494	0.732	0.579	0.649	0.449	0.274

(continued)

Table 16.6 (continued)

	Frequency							
	0.02 (314 week)	0.13 (48 week)	0.26 (24 week)	0.52 (12 week)	0.78 (8 week)	1.57 (4 week)	2.09 (3 week)	3.13 (2 week)
$BTC^- \neq> RH^-$	2.716	3.052	3.266	2.394	1.847	0.604	2.004	3.194
$BTC^- \neq> RH^+$	1.178	1.168	2.063	3.054	3.004	2.177	2.957	3.761
$BTC^+ \neq> RH^-$	0.485	0.543	1.468	1.926	1.987	1.752	1.581	1.544
$BTC^+ \neq> RH^+$	2.854	2.541	1.551	1.138	0.822	0.093	0.807	1.332
$BTC^- => RU^-$	0.836	1.044	2.513	3.120	3.140	2.616	2.416	2.586
$BTC^- \neq> RU^+$	4.119	3.943	3.252	3.161	3.373	3.297	1.156	0.169
$BTC^+ \neq> RU^-$	0.361	0.382	0.453	0.497	0.506	0.407	0.170	0.120
$BTC^+ \neq> RU^+$	2.075	1.553	0.328	0.195	0.352	1.108	2.164	2.295
$BTC^- => S^-$	0.923	1.071	1.256	0.836	0.680	0.954	0.839	0.598
$BTC^- \neq> S^+$	0.662	0.426	0.340	1.216	1.475	1.423	1.472	1.589
$BTC^+ \neq> S^-$	3.626	3.518	3.306	3.147	2.725	0.175	0.504	1.202
$BTC^+ \neq> S^+$	0.417	0.420	0.638	0.922	1.212	2.390	2.011	1.573

The symbol A $\neq>$ B indicates that there is no causality from variable A to variable B. Optimal lag length in VAR models were determined based on the HJC information criteria. *%1, **%5, ***%10 indicate significance level

Wald values written in bold indicate significant causality between the variables at the stated frequency their industrial usage. Another alternative for investors is cryptocurrencies. Even though they are not developed as an investment vehicle, in recent years they are included in the portfolios as a financial asset. Especially since they have lower correlations with the other asset classes, it is very popular in the finance world. But, in order to invest in these assets and to provide the expected benefit, the relationships between the assets should be analyzed appropriately, and in the analyses the asymmetric structure of the relationships and time-varying effects should be considered. For this reason, in the study it is aimed to determine the relationships between Bitcoin and gold, iridium, osmium, palladium, platinum, rhenium, rhodium, ruthenium and silver in detail.

In line with the purpose of the study, the relationships between the mentioned variables were investigated by asymmetric causality in frequency domain test besides traditional econometric methods. Based on the analyses, it is concluded that there are causality relationships between Bitcoin and precious metals in different asymmetric structure and dimensions. The most important finding of the study is that there are causalities from all the precious metals to Bitcoin even in different frequencies. On the other hand, increases in rhenium and ruthenium prices are cause of increases in Bitcoin prices in all frequencies. However, it is determined that Bitcoin prices are cause of only iridium, osmium and rhenium prices in different frequencies and asymmetric structures, and there are causalities from increases in Bitcoin price to decreases in iridium prices in almost all frequencies.

The results indicate that there are generally mid and long term causality relationships between Bitcoin and precious metals. In this context, it may be suggested investors to invest in these assets for short investment horizon, and to consider the relationships between Bitcoin and gold, palladium, platinum, rhenium and ruthenium portfolio allocation decisions. In further studies, the dynamic correlation between Bitcoin and precious metals may be determined, and portfolio implications may be examined in order to determine to proportion of Bitcoin and precious metals in the portfolio.

Appendix 1

Table 16.7 Ng-Perron unit root test results

	İntercept				Intercept and trend			
	MZa	MZt	MSB	MPT	MZa	MZt	MSB	MPT
BTC	−5.4786	−1.5876	0.2898	4.6725	−11.837	−2.4278	0.2051	7.7259
G	−0.0380	−0.0368	0.9664	52.251	−2.7508	−1.0164	0.3695	28.480
IR	−0.7151	−0.4122	0.5763	19.941	−0.0949	−0.0551	0.5812	75.096
OS	−0.0769	−0.0574	0.7467	33.701	−5.1227	−1.5643	0.3054	17.640
PD	2.5845	0.8490	0.3285	16.163	−1.5833	−0.4947	0.3125	27.784
PT	0.5540	0.4994	0.9015	53.089	−12.230	−2.4397	0.1995	7.6378
RE	0.1216	0.1302	1.0709	65.123	−1.9670	−0.8546	0.4344	38.122
RH	−0.1005	−0.0812	0.8085	38.095	0.9440	0.7752	0.8212	155.77
RU	−0.2902	−0.2081	0.7172	30.102	0.5593	0.4063	0.7265	119.20
S	0.3700	0.4499	1.2158	87.044	−2.4748	−0.9870	0.3989	32.067
BTC^-	4.8186	4.5074	0.9354	108.73	1.4571	1.0252	0.7036	125.55
G^-	1.1829	9.0182	7.6239	3815.1	0.6659	0.6268	0.9414	193.05
IR^-	0.6179	1.0601	1.7155	176.57	−1.1502	−0.5686	0.4944	50.875
OS^-	1.6364	11.012	6.7292	3260.6	−2.9343	−1.2049	0.4106	30.880
PD^-	1.5452	10.291	6.6599	3137.4	−9.3099	−2.1246	0.2282	9.9295
PT^-	1.2580	11.003	8.7468	5099.8	0.6057	0.5466	0.9024	177.15
RE^-	1.0285	3.8990	3.7908	917.53	−0.0495	−0.0313	0.6316	86.594
RH^-	1.2961	5.4083	4.1728	1175.3	−1.5995	−0.7548	0.4719	44.747
RU^-	0.6790	2.9183	4.2981	1087.3	0.5891	0.7902	1.3413	373.01
S^-	1.0579	7.1462	6.7550	2916.9	0.6248	0.7842	1.2551	330.03
BTC^+	3.5663	3.3957	0.9522	96.556	0.5472	0.3664	0.6695	103.26
G^+	1.3571	10.015	7.3796	3707.5	−0.5494	−0.2866	0.5217	59.740

IR^+	2.6062	3.4175	1.3113	153.82	−1.0058	−0.4563	0.4537	45.984
OS^+	1.5327	10.845	7.0757	3531.8	−3.5026	−1.2876	0.3676	25.400
PD^+	2.0535	14.372	6.9990	3812.0	1.6903	0.8398	0.4969	71.524
PT^+	1.2672	8.7859	6.9333	3213.1	0.1773	0.1194	0.6737	99.502
RE^+	1.2912	8.0219	6.2129	2594.3	−0.0832	−0.0728	0.8752	152.21
RH^+	2.4150	7.2736	3.0118	757.37	0.2933	0.1631	0.5562	73.470
RU^+	2.8530	2.6985	0.9458	86.459	−0.8256	−0.3901	0.4725	49.818
S^+	1.1807	7.4429	6.3038	2609.3	0.3686	0.3071	0.8334	148.44

The symbols + and − indicates the series consisting of positive and negative components

Appendix 2

Table 16.8 Results of multivariate normality and multivariate ARCH test

	Multivariate normality	Multivariate ARCH
BTC, G	<0.0001	163.21*
BTC^-, G^-	0.5217	178.80*
BTC^-, G^+	<0.0001	190.82*
BTC^+, G^-	<0.0001	158.52*
BTC^+, G^+	0.5551	170.76*
BTC, IR	<0.0001	417.63*
BTC^-, IR^-	0.9539	79.453*
BTC^-, IR^+	0.6932	82.707*
BTC^+, IR^-	0.7882	92.685*
BTC^+, IR^+	0.8406	95.793*
BTC, OS	<0.0001	87.970*
BTC^-, OS^-	0.6976	89.214*
BTC^-, OS^+	<0.0001	94.522*
BTC^+, OS^-	0.4921	91.513*
BTC^+, OS^+	<0.0001	97.462*
BTC, PD	<0.0001	103.73*
BTC^-, PD^-	0.2826	131.44*
BTC^-, PD^+	<0.0001	133.82*
BTC^+, PD^-	0.3227	128.89*
BTC^+, PD^+	<0.0001	132.19*
BTC, PT	<0.0001	100.73*
BTC^-, PT^-	0.7246	156.17*
BTC^-, PT^+	0.5183	155.28*
BTC^+, PT^-	0.6510	153.56*
BTC^+, PT^+	0.0940	151.89*
BTC, RE	<0.0001	520.91*
BTC^-, RE^-	<0.0001	66.712*
BTC^-, RE^+	<0.0001	64.768*
BTC^+, RE^-	0.3852	79.387*
BTC^+, RE^+	<0.0001	76.394*
BTC, RH	<0.0001	553.25*
BTC^-, RH^-	0.9923	66.712*
BTC^-, RH^+	0.0074	64.768*
BTC^+, RH^-	0.0852	79.387*
BTC^+, RH^+	0.9692	76.394*
BTC, RU	<0.0001	685.12*
BTC^-, RU^-	0.9795	330.76*
BTC^-, RU^+	0.9144	325.26*
BTC^+, RU^-	0.6850	335.42*
BTC^+, RU^+	0.8385	336.16*

(continued)

Table 16.8 (continued)

	Multivariate normality	Multivariate ARCH
BTC, S	<0.0001	205.56*
BTC^{-}, S^{-}	0.8274	164.97*
BTC^{-}, S^{+}	0.0010	172.84*
BTC^{+}, S^{-}	0.0123	157.90*
BTC^{+}, S^{+}	<0.0001	165.76*

*%1 significance level

References

Ali, R., Barrdear, J., Clews, R., & Southgate, J. (2014). Innovations in payment technologies and the emergence of digital currencies. *Bank of England Quarterly Bulletin, Q3.*

Aloui, C., Hkiri, B., Lau, C. K. M., & Yarovaya, L. (2016). Investors' sentiment and US Islamic and conventional indexes nexus: A time—frequency analysis. *Finance Research Letters, 19*, 54–59. https://doi.org/10.1016/j.frl.2016.06.002

Arouri, M., Uddin, G. S., Kyophilavong, P., Teulon, F., & Tiwari, A. K. (2014). *Energy utilization and economic growth in France: Evidence from asymmetric causality Test (No. 2014-102).* Retrieved from http://v6.ipag.fr/wp-content/uploads/recherche/WP/IPAG_WP_2014_102.pdf

Baek, C., & Elbeck, M. (2015). Bitcoins as an investment or speculative vehicle? A first look. *Applied Economics Letters, 22*(1), 30–34. https://doi.org/10.1080/13504851.2014.916379

Baur, D. G., Dimpf, T., & Kuck, K. (2018). Bitcoin, gold and the US dollar—A replication and extension. *Finance Research Letters, 25*, 103–110. https://doi.org/10.1016/j.frl.2017.10.012

Baur, D. G., Hong, K., & Lee, A. D. (2018). Bitcoin: Medium of exchange or speculative assets? *Journal of International Financial Markets Institutions and Money, 54*, 177–189. https://doi.org/10.1016/j.intfin.2017.12.004

Baur, D. G., & McDermott, T. K. (2010). Is gold a safe haven? International evidence. *Journal of Banking & Finance, 34*(8), 1886–1898. https://doi.org/10.1016/j.jbankfin.2009.12.008

Beckmann, J., Berger, T., & Czudaj, R. (2015). Does gold act as a hedge or a safe haven for stocks? A smooth transition approach. *Economic Modelling, 48*, 16–24. https://doi.org/10.1016/j.econmod.2014.10.044

Bouri, E., Jalkh, N., Molnár, P., & Roubaud, D. (2017). Bitcoin for energy commodities before and after the December 2013 crash: Diversifier, hedge or safe haven? *Applied Economics, 49*(50), 5063–5073. https://doi.org/10.1080/00036846.2017.1299102

Bouri, E., Molnár, P., Azzi, G., Roubaud, D., & Hagfors, L. I. (2017). On the hedge and safe haven properties of Bitcoin: Is it really more than a diversifier? *Finance Research Letters, 20*, 192–198. https://doi.org/10.1016/j.frl.2016.09.025

Bredin, D., Conlon, T., & Potì, V. (2015). Does gold glitter in the long-run? Gold as a hedge and safe haven across time and investment horizon. *International Review of Financial Analysis, 41*, 320–328. https://doi.org/10.1016/j.irfa.2015.01.010

Breitung, J., & Candelon, B. (2006). Testing for short and long-run causality: A frequency-domain approach. *Journal of Econometrics, 132*(2), 363–378. https://doi.org/10.1016/j.jeconom.2005.02.004

Brière, M., Oosterlinck, K., & Szafarz, A. (2015). Virtual currency, tangible return: Portfolio diversification with Bitcoin. *Journal of Asset Management, 16*(6), 365–373. https://doi.org/10.1057/jam.2015.5

Brito, J., & Castillo, A. (2013). *Bitcoin: A primer for policymakers*. Arlington: Mercatus Center at George Mason University.

Carrick, J. (2016). Bitcoin as a complement to emerging market currencies. *Emerging Markets Finance and Trade, 52*(10), 2321–2334. https://doi.org/10.1080/1540496X.2016.1193002

Chuen, D. L. K., & Wang, L. G. (2018). Cryptocurrency: A new investment opportunity? *Journal of Alternative Investment, 20*(3), 16–40. https://doi.org/10.3905/jai.2018.20.3.016

Ciner, C. (2011). Information transmission across currency futures markets: Evidence from frequency domain tests. *International Review of Financial Analysis, 20*(3), 134–139. https://doi.org/10.1016/j.irfa.2011.02.010

Conover, C. M., Jensen, G. R., Johnson, R. R., & Mercer, J. M. (2009). Can precious metals make your portfolio shine? *Journal of Investing, 18*(1), 75–86. https://doi.org/10.3905/JOI.2009.18.1.075

Corbet, S., Meegan, A., Larkin, C., Lucey, B., & Yarovaya, L. (2018). Exploring the dynamic relationships between cryptocurrencies and other financial assets. *Economics Letters, 165*, 28–34. https://doi.org/10.1016/j.econlet.2018.01.004

Doornik, J. A., & Hansen, H. (2008). An omnibus test for univariate and multivariate normality. *Oxford Bulletin of Economics and Statistics, 70*, 927–939. https://doi.org/10.1111/j.1468-0084.2008.00537.x

Dyhrberg, A. H. (2016). Bitcoin, gold and the dollar—A GARCH volatility analysis. *Finance Research Letters, 16*, 85–92. https://doi.org/10.1016/j.frl.2015.10.008

Faisal, F., Tursoy, T., & Berk, N. (2018). Linear and non-linear impact of internet usage and financial deepening on electricity consumption for Turkey: Empirical evidence from asymmetric causality. *Environmental Science and Pollution Research, 25*(12), 11536–11555. https://doi.org/10.1007/s11356-018-1341-7

Geweke, J. (1982). Measurement of linear dependence and feedback between multiple time series. *Journal of the American Statistical Association, 77*(378), 304–313. https://doi.org/10.1080/01621459.1982.10477803

Giudici, P., & Hashish, I. A. (2018). What determines Bitcoin exchange prices? A network VAR approach. *Finance Research Letters, 28*, 309–318. https://doi.org/10.1016/j.frl.2018.05.013

Glaser, F., Haferhorn, M., Weber, M. C., Zimmarmann, K., & Siering, M. (2014, April 15). *Bitcoin-asset or currency? Revealing users' hidden intentions*. ECIS.

Gozgor, G. (2015). Causal relation between economic growth and domestic credit in the economic globalization: Evidence from the Hatemi-J's test. *The Journal of International Trade and Economic Development, 24*(3), 395–408. https://doi.org/10.1080/09638199.2014.908325

Granger, C. W., & Yoon, G. (2002). *Hidden cointegration* (Economics working paper, (2002-02)). University of California. https://doi.org/10.2139/ssrn.313831

Guesmi, K., Saadi, S., Abid, I., & Ftiti, Z. (2018). Portfolio diversification with virtual currency: Evidence from Bitcoin. *International Review of Financial Analysis, 63*, 431–437. https://doi.org/10.1016/j.irfa.2018.03.004

Hacker, R. S., & Hatemi-J, A. (2005). A test for multivariate ARCH effects. *Applied Economics Letters, 12*(7), 411–417. https://doi.org/10.1080/13504850500092129

Halkos, G. E., & Tzeremes, N. G. (2013). Renewable energy consumption and economic efficiency: Evidence from European countries. *Journal of Renewable and Sustainable Energy, 5*(4), 1–13. https://doi.org/10.1063/1.4812995

Hatemi-j, A. (2012). Asymmetric causality tests with an application. *Empirical Economics, 43*(1), 447–456. https://doi.org/10.1007/s00181-011-0484-x

Hillier, D., Draper, P., & Faff, P. (2006). Do precious metals shine? An investment perspective. *Financial Analysts Journal, 62*(2), 98–106. https://doi.org/10.2469/faj.v62.n2.4085

Hong, K. (2017). Bitcoin as an alternative investment vehicle. *Information Technology and Management, 18*(4), 265–275. https://doi.org/10.1007/s10799-016-0264-6

Hood, M., & Malik, F. (2013). Is gold the best hedge and a safe haven under changing stock market volatility? *Review of Financial Economics, 22*(2), 47–52. https://doi.org/10.1016/j.rfe.2013.03.001

Hosoya, Y. (1991). The decomposition and measurement of the interdependency between second-order stationary processes. *Probability Theory and Related Fields, 88*(4), 429–444. https://doi.org/10.1007/BF01192551

İçellioğlu, C. Ş., & Öztürk, M. B. E. (2018). Bitcoin ile seçili döviz kurları arasındaki ilişkinin araştırılması: 2013–2017 dönemi için Johansen Testi ve Granger Nedensellik Testi. *Maliye ve Finans Yazıları, 109*, 51–70.

Iqbal, J. (2017). Does gold hedge stock market, inflation and exchange rate risks? An econometric investigation. *International Review of Economic and Finance, 48*, 1–17. https://doi.org/10.1016/j.iref.2016.11.005

Ji, Q., Bouri, E., Gupta, R., & Roubaud, D. (2018). Network causality structures among Bitcoin and other financialassets: A directed acyclic graph approach. *The Quarterly Review of Economics and Finance, 70*, 203–213. https://doi.org/10.1016/j.qref.2018.05.016

Klein, T., Thu, H. P., & Walther, T. (2018). Bitcoin is not the new gold—A comparison of volatility, correlation, and portfolio performance. *International Review of Financial Analysis, 59*, 105–116. https://doi.org/10.1016/j.irfa.2018.07.010

Lucey, B. M., & Li, S. (2015). What precious metals act as safe havens, and when? Some US evidence. *Applied Economics Letters, 22*(1), 35–45. https://doi.org/10.1080/13504851.2014.920471

Malović, M. (2014). Demystifying Bitcoin: Sleight of hand or major global currency alternative? *Economic Analysis, 47*(1–2), 32–41. Retrieved from https://www.library.ien.bg.ac.rs/index.php/ea/article/view/283

McCown, J. R., & Shaw, R. (2017). Investment potential and risk hedging characteristics of platinum group metals. *The Quarterly Review of Economics and Finance, 63*, 328–337. https://doi.org/10.1016/j.qref.2016.06.001

Mensi, W., Hammoudeh, S., & Kang, H. (2015). Precious metals, cereal, oil and stock market linkages and portfolio risk management: Evidence from Saudi Arabia. *Economic Modelling, 51*, 340–358. https://doi.org/10.1016/j.econmod.2015.08.005

Nguyen, D. K., Sousa, R. M., & Uddin, G. S. (2015). Testing for asymmetric causality between US equity returns and commodity futures returns. *Finance Research Letters, 12*, 38–47. https://doi.org/10.1016/j.frl.2014.12.002

Peng, Y., Albuquerque, P. H. M., Camboim de Sá, J. M., Padula, A. J. A., & Montenegro, M. R. (2018). The best of two worlds: Forecasting high frequency volatility for cryptocurrencies and traditional currencies with Support Vector Regression. *Expert Systems with Applications, 97*, 177–192. https://doi.org/10.1016/j.eswa.2017.12.004

Ranjbar, O., Chang, T., Nel, E., & Gupta, R. (2017). Energy consumption and economic growth nexus in South Africa: Asymmetric frequency domain approach. *Energy Sources, Part B: Economics, Planning and Policy, 12*(1), 24–31. https://doi.org/10.1080/15567249.2015.1020120

Rehman, M. U., & Apergis, N. (2018). Determining the predictive power between cryptocurrencies and real time commodity futures: Evidence from quantile causality tests. *Research Policy, 61*, 603–616. https://doi.org/10.1016/j.resourpol.2018.08.015

Shahbaz, M., Van Hoang, T. H., Mahalik, M. K., & Roubaud, D. (2017). Energy consumption, financial development and economic growth in India: New evidence from a nonlinear and asymmetric analysis. *Energy Economics, 63*, 199–212. https://doi.org/10.1016/j.eneco.2017.01.023

Sontakke, K. A., & Ghaisas, A. (2017). Cryptocurrencies: A developing asset class. *International Journal of Business Insights and Transformation, 10*(2), 10–17.

Taylor, N. J. (1998). Precious metals and inflation. *Applied Financial Economics, 8*(2), 201–210. https://doi.org/10.1080/096031098333186

Trabelsi, N. (2018). Are there any volatility spill-over effects among cryptocurrencies and widely traded asset classes? *Journal of Risk and Financial Management, 11*(66), 1–17. https://doi.org/10.3390/jrfm11040066
Yermack, D. (2014). *Is Bitcoin a real currency? An economic appraisal* (NBER working paper, no 19747).

Melik Kamisli received his MS in Business Administration Master Program from Osmangazi University in 2008. He was hired as a lecturer by Bilecik University in 2009. He received his Ph.D. in Finance program from Anadolu University in 2015 and became an assistant professor at Bilecik University in 2015. He currently gives international finance, investment decisions, and portfolio analysis, financial crisis and risk management courses in Finance and Banking Programme.

Part IV
Crypto Currency Taxation in Emerging Markets

Chapter 17
Effective Taxation System by Blockchain Technology

Habip Demirhan

Abstract One of the basic functions of a government is to deliver public services to citizens. Service delivery of this kind requires public expenditure. Hence, governments require resources to finance their expenditure. Although there are a number of different methods available to fund public expenditure, the most important one is taxation. However, governments incur costs when collecting taxes. It is, therefore, important for a government to ensure the efficiency of its tax collection system and to collect taxes in such a way that only minimal costs are incurred. Providing transparent, controllable, secure, and real-time information is vital in terms of ensuring the effectiveness of a tax collection system. Changes and developments in information and communication technologies have prompted the public sector to identify new ways to collect taxes. In recent years, discussions regarding the applicability of blockchain technology (or, more commonly referred to as crypto coins), for the public sector have emerged. In this study, the applicability of blockchain technology for use in a tax system is discussed. The properties and benefits of different blockchain technologies are analyzed in terms of both data and transparency. It has been concluded that blockchain technology could be applied in a number of areas to reduce the administrative tax burden and the costs associated with tax collection. This study, therefore, attempts to explain the applicability of blockchain technology in relation to taxation, and it clarifies (1) how blockchain technology represents a new approach to taxation, (2) how blockchain technology reduces tax expenditure, (3) how blockchain technology increases both transparency and accountability, (4) how tax evasion can be reduced using blockchain technology, and (5) how blockchain technology can reduce the administrative tax burden.

H. Demirhan (✉)
Faculty of Economics and Administrative Sciences, Department of Public Finance, Hakkari University, Hakkari, Turkey
e-mail: habipdemirhan@hakkari.edu.tr

© Springer Nature Switzerland AG 2019

U. Hacioglu (ed.), *Blockchain Economics and Financial Market Innovation*, Contributions to Economics, https://doi.org/10.1007/978-3-030-25275-5_17

17.1 Introduction

The rapid development of information and communication technologies (ICTs) that has taken place in recent decades has brought about new approaches in the field of public administration. Indeed, since the early 1980s, public administration has been increasingly influenced by ICTs. Concepts such as accountability, transparency, effectiveness, and productivity, which the private sector was already familiar with, have also become important tools for the public sector. It is certainly true that the accession of reliable information has become increasingly important in recent times. Further, the development of ICTs has increased the number of digital public services (e-governments).

Tax is the most important source of public funding for any government. For this reason, tax collection is always an important issue. The government always aims to collect tax effectively, although tax collection is really a double-edged sword. On the one hand, tax authorities seek to improve the tax collection process; while on the other hand, taxpayers attempt to minimize their tax bases. Importantly, deficiencies in tax collection are not unilateral.

The idea that the integration of blockchain technology into the tax system would be a positive development is due the various features that such technology offers. The key blockchain features that are to affect the tax system include (PwC, 2017):

- Transparency: Blockchain provides a transparent structure by allowing the provenance of transactions to be traced.
- Controllability: The permissions accession network is restricted to identified users.
- Security: Once digital data has been entered, nobody can alter or tamper with the digital ledger. It is easy to trace fraud using blockchain technology. Thus, blockchain helps to prevent fraud.
- Real-time information: Everyone in the network can see updated information at the same time. In other words, if information is updated, then the update is available to everyone.

In this chapter, we will analyze the effects of blockchain technology on taxation. In order accomplish this, we will explain the history and features of blockchain within the context of taxation, as well as how blockchain technology will affect both different tax types and the tax audit mechanism.

17.2 The History of Blockchain Technology

In 2008, Satoshi Nakamoto, whose true identity remains unknown, published a white paper entitled "Bitcoin: A Peer-to-Peer Electronic Cash System." In the paper, he claimed to have created a solution to the problem of double-spending digital currency via distributed databases, which combined cryptography, game theory, and computer science (Nakamoto, 2008). The increasing value of cryptocurrencies, especially Bitcoin, has caused researchers and investors alike to consider the logic of the associated process. Blockchain technology relies on the process being characterized by trust and openness. Although discussions concerning blockchain have increased in recent years, studies of blockchain actually date back to the 1990s. The logical concept of blockchain was first mentioned in 1991 in an article entitled "How to Tine-Stamp a Digital Document Block," which was written by Stuart Haber and W. Scott Stornetta. In their article, Haber and Stornetta described the concept of a cryptographically secured network of blocks. According to them, the customer sends a document to the timestamp server to receive a timestamp, and the server then signs the document with the current timestamp. The server also links the document to any previous document (Haber & Stornetta, 1992). This model of recording digital documents using the timestamp method has two important features.

> First, one must find a way to time-stamp the data itself, without any reliance on the characteristics of the medium on which the data appears, so that it is impossible to change even one bit of the document without the change being apparent. Second, it should be impossible to stamp a document with a time and date different from the actual one (Haber & Stornetta, 1992, p. 1).

Although cryptographic studies have been conducted since 1991, it was the invention of Bitcoin that really maximized the recognition of blockchain technology. Nakamoto's (2008) paper claimed that such technology provided a solution to the problem of double-spending in digital currency by using a peer-to-peer network. The main purpose of the study described in Nakamoto's (2008) paper was hence to create a peer-to-peer digital currency that allowed people to spend directly without having to go through a financial institution. Such an invention represented a major innovation that enabled users to operate without the need for a third party to establish trust.

Nakamoto (2008) intended blockchain to serve as a transaction book for the Bitcoin cryptographic currency. One of the key features of blockchain is that it has a self-governing autonomous structure rather than a central structure. In other words, blockchain technology is based on "Decentralized Distributed Ledger Technology". In this system, the data is entered. After the data entry, the accuracy of the data is determined and the data entered blocks are linked to each other to form a chain. Thus, data in each block is linked to each other and saved. As a result, the process forms a structure that cannot be recycled and deleted. The infrastructure provided by such a system offers opportunities for many areas in the future. In particular, it is stated that smart contracts with block chain will be made in the future and thus, the transactions between people will be more reliable and faster. Furthermore, it is argued that the violation of the rights arising from the contract between the persons will be eliminated by the smart contracts. The researchers state that the blockchain technology will cause a serious change in the field of taxation in the coming years.

17.3 The Features of Blockchain Technology

Blockchains have numerous features that render it flexibly usable in numerous fields. In this context, we will try to explain some features that make it important in using transfer pricing.

Blockchain as a Data Structure Blockchains are a customized version of a linked list structure. In the standard single-link structure, each element of the list points following elements by a pointer routine. Thus, all elements from the starting element of the list to the tail element are interconnected (Cormen, Leiserson, Rivest, & Stein, 2009). In the blockchain structure, each element (block) not only points to the next block but it also hides the hash value of that block. That is, the block chain is a special linked list structure created with hash-pointers (Narayanan, Bonneau, Felten, Miller, & Goldfeder, 2016). The hash pointer structure does not allow changes. In other words, it can be understood easily if such a change is made, given that the hash value of the newly added block is different from the value indicated by the hash pointer that pointing to the new block. This feature is one of the important factors that make the block chain a safe structure.

Immutability and Tamper Detection According to Webster's Dictionary of the English Language, the word *immutable* refers to being *not capable of or susceptible to change.* When this considered in the context of blockchain, it means "*the inability of a block to be deleted or modified once it is in blockchain*". In other words, a verifiable, traceable, and immutable log system is required in order to establish "trust" among participants (Pourmajidi & Miranskyy, 2018). Once the transaction in the blockchain has received a sufficient level of validation, cryptography then ensures its irreversibility and irreplaceability. Thus, the information cannot be edited or deleted. Block chain technology uses computational algorithms and approaches in order to ensure irreversibility of stored records (Iansiti & Lakhani, 2017). The information in blocks cannot be tampered. Without corrupting the chain, it is impossible to change and/or to alter the information in a block, and if it is corrupted, it becomes visible within all nodes. Once digital data are entered, nobody can alter or tamper with the digital ledger. It is easy to trace fraud using blockchain technology. Thus, blockchain helps prevent fraud.

Data Protection Even though more secure financial systems can be hacked, numbers of computers called nodes that confirm the transaction on the blockchain network provides the security of blockchain. There is no a single way of shutting down the system. Blockchain is a data structure that is sorted by time and that is constantly growing. The blocks hold the transaction(s) and the address of the previous block. The recorded transactions cannot be changed. Thus, blockchain is mostly defined as a ledger that the list of transactions is recorded.

Decentralized Ledger Technology Blockchain technology is a distributed database structure (Iansiti & Lakhani, 2017; Xu et al., 2016). The database is distributed between nodes. While some researchers view them as being computers, others view them as a part of the system as a general definition. Nodes in the block technology system are accessible to the entire database; however, a single node cannot control the data stored in blockchain. All of the nodes verify every new record entry in the blockchain (transactions) in the system without any intermediaries. In other words, the blockchain's decentralized ledger feature eliminates the necessity of third parties in transactions. Each user can directly access to the system form the web and store their assets. The system does not need any governing authority and each user has direct control over transactions. Thus, the blockchain's decentralized structure gives people their rights back on their assets.

Peer to Peer Network Instead of using any centralized structure for communication between the parties, individual nodes transmit and store information to each other in a peer-to-peer network (Nakamoto, 2008). Due to the consensus between nodes in the block chain system, there is no need of intermediaries. All of the nodes store the information in the blockchain is stored in the form of BitShares (Iansiti & Lakhani, 2017). Some authors claim that the block chain is not stored by all nodes but can be used by all nodes (Nakamoto, 2008). They also note that new processes do not have to reach all of the blocks, but that the process must reach a sufficient number of nodes so that it can be included in a block over time.

Transparency The concept of transparency in block chain technology is realized when all of the blocks can oversee all operations, which in turn means that it is more transparent as opposed to a centralized structure managed by a third party. The block chain allows the digitized information to be recorded on different nodes on a network so that the recorded information can be validated and transparent by sharing with other nodes. All users of the system can see the transactions. Consequently, Blockchain provides a transparent structure by allowing the provenance of transactions to be traced.

17.4 Blockchain and Smart Contracts

One of the best things about the blockchain is that it saves you time and conflict because it is a decentralized system, meaning it exists between all permitted parties. In other words, there are no intermediaries to pay, thus making it faster, undeniable, more secure and cheaper than traditional systems. In 1997, Szabo has introduced the term "smart contract". A smart contract is a computer code running on top of a blockchain containing a set of rules under which the parties of that contract agree to interact with one each other. The agreement is automatically enforced, while the pre-defined rules are met. The smart contract code is a simplest form of a decentralized automation that facilitates, verifies, and enforces the negotiation or performance of an agreement or transaction. If a smart contract is deployed on a blockchain network, it can send message to the other contracts. This message is composed by the address of sender, the address of recipient, value of transfer, and a data field containing the input data to the recipient contract to other contracts (Karamitsos, Papadaki, & Al Barghuthi, 2018).

Smart contracts are pieces of software that extend block chain's utility from simply keeping a record of financial transaction entries to automatically implementing terms of multiparty agreements (Ream, Chu, & Schatsky, 2016). An agreement of this sort would mean that the related parties maintain separate databases; these kinds of smart contracts were impossible prior to the advent of blockchain technology. "*With a shared database running a blockchain protocol, the smart contracts auto-execute and all parties validate the outcome instantaneously and without need for a third-party intermediary* (Ream et al., 2016, p. 2)". The smart contracts can be a worthwhile option where frequent transactions occur among a network of parties, and whereby counterparties perform manual and duplicative tasks for each transaction. The blockchain act as a shared database in order to provide a secure, single source of truth. Smart contracts moreover automate approvals, calculations, and other transacting activities that are prone to lag and error. The benefits of blockchain-based smart contracts are:

- **Speed and Real-time Updates**: The smart contracts can increase the speed of a wide variety of business processes.
- **Accuracy**: Automated transactions are less prone to manual error.
- **Lower execution risk**: The execution is managed automatically by the network rather than an individual party. Therefore, the decentralized process of execution virtually eliminates the manipulation, nonperformance, or errors risk.
- **Fewer intermediaries**: Smart contracts can eliminate or reduce reliance on third party intermediaries that provide "trust" services such as escrow between counterparties.
- **Lower Cost**: This process requires less human intervention and fewer intermediaries, and will therefore reduce costs.
- **New business or operational models**: Given that smart contracts provide a low-cost way of ensuring that the transactions are reliably performed as agreed upon, they will enable new kinds of businesses spanning peer-to-peer renewable energy trading to automated Access to vehicles and storage units.

Smart contracts can be either public or private contracts. The usage of smart contracts in transfer pricing will build trust between tax authorities and MNEs. Thus, the transactions between associated parties will be tracked easily and tax evasion will be prevented.

17.5 Blockchain Technology and Taxation

A state needs funding in order to perform its functions. In other words, public expenditures require public finance. Undoubtedly, tax remains the most important and the most powerful source of funding, as has been demonstrated historically. Taxes are the economic values that the state receives from its natural and legal citizens for the financing of public expenditures on the basis of compulsory contribution. Furthermore, taxes are not only a source of funding for public expenditures but also an important instrument of fiscal policy. The government may also have the opportunity to direct its national economy through this policy. For this reason, one of the highest priorities of a government is to collect maximum taxes. In other words, ensuring the optimization of tax collection is among the most vital tasks of the state. The improvement of information and communication technologies brings opportunities for states to deliver services on the digital platform.

The blockchain technology with its features will give an opportunity to the tax authority to control the taxpayers effectively. In this part of our studies we will try to clarify the using of blockchain technology in terms of tax types.

17.6 Payroll Tax

Payroll tax is a tax that imposed on employers or employees, and is usually calculated as a percentage of the salaries that paid to staffs by employers. There are two categories of payroll taxes; (1) deductions from an employee's wage (2) taxes paid by the employer based on the employee's wage. In recent years, many developed countries digitalized their payroll tax systems. There are two main advantages of using blockchain technology in the field of payroll taxation. First, collection of payroll taxes with blockchain technology will ensure the tax security for both employers and employees. Second, blockchain technology will speedup tax collection for the states (Johnston & Lewis, 2017). In other words, using blockchain technology will reduce transaction costs both for the state and employers.

In traditional payroll taxation, employers act as a government agent. They withhold taxes from the payments to the employee's earnings. Blockchain technology by embedding smart contracts will allow removal of an intermediary. In this system, the employer will make the gross payment plus social security contributions into the system, and will not need to do any other transaction. Then, in a system that will probably be open only to the tax office and perhaps to banks, the use of smart contracts will allow the calculation of the correct tax and social security, and this will be matched to the payment, which in turn will be transferred to the worker (WU Net Team, 2017). The transactions done by blockchain are tracked in a completely decentralized environment. That feature allows transaction history to be transparent and indisputable. This system can reduce and discrepancies and increase timesaving in a highly regulated payroll industry. Among many other technologies and methods, blockchain is also being explored as a part of the solution for the real-time payments. The system can be done, for example, by embedding smart contracts that fully automate the process. This could be done in the following steps (WU Net Team, 2017):

1. Insert of gross amount of salary by the employer.
2. Match of tax data with the payment by smart contract technology within the blockchain system and calculation of the correct tax and social security amounts.

3. Transferring of net salary automatically to the employee's amount and calculated tax to the government.

As a result, we can list the advantages of using blockchain in payroll system as below:

- Payroll related payments with blockchain databases have a potential for faster transaction speeds and a more efficient cash flow.
- Blockchain technology offers a very secure environment for payroll related payments.
- Blockchain technology offers the ability for tax authority professionals to authenticate employment history.

17.7 Taxation of Value Added Tax

Value added tax (VAT) is an indirect tax levied on consumption. In some countries it is also known as a goods and services tax (GST). The VAT is placed on a product whenever value is added at each stage of the supply chain, from the production to the point of sale. In almost all countries, VAT is the most important factor of tax administrations and the largest contributor to the governmental budgets. For this reason, "*an effective VAT collection is very important for the tax authorities in order to gain more revenue and shorten budget gap* (Frankowski, Baranski, & Bronowska, 2017, p. 12)."

Businesses play a big role in the current VAT system in order to access and collect tax due, and in order remit it to the government. This system includes big risks in and of itself. When businesses get into financial trouble, the system involves major risks of non-payment, especially which means large-scale payment fraud and a high compliance burden for business (Ainsworth & Shact, 2016). Blockchain technology will help to record transactions on a distributed ledger and pay through smart contracts. Thus, the VAT will be calculated correctly, and "*the tax due could be split from the payment as it is made by the costumer and be sent directly to the government* (WU Net Team, 2017, p. 8)." As a result, the system will reduce transaction costs significantly as well as will reduce the risks for fraud. In current system, the VAT transaction is processed in the following steps (Frankowski et al., 2017):

1. The company issues a VAT invoice.
2. The client pays the bills, including VAT.
3. The company records the information about the payment in its system.
4. The company pays their suppliers bill ex by bank transfer.
5. The company calculates VAT due to the tax authorities and fills a tax return (quarterly, monthly, yearly).

In blockchain system the steps are minimized for VAT transactions. There are two steps in blockchain based VAT transactions. At first step, the client pays the voice to the company. During this step, at same time, blockchain smart contracts calculate the invoice VAT and divide it into the non-VAT and VAT parts. The VAT is paid directly to the tax authority by smart contracts, and the non-VAT part is transferred to the company's account using a smart contract. During the second step, the company pays the suppliers invoice via smart contract. The company fills in the needed amount and the smart contract performs the payments. At the same time, the amount due is sent to the supplier, and the smart contract calculates VAT and sends it to the tax authorities. Thus, many transactions in the VAT tax chain are eliminated by the blockchain. Payment tools are diversified within the context of advanced technology. It has been stated that, in the coming years, bitcoin-like VATCoins will be used as tax payment tool (Frankowski et al., 2017). According to this;

- In countries which accept VATCoin and will be integrated to this system, VAT payments will be made by using VATCoin only in terms of smart contract written invoice documents.
- As a rule of thumb, VATCoins cannot be converted into cash or any other currency, however, VATCoins can be used a cash exchange tool among countries integrated to the system.
- VATCoins paid in both input and output taxes will be recorded in real time and added to the blockchain.

To summarize, there are many benefit of using blockchain technology in VAT payment. The system reduces the administrative burden of companies significantly. Thus, the system saves time and accounting services costs. VAT transactions are conducted in real time, what it more, all transactions executed by smart contracts are tamper proof and transparent. The system reduces the risk of fraud and mistakes.

17.8 Taxation of Transfer Pricing

Transfer pricing is defined as the pricing applied between related parties of an intercompany transaction (e.g. parent companies, divisions, branches, subsidiaries, and sales of goods and services or other similar commercial transactions within the same business organization). The pricing is mostly used to increase global profit after tax income. When the related parties are required to transact with each other, a transfer price is used to determine the costs. Transfer prices directly affect the allocation of taxable income. Hence, the transfer pricing of companies can directly affect the its after tax income to extend that rates differ across national jurisdictions. The companies mostly aim to reduce taxes on foreign trades, to receive more tax refund in export, to hide or reduce the tax deduction base, distributing head office expenses in a way to reduce tax burden and to reduce the tax burden indirectly by exceeding foreign trade and profit transfer restrictions. In tax literature, transfer pricing is defined a way of concealed gain distribution. In other words, if entities purchase or sell goods or services at a price that they determine against arm's length principle with their related parties, earnings are deemed to be wholly or partially concealed gain distribution through transfer pricing. The arm's length principle is an internationally accepted standard adopter for transfer pricing between related parties. The principle requires that transfer prices between related parties are equivalent to prices that unrelated parties would have charged in the same or similar circumstances. Article 9 of the OECD Model Tax Convention is dedicated to Arm's Length Principle. It states that the transfer price between the corporate entities should be function like two different entities. The primary of the objective of concealed gain distribution through transfer pricing method is to prevent the loss of treasure (OECD, 2014). The main purpose of concealed gain transfer ban is the protection of shareholders' rights. Nevertheless, companies for the reasons mentioned above are to try and find new methods to get concealed gain through transfer pricing. For this reason, transfer pricing is an important issue between companies and tax authorities. The companies try to increase their global profit after tax income by concealed gain distribution methods through transfer pricing while tax authorities try to not let companies apply a price against arm's length principle. According to data released by United Nations, intra firm trade makes up around 30% of global trade altogether. The laws regulating transfer pricing are different for each country, and require that cross border transactions between related parties comply with arm's length price. Simply put, this price should mirror the proposed or applied price between non-related parties in an open market.

The rapid advancement of Information and Communication Technologies has changed the service delivery structure both in the public and private sectors. The new technology advancement provides opportunities for both sides. Blockchain has become the new buzzword in recent years. The financial services industry has already begun using it to facilitate transactions, speed up trade settlement, cut out the need for intermediaries, and increase the traceability through transaction chains (PwC, 2017). It has been stressed that, with its properties, Blockchain can help reduce transfer-pricing complexity. In other words, the use of Blockchain technologies will help both companies and public sector to monitor transfer-pricing transactions. Tax authorities try to maximize revenues for the states while companies try to maximize their profit. These two goals are in clash with one other. For this reason, there are disputes between tax authorities and taxpayers. In other words, disputes between tax authorities and companies as taxpayers may arise in many areas. Also, increasing documentation requirements resulting from international developments and specific transfer pricing inspections based on risk analysis make it mandatory for companies to manage their transfer pricing policies strategically, in conformity with both the relevant legislation and their operational structures.

To summarize, transfer pricing is an arm's length charge between related parties (e.g. a parent corporation) and a controlled foreign corporation. For this reason, transfer pricing has a potential area of high-tax-compliance risk for multinational corporations and carries important implications for tax planning and financial reporting. Determining an arm's-length transfer price typically requires identifying where value is created and transferred. It also requires analyzing such factors as assets used, risks assumed, and functions of the respective parties, and requires correctly applying an appropriate economic method provided in Treasury regulations and other guidance. Transfer pricing issues often give rise to uncertain tax benefits. Transfer pricing forces tax authorities to find new way to challenge with it. In recent years, Blockchain technology has become popular for many transactions.

Traditional transfer pricing depends upon conventional intra-firm agreements (non-digital agreements). These agreements are executed manually. The transactions documents have a high risk of falsification. It is possible to tamper records and documents easily (Frankowski et al., 2017; Parekh, 2017). In blockchain-based transfer pricing, it is easy to track the flow of transactions as well as track the identity of all involved parties by a blockchain distributed ledger. The agreements are written into self-executing a smart contract. All transactions are time-stamped and cryptographically sealed. Thus, the possibility of tampering the transactions eliminated. A blockchain stores each piece of information. Any party who has access to the blockchain can see this information flow. There are specific determined conditions, and the payments are executed by smart-contracts if they meet these conditions.

17.9 Tax Audit and Blockchain

Auditing has deep-rooted principles and its history date backs to ancient times. According to the Turkish Language Association (TDK), auditing refers to research, review and supervision conducted in order to understand whether a task is being carried out, inspected, controlled, and examined or not (TDK, 2018). Auditing constitutes an independent review of an asset, liability, activity, organization, or set of financial statements (O'Reagan, 2004). An audit is usually done to support or reject a defined audit objective, and it normally leads to an audit opinion on the subject of the review. The evidences related to monitoring and reporting of public finance date as far back as the Roman and Chinese Empires. According to some researchers, the history of auditing goes back to the ancient Mesopotamia (Sawyer & Vinten, 1996). A tax audit is one of the most important auditing mechanisms of finance. Tax auditing is defined as the examination of accounts and transactions of taxpayers related to taxable events through the staff of the tax offices, who are deemed experts in the relevant field (Pehlivan, 1986). Tax, for a state, is not only a public finance instrument but also a fiscal policy tool. The purpose of a public finance instrument for a state is to maximize tax revenues on the basis of laws. The main objective of the tax audit is to extend the tax base and to minimize tax loss and tax evasion.

In tax literature, the difference between the total amount of taxes (i.e. collectible tax potential and taxes paid voluntarily) is called a compliance gap. The main objective of the tax audit is to minimize this gap. The reasons leading to ineffectiveness in auditing mechanisms can be eliminated through blockchain applications. For example, for tax authorities, there is would not be any auditing supervisor–related problems in blockchain technology. Furthermore, data entered into a block will not be reversible and cannot be deleted by the taxpayers, and will become functional with the approval of the tax authority. In other words, the taxpayers' transactions will be recorded in virtual accounts. Thus, traditional bookkeeping will be eliminated. It is necessary to explain the blockchain process mechanism in order to understand the issue. Blockchain is an information book replicated on computers as a result of peer-to-peer network participation. The communication mechanism within the network uses cryptography (encryption) to identify the person receiving and the person sending the information. When one of the peers wants to add data to the book, they provide the accuracy and reliability of this information to be added to a block. It is necessary to have the approval of the other peer for the integration of the data during the data entry stage. Approved data is added to a block to form a chain of blocks and linked to other blocks, like a digital fingerprint, in order to form a blockchain. In this manner, peers are allowed to share information and perform transactions safely with each other without an intermediary. Moreover, nobody can change the information put on the blockchain because it is impossible to change the information stored in a blockchain without changing the root of a chain. There are functions known as "hash" on the base of each chain. They have a certain size regardless of the size of the data entered. Each data entry produces a new output, which means the produced output belongs to the data entered. Therefore, the outputs of different data will not have a chance to match.

Auditors generally work independently apart from their institutions in terms of strict regulations, professional standards and codes of conduct. As a result of the use of the blockchain in the tax audit, auditors will also concentrate their performance on matters not provided and their performance will be used effectively. In other words, blockchain technology will mean a serious change for auditors. The audit is not only the control of a transaction, but also inspects the detail of the monetary amount, as well as how it is recorded and classified. When a transaction is checked, it is examined to see whether it is a sale, an expenditure, a payment to a creditor, or a value created transaction. The auditors will save more time in focusing on these issues surrounding blockchain technology. Blockchains also offer important opportunities for accountants. Accountants are considered as experts in the field of record keeping, alongside in the implementation of complex rules, business logic and setting standards. The blockchain and smart contracts will transform the part of the accounting related to transaction security and transfer of ownership rights in the future. It will focus more on how transactions will be accounted for and taken into account. It will focus on which points should be explained in terms of reducing the need for reconciliation and dispute management, combined with increased precision of rights and obligations. Many current accounting department transactions and processes can be optimized with modern technology like blockchains, smart contracts, and learning machines, which in turn will increase the efficiency and value of the accounting function. As a result, the range of necessary skills in accounting will transform (ICAEW IT Faculty, 2017). Some work such as reconciliations and resource assurance will diminish or disappear, whereas other areas like technology, consulting and other value-added activities will expand. The issues that the auditor will focus on will also change if a company is properly audited through a significant block-chain-based transaction.

17.10 Conclusion

The accelerated development of ITCs has been affecting the public sphere in recent years. Blockchain technology brings new approaches and opportunities to both the public and private sectors. The states have started to establish digital databases for the future, especially so as countries try to find new methods in order to maximize their tax bases. Blockchain has the potential to decrease tax evasion and increase tax compliance and tax awareness. A blockchain provides transparent and secure transactions, as well as provides immediate access to real time information. The integration of blockchain technology into tax systems is a new phenomenon that as yet to been fully established. The states should establish databases for the development of this model. In a blockchain mechanism, the peers are taxpayers and tax authorities. In such a system, the taxpayers will carry out their transactions through a digital ledger while the tax authorities will improve the validity of the data entry through the same thing.

Thus, data entry will continue and the blocks will be interlinked in order to form a chain. In the end, a tax return will be provided, and tax assessment will be calculated by considering distributed ledgers.

Research shows that cryptocurrencies are one powerful application of blockchain technology. Nevertheless, blockchain has a potentially significant application in other areas. For example, global banks are experimenting with blockchains in order to streamline processes and reduce costs. They also use them to create a new competitive advantage through modified business models. Moreover, accounting firms facilitating innovation and application. At present, blockchain technology is relatively new. Despite it owes its reputation to the bitcoin, blockchain has the potential to be used in many areas and could revolutionize business. Research reveals that that it therefore could be a game changer for transfer pricing by providing tax authorities a more reliable audit trail.

References

Ainsworth, R. T., & Shact, A. (2016). *Blockchain (distributed ledger technology) solves VAT fraud* (Boston University School of Law and Economic Research Paper No. 16-41).

Cormen, T. H., Leiserson, J. E., Rivest, R. L., & Stein, C. (2009). *Introduction to algorithms* (3rd ed.). Cambridge: MIT Press.

Frankowski, E., Barański, P., & Bronowska, M. (2017). *Blockchain technology and its potential in taxes*. Deloitte. Accessed December 21, 2018 from https://www2.deloitte.com/content/dam/Deloitte/pl/Documents/Reports/pl_Blockchain-technology-and-its-potential-in-taxes-2017-EN.PDF

Haber, S. A., & Stornetta Jr., W. S. (1992). How to time-stamp a digital document. *Journal of Cryptology, 3*(2), 99–111.

Iansiti, M., & Lakhani, K. (2017). The truth about blockchain. *Harvard Business Review, 95*(1), 118–127.

ICAEW IT Faculty. (2017). *Blockchain and the future of accountancy*. Accessed December 13, 2018 from https://www.icaew.com/-/media/corporate/files/technical/informationtechnology/technology/blockchain-and-the-future-of-accountancy.ashx
Johnston, S. S., & Lewis, A. (2017). New frontiers: Tax agencies explore blockchain. *Tax Notes International, 86*(9), 16–19.
Karamitsos, I., Papadaki, M., & Al Barghuthi, N. B. (2018). Design of the blockchain smart contract: A use case for real estate. *Journal of Information Security, 9*, 177–190.
Nakamoto, S. (2008). *Bitcoin: A peer-to-peer electronic cash system*. Accessed December 20, 2018 from https://bitcoin.org/bitcoin.pdf
Narayanan, A., Bonneau, J., Felten, E., Miller, A., & Goldfeder, S. (2016). *Bitcoin and cryptocurrency technologies: A comprehensive introduction*. New Jersey: Princeton University.
O'Reagan, D. (2004). *Auditor's dictionary: Terms, concepts, processes, and regulations*. Hoboken, NJ: Wiley.
OECD. (2014). *2014 Update to the OECD model tax convention*. Accessed December 23, 2018, from http://www.oecd.org/tax/treaties/2014-update-model-tax-concention.pdf
Parekh, P. (2017, November). *Blockchain technology: Possible future of digital transfer pricing*. Accessed December 14, 2018 from https://indiataxinsightsblog.ey.com/2017/11/10/blockchain-technologypossible-future-of-digital-transfer-pricing/
Pehlivan, O. (1986). Tax audit and efficiency in tax audit (Vergi Denetimi ve Vergi Denetiminde Etkinlik). *Vergi Dünyası, 62*, 35–42.
Pourmajidi, W., & Miranskyy, A. (2018). Logchain: Blockchain-assisted log storage. *arXiv preprint arXiv*:1805.08868.
PwC. (2017, February). *How blockchain technology could improve tax system*. Accessed December 9, 2018 from https://www.pwc.co.uk/issues/futuretax/assets/documents/how-blockchain-could-improve-the-tax-system.pdf
Ream, J., Chu, Y., & Schatsky, D. (2016). *Upgrading block chains: Smart contract use cases in industry*. Brussels: Deloitte University Press.
Sawyer, L. B., & Vinten, G. (1996). *Manager and the internal auditor: Partners for profit*. Hoboken, NJ: Wiley.
Turkish Language Association. (2018). Accessed December 26, 2018, from http://www.tdk.gov.tr/index.php?option=com_bts&view=bts&. kategori1=veritbn&kelimesec=88169
WU Net Team. (2017, March 15–16). *Blockchain: Taxation and regulatory challenges and opportunities*. WU Global Tax Policy Center of Vienna University of Business and Economics. Accessed December 21, 2018, from https://www.wu.ac.at/fileadmin/wu/d/i/taxlaw/institute/WU_Global_Tax_Policy_Center/Tax___Technology/Backgrd_note_Blockchain_Technology_and_Taxation_03032017.pdf
Xu, X., Pautasso, C., Zhu, L., Gramoli, V., Ponomarev, A., Tran, A. B., & Chen, S. (2016). *The blockchain as a software connector*. In 13th Working IEEE/IFIP Conference on Software Architecture (WICSA), pp. 182–191.

Habip Demirhan is Assistant Professor of Public Finance at Faculty of Economics and Administrative Science of Hakkari University. Dr. Demirhan graduated from Faculty of Political Sciences of Ankara University in 2005. During 2006–2010, he worked in local governments as local as project manager and foreign relations advisor. He wrote and coordinated many European Union projects. Shortly after that, he started his academic career at Hakkari University as a Research Assistant. Dr. Demirhan received both his MSc and PhD from Social Science Instute of Dokuz Eylul University. He has been working as Assistant Professor of public finance since September 2017. Dr. Demirhan's research and teaching interests are primarily in the Budget and Fiscal Issues. He has various articles, book chapters and studies in the field of public finance.

Chapter 18
The Size and Taxation of Cryptocurrency: An Assessment for Emerging Economies

Erdoğan Teyyare and Kadir Ayyıldırım

Abstract Cryptocurrency started to be used in many countries as soon as its system became widespread. Countries have begun to create their own cryptocurrencies as well. Along with this increase in the dimensions of cryptocurrencies across the world, some regulatory needs have arisen. Many countries have been regulating their legal and economic infrastructures to cover cryptocurrencies, and many others have been on a quest to do so. While developed countries have been taking certain steps concerning cryptocurrencies, developing countries do not have a well-established theoretical framework for cryptocurrencies. Furthermore, inquiry and debate are still ongoing regarding the legal dimensions of these circulating currencies as well as how they are to be handled and taxed within the economy. This chapter discussed the dimensions of cryptocurrencies in developing countries and shows the debates on taxation of these currencies. Since countries' tax systems and taxable incomes may differ, the income category under which cryptocurrencies and income arising from such currencies will be treated and how they will be taxed are still being debated. In this regard, our aim was to determine the current situation in certain developing countries including Turkey and to put forward some policy recommendations regarding taxation.

18.1 Cryptocurrency and Its Dimensions in Developing Economies

The concept of cryptocurrency is used to express digital currencies with certain characteristics. In general, there is no agreed-on definition for cryptocurrencies in different categories under the same roof of virtual currencies. As a result, it is clear that the cryptocurrency market has not yet completed its development and the legal regulations on the issue have not been fully identified (FATF, 2014; IMF, 2016).

E. Teyyare (✉) · K. Ayyıldırım
Department of Public Finance, Faculty of Economics and Administrative Sciences, Bolu Abant Izzet Baysal University, Bolu, Turkey
e-mail: erdoganteyyare@ibu.edu.tr; ayyildirimkadir@ibu.edu.tr

© Springer Nature Switzerland AG 2019

U. Hacioglu (ed.), *Blockchain Economics and Financial Market Innovation*, Contributions to Economics, https://doi.org/10.1007/978-3-030-25275-5_18

Cryptocurrencies refer to the diffuse, open source, mathematically-based and peer-to-peer network currencies. There is neither a centralized management authority nor centralized control and supervision for these currencies. In cryptocurrencies, cryptographic principles are used to develop a secure information economy that can operate in a dispersed fashion without being connected to a center (FATF, 2014). According to some definitions, it is a virtual currency issuance system that provides a virtual payment tool for goods and services without a reliable central authority and works as a standard currency (Farell, 2015: 3).

There are legal shortcomings in the definition and operation of cryptocurrencies, both at the international and national levels. The lack of regulatory rules for cryptocurrencies provides them with areas where they can maintain their existence autonomously and create a perception that they are autonomous structures that are independent of legal arrangements. But this situation is temporary because both international organizations and the relevant institutions of many states continue to seek legal arrangements for these structures. There are some uncertainties in many aspects of legal cryptocurrencies in some countries and illegal ones in some other countries. Some countries have prepared the necessary legal and technological infrastructure for cryptocurrencies and their technologies while some countries are still searching. The characteristics of cryptocurrencies and technological requirements create a gap in many areas in terms of national economies. This gap leads to a number of challenges and opportunities. In terms of economic authorities and policymakers, the infrastructure provided by economic theories in order to fill this gap and make optimal choices is limited (Chiu & Koepl, 2017: 2).

It can be said that, on the basis of the search for legal regulation on cryptocurrencies, the idea that providing a healthy reflection of the impact of this technological innovation on both the national and international dimensions to social, political, economic, and financial life and preventing its abuse is essential because cryptocurrencies are non-centralized systems, the existence of which are created by individuals or institutions that are outside of public authority and control and are part of a system in which anybody who wants to can participate, hide their identity, and move money without any obstacles and controls. These characteristics provide suitable environments for the realization of many different illegal activities regarding which the international community is sensitive such as laundering money from illegal activity, illegal trade, illegal betting, tax evasion, and the financing of terrorism (Narayanan, Bonneau, Felten, Miller, & Goldfeder, 2016: 204).

Many economic activities performed with cryptocurrencies can lead to a significant loss of income in terms of national economies. The ecosystems created by the cryptocurrency systems in their environment and various activities such as the block chain architecture[1] that is accessible to the users, mining,[2] etc. create uncertainties in

[1]The block chain is a general ledger that keeps track of all the crypto-coin transactions taken place and that enlarges as the blocks are being attached to each other.

[2]Mining is a transaction registration service. Miners ensure to collect the newly published transactions and make them blocks and they ensure the block chain to be consistent, complete and

the establishment of a legal infrastructure and regulation of for taxation purposes. How tax a new activity for which no legal infrastructure has been able to be established arises as another problem. In order to solve these problems, many countries have stepped up efforts to create their own digital money to replace cryptocurrencies.

Cryptocurrencies contain both opportunity and risk for emerging economies. The idea that the negativities arising from the problems such as speculative attacks and excessive depreciation of the national currency that are experienced by many emerging economies can be reduced by means of crypto money, which leads emerging economies to look positively on cryptocurrencies. Uncertainties and inadequacies about cryptocurrencies in terms of both legal and economic infrastructure confront emerging economies with a bigger problem. The difficulties for the detection and monitoring of problems such as uncertainties in the legal and economic fields and the probability of misuse for illegal purposes including fraudulent tax losses and evasions enlarge the problem in terms of emerging economies.

While advanced economies have clear regulation of their legal, economic, and technological infrastructures, emerging economies are still grappling with these issues. Although developing countries do not have much activity in cryptocurrencies, they are not entirely outside this system. Some countries even have created their own cryptocurrencies. The crypto coin market, which started in 2009 with Bitcoin, has become active in many countries of the world. Cryptocurrencies, which have attracted attention with the increase in their values over time, have found an increasing usage area in terms of volume and variety. In the crypto-coin markets around the world, more than 2000 different types and values of tools can be used. In addition, platforms and markets with crypto-currency transactions worldwide number around 16,000. The value of these instruments in these markets, however, fluctuates at a constant level of $100–150 billion as of the end of 2018 (Coin Market Cap, 2018).

In emerging economies, crypto coin markets and transaction volumes are different. Graph 18.1 shows the number of cryptocurrency markets in emerging economies. Accordingly, India comes first with nine markets is followed by the Philippines, Mexico, Zimbabwe, Brazil, and Vietnam with three markets each.

Cryptocurrencies, which were first used in India in 2012 and have a very small usage area, started to grow rapidly due to the crypto coin swap platforms that began to emerge in the country. With a decision taken on 8 November 2016, 86% of the paper money in the country was withdrawn from circulation, and the people started to turn to cryptocurrencies. With these developments, the small-scale cryptocurrency markets displayed an extraordinary rate of growth. Although it has a large population, India contributes only 2% to the total global crypto coin markets (Jani, 2018: 3).

---

irreversible by consistently verifying it. Verification of the transactions is carried out by several miners who provide the computer power of the crypt coin network. With the algorithm used to solve a cipher puzzle as part of the verification process, miners ensure that each block contains the cipher extraction of the previous block. In this way, the blocks are connected to each other and by forming a chain; they take the name of block chain (Üzer, 2017: 31–32).

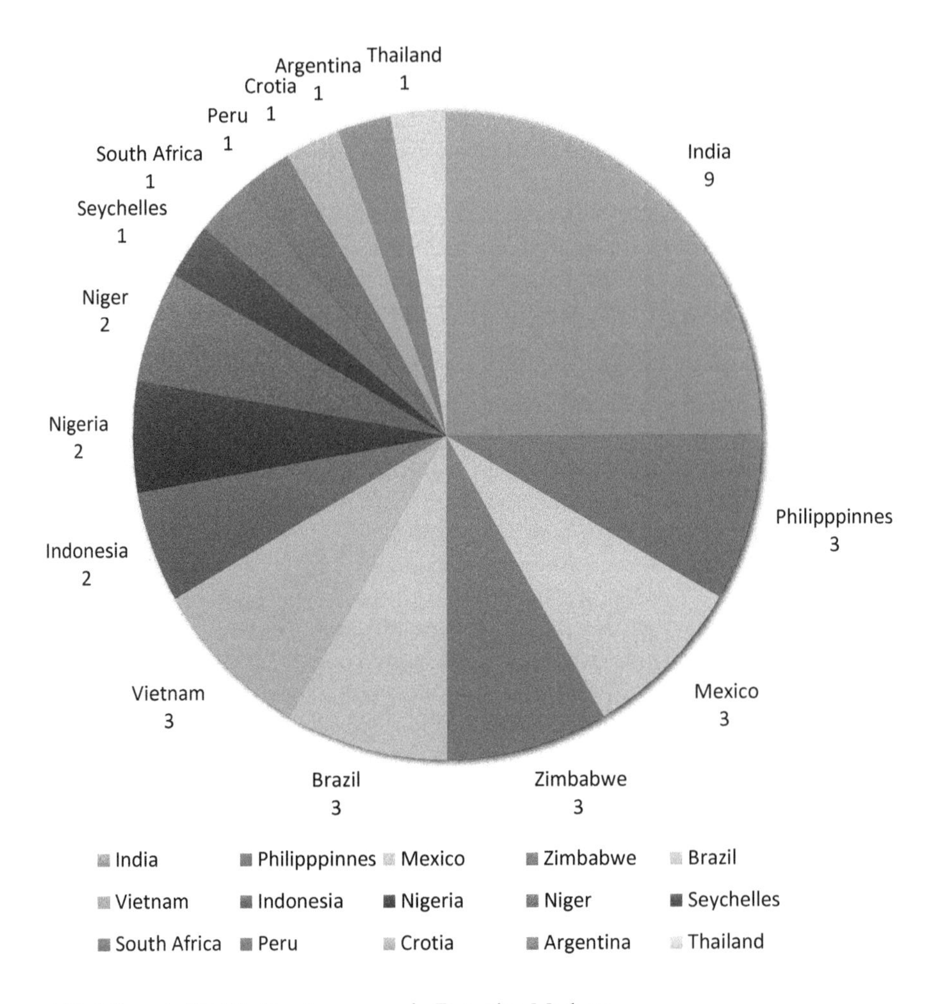

Allied Crowds (2018), Cryptocurrency in Emerging Markets

Graph 18.1 Cryptocurrency markets by country

This figure would be considered small for emerging economies but large for developed economies.

Many emerging economies see cryptocurrencies solution to problems such as decline in the value of national currency and speculative attacks and as alternative trade opportunities. However, some emerging economies are cautious about cryptocurrencies, and some countries even prohibit certain transactions. For example, China began to prohibit certain transactions in the crypto coin market in 2017. South Korea and Thailand are conducting studies with a view towards taking regulatory measures. According to Vietnamese law, tax cannot be levied on earnings in cryptocurrencies because they cannot be considered as assets. At the end of 2017, it was decided that cryptocurrencies were not a legal means of payment, and

Table 18.1 Status of Governments on Cryptocurrency around the World

Country	Status
South Korea	Neutral
China	Hostile
Thailand	Neutral
Vietnam	Neutral—Hostile
Iran	Friendly
Russia	Friendly
Poland	Friendly—Neutral
Venezuela	Friendly
Brazil	Hostile
Mexico	Friendly
South Africa	Friendly

regulatory efforts were made by prohibiting its use in the market. Iran has a positive approach to cryptocurrencies, and they are also working on a project to create their own crypto coin. Russia is another country that is looking positively on cryptocurrencies, but research and development activities in various fields have continued. While Poland has been working on cryptocurrencies, it has taken cautionary measures against its citizens. In Venezuela, cryptocurrencies are important as a way out of the sanctions imposed by the US and the negative consequences of the depreciation of the national currency. Venezuela also took their place in this market by releasing their own cryptocurrencies. Brazil sees cryptocurrencies as a balloon and does not favor the use of crypto coins. Mexico is one of the most moderate countries in terms of cryptocurrencies in the Latin American region. In order to expand its use and prevent negativity, regulatory efforts have continued. South Africa has a positive outlook, and the necessary tests are being carried out for the use of cryptocurrencies (Jani, 2018: 11–15). The viewpoints of the various countries regarding cryptocurrencies are given in Table 18.1.

Compared to developed countries, crypto coin swap platforms and their usage in Turkey has remained at very low levels. The swap platform in Turkey that allows purchase and sale using cryptocurrency is very limited. As of the end of 2018, crypto coin was traded on four platforms. The transaction volumes of these platforms are also relatively low compared to developed countries. The transactions on daily basis do not exceed a few million dollars. By the end of 2018, the first crypto coin ATM was put into operation. Many factors such as the lack of recognition of existing systems, deficiency in reliability and lack of depth in the financial markets, and the fact that the legal and institutional framework have not yet matured are probable reasons for this situation.

18.2 Taxation of Cryptocurrencies in Developing Countries

In developing countries, a number of legal and administrative regulations and/or evaluations are made on cryptocurrencies. The common questions of these regulations and assessments is whether cryptocurrencies will be legally recognized; what this legal definition might be; and, following the legal definition, how the income to be obtained from crypto coin would be taxed. Because developing countries as well as developed countries do not remain indifferent to crypto coin, all countries are not only trying to bring these currencies into a legal status but also to make explanations and warnings so that their citizens are aware of these currencies.

In this context, for example, in the ninth meeting of the Financial Stability Committee in Croatia, the use of virtual currencies such as Bitcoin, Ethereum, Ripple, and Litecoin was discussed, and, compared to their regulated counterparts, it was demonstrated that they carry a number of significant risks such as the digital money being stolen or people being swindled in the buying and selling process. In addition, while emphasizing that cryptocurrency values are subject to excessive volatility, on the one hand, it is emphasized that those who hold these cryptocurrencies are also aware of the possibility of being taxed on the other hand.[3]

Similarly, in the guidance published by the Ministry of Finance in Poland on April 4, 2018, the tax effects of the trade in crypto coin were discussed. In this context, it was stated that the crypto coin trade is subject to income tax and will be taxed in the tax bracket of 18–32% (Global Legal Research Center (GLRC), 2018: 52). In addition, the sale or purchase of these cryptocurrencies between individuals will be accepted as transferring the ownership right and will be subject to taxation at the ratio of 1% (Krasuski, 2018). However, in Poland, the taxation of all kinds of crypto coin transactions carried out by people who were engaged in this business without considering whether or not they gained any income from this transaction was found to be distracting, and thus the Minister of Finance stated that they would work on the taxation of virtual currencies under more suitable circumstances (Krasuski, 2018).

In South Africa, there is currently no specific law or regulation on the use or trade in cryptocurrencies. However, in the statement made by the South Africa Revenue Administration on March 6, 2018, it was explained that crypto coin would not be accepted as "money" but as "assets". In addition, in this statement, the tax status of the cryptocurrencies was also clarified. Accordingly, the Administration stated that the current income tax rules would be applied to cryptocurrencies and the taxpayers are expected to present their income from crypto coin as part of their income to be taxed (South African Revenue Services (SARS), 2018). In this respect, it was stated that taxpayers in the country would be obliged to declare their income from all crypto coin transactions; otherwise, the accessory public receivables such as tax penalty and interest as well as the primary public receivables of the relevant taxpayer would be collected (SARS, 2018).

Similarly, in Romania, although cryptocurrencies were not legally recognized, the National Revenue Administration announced that the income derived from crypto coin transactions would be subject to income tax (GLRC, 2018: 53). In addition, the Administration stated that people who earn income through this declaration should, on the one hand, declare this income, but, on the other hand,

[3]See Press Release of the Financial Stability Committee on the 9th session of 18.12.2017.

as a result of their research, they stated that they have not been able to identify any person who had earned income from crypto coin until that time (Ionaşcu, 2018).

Conversely, in Belarus, with the Presidential Decree regarding the field of digital economy which entered into force on March 28, 2018, it was made clear that it is not considered risky for individuals to buy, sell, swap, donate, legate, or dispose of cryptocurrencies in any way or for the mining of it by individuals since the purchase, sale, swapping, and mining of cryptocurrencies are legally permitted. Additionally, the income obtained from mining or virtual money transactions by individuals is exempted from tax until 2023 (GLRC, 2018: 60–61).

In Argentina, after recent regulations in the income tax law, the profits arising from the sale of crypto coin will be considered income, and, in this context, such profits will be subject to income tax (GLRC, 2018: 7). Likewise, in Bulgaria, in the taxation of gains from the sale of cryptocurrencies, the provisions applied to the gains obtained from the sale of "financial assets" will be valid, and, in this context, the income obtained from crypto coin should be declared as an annual declaration, and the tax must be paid (Marinov, 2017).

Although the cryptocurrencies in India are not legal means of payment, transactions on many platforms and in many markets can also be made. The legal framework for the taxation of cryptocurrency is not fully drawn up, but the requirement for taxation is accepted. However, the discussion and research regarding the definition of cryptocurrencies and under exactly which framework of gain they will be evaluated are continuing (Abraham, 2018).

China has a rather rigid attitude towards cryptocurrencies and does not see them as a legal means of payment. In 2013, the Popular Bank of China prohibited the crypto coin transactions of financial institutions, and, in 2017, they went even further and prohibited the crypto coin markets. Further studies regarding taxation in China are planned for the future (Peaster, 2018).

Finally, in Russia, the draft text of the regulation published by the Ministry of Finance on January 20, 2018, crypto coin mining has been dealt with as an entrepreneurial activity, and it was stipulated that those who deal with mining would be subject to tax if they exceeded the energy consumption amounts to be determined by law for three consecutive months (GLRC, 2018: 76).

18.3 Legal Status of Cryptocurrencies According to Legislation in Turkey

In Turkey, as well as in other developing countries, some studies on the recognition of the legal status of cryptocurrencies and taxation of the income to be obtained from these currencies have been made. In addition, many other countries have warned their citizens about relevant institutions and organizations regarding the use of cryptocurrencies in Turkey, and they have made a series of announcements so that the citizens do not face disadvantages.

In this context, it was announced by the Banking Regulation and Supervision Agency (BRSA) that bitcoin and its counterparts cannot be considered as electronic money due to their current structure and operation under Law No. 6493; for this reason, their supervision and control are impossible. In addition, it was stated that, because the parties' identities could not be known in transactions carried out with bitcoin and similar cryptocurrencies, an environment of illegal activities might arise from the use of these cryptocurrencies.[4] Furthermore, these currencies are open to certain risks such as the market value being too volatile, the possibility digital wallets being stolen, lost, or improperly used without the owners' knowledge, as well as the fact that the risks caused by operational errors or abuse by malicious vendors are irreversible.[5]

Furthermore, the Financial Crimes Investigation Board (FCIB) has determined the transaction of "money transfer for the purchase of bitcoin to the intermediary institutions who sell bitcoin from their bank accounts" as a suspected transaction related to the field of banking; in this context, the Board stated that banks have the liability of taking notice and the Postal and Telegraph Organization Co (PTT). which is limited to banking activities, should inform the FCIB of such transactions.[6]

As a result, there is no regulation in the Turkish tax legislation related to bitcoin and similar cryptocurrencies. However, persons representing the tax administration have stated that a number of studies have been carried out to define the legal status of cryptocurrencies, and they should be taxed in this context. However, it is considered to be necessary to evaluate the status of the cryptocurrencies against the current legislation, and the necessary evaluations are made in the following section.

18.4 A General Assessment in Terms of Taxation of Cryptocurrencies in Turkey

There is no provision explicitly set forth in the legislation regarding the taxation of cryptocurrencies in the Turkish tax legislation. However, it is known that research on taxation of the income obtained from crypto coin transaction (specific to bitcoin) has been made. In this context, firstly, it is emphasized how to legally define cryptocurrencies. In this respect, it must be stressed that bitcoin is identified as a "commodity", defined as "securities," and named as "money."

However, securities in accordance with Article 3/(1)-o of Capital Market Law (CPL) express that shares and other similar assets including warehouse certificates

[4]See Press Release of BRSA (25.11.2013-2013/32) https://www.bddk.org.tr/ContentBddk/dokuman/duyuru_0512_01.pdf (11.01.2019).

[5]See Press Release of the Financial Stability Committee of Republic of Turkey Ministry of Treasury and Finance (11.01.2018-2018/01).

[6]See http://www.masak.gov.tr/userfiles/file/rehber1_1/MSK-RHB-%C5%9E%C4%B0B-001-1_1.pdf (12.01.2019).

and debt instruments or securitized assets and income-based debt instruments and warehouse certificates for such securities with the exception of money, checks, policies and bills substantially represent the participating nature or right to claim. For the issuance of securities, an underlying asset or receivable is required. Therefore, it can be said that the cryptocurrencies like bitcoin do not have the possibility to be accepted as securities because they are not based on any real assets or receivables in production and supply (Şahin, 2018: 30–31).

Similarly, the experts of the Central Bank hold the opinion that cryptocurrencies cannot be regarded as money (specific to bitcoin), and the cryptocurrencies like bitcoin should be considered assets rather than currency in their current state in accordance with the fact that money must be printed by the state, in other words, by a central bank of the state and in exchange for a value even if it is fictive (TCMB, 2018: 61).[7] In addition, the Central Bank holds the opinion that it is not possible to accept cryptocurrencies as electronic money because they are not circulated in accordance with the law and not subject to regulations but are rather mostly controlled by the promoters (TCMB, 2018: 60).

Then, according to Article 35/(1)-ç of Law No. 6493, electronic money is defined as "the monetary value issued in exchange for the funds accepted by the issuing institution, electronically stored, used to carry out the payment transactions defined in this Law, and accepted as a means of payment by the individuals and legal entities other than the electronic money issuing institution." In Article 3/(1)-d of the same Law, the electronic money institution represents the-legal entity that is authorized to issue electronic money under Law No. 6493. However, in accordance with Article 13 of this Law, banks, electronic money institutions, payment institutions and the PTT are defined as payment service providers. In addition, the payments will be made by using the currency agreed upon by the parties, provided that they comply with the regulations set out in Decision No.32 in Protection of the Value of Turkish Currency (Kükrer, 2016: 589). Therefore, it is natural that cryptocurrencies will not be accepted as electronic money according to the relevant legislation.

In view of these opinions, it can be concluded that the acceptance of cryptocurrencies as a commodity is more likely, as in the case of Canada. Commodities, which is the plural form of the word "meta" in Arabic, which means goods, refers to tradables. However, in the doctrine, based on the description of commodity, there are also opinions that it is not correct to accept everything that can be bought and sold as a commodity and that cryptocurrencies such as bitcoin cannot be considered commodities (Şahin, 2018: 31), but it would be more convenient to accept them as a digital commodity (Polat, Yusufoğlu, & Çakır, 2018: 51).

[7] Also see "Bitcoin'e vergi geliyor (Maliye, SPK ve MB kripto paraları inceliyor)", https://www.ntv.com.tr/teknoloji/bitcoine-vergi-geliyor-maliye-spk-ve-mb-kripto-paralari-inceliyor, riHhA6CgEUqJgGXs7RWlgA?_ref=infinite (11.01.2019).

18.4.1 The Importance of Creating a Legal Description of Cryptocurrencies

In Turkey, the taxpayer who performs a taxable transaction pays the tax required to be paid within the process of taxation including the taxpayer, tax assessment, notification, and accrual and collection stages. The state collects the relevant tax from the taxpayer within the scope of the taxation authority. This power, which is part of the condition of existence of a state, represents the legal and de facto power of taxation based on the sovereignty of the state over its citizens (Çağan, 1982: 3).

Undoubtedly, this power originates from the Constitution (Çağan, 1982: 95). Then, in the first paragraph of Article 73 of the Turkish Constitution, it is ensured that everyone is obliged to pay taxes in proportion to their financial power in order to cover public expenditures. On the other hand, this power shall be used only by the legislature in accordance with the provisions of the Constitution. As a matter of fact, it is ensured in the third paragraph of Article 73 of the Constitution that the financial obligations such as taxes, duties, and similar financial obligations shall be statutory, changed, or abolished by law. This provision, which is called the principle of the legality of taxes,[8] requires that a tax should only be the product of the parliament (under the principle of no taxation without representation), regulated by law (the principle of no taxation without law), and include the basic elements in the relevant law (the principle of legal certainty) (Güneş, 2011: 14).

The principle of the legality of taxes necessitates that the laws contain these basic elements and that the obligation and method relations arising out of the tax are regulated by law. In this respect, in order for a tax to be enacted by law, it is deemed to be in compliance with the principle of the legality of the tax, the subject of the tax,[9] the taxpayer,[10] the taxable event,[11] the tax base[12] and rate,[13] the exception, and the exemption and the deduction of the tax; and all regulations about tax obligation and tax method must be included in the laws (Güneş, 2011: 134).

Therefore, the taxation of cryptocurrencies and their transactions depends on the legal definition of these funds. In this way, which tax will cover the income obtained from the crypto coin will be clarified in terms of what the subject of the tax and the event that gave rise to the tax will be, who the taxpayers are, and at which rate and

[8]For a comprehensive study, see Güneş, G (2011). Verginin Yasallığı İlkesi. Third Press, İstanbul: On İki Levha.

[9]The subject of the tax refers to the actual or legal situation on whether the tax is levied or the tax is received (Güneş, 2011: 137; Şenyüz, Yüce, & Gerçek, 2018a: 86).

[10]The taxpayer refers to the person carrying the subjective and objective conditions of the taxable event (Başaran Yavaşlar, 2013: 39–40).

[11]The taxable event refers to the events or facts that the legislator foresees to be taxed if occurs (Akkaya, 2002: 19).

[12]The tax base refers to the technical-physical or economic-monetary magnitude in which the tax subject is reduced in order to calculate the tax receivable (Turhan, 1993: 44).

[13]The tax rate is the percentage of the basis paid or reserved as tax (Oktar, 2014: 102).

how they will be taxed. In other words, since a different taxation technique is predicted for each income item, a number of differences in a number of formal obligations, exceptions, and exemption provisions related to the relevant income item is also likely to come into the agenda (Öncel, Kumrulu, & Çağan, 2014: 241–242). Therefore, it is important to determine the legal nature of the cryptocurrencies to prevent uncertainty in these respects in terms of the principles of legal security and certainty.

18.4.2 Evaluation of Cryptocurrencies for Income Tax

The income of the individual is subject to income tax in accordance with Article 1 of the Personal Income Tax (PIT). According to this Law, income is accepted as the net amount of earnings and revenues that an individual has acquired in a calendar year. It is considered as the income items limitation in Article 2 of the PIT. In this context, income and revenues accepted as income consist of business profits, agricultural profits, salaries and wages, incomes from independent personal services, income from immovable property and rights, and income from capital investment and other income and earnings without considering the source of income. Therefore, in order that any economic value can be taxed as income, this value should be included in any of these income items determined in the law (Öncel et al., 2014: 241).

In addition, there are two main theories of income taxation: the theory of resources and the net increase theory. According to the first of these, income refers to the sum of economic values that arise from the combination of labor or property with human resources and labor and capital and which are continuous in a certain period (Başaran Yavaşlar, 2011: 77; Turgay, 1967: 5). According to the second of these theories, regardless of the origin of the source, the increase in purchasing power (Öncel et al., 2014: 240) in a certain period of time–in other words, due to any increase in the value of the positive difference (Turgay, 1967: 5)–is considered as revenue. It is clear that the Personal Income Tax has adopted both theories in this context. Then, the first six of the seven income items listed in Article 2 of the PIT according to the resource theory are considered as income, and the last one is considered as income according to the net increase theory.

In this context, for the taxation of the income derived from cryptocurrencies, they are required to be accepted as income in accordance with any of these theories and included in any of the income items listed in Article 2 of the PIT.

18.4.3 Evaluation of Cryptocurrencies in Terms of Commercial Gains

In accordance with Article 37 of the PIT, the income from all kinds of commercial and industrial activities will be considered as business profits and will be subjected to income tax. The law does not define what the commercial and industrial activity is. However, "all activities carried out with the use of labor and capital elements in a continuous organization and outside the agricultural and self-employment activities within the scope of the Law as commercial activity and value creation activities as a result of manufacturing and production works" are defined as industrial activity (Şenyüz, Yüce, & Gerçek, 2018b: 14).

Here, the multiplicity of one of the labor and capital elements to the other, for example, of the labor being greater than capital or of the capital than the labor, has no effect on consideration of the activity as commercial activity and the taxation of income as business profits (Özbalcı, 2012: 290). On the other hand, in order to qualify the income obtained from an activity as a commercial profit, it is necessary to have continuity of the activity. In other words, the income obtained as a result of the activity by the intent of the continuity will be considered as business profit (Özbalcı, 2012: 290). Otherwise, the income to be derived from such activity will be taxed as incidental profit in accordance with Article 82 of the PIT. The existence of subjective continuity of intention and aspiration here may be based on a number of objective indications such as opening a workplace, recruiting staff, advertising, registering in a trade register, or a number of objective signs, as well as it can be based on a number of indications such as the multiplicity of the treatments carried out without them (Özbalcı, 2012: 291; Turgay, 1967: 258).

In the light of the relevant legislation provisions and the explanations, if the acceptance of cryptocurrencies as commodity and the continuous sale of these funds in an organization or without the explicit existence of an organization, it is natural to accept this activity as commercial activity and the profit obtained from this activity as business profit and to subject them to personal income tax. Also, in the event that there are dealings with bitcoin mining continuously and within an organization and there is an increase in the purchasing power as well as the cost of the bitcoins to be obtained from it, bitcoin mining will be considered as a commercial activity and the gain to be obtained in this way will be considered as commercial gain (Ateş, 2014: 137; Polat et al., 2018: 53). The presence of the organization in Bitcoin mining may be based on a number of objective indications and/or signs, such as the presence of one or more high-processor computers, involvement in mining pools, and high amount of energy consumption (Karakoyun, 2018: 76). Additionally, in the event that the cryptocurrencies are accepted as a security and they are engaged in the trading of these funds on their own behalf and account, it is thought that the profit of this activity shall be considered as business profit in accordance with Article 37/(2)-5 of the PIT, and it will be subjected to tax.

Furthermore, in the event that the cryptocurrencies are accepted as commodity and their purchase and sale are carried out in a non-continuous manner, the profit to be obtained will be considered as incidental profit in accordance with the provision that acceptance of the profits obtained from the performance of the commercial transactions incidentally or from intervening in the transactions in this quality are incidental profits. However, in the event that such a profit is obtained via bitcoin, the

27,000 TL[14] portion of the total income is exempted from the income tax in accordance with this article of the Law. Therefore, if the total earnings with the cryptocurrency is below this amount, they will not pay any tax. However, if the incidental gain obtained in this way is above the exemption amount, the gain above the amount of the exception will be declared with an annual declaration in accordance with Article 85 of the PIT, and the income tax will be paid in proportion to the amount specified in the Law. In determining the net amount of the income, it is also necessary to reduce the costs incurred from the sales price and sales expenses in accordance with Article 82/(4)-1 of the PIT.

For example, suppose that bitcoin, which is accepted as a crypto coin, is considered a commodity and that an individual (A) buys a bitcoin of $3698[15] on 15.09.2017 and sells this bitcoin at $7278[16] on 15.11.2017. In this case, the profit on the date of sale is 28,264,840 TL. When we deduct the cost of 12,692,275 TL from the sales price, we reach the net amount to be taxed at 15,572,565 TL. However, since this amount does not exceed the amount of 24,000 TL for the year 2017,[17] no tax return will be made, and no tax will be paid in 2018 due to this profit.

18.4.4 *Evaluation of Cryptocurrencies from the Perspective of Income from Capital Investments*

In Article 75 of the PIT, income from capital investment is described as "a dividend, interest, rent, and similar incomes obtained as a result of the capital composed of cash capital or money except for the commercial, agricultural, or occupational activities of its owners". Also, it is stipulated in Article 75/(2) of the PIT that, no matter what the source is, some of the assets will be accepted as the income from capital investment. In addition, according to Article 3/(1)-o of CML, securities describe shares, other similar assets, and repository certificates related to such shares, debt instruments, or debt securities based on securitized assets and revenues and warehouse certificates for such assets excluding money, checks, policies, and bonds.

In the event that Bitcoin is accepted as a security and an interest, and similar income is obtained from it, this will be deemed to be inheritance capital in accordance with Article 75 of the PIT. It is natural that the acceptance of such payments to

[14]This amount is effective from 01.01.2018. For the year of 2017, this amount is determined to be 24,000 TL.

[15]The exchange rate of the (Central Bank of the Republic of Turkey) CBRT is assumed to be 3.4322 TL.

[16]The exchange rate of the CBRT is assumed to be 3.8836 TL.

[17]Since the taxable event occurred in 2017, the exemption amount determined for 2017 will be taken into consideration.

be obtained through Bitcoin as securities of movable capital depends on the acceptance of Bitcoin as a capital market instrument by the CML. In order to be accepted as income from capital investment according to Article 75/(2)-17 of PIT, it is stipulated that the said income "should be obtained from all capital market instruments issued in accordance with the provisions of the CML except for the ones listed in the Law."

In addition, in the doctrine, the withholding procedure (stoppage) included in the temporary Article 67 of the PIT is not applied, since it is due to the provision of this income through the banks and intermediary institutions, but the difference for the Bitcoin returns in the world is that it can be obtained through the stock exchange and the sites, and it can be done without a bank and intermediary institution (Şahin, 2018: 34). However, if it becomes possible to obtain an interest and a similar increase over bitcoin through a stock broker based in Turkey and on condition that it is stipulated in legal legislation, it is natural to withhold in the ratio specified in the Law in accordance with Temporary Article 67 of the PIT over this income.

On the other hand, in the event that bitcoin is accepted and disposed of as a security or other capital market instrument, the income obtained due to such disposal will be accepted as gains from appreciation and be subjected to the tax in this context in accordance with the provision that "the profits to be obtained from the disposal of securities or the other capital market tools are gains from appreciation except for the equity share acquired gratuitously that belongs to fully responsible institutions and kept in hand for more than two years," as written in Article 80/(1)-1 bis of the PIT. The mention of disposal mentioned here expresses the sale of the goods and rights written in this article, transfer and assignment, exchange, and nationalization; nationalization of the trade in exchange for a vise means to be placed as capital for trade companies in accordance with Article 80/(2) bis of the PIT. The profits obtained after disposal of Bitcoin in the aforementioned way express the sale of the goods and rights written in this article, the transfer and assignment, exchange, exchange, nationalization and nationalization of the goods and rights mentioned in this article, and using them as capital for trade companies. Therefore, gains resulting from the disposal of bitcoin in the aforementioned way will be considered to be gains from appreciation. In this way, all of the profits will be subjected to income tax in accordance with the provision that "of the gains from appreciation, 11,000[18] is exempted from income tax except for those provided from the disposal of gains from appreciation, securities, and the other capital market tools" written in Article 80/(3) bis of the PIT; in other words, any part of it cannot be exempted from income tax. On the other hand, it can be said that the gains from appreciation arising from the disposal of securities and other capital market instruments will be subject to withholding within the scope of Temporary Article 67 of the PIT.

Here, the question of how to calculate the gains from appreciation will arise. In Article 81 bis of the PIT, it is expressed that the net amount in the gains from appreciation "will be found by deducting cost values of the goods and rights

[18]The mentioned monetary amount is effective from 01.01.2018.

disposed and the expenses held by the seller and the taxes and charges from the amount of all kinds of expenses provided by money given in exchange for disposal and can be represented by money." It is ensured in the article that "in the case that cost value cannot be detected by the taxpayers, the cost is to be detected by the valuation commission in accordance with the Tax Procedure Law (TPL)" will be based. Furthermore, in the case of the disposal of securities in the article, if the acquisition cost cannot be confirmed, it is stated that the nominal value written in Article 266 of the TPL will be accepted as the acquisition cost. According to this, in the case that the bitcoin is accepted as security, it is considered to be an acquirement cost if it is promoted; if not, its nominal value is considered and, in the case that it is accepted as another capital market instrument, its cost value is considered if it can be detected; if not, the cost determined by the appraisal commission will be considered in the determination of the net amount of the gains from appreciation (Şahin, 2018: 33).

Finally, it should be noted that, if no regulation is made in accordance with the provisions of the current legislation, in other words, as it is explained above, the acceptance of income to be obtained from the disposal of bitcoins as gains from appreciation, it does not seem possible to accept the income to be obtained from the disposal of bitcoin as gain from appreciation or to be subjected to income tax as other incomes and earnings without considering the source of income.

18.4.5 Evaluation in Terms of Value-Added Tax

In accordance with Article 1/(1) of the law of Value Added Tax (VAT), the transaction subject to value-added tax will be subject to tax if carried out in Turkey. The transactions that are included under the scope of value-added tax are stated in the Law as "delivery and services within the scope of commercial, industrial, agricultural, and self-employment activities ... importing of all kinds of goods and services ... [and] delivery and services arising from other activities".

In the event that bitcoin trading or bitcoin mining is considered as a commercial activity, since these transactions shall be considered within the scope of "delivery and services within the scope of commercial, industrial, agricultural, and self-employment activities" written in Article 1/(1)-1 of the VAT law, as a result of these transactions, the arising value-added tax is natural. Likewise, in the case of the commissioning of bitcoin trading and if commission is taken for this reason, it will be necessary to calculate value-added tax for the service fee of the institution providing intermediation service (Cebecioğlu, 2018). On the other hand, in the event that the bitcoin gain is accepted as either an improper gain or an increase in value, in other words, in the case of transactions that fall within the scope of other earnings and revenues related to bitcoin, value added tax will not be generated.

If bitcoin is considered as a commodity, it can be expressed with an example as follows:

- The process of selling bitcoin by a person (X) who is non-taxpayer and does not buy and sell a bitcoin continuously to his friend (Y), a person who is non-taxpayer like himself and wants to own a bitcoin for investment purposes, will not incur VAT.
- The process of selling bitcoin by a person (Z), who is non-taxpayer, to a person (T), who buys and sells bitcoin on a regular basis, will not incur VAT.
- The process of selling bitcoin by a person (Z), who is non-taxpayer, to a person (T), who buys and sells bitcoin on a regular basis,— in other words, the process of delivering bitcoin from person (Z) to person (T)—will incur VAT.
- Providing bitcoin purchase services by an institution (A) that provides brokerage services to Bitcoin and obtains commission fees in return for services to person (B), who is a taxpayer or not, will give rise to VAT.

In the case that Bitcoin is regulated as money or securities or other capital markets, the Bitcoin stock market is accepted as having the status of stock markets established in Turkey, over the transactions related to Bitcoin delivery, it is thought that the calculation of value-added tax will not come into question when considering the provision related "...foreign exchange, money...stocks, bonds...capital market instruments processed in the stock markets established in Turkey... to keep their delivery exempt from value added tax".[19]

18.5 Conclusion

In terms of the worldwide monetary volume and transactions, cryptocurrencies are still limited. This is more evident in emerging economies. The lack of confidence in cryptocurrencies, excessive fluctuations, and lack of a centralized structure, legal deficiencies, and the potential for illegal use are the reasons for this situation.

There is uncertainty about the taxation of cryptocurrencies worldwide. It is seen that the legal framework for taxation, except in a few developed countries, has not been fully established. The primary questions discussed here are:

- How are cryptocurrencies defined in the country's tax system?
- What type of earnings will be evaluated?
- Which transactions will be taxed?
- Will losses be taken into consideration in taxation?

The uncertainties over cryptocurrencies, such as the variety of transactions, the possibilities for alternative use, and the difficulty in identification of such, bring about difficulties in regulating the taxation of cryptocurrencies. Emerging economies are facing a much greater risk than developed economies because of both

[19]For a similar opinion, see Taşdöken S (2016). Dijital Para Bitcoin'in KDV'si. Vergi Dünyası, 35 (417), 118–121.

institutional weaknesses and deficiencies in technological infrastructure. In emerging economies, the search for taxation and legal status of cryptocurrencies continues.

When the evaluations of the cryptocurrencies in both developed and developing countries are examined, it is seen that the general discussion need to be shaped within the framework of whether the legal status of cryptocurrency needs to be defined; and if required, how it needs to be defined, revising the tax system in accordance with this definition.

In Turkey, the legal status of cryptocurrencies as commodity, security, or money must be defined and, in this context, we attempted to work on this matter. The common sense view is that it would be appropriate to define cryptocurrencies as commodities. In fact, since it is clear that each definition would bring different taxation regime, we conclude that the legal definition should be made by taking into consideration all possible results. The legal definition and the revision of the tax legislation are both related to the principle of legality, the state of law, and the principle of legal security as well as the principle of certainty.

On the other hand, the lack of transparency of transactions related to crypto coin, in other words, the fact that the operations of the crypto coin exchanges and crypto coin cannot be audited by the administrative institutions in the country makes it difficult to fully comprehend the income from the crypto coin. As a result, defining the legal status and adapting tax legislation does not mean that the income from crypto coin will always be fully taxed.

It is not known where cryptocurrencies will evolve and how they will produce new instruments. In this respect, it is very challenging to fully prepare the conceptual framework for taxation.

References

Abraham, R. (2018). *How to pay tax on cryptocurrency assets in India*. Retrieved from https://www.moneycontrol.com/news/business/cryptocurrency/how-to-pay-tax-on-cryptocurrency-assets-in-india-2820361.html

Akkaya, M. (2002). *Vergi hukukunda ekonomik yaklaşım*. Ankara: Turhan Kitabevi.

Allied Crowds. (2018). *Cryptocurrency in emerging markets*. Retrieved from https://alliedcrowds.com/static/reports/AlliedCrowds-Cryptocurrency-in-Emerging-Markets-Directory.pdf

Ateş, L. (2014). Bitcoin: sanal para ve vergileme. *Vergi Sorunları Dergisi, 37*(308), 131–141.

Başaran Yavaşlar, F. (2011). *Gelir vergilendirmesinin temelleri*. Ankara: Seçkin Yayıncılık.

Başaran Yavaşlar, F. (2013). *Vergi ödevi ilişkisinin tarafları üzerinden Alman vergilendirme usulü*. Ankara: Seçkin Yayıncılık.

Çağan, N. (1982). *Vergilendirme yetkisi*. İstanbul: Kazancı Hukuk Yayınları.

Cebecioğlu, E. (2018). *Kripto paraların (bitcoin) uluslararası boyutuyla vergisel açıdan incelenmesi*. Retrieved from http://www.pkfistanbul.com/kripto-paralararin-bitcoin-uluslararasi-boyutuyla-vergisel-acidan-incelenmesi/

Chiu, J. & Koeppl, T. (2017). *The Economics of Cryptocurrencies - Bitcoin and beyond*. https://doi.org/10.2139/ssrn.3048124

Coin Market Cap. (2018). https://coinmarketcap.com/

Farell, R. (2015). *An analysis of the cryptocurrency industry*. Wharton Research Scholars 130. Retrieved from https://repository.upenn.edu/cgi/viewcontent.cgi?article=1133&context=wharton_research_scholars

Financial Action Task Force (FATF). (2014). *Virtual currencies—Key definitions and potential aml/cft risks*. Retrieved from https://www.fatf-gafi.org/media/fatf/documents/reports/Virtual-currency-key-definitions-and-potential-aml-cft-risks.pdf

GLRC. (2018). *Regulation of cryptocurrency in selected jurisdictions*. https://www.loc.gov/law/help/cryptocurrency/regulation-of-cryptocurrency.pdf

Güneş, G. (2011). *Verginin yasallığı ilkesi*. İstanbul: On İki Levha Yayıncılık.

International Monetary Fund (IMF). (2016). *Virtual currencies and beyond: Initial considerations*. Retrieved from https://www.imf.org/external/pubs/ft/sdn/2016/sdn1603.pdf

Ionaşcu, D. (2018). *Românii cu bitcoin, datori la fisc. trebuie plătit impozit pe venit şi contribuţii sociale, deşi monedele virtuale nu sunt reglementate în România*. https://www.libertatea.ro/stiri/exclusiv-romanii-cu-bitcoin-datori-la-fisc-trebuie-platit-impozit-pe-venit-si-contributii-sociale-desi-monedele-virtuale-nu-sunt-reglementate-in-romania-2163395

Jani, S. (2018). *The growth of cryptocurrency in India: Its challenges & potential ımpacts on legislation*. Vadodara: Faculty of Management Studies, Parul University. Retrieved from https://www.researchgate.net/publication/324770908_The_Growth_of_Cryptocurrency_in_India_Its_Challenges_Potential_Impacts_on_Legislation

Karakoyun, F. (2018). Sanal paralar ve bitcoin'in vergilemesinde yaklaşımlar ve Türkiye perspektifinden değerlemesi. *Vergi Sorunları Dergisi, 41*(363), 67–84.

Krasuski, K. (2018). *Crypto traders protest Poland's tax decision*. https://www.bloomberg.com/news/articles/2018-04-09/crypto-traders-protest-as-poland-wants-tax-from-all-transactions

Kükrer, C. (2016). İnternet Ortamında Bazı Faaliyetlerin Kavramsal Tanımları ve Türk vergi sistemi karşısındaki durumlarının değerlendirmesi. *Yönetim Bilimleri Dergisi, 14*(27), 583–604.

Marinov, M. (2017). *Legal and tax treatment of bitcoin in Bulgaria*. https://www.ruskov-law.eu/bulgaria/article/legal-tax-treatment-bitcoin.html

Narayanan, A., Bonneau, J., Felten, E., Miller, A., & Goldfeder, S. (2016). *Bitcoin and cryptocurrency technologies: A comprehensive introduction*. New Jersey: Princeton University Press.

Oktar, S. A. (2014). *Vergi hukuku*. İstanbul: Türkmen Kitabevi.

Öncel, M., Kumrulu, A., & Çağan, N. (2014). *Vergi hukuku*. Ankara: Turhan Kitabevi.

Özbalcı, Y. (2012). *Gelir vergisi kanunu yorum ve açıklamaları*. Ankara: Oluş Yayıncılık.

Peaster, W. M. (2018). *Bitcoin, cryptocurrency and taxes: What you need to know*. Retrieved from https://blockonomi.com/cryptocurrency-taxes/

Polat, A., Yusufoğlu, A., & Çakır, M. (2018). Bitcoin özelinde sanal paraların Türk vergi sistemi karşısındaki durumunun incelenmesi. *Vergi Dünyası, 38*(445), 43–56.

Şahin, F. S. (2018). Kripto para: alternatif sanal para. *Vergi Dünyası, 38*(443), 25–37.

SARS. (2018). *Sars's stance on the tax treatment of cryptocurrencies*. http://www.sars.gov.za/Media/MediaReleases/Pages/6-April-2018%2D%2D-SARS-stance-on-the-tax-treatment-of-cryptocurrencies-.aspx

Şenyüz, D., Yüce, M., & Gerçek, A. (2018a). *Vergi hukuku (genel hükümler)*. Bursa: Ekin Yayınevi.

Şenyüz, D., Yüce, M., & Gerçek, A. (2018b). *Türk vergi sistemi*. Bursa: Ekin Yayınevi.

Taşdöken, S. (2016). Dijital para bitcoin'in KDV'si. *Vergi Dünyası, 35*(417), 118–121.

TCMB (Türkiye Cumhuriyet Merkez Bankası). (2018). Dijital para ve merkez bankaları, İktisat ve Toplum Dergisi. *Sayı, 8*(95), 60–63.

Turgay, R. (1967). *Gelir vergisi kanunu ve tatbikatı*. İstanbul: Güven Basım ve Yayınevi.

Turhan, S. (1993). *Vergi teorisi ve politikası*. İstanbul: Filiz Kitabevi.

Üzer, B. (2017). *Sanal para birimleri*. Uzmanlık Yeterlik Tezi. Ankara: Türkiye Cumhuriyet Merkez Bankası Ödeme Sistemleri Genel Müdürlüğü.

Erdogan Teyyare is an Assistant Professor in the Faculty of Economics and Administrative Sciences at Bolu Abant Izzet Baysal University in Bolu (Turkey). He received a Ph.D. in Economics at Zonguldak Bulent Ecevit University, Institute of Social Sciences, Turkey. His research interests focus on public economics, fiscal policy, institutional economics, and public finance. He has several published book chapters and articles. He took part in different international conferences.

Kadir Ayyıldırım is an Assistant Professor in the Faculty of Economics and Administrative Sciences at Bolu Abant Izzet Baysal University in Bolu (Turkey). He received a Ph.D. in Fiscal Law at Marmara University, Institute of Social Sciences, Turkey. His research interests focus on Tax Law, Tax Procedural Law, and Administrative Procedural Law. He has a published book entitled "Notification in Turkish Tax Law (in Turkish)" and also several articles.

Chapter 19
Accounting and Taxation of Crypto Currencies in Emerging Markets

Ali Kablan

Abstract With the development of the Internet and e-commerce, the fact that cryptocurrencies are transferred between accounts using crypts and not controlled by central banks has remained on the agenda in recent years, and the increase in the number of people and entities investing heavily particularly in cryptocurrency production has forced governments to take certain steps to regulate this structure. In many countries, the trading volumes of cryptocurrencies, which became popular with Bitcoin, have reached a significant level. As a result of the increasing interest in such currencies and their widespread usage, it is also necessary to analyze the taxing, legal and accounting aspects of these currencies. Thus, this study aimed to discuss the accounting and taxation of cryptocurrencies, whose use has increased rapidly in recent years. To that end, the study first addresses the question of why cryptocurrencies are needed in the first place, and then, attempts to introduce blockchain technology, used to create new cryptocurrencies, and Bitcoin, the most renowned and valuable cryptocurrency of this technology. The study then makes recommendations on the accounting and taxation of cryptocurrencies by providing examples of accounting records. Finally, the study recommends that common definitions are made for these new assets throughout the world, and globally accepted international cryptocurrency standards for the accounting and taxation of these currencies are established and implemented.

19.1 Introduction

Internet technologies, which have changed human life rapidly, have influenced many fields including the field of economy creating new jobs and sectors. Although these technologies have been developing rapidly, the methods used today for the exchange and transfer of money are still insufficient and costly. For example, the EFT method used to transfer money domestically, is only possible on weekdays during

A. Kablan (✉)
Uzunköprü School of Applied Sciences, Trakya University, Edirne, Turkey
e-mail: alikablan@trakya.edu.tr

© Springer Nature Switzerland AG 2019

U. Hacioglu (ed.), *Blockchain Economics and Financial Market Innovation*, Contributions to Economics, https://doi.org/10.1007/978-3-030-25275-5_19

determined times, apart from these times or at the weekends it is not possible to use this method. Furthermore, this method requires various commissions to be paid. The SWIFT method, which can be used to transfer money abroad, can be difficult, costly and time-consuming. The delays in transactions and the fact that it is costly leaves people unsatisfied who want to do money transfers in a short time and with minimum cost possible. People no longer want to come across time and place restrictions when transferring money (Serçemeli, 2018: 34).

Block chain technology, in other words distributed ledger technology, is described as the most important technological development ever following the Internet and it draws attention with its expanded application potential. Bitcoin cryptocurrency was first mentioned in a study published by a person or group under the nickname Satoshi Nakamoto, whose identity is still unknown, and since then block chain technology has been developed rapidly, especially in accounting and finance.

The process of asset handover has always been on the agenda as a problem to be solved ever since human beings met with trade. Different alternatives have been used for this process throughout history. Together with the advancement of communication technologies, first the Internet and now blockchain technology have emerged, both presenting a better solution than their predecessors. Blockchain technology is an important infrastructure technology that is used in many different areas and situations from notaries and counting votes to supply chains and currencies (Underwood, 2016: 15).

Both the opportunities of technology and the changing needs of people have led to the creation of safer and quicker alternative currencies. Today, people can exchange, save or transfer their assets quickly and transparently without the requirement of a central authority by means of technologies such as blockchain technology. These assets, which are created by using this technological infrastructure, are called cryptocurrency (Serçemeli, 2018: 35).

Bitcoin, which is the most popular cryptocurrency used today, was first used as a mean of payment in 2009. However, it first came about in 2008 in an article titled "Bitcoin: A Peer-to-Peer Electronic Cash System" published by Satoshi Nakamoto, whose real name is unknown. The popularity and usage area of Bitcoin, which is a network that operates directly between users without the need of a medium, have increased over the years and it has undergone many changes both on national and international levels (Dizkırıcı & Gökgöz, 2018: 93).

The increase in the interest of these currencies and their widespread usage necessitates revisions in taxing, accounting and auditing. In this study, the cryptocurrencies that have been known and used for many years have been analyzed with reference to Bitcoin and other cryptocurrencies (altcoin). Within the scope of this study, the usages and legislative regulations of Bitcoin around the world have been analyzed and suggestions regarding Bitcoin accounting process have been put forward.

19.2 Crypto Currencies

In today's digital era, payments, money transfers and trade are predominately carried out on the internet. This is the result of the speed of digitalization and the rise of digital businesses, which have also reinforced global dependence (Khalilov, Merve, Mücahit, & İrfan, 2017).

In an extensive research conducted in 2015 regarding the payment methods of the digital era and mobile banking with 14,829 participants across Europe, Australia and the U.S.A, the following results were obtained: (ING International Survey, 2015)

- More than 65% of the participants stated that they had already been using mobile banking or would probably use it within a year,
- 48% checked their accounts using mobile banking,
- In the past year, 58% of the participants from Europe bought goods and services via mobile devices
- The young population is more willing to use mobile devices for shopping,
- More than 50% of the participants used fewer physical assets this year compared to the previous year and 84% will use much fewer physical assets in the near future,
- 51% of the participants from Europe will use mobile payment applications this year,
- People used mobile payment applications because they are quick and easy.

The Ali Baba Group made online transactions of more than $14 billion on 11 November 2015, 68% of which were made via mobile devices (Schwab, 2017: 64). This shows that people are eager to make transactions by e-commerce using mobile devices instead of physical money.

Such developments prompted people to seek innovations in money. Therefore, firstly virtual currencies were created, which failed due to various reasons including double spending. The development of cryptology technologies have enabled the control of double spending and spenders. These solutions, which were created via cryptology technologies, were based on the blockchain infrastructure (Serçemeli, 2018: 36).

The term cryptocurrency signifies alternative values to money that can be used in e-commerce on a global scale in today's widespread and enhanced Internet environment. Due to their digital and virtual structures, as global mediums, cryptocurrencies are not subject to central bank procedures, restrictions, audits and guarantees. As the result of developed technology, the usage of cryptocurrency has increased rapidly around the world. According to Çarkacıoğlu (2016: 8, 9), cryptocurrencies generally possess the following features:

- In contrast to central electronic money and banking systems, cryptocurrencies are non-central currencies and are controlled using blockchain databases.
- Cryptocurrencies are created at specified rates in the establishment phase of the non-central crypto systems with methods that are accessible to and known by the

public. The amount, asset source and timing of cryptocurrencies submitted to circulation are determined during the establishment phase of the crypto system.
- In many cryptocurrency systems, currency creation decreases over time in order to remain stable in circulation.
- There is no third medium in crypto system to establish confidence. The safety, integrity and accuracy of its global ledger are realized by miners (creating limited assets and earning assets on the systems by organizing algorithms) that do not rely on each other. Although the system is reliable, parties do not have to trust each other.

Today, cryptocurrencies are classified as Bitcoin and alternative coins. All cryptocurrencies that have been developed after the Bitcoin are called Altcoin, meaning alternative coin, due to the fact that they have emerged as an alternative to Bitcoin and were inspired from Bitcoin.

There are a total of 2104 altcoins including Ripple, Ethereum, Litecoin, Namecoin, Anoncoin, Darkcoin, Counterparty, Nextcoin, Dogecoin, Bitshares in circulation, with new altcoins coming into circulation every day (Aslantaş, 2016: 360). The existing marketing values of cryptocurrencies including Bitcoin are $124,388,723,872 (https://coinmarketcap.com).

There are four functions of money which are to operate as a medium of exchange, a unit of account, a medium of saving and a way to pay debts. In order for an asset to be considered as money, it must persuade people that it has these functions (Boyes & Melvin, 2013: 294). It must also be acceptable, standard, enduring, divisible, portable, while not being rare (Erdem, 2008: 2). Whether the cryptocurrencies carry out these functions and thus can be considered as money or not is still debated, however their usage continues to increase rapidly.

Bitcoin and other Altcoin currency platforms can be used in money transactions independent from the banking sector and any kind of authority, and can also provide financial services outside of banks for less by using only technology and establishing trust. For this reason, cryptocurrencies can be destructive for the financial systems in the future (Hayes, 2017: 2).

19.2.1 Bitcoin (BTC)

Bitcoin is a peer-to-peer electronic cash payment system that was first revealed in an article by Satoshi Nakamoto, whose real name and identity is unknown (Nakamoto, 2008).

Bitcoin was introduced via e-mail sent to a mail group by Satoshi Nakamoto in November 2008. The first Bitcoin software was brought out in 2009. There is no information regarding the identity of Satoshi Nakamoto. It is claimed that Nakamoto quit the system towards the end of 2010, because he had earned enough from Bitcoin and was busy with different professions as of 2011. It is certain that the email sent by Nakamoto was from Japan but whether Nakamoto is Japanese or not is not known.

There have been many assumptions as to who Nakamoto is. A student in cryptography whose group has three members has been said to be Dorian Nakamoto, although he refused the claims and filed a lawsuit (Aslantaş, 2016: 354).

The interest in Bitcoin when coupled with curiosity regarding person(s) who built the system leads to interesting outcomes. This curiosity even led to the creation of a comic book of Nakamoto. The comic book titled "On the track of Satoshi Nakamoto" published by a Spanish promoter received a lot of interest from the Bitcoin community (http://www.bitcoinhaber.net/2014/12/bitcoin-cizgi-roman-oldu.html 15.01.2019).

In the article published by Nakamoto, it is stated that Bitcoin currency makes online payment possible via a peer-to-peer cash electronic system without the intervention of any financial corporation. In the process of the exchanging of money, when the sender makes a transaction, s/he adds the summary of the previous transaction using her/his own digital signature and signs it off to the recipient by adding this signature. Bitcoin can be shared by the users as wished via transfer in a computer network system in a virtual environment and each electronic transaction is signed with the system's digital signature, which is recorded in the central database. Therefore, the only thing required to make transactions is the Internet as opposed to the government, organizations or bank mediums. Unlike other payment methods, there is almost no possibility of the sellers charging unexpected costs or prices (account maintenance fee, subscription fee etc.) In this case, when compared with its other equivalents, the Bitcoin software system generates minimum cost and provides maximum speed. Bitcoin owners can also freely carry out international transactions by paying low cost to miners. Furthermore, by means of the crypto mechanism of the system, the privacy of all transactions is guaranteed (Aslantaş, 2016: 355).

When Bitcoin is bought and sold, there is a virtual ledger called blockchain which records how much Bitcoin there is and keeps records of the parties based on their Bitcoin addresses before each transaction. All transactions regarding Bitcoin, transaction steps and transaction pasts are tracked in blockchain and all new transactions are added to the end of the confirmed transaction chain (Hepkorucu & Genç, 2017: 49). After a Bitcoin is sold, the output record from the crypto address in the wallet of the previous owner and the input record for the same amount are made automatically to the buyer's address after a 10 min confirmation period. At the end of the transfer, the seller no longer has any rights over the Bitcoin. Meanwhile, everybody can see how much Bitcoin has been sent to which address by means of this transaction (Aslantaş, 2016: 356).

Bitcoin transactions are made according to the following steps: A person who wants to buy Bitcoin submits his/her address to the seller The seller adds this address to the buyer section of the transaction and after receiving conformation via a private password, the transaction is completed, which is put forward for all users to see. The reason why it shares this record of the completed transaction with all users is to prevent changes in the transaction. In order to prevent the usage of the same Bitcoin in different places at the same time, all Bitcoin users have a copy of the transactions. A person who wants to reuse the same Bitcoin needs more processor power than all the Bitcoin users. It is not possible to make changes to a Bitcoin transaction, as each Bitcoin transaction is added the Bitcoin itself. A person who wants to change the data needs entry to the previous owner's account and due to the privacy of user identities this is not possible. However, the user can save his/her own private password (www.okanacar.com).

Theoretically, Bitcoin can be created by anyone according to Atik et al. (2015: 249). It is calculated by solving various mathematical problems that require complicated transactions, of different difficulty levels using Bitcoin Miner Software which is submitted to miners who participate in the creating process. The first person to find a solution for the problem is rewarded with a certain amount of Bitcoin. Everybody in the network (P2P network) can reach this information and new and more difficult problems are submitted for the miners to solve on the related software. Miners are able to create Bitcoin the amount of which decreases with their forecasts. The amount of Bitcoin in the circulation decreases over time. The upper limit of Bitcoin emission volume is 21 million, which is the maximum amount created by miners.

The first Bitcoin transaction was made by the system establishers in January 2009. The first person to make a transaction using Bitcoin was Laszlo Hanyecz who bought two slices of pizza in exchange for 10,000 BTC on 22 May 2010. At that time, the pizza that costed $25 has come to be known as the most expensive pizza ever bought in the world (Koçoğlu, Çevit, & Tanrıöven, 2016: 79).

As understood from the explanations, Bitcoin has more advantages than conventional payment systems. These advantages are as follows: (Serçemeli, 2018: 49)

- Bitcoin transfers can be made from anywhere that has internet,
- It is not necessary to physically be in a place,
- Transfers are very quick,
- It can easily be transported and stored,
- It is not affected by inflation,
- Unlike conventional methods, it does not charge any expenses or costs,
- It is not affected by the social-economic situations of countries,
- There are no restrictions to payment amounts. Even very low payments can be made,
- The payment transaction expenses of companies are eliminated,
- It is a transparent and clear system,
- It is divisible,
- Just like gold it cannot be imitated which protects its value.

19.3 Blockchain Technology

Blockchain draws attention as the technology that is based on the most popular currency, Bitcoin. However, blockchain is much more than a digital currency (Rosenberg, 2017). In November 2008, with the article published by Satoshi

Nakamoto, Bitcoin was suggested as a peer-to-peer cash system that is not central and does not require third parties. Different from conventional currencies, Bitcoin is known to use blockchain technology to store and save transactions on a peer-to-peer network (Raiborn & Sivitanides, 2015: 26, 27). Nakamoto used the description 'a chain including digital signatures' for cryptocurrencies in his article published in 2008 (Nakamoto, 2008: 2). Bitcoin is a distributed digital currency which does not require any corporations or authorities, is not supported by the government or legal institutions and is not dependent of any commodity. It works on a blockchain with a structure based on cryptography and peer-to-peer network (Grinberg, 2011: 160). Each Bitcoin transaction is saved in the blockchain which is a distributed ledger. Blockchain keeps the record of all of the Bitcoin transactions (Vranken, 2017: 2).

Blockchain is a ledger whose content cannot be changed or records cannot be deleted (Byström, 2016: 3). In brief, it is a chronologic database that includes recorded transactions in a network via computers (Peters & Panayi, 2016: 3). A blockchain distributed database is such an amplified database that even the creators cannot make a change on data and it keeps data records continuously. Each area in the system that includes data is called a block. In order to create new block, previous blocks must be completed. These blocks that include data come together to create a chain, which is called a blockchain (Fanning & Centers, 2016: 53).

The most basic feature of blockchain is that no corporation, institution or authority has a regulative effect on its system. Today, exchanging anything that has value between corporations or persons requires a medium or regulator in order to be completed correctly. For example, a medium corporation or regulative authorities are required for share certificate exchange transactions, similarly when buying a house, notaries and land offices for title deed exchanges and a bank for money exchange are required. The blockchain technology is an infrastructure that facilitates transactions quickly without the need of any mediums or regulators (Kosba, Miller, Shi, Wen, & Papamanthou, 2016: 839).

The blockchain technology, also called Internet of Values, ensures the exchange of transactions without any medium and is also safer than the methods used today that require mediums and regulatory authorities. In the systems working with the distributed database conception, each record carries a time mark. If one block is changed by foreign intervention, the other blocks will not be affected by this change, thus it is thought that the actual data will be protected (Özdoğan & Kargın, 2018: 163).

According to Johann Palychata, head of Blockchain at BNP Paribas, Bitcoin blockchain is as important as the invention of steam machines or internal combustion engines as it has the potential to turn into something beyond the finance environment (Crosby, Pradan, Sanjeev, & Vignesh, 2016: 8).

Blockchain technology provides users with advantages in various fields, the most important being the advantage to submit distributed ledger structure. Blockchain records all the transactions made by the users in a single ledger and provides this service without decentralized transaction periods. In addition, this technology enables all transaction copies to be kept in the chain. Even if users are not online, they can easily reach the copies of the transaction they have made by means of ledger. Since each blockchain saves copies of previous transaction in the chain, it is possible to obtain copies of retroactive transactions (Deloitte Report, 2017: 4). The system's transparency does not enable transactions to be deleted or changed which is an important advantage (Ovenden, 2017). This also influences accountability, as the mentioned advantage naturally provides a decrease in costs and an increase in the efficiency of regulations and legislative harmonization in enquiries. Therefore, due to the distributed ledger structure of blockchain technology, all assets used in economic transactions can be tracked safer, easier and faster. Transactions are gathered in local ledgers by blockchain technology and the copies are then transmitted to many computers around the world. The records can be accessed easily in blockchain which are organized according to date starting from the first transaction (Rosenberg, 2017).

With the help of public-key cryptographs, blockchain technology also removes the double spending problem. Public keys are the addresses hiding in blockchains and created in cryptographic. Each asset is associated with one address which is used in the transactions made in the crypto economy trade assets from one address to another. A striking feature of blockchain is that public keys are never related to real world identities. Although transactions can be seen by everybody, they are made without mentioning the identity of the people involved (Pilkington, 2016: 226). Distributed conformity and anonymity are the most significant features of blockchain technology (Crosby et al., 2016: 8).

The potential of blockchain is based on its many valuable features. The basic features of blockchain technology are given below: (Schatsky & Muraskin, 2017):

- *Reliability and Accessibility*: Due to being used by a large community, there is no room for failure and it is designed to resist against interruptions and attacks. If errors occur in participatory network, other users can provide information and carry out operations.
- *Transparency*: The transactions in blockchain is made so that all participants can see, which increases audit and safety.
- *Invariability*: Making changes in the chains is almost impossible without being detected, which increases reliability while decreasing the possibility of fraud.
- *Unrecoverability*: It is possible to easily carry out administrative affairs and increase the accuracy of records which will be unrecoverable.
- *Digital*: Almost all assets and documents can be coded and limited or presented as bookkeeping entries. This means that blockchain technology has more expanded applications than the existing application fields.

In 2015, Estonia created the first virtual BitNation using blockchain based on identification technology for the identity cards of citizens. It is also the first government to ever use blockchain technology (Schwab, 2017: 171).

Bitcoin usage is rather widespread in developed countries such as Japan, Canada, U.S.A, Germany and France. In also Japan and Ireland, there are even Bitcoin ATMs. The first Bitcoin ATM in the world was used in Canada. In Turkey, the first Bitcoin ATM was launched in Istanbul Atatürk Airport (Aslantaş, 2016: 357). In addition, a firm producing 3D printers in Turkey, 3Dörtgen, is the first firm to pay its employee's salaries with Bitcoin (Dizkırıcı & Gökgöz, 2018: 98).

The American NASDAQ tested the blockchain technology in share certificate transfers between the companies in its special market before making them public. By using Colored Coin technology, confirmation period is no longer needed in the transfer of digital assets and more transaction volume per second can be made at an almost nonexistent level of the transaction fee (NASDAQ, 2016).

Similarly, the Australia Stock Exchange also made preparations to use blockchain technology as a post processing period platform. It made partnership investments in companies for blockchain technology, with the aim to decrease risks and abbreviate the periods of post share certificates (ASX, 2016).

R3 distributed ledger platform, established in 2015, supported by a consortium including 80 institutions related with the banking and finance sectors, such as Bank of Montreal, BNY Mellon, CIBC, Commerzbank, Commonwealth Bank of Australia, ING, Macquarie, Mitsubishi UFJ Financial Group, Mizuho Bank, Nordea, RBC, Société Générale, State Street, TD Bank, UniCredit, Wells Fargo, aims to prepare the finance sector for blockchain technologies and adopt these technologies in the future (WEF, 2017).

VISA, which is the leading payment institution in the world, has started to test blockchain technology to use it in transfers made between banks. In line with this aim, it has requested that the system be tried by sending invitation to various banks (VISA, 2016).

In a report published in 2016 by Bain & Company, which is an international consultancy management firm, it was highlighted that distributed ledger technology simplifies transactions by removing any medium and keeping the records protected against interventions and thus has the potential to increase payment speed, transparency and efficiency.

According to survey results by PWC, representatives of the financial services sector, it is expected that blockchain will be a part of the production system by 2020 (PWC, 2017).

Amazon Web Services, which is a company of Amazon, announced in a conference that blockchain could be used for keeping health records, sharing confidential information, smart contracts and corporate management applications as well as finance and they announced that they had started taking actions in this area (AWS, 2016).

Similarly, IBM together with Sichuan Heja which is a company based in China announced their support towards to the blockchain Yijian which is aimed to be used in the improvement of supply applications in the pharmaceutical industry. With this blockchain, it is expected that this sector will become transparent by observing the movements of medicines in the supply chain and companies with low credit scores will acquire better scores (Mansfield-Devine, 2017).

According to the results of a survey carried out by the World Economic Forum with information and communication specialists and senior executives, it was seen that at least 10% of global gross revenue would be stored in blockchain platforms by 2027. (Espinel, O'Halloran, Brynjolfsson, & O'Sullivan, 2015: 24). Today, blockchain technology is predominantly used for cryptocurrencies. In addition to this, it is expected that the same technology will be used in many business applications such as share certificates and other valuable paper transactions, audit periods, accounting records and contracts (Özdoğan & Kargın, 2018: 164).

19.4 The Effects of Blockchain and Crypto Currencies on Accounting and Accounting Audit

Financial transactions create the scope of accounting. Due to their properties, transactions made using cryptocurrencies are also included in the scope of accounting and should take place in the accounting information system. The basic function of accounting which is an information system is to give the required financial information to personals in businesses. As in each transaction, accounting is responsible for giving accurate and complete information to internal or external users in businesses regarding cryptocurrencies. Therefore, cryptocurrencies should also be recorded and reported (Raiborn & Sivitanides, 2015: 33).

The need to keep regular entries in accounting dates back to centuries ago. The first data showing the existence of regular accounting in history goes back to thirteenth century. Sombart (2008: 138) stated that a study published in 1202 by Leonardo Fibonacci and Liber Abacci is the first important advancement in accounting. In the following years, the first part of subsection 9 in section 11 of the book titled Summa de Arithmetica, Geometria, Proportinoni et Proportionalita, (1494) by Fra Luca di Borgo whose real name is Luca Pacioli, mentions an accounting model that is based on double-entry. This book is the first printed material to contain the double-entry system.

Summa which is by Pacioli consists of 36 short sections. One of these sections is titled "Particularis de Computis et Scriptus- Accounts in Accounting and Explanation on Record" and includes explanations related to the double-entry system (Isaac, 1947: 23). In addition to this, two ledgers are also mentioned in Summa. The first of these ledgers is Manudo or Giornale in which transactions are kept according to a time sequence and the second described as the main ledger is Quaderno in which each transaction is recorded twice. The second ledger which is double-sided shows an innovation in the accounting entry system. This ledger enables the balance between credit and debit at any time. If the balance is not found, which indicates a mistake and it requires to detect it (Braudel, 2004: 512–514). The most important feature of Summa is that it is the first printed work to show accounting transactions that accountants systematically applied during that era.

Subsequently, the double-entry system has continued to be used with improvements. In conclusion, the system, which has been expanded and developed since the Renaissance, still survives with respect to accounting applications today. However, developments in digital assets and blockchain technology that have recently started to be used today are the developed versions of this old entry system (Uçma & Kurt, 2018: 469).

The most important innovation provided by blockchain technology in accounting applications is shown as the transformation of the double-entry system to a triple-entry system, which is presented as the unique feature of the system. It is known that the double-entry system is used in accounting applications. In other words, in the reporting period of economic transactions of businesses, an entry system consisting of asset and fund accounts that work double side at the same time is used. This means both sides of the economic transactions of businesses are recorded (Uçma & Kurt, 2018: 472, 473). However, according to Potekhina and Riumkin (2017: 11–13), as blockchain technology has a ledger property for Bitcoin transactions, this significantly influences the conventional existing system because Bitcoin transactions are presented as the data from classified, stored and saved in real accounting system. This presentation is made by software showing the transfer of digital money, financial assets and other digital documents in based on a blockchain between two persons or more in real-time instantaneously. As they are cryptographic, the blocks of the transactions made allow financial statement components to be accessed at any time. In other words, just as any transaction is made; it can be given access on blockchain easily. This enables the real-time access of accounting applications for both internal and external users in businesses. In addition, Potekhina and Riumkin (2017: 11–13) explained the developing accounting periods based on blockchain with these factors below:

- *Transparency*: Transactions can be seen in real-time.
- *Irrecoverableness*: Transactions made in the chain are irrecoverable, deleting or changing them is not possible.
- *Accessibility*: All data on the chain are easily accessible by a large community of partners.

Comprising the factors counted above, the distributed ledger structure mentioned in accounting applications and created based on blockchain technology leads to the use of this ledger as the third side in addition to the traditional double-entry system. By this way, a shared blockchain distributes a ledger so that the related groups can access the transaction records that take place as third side in this structure, which is data stored in a transparent, accessible and irrecoverable way. Meanwhile, the parties that made the transactions confirm their integrity in the shared ledger, which provides much more assurance than in realistic presentations of financial information (Wunsche, 2016: 17, 18). For this reason, various writers use the term 'triple-entry system conception' in order to explain accounting based on blockchain.

The triple-entry system is more developed than the double-entry system and is expressed as a cryptographic system in which blockchain takes place as a third side. Economic transactions made in the blockchain processes are expressed as data completely based on automation and it is important to submit information that the third party requests in non-central distributed ledger system safely (Potekhina & Riumkin, 2017: 13). For example, when a seller receives assets in cash as the exchange of sold goods or services, this is recorded on the debit side according to the double-entry system. In the same transaction, when a buyer pays cash for the exchange of goods or services, this is recorded on the credit side. For example, economic transactions are showed in separate ledgers with each side recorded as double sided. However, based on blockchain, parties in same transaction take place in same ledger as interconnected accounting records set instead of being separate, which creates a third side. In other words, the same economic transaction kept in the blockchain with a distributed ledger structure creates the third side of the double-entry in a single block by means of distributed ledger structure (Uçma & Kurt, 2018: 474).

In this way, blockchain technology is the next step of accounting. Instead of an entry system based on the separate commercial documents of businesses, it comprises an ecosystem or extensive entry system that enables the direct access into the entry system to be used jointly by all businesses and provides to be in place of transactions made on interconnected chain that accounting records are. The mentioned ledger structure also enables the concept of 'World Wide Ledger-WWL' which blockchain technology brings to economic life to become widespread. This ledger structure is a certifiable, controllable and searchable blockchain accounting system application. This system ensures that all the data of businesses is published on an international level and becomes accessible for regulators and key partners. Furthermore, the system allows all partners and regulatory authorities to access entry based on exact transparency (Potekhina & Riumkin, 2017: 13).

According to Uçma and Ganite (2018: 474), blockchain along with the global ledger structure has a potential to directly assist today's accounting applications. It provides integration on the chain by typical accounting procedures in a gradual way, starting from the reliability of the integrated records and it also provides observable records in real-time for independent audit periods that occur at the next step of accounting applications. As for the final stage which is provided by this technology, audits based on complete automation turn into reality.

When businesses present their financial statements after completing the audit process for tax or regulatory aims, this means that they have carried out the audit actions based on paper documents. As it is not possible to delete or change transactions in blockchain technology, showing the observable audit materials and the audited transaction processes is possible (Simon, Kasale, & Manish, 2017: 9). Since this provides digital and integrated data transfer by stepping back from paper document audits, it makes easier applications of continuous audit concepts which have advantages in many ways (Uçma and Ganite, 2018: 474).

In a conventional accounting system, transaction and audit processes are carried out as follows (Byström, 2016: 4):

- Financial reports are prepared by the accounting departments of businesses,
- An auditor expresses an opinion regarding the accuracy and reliability of the prepared statements,
- Investors, creditors and other external partners trust in both the opinion of the auditor and that the business is not provided with misleading information.

- In an accounting system using blockchain technology, transaction and audit processes are carried out as follows: (Byström, 2016: 4)
- All monetary transactions of businesses are recorded on blockchain together with time marks of each transaction,
- All accounting records of businesses can simultaneously be observed by all authorized partners and all partners requesting financial statements can see the real-time financial statements of the business.

As it can be understood, blockchain technology enables the improvement of information quality and increases transparency by presenting the accounting transactions to the partners more reliably and in real-time. Thus, the duration of audit operations decreases, the reliability of audit results increases and important time and cost advantages are ensured.

Distributed ledger structure provided by blockchain technology will be able to turn into real-time consolidates financial statements easily and quickly and delays in the reports end of month will remove. As a result, regulatory authorities will be able to access businesses' real-time data (Wunsche, 2016: 17, 18). The presentation of real-time information also enables financial statements to be debated in their presentation times and even if quarterly or yearly reporting periods are adopted for decision makers, it is possible that an authorized manager in the businesses can observe and track all transactions in real-time with the system (Potekhina & Riumkin, 2017: 17).

The effects of cryptocurrencies and blockchain technology on accounting science make it necessary to develop accounting applications related to these cryptocurrencies. On an international level, professional societies have been carrying out studies in this field and publishing acceptable regulations for other countries. For example, because Bitcoin has asset properties serving an economical purpose and are created at the end of past transactions, the Financial Accounting Standards Board (FASB) states that it should be accepted as an asset for financial reporting (Uçma & Kurt, 2018: 475). However, according to Raiborn and Sivitanides (2015: 27–30) in order to consider Bitcoin as an asset, there are still uncertainties about the type of asset (cash, tangible asset, investment etc.) it should be accepted as. This has led to the adoption of different reporting types among countries. As is known, in order to be accepted as cash, an asset is required to be a currency that is used for exchange. In this sense, Bitcoin is considered cash and should be recognized in accounting system like all other assets. However, as it is not accepted in the transactions of businesses, it does not entirely meet the features of cash and is treated as a different asset in many countries. For example, Finland and China accept Bitcoin as cash, while Norway, Germany and Korea do not. In 2014, the American National Income Administration announced that it had accepted Bitcoin and other cryptocurrencies as assets providing operational fund gains.

According to Wunsche, 2016: 18) accounting applications based on blockchain possess a suitable infrastructure to process transactions that have been made in businesses. This infrastructure consists of the following:

- *Cash*: Cash flows based on digital assets will able to be explained directly without requiring the verification records of third parties.
- *Receipt and Payment*: Thanks to embedded devices in the form of smart contracts, all receipts and payment transactions will able to be stored on the chain automatically.
- *Stocks*: Stocks will be recorded based on the asset transfers in smart contracts and will be activated with the "buy" command in the stock management system of the buyer.
- *Intangible Assets*: Intangible assets will be transferred to smart rights contracts based on the intellectual property rights. Conflicts on property rights will be removed thanks to the time mark of blockchain.
- *Capital*: These accounts will be monitored on blockchain and transferred as in digital asset flows. In addition, blockchain ledgers will be easily accessible and transferable.
- *Credits*: These accounts will be digitally recorded with smart credit contracts on blockchain. Once the credit accounts are submitted in the form of smart credit contracts, they will be able to be monitored until the debts have been payed.

19.5 Recognition of Crypto Currencies in Accounting

Although there are no generally accepted applications and regulations of cryptocurrencies in accounting yet, the transactions related to the recording, valuation and reporting of cryptocurrencies will be able to be carried out by made inferences within the frame of general accepted accounting principles, basic concepts of accounting, accounting and financial reporting standards and operational system of accounting.

Bitcoin can be evaluated in the following six situations regarding accounting.

1st Situation; The medium of payment,
2nd Situation; Foreign currency,
3rd Situation; Valuable mine,
4th Situation; Marketable security,
5th Situation; Stock,
6th Situation; Intangible asset.

1st Situation

In order to consider Bitcoin as a medium of payment, a new account must be opened in the group of Liquid Assets or the Bitcoin received in the exchange for commercial transaction can be recorded in a cash account.

Example Firm X has sold a good for $250,000 that cost $100,000. The buyer has paid its cost as Bitcoin.

Cash	250,000	
Bitcoin		
Sales		250,000
Good sales record in the exchange for Bitcoin		
Cost of trade goods sold	100,000	
Trade goods		100,000
Cost record of sales		

2nd Situation

Businesses can consider Bitcoin as a different currency to national currencies and accept it in exchange for their sales or can buy Bitcoin directly from the bitcoin stock exchange, at the result the received Bitcoin should be evaluated as a foreign asset and should be recorded in the form of 'bitcoin cash' in an adjunct account under a cash account like other foreign assets. In accordance with the "substance over form" concept of the Accounting Basic Conceptions, transactions made as foreign assets should take place in the records of exchange to national currency. In addition, existing Bitcoins should be evaluated at the end of each period like other foreign assets. The International Accounting Standard (IAS) 21 Recognition Standard of Foreign Asset Transactions states that the effective rate in foreign asset transactions is the existing rate during the delivery moment, namely the spot rate (IAS 21 artc. 8). However it has not clearly indicated which institution's rate the spot rate belongs to. Furthermore, if there is not a lot of fluctuation between the rates, it has been indicated that averages can also be accepted (Örten, Kaval, & Karapınar, 2013: 331). Bitcoin can be shown in Liquid Assets when accepted as a currency of Exchange and be recorded as Foreign Exchange Gain and Foreign Exchange Losses.

Example With the aim of benefitting from short-term price movements, Firm X buys Bitcoin for $100,000.

(a) If sold for $120,000

Cash	100,000	
Bitcoin		
Cash		100,000
$		
Buying of Bitcoin		
Cash	120,000	
$ *Cash*		100,000
Bitcoin		
Foreign exchange gain		20,000
Sale of Bitcoin by gaining profit		

(b) If sold for $80,000

Cash	100,000	
Bitcoin		
Cash		100,000
$		
Buying of Bitcoin		
Cash	80,000	
$		
Foreign exchange losses	20,000	
Cash		100,000
Bitcoin		
Sale of Bitcoin by making loss		

3rd Situation

By accepting Bitcoin as a valuable mine, shown in the group of Liquid Assets, it can be recorded as Foreign Exchange Gain, Foreign Exchange Losses.

Example With the aim of benefitting from short-term price movements, Firm X buys Bitcoin for $100,000 sells it for $120,000.

Other liquid assets	100,000	
Bitcoin		
Cash		100,000
TL		
Buying of Bitcoin		
Cash	120,000	
TL		
Other liquid assets		100,000
Bitcoin		
Foreign exchange gain		20,000
Sale of Bitcoin by gaining profit		

4th Situation

Bitcoin can be recorded as Gains on Marketable Securities Sales and/or Losses on Marketable Securities Sales when accepted as Marketable Securities.

Example In order to benefit from short-term price movements, Firm X buys Bitcoin for $100,000.

(a) If sold for $120,000

Other marketable securities	100,000	
Bitcoin		
Cash		100,000
$		
Buying of Bitcoin		
Cash	120,000	
$ *Other marketable securities*		100,000
Bitcoin		
Gains on marketable securities		20,000
Sales		
Sale of Bitcoin by gaining profit		

(b) If sold for $80,000

Other marketable securities	100,000	
Bitcoin		
Cash		100,000
$		
Buying of Bitcoin		
Cash	80,000	
$		
Losses on marketable securities sales	20,000	
Other marketable securities		100,000
Bitcoin		
Sale of Bitcoin by making loss		

(c) If the value of Bitcoin is $80,000 at the end of the period

Provision expenses	20,000	
Provision for decrease in		
Value of		
Marketable securities (−)		20,000
Valuation record of Bitcoin at the end of period		
Provision Separation		

Furthermore, the long-term investment in a cryptocurrency can be made at the first public offering of cryptocurrencies. In such cases, it can be considered that the received cryptocurrency falls under Financial Fixed Assets.

Example Mr. Ali buys X Cryptocurrency that is valued at $100,000 from its first public offering with the aim of investing long-term.

Other financial assets	100,000	
Bitcoin		
Cash		100,000
$		
Buying of Bitcoin with the aim of investment		

5th Situation

For Bitcoin miners, Bitcoin is like a physical production on a digital environment. In such cases, accounting entries can be organized just like the accounting processes used by manufacturers. The created Bitcoins can be seen in the account of Other Inventories in the Balance Sheet.

Example Mr. Ali buys technic equipment for $100,000 with the aim of creating Bitcoin and begins the creation process. In the course of Bitcoin creation spends $50,000 on electricity. Later, the created Bitcoins are sold for $200,000, the accounting entry would be as follows:

Direct raw materials and supplies expenses	100,000	
General production expenses	50,000	
Cash		150,000
Taking of costs into cost accounts		
Other inventories	150,000	
Bitcoin		
Reflection account for		
Direct raw		
Materials and supplies		100,000
Reflection account		
For general		
Production expenses		50,000
Taking of costs into other inventories account		
Cash	200,000	
Sales		200,000
Record of Bitcoin sale		
Cost of other sales	150,000	
Other inventories		150,000
Bitcoin		
Record of Bitcoin cost		

Bitcoin miners get commission either by solving complicated problems for the creation of Bitcoin or by confirming the accuracy of transaction like notaries in Bitcoin transfer transactions. The obtained income for confirming the accuracy of transactions by mediating Bitcoin transfers can be seen in the Commission Income account.

Example Mr. Ali confirms the accuracy of transactions by mediating Bitcoin transfers and for each transaction he gets $200.

Cash	200	
Commission income		200

6th Situation

When Bitcoin is considered as an Intangible Asset, it should be reported in the Balance Sheet between Long Term Assets.

Example If Firm X buys Bitcoin for $100,000 they must account them as Intangible Assets;

Intangible assets	100,000	
Cash		100,000
Record of Bitcoin as intangible assets		

19.6 Legislative Regulations Related to Cryptocurrency

In order to tax cryptocurrencies, they first must be legally accepted in accordance with the related laws. First, cryptocurrencies are required to be recognized as assets in accordance with laws, subsequently they can be taxed. For this process the outlooks of various countries on cryptocurrencies taken into consideration and applied to the taxing of the cryptocurrencies.

As cryptocurrencies, which can be carried out at the speed of light as they have no physical existence and are called the perfect asset, they are not subjected to taxation, do not circulate within the frame of determined rules, and are not affected by the volatility movements on their prices, many national central banks and councils of bank audit-regulations have placed bans on using crypto assets and warned users against the possible risks that they can encounter. They have voiced concerns such as, the fact that cryptocurrencies are not subjected to any authority, do not require any formal agents or inspections, are not influenced by the economic situation of any countries, their prices are only determined by user transactions can justify the institutions' warnings in the future (Kuzu, 2018: 42).

When the legal positions of cryptocurrencies in EU countries is examined, it can be seen that only four countries out of the 28 EU countries chose to recognize these assets. Thirteen countries, namely Bulgaria, Ireland, Greece, France, South Cyprus, Lithuania, Latvia, Hungary, Netherland, Austria, Portugal, Romania, Slovakia have not taken any position against Bitcoin, while 11 of the countries have only emphasized what Bitcoin is not. Belgium, Croatia, Finland, Italia, Luxemburg, Malta and Poland have accepted these assets as neither legal nor electronic. The Czech

Republic has claimed that Bitcoin is not a banknote, coin, written or electronic money. Denmark has stated that Bitcoin has no real commercial value compared with gold and silver. Spain has claimed that such assets cannot be accepted as legal currency because they are not created by the asset authority of any country. Slovenia, on the other hand, has claimed that these assets are not acceptable and were risky as they could be used for money laundering (Kowalski, 2015: 149). Although these countries do not accept cryptocurrencies and do not consider any tax regulations for them, it can be said that avoiding this ecosystem is extremely difficult.

The EU countries that legally accept cryptocurrencies are Germany, Estonia, Sweden and the United Kingdom. In addition, the Germany Finance Ministry has stated that Bitcoin is a legal currency and is considered as a financial medium or an accounting unit, and has accepted it that Bitcoin can be used by businesses in special transactions with the permission the Federal Financial Authorization. The Estonia Central Bank and Finance Ministry have accepted cryptocurrencies as alternative payment methods, that the buying or selling of Bitcoin is not illegal and that they are an alternative payment medium for entrepreneurs, however they do not accept them as a currency. In Sweden, the Tax Office has not accepted Bitcoin as a currency due it not depending on any central bank and claims that Bitcoin should be classified as "a different type of asset". The Bank of England has shown a very theoretical approach towards Bitcoin. It has considered cryptocurrencies can be transferred like money (special money for persons with internet devices) (Kowalski, 2015: 149).

The legal positions of cryptocurrencies in countries outside of the EU are given in Table 19.1 (Serçemeli, 2018: 56).

In Turkey, in a press briefing held on 25 November 2013, the Banking Regulation and Supervision Agency of Turkey informed the public of their opinion stating that "In accordance with the law no 6493, the supervision and audit of Bitcoin cannot be carried out and there is a risk of hacking and fraud." although no bans were introduced (www.bddk.org.tr).

As for the Central Bank of the Republic of China stated that Bitcoin is not a real asset and has no legal status. Bitcoin was forbidden in China in 2014. However, in 2015 Bitcoin mining started to be carried out in China and in January 2016 the Central Bank of the Republic of China announced that had started to research ways to improve their own digital currency. Similarly, Bitcoin is also forbidden in Russia. In Australia, a bill was brought out towards the end of 2015 to make Bitcoin equivalent to other currencies (Aslantaş, 2016: 358).

19.7 Taxation of Crypto Currencies

Cryptocurrencies reflect the conventional characteristics of tax havens, where gains are not subjected to taxation and the identities of tax payers are unknown. For this reason, cryptocurrencies will probably overcome the detection of governments regarding tax evasion. This shows that the necessary innovator policies should be put in place (Omri, 2013: 921–927).

Table 19.1 Legal position of cryptocurrencies

Country	Legal position
Hong Kong	Permitted
Indonesia	Permitted
Israel	Permitted
Iran	Permitted
Jamaica	Permitted
Japan	Permitted
South Korea	Permitted
Lebonon	Permitted
Mexico	Permitted
Malaysia	Permitted
New Zealand	Permitted
Philippines	Permitted
Singapore	Permitted
Turkey	Permitted
Taiwan	Permitted
Ukraine	Permitted
U.S.A.	Permitted
South Africa	Permitted
Bangladesh	Controversial
India	Controversial
Jordan	Controversial
Kazakhstan	Controversial
Russia	Controversial
Thailand	Controversial
Dominican Republic	Not permitted
Ecuador	Not permitted
Iceland	Not permitted

Placing tax on the income earned from Bitcoin is for the good of the government Brazil, Canada, Finland, Bulgaria and Denmark have made regulations regarding the taxation of Bitcoin usage. Singapore considers Bitcoin as an asset or product and even collects value-added-tax from local shopping done with Bitcoin. The highest number of cryptocurrency trade and the leader of the world in Bitcoin trade volume is the US. Many countries in the world are waiting for the position that the US will take towards legislative regulation of cryptocurrencies to see the results (Çarkacıoğlu, 2016: 56).

There are two turning points that are expected to occur by 2025. The first is the estimation that artificial intelligence make up 30% of company audits and the second is governments starting to collect tax from blockchain for the first time (Schwab, 2017: 36).

19.8 Conclusion

The accounting profession has entered a new configuration period due to transforming technologies such as artificial intelligence, automatic data analyzes, cryptocurrencies and blockchain. The basic functions of the accounting profession such as reporting, tax and audit have also been influenced by these changes (Drew, 2018).

In the present study, the transformation period of the double-entry system, which has a conventional application field in seven centuries with blockchain technology, was particularly emphasized. The distributed ledger structure based on blockchain brings the triple-entry system to the agenda and suggests that the existing accounting and audit actions are unnecessary. This does not mean that the applications of the profession will be removed, it merely provides the creation of new roles in this new technological system. The new system provides a suitable environment for the members of the accounting profession to take more care in issues of planning and valuation instead of keeping records.

The new system is being adopted by businesses that accept Bitcoin as a medium of payment and make transactions in the stock exchange of bitcoin-national currency. Therefore, the accounting, taxation and audit of cryptocurrencies, which are the basics of blockchain, is extremely important. When the usage and the usage areas of cryptocurrencies and their recognition in accounting are examined, in accordance with the "substance over form" concept, accounting record transactions should be made according to the evaluation of the payment medium of Bitcoin, foreign currencies, Valuable Mine, Marketable Security, Stock and Intangible Asset.

Distributed ledger technology, namely blockchain or IoT, which expands the potential usage areas ranging from financial markets to all firm functions and to even all transactions including any exchange value provides activity, transparency and more reliability especially in accounting and finance applications. Although the present legal gaps may seem unimportant for now, they may cause serious problems in the future. So, the arrangements of regulations to fit the system's purpose is of great importance even though the system aims to eliminate centralization. The International Federation of Accountants (IFAC) suggests that financial reporting should be adapted to the new technology and study how suitable the audit period will be realized in changing conditions.

When blockchain technology is accepted especially in the fields of accounting and auditing, there will be a strong need for the revision of policies and standards implemented in these fields and for new standards suitable to blockchain.

Just as there are standards for accounting, financial reporting and audit which are applied commonly, it is suggested that to globally accepted cryptocurrency standards should also be created and put into practice.

References

Aslantaş, B. A. (2016). Kripto Para Birimleri, Bitcoin ve Muhasebesi. *Çankırı Karatekin Üniversitesi Sosyal Bilimler Enstitüsü Dergisi, 7*(1), 349–366.

Asx. (2016). Retrieved August 28, 2017, from http://www.asx.com.au/documents/about/ASX-Selects-Digital-Asset-to-Develop-Distributed-Ledger-Technology-Solutions.pdf

Atik, M., Yaşar, K. S., Yılmaz, B., & Sağlam, F. (2015). Kripto Para: Bitcoin ve Döviz Kurları Üzerine Ektileri. *Bartın Üniversity, The Journal of Faculty of Economics and Administrative Sciences, 6*(11), 247–261.

Aws. (2016). *AWS Re: Invent 2016: Blockchain on AWS: Disrupting the norm (GPST301).* Retrieved August 25, 2018, from www.slideshare.net/AmazonWebServices/aws-reinvent-2016-blockchain-on-aws-disrupting-the-norm-gpst301

Bain & Company. (2016). *Distributed ledgers in payments: Beyond the bitcoin hype.* Retrieved September 10, 2018, from http://www.bain.com/publications/articles/distributed-ledgers-in-payments-beyond-bitcoin-hype.aspx

Boyes, W., & Melvin, M. (2013). *Ekonominin Temelleri.* 5. Basımdan Çeviri Editörü: Erdinç Telatar. Ankara: Nobel Yayıncılık.

Braudel, F. (2004). *Maddi Uygarlık: Mübadele Oyunları.* Çeviren: Mehmet Ali Kılıçbay, İkinci Baskı. Ankara: İmge Kitabevi.

Byström, H. (2016). *Blockchains, real-time accounting and the future of credit risk modeling.* Working Paper/Department of Economics, School of Economics and Management, Lund University (4), pp. 1–11.

Çarkacıoğlu, A. (2016). Kripto-Para Bitcoin. Sermaye Piyasası Kurulu Araştırma Raporu, Ankara.

Crosby, M., Pradan, P., Sanjeev, V., & Vignesh, K. (2016). Blockchain technology: Beyond bitcoin. *Applied Innovation, 2*, 6–10.

Deloitte Report. (2017). *Blockchain technology and its potential impact on the audit and accounting.* Retrieved January 20, 2019, from https://www2.deloitte.com/us/en/pages/audit/articles/impact-of-blockchain-in-accounting.html

Dizkırıcı, A. S., & Gökgöz, A. (2018). Kripto Para Birimleri ve Türkiye'de Bitcoin Muhasebesi. *Journal of Accounting, Finance and Auditing Studies, 4*(2), 92–105.

Drew, J. (2018). *How AI, blockchain, and automation will reinvent accounting.* Retrieved January 10, 2019, from https://www.journalofaccountancy.com/podcast/ai-blockchain-automation-reinventing-accounting.htm

Erdem, E. (2008). Para Banka ve Finansal Sistem, 2. Baskı. Ankara: Detay Yayıncılık.

Espinel, V., O'Halloran, D., Brynjolfsson, E., & O'Sullivan, D. (2015, September). *Survey report: "Deep shift: Technology tipping points and societal impact"*. World Economic Forum.

Fanning, K., & Centers, D. P. (2016). Blockchain and its coming impact on financial services. *Journal of Corporate Accounting and Finance, 27*, 53–57. https://doi.org/10.1002/jcaf.22179

Grinberg, R. (2011). Bitcoin: an innovative alternative digital currency. *Hastings Science and Technology Law Journal, 4*, 160–208. Retrieved August 20, 2018, from https://ssrn.com/abstract=1817857

Hayes, A. S. (2017). Cryptocurrency value formation: An empirical study leading to a cost of production model for valuing bitcoin. *Telematics and Informatics, 34*(7), 1308–1321.

Hepkorucu, A. & Genç, S. (2017). Finansal Varlık Olarak Bitcoin'in İncelenmesi ve Birim Kök Yapısı Üzerine Bir Uygulama. *Osmaniye Korkut Ata Üniversitesi Osmaniye Korkut Ata University İktisadi ve İdari Bilimler Fakültesi Dergisi, 1*(2), 47–58.

Ing International Survey. (2015). *The rise of mobile banking and the changing face of payments in the digital age.* Retrieved September 19, 2017, from https://www.ezonomics.com/ing_international_surveys/mobile_banking_2015/

Isaac, A. (1947). *İşletme İktisadı*, Cilt:1, Çeviren: Dr. Orhan Tuna. İstanbul Üniversitesi Yayınları No: 117, İktisat Fakültesi Yayınları No: 8, İsmail Akgün Matbaası, Gözden Geçirilmiş ve Genişletilmiş İkinci Baskı, İstanbul.

Khalilov, K., Merve C. Mücahit, G., & İrfan, K. (2017). *Bitcoin ile Dünya ve Türkiye'deki Dijital Para Çalışmaları Üzerine Bir İnceleme*. 19 Akademik Bilişim Konferansı. Aksaray: Aksaray Üniversitesi.

Koçoğlu, Ş., Çevit, Y. E., & Tanrıöven, C. (2016). Bitcoin Piyasalarının Etkinliği, Likiditesi ve Oynaklığı. *İşletme Araştırmaları Dergisi, 8*(2), 77–97.

Kosba, A., Miller, A., Shi, E., Wen, Z., & Papamanthou, C. (2016). *Hawk: The blockchain model of cryptography and privacy-preserving smart contracts*. In: 2016 IEEE symposium on security and privacy (SP), pp. 839–858.

Kowalski, P. (2015). Taxing bitcoin transactions under polish tax law. *Comparative Economic Research, 18*(3), 139–152.

Kuzu, S. (2018). *Alternative investment medium and crypto assets as being hedge instrument*. Istanbul: Kriter Publisher. isbn:978-605-2228-92-0.

Mansfield-Devine, S. (2017). Beyond bitcoin: Using blockchain technology to provide assurance in the commercial world. *Computer Fraud and Security, 2017*(5), 14–18. Retrieved January 15, 2019, from https://doi.org/10.1016/S1361-3723(17)30042-8

Nakamoto, S. (2008). *Bitcoin: A peer-to-peer electronic cash system*. Retrieved February 24, 2018, from http://www.bitcoin.org/bitcoin.pdf

Nasdaq. (2016). Retrieved August 20, 2018, from http://www.nasdaq.com/article/colu-announces-colored-coins-and-lightning-network-integration-cm710111

Omri, M. (2013). Is cryptographic currency an outstanding tax heaven? *Michigan Law Review, First Impressions, 38*. Çev. Gürlek Keleş (2016), S. P. İÜHFM, LXXIV(2).

Örten, R., Kaval, H., & Karapınar, A. (2013). *Türkiye Muhasebe Finansal Raporlama Standartları Uygulama ve Yorumları* (Vol. 7). Ankara: Baskı, Gazi Kitabevi.

Ovenden, J. (2017). *Will blockchain render accountants irrelevant?* Retrieved December 5, 2018, from https://channels.theinnovationenterprise.com/articles/will-blockchain-render-accountants-irrelevant

Özdoğan, B., & Kargın S. (2018). Blok Zinciri Teknolojisinin Muhasebe ve Finans Alanlarına Yönelik Yansımaları ve Beklentiler. *Muhasebe ve Finansman Dergisi, 80*, 161–176. issn:2146-3042. https://doi.org/10.25095/mufad.465928.

Peters, G. W., & Panayi, E. (2016). Understanding modern banking ledgers through blockchain technologies: Future of transaction processing and smart contracts on the internet of money. In: *Banking beyond banks and money* (pp. 239–278). Cham: Springer.

Pilkington, M. (2016). *Blockchain technology: Principles and applications*. Research handbook on digital transformations. isbn:9.781.784.717.759. https://doi.org/10.4337/9781784717766.

Potekhina, A., & Riumkin, I. (2017). *Blockchain—A new accounting paradigm: Implications for credit risk management*. Retrieved January 10, 2019, from https://umu.diva-portal.org/smash/get/diva2:1114333/FULLTEXT01.pdf

Pwc. (2017). *Global FinTech report 2017*. Retrieved August 20, 2018, from https://www.pwc.com/jg/en/publications/pwc-global-fintech-report-17.3.17-final.pdf

Raiborn, C., & Sivitanides, M. (2015). Accounting issues related to bitcoins. *Journal of Corporate Accounting and Finance, 26*(2), 25–34.

Retrieved March 3, 2018, from https://coinmarketcap.com/

Retrieved January 15, 2019, from http://www.bitcoinhaber.net/2014/12/bitcoin-cizgi-roman-oldu.html

Rosenberg, E. (2017). *How blockchain is going to change accounting forever*. Retrieved October 15, 2018, from https://due.com/blog/blockchain-to-change-accounting-forever/

Schatsky, D., & Muraskin, C. (2017). *Beyond bitcoin blockchain is coming to disrupt your industry*. January 15, 2019, from https://www2.deloitte.com/tr/tr/pages/technology-media-andtelecommunications/articles/trends-blockchain-bitcoin-security-transparency.html

Schwab, K. (2017). *Dördüncü Sanayi Devrimi*. Çev.: Zülfü Dicleli. İstanbul: Optimist Yayınevi.

Serçemeli, M. (2018). Kripto Para Birimlerinin Muhasebeleştirilmesi ve Vergilendirilmesi. *Finans Politik & Ekonomik Yorumlar, 639* (Mayıs), 33–66.

Simon, A. D., Kasale, S., & Manish, P. M. (2017). Blockchain technology in accounting and audit. *IOSR Journal of Business and Management.*

Sombart, W. (2008). *Burjuva: Modern Ekonomi Dönemine Ait İnsanın Ahlaki ve Entelektüel Tarihine Katkı.* Çeviren: Oğuz Adanır. Ankara: Doğu Batı Yayınları.

Uçma, T., & Ganite, K. (2018). Blockchain technology in accounting and auditing. *Suleyman Demirel University, The Journal of Faculty of Economics and Administrative Sciences, 23*(2), 467–481.

Uçma, T. U., & Kurt, G. (2018). Muhasebe ve Denetiminde Blok Zinciri Teknolojisi. *Süleyman Demirel Üniversitesi İktisadi ve İdari Bilimler Fakültesi Dergisi, 23*(2), 467–481.

Underwood, S. (2016). Blockchain beyond bitcoin. *Communications of the ACM, 59*(11), 15–17.

Visa. (2016). Retrieved September 10, 2018, from http://investor.visa.com/news/news-details/2016/Visa-Introduces-International-B2B-Payment-Solution-Built-on-Chains-Blockchain-Technology/default.aspx

Vranken, H. (2017). Sustainability of bitcoin and blockchains. *Current Opinion in Environmental Sustainability, 28*, 1–9.

Wef. (2017). *Realizing the potential of blockchain, a mutlistakeholder approach to the stewardship of blockchain and cryptocurrencies.* World Economic Forum White Paper. Retrieved August 20, 2018, from http://www3.weforum.org/docs/WEF_Realizing_Potential_Blockchain.pdf

Wunsche, A. (2016). *Technological disruption of capital markets and reporting?* Retrieved January 10, 2019, from https://www.cpacanada.ca/-/media/site/business-and-accounting-resources/docs/g10157-rg-technological-disruption-of-capital-markets-reporting-introduction-to-blockchain-october-2016.pdf

www.bddk.org.tr.

Ali Kablan, after receiving a bachelor's degree in Accounting and Finance Education at the Faculty of Commerce and Tourism Education, Gazi University, earned both his master's degree and doctoral degree in the Department of Business Administration-Accounting in Istanbul University. He has published many articles and contributed chapters in books on "Accounting—Finance". He wrote his master's thesis on "Accounting of Derivatives" and a doctoral thesis on "Accounting System at Municipalities". He continues his research as Assistant Professor at Trakya University.

Chapter 20
Cryptocurrency and Tax Regulation: Global Challenges for Tax Administration

Gamze Öz Yalaman and Hakan Yıldırım

Abstract This chapter investigates whether the government should taxed cryptocurrency or not by using game theoratical framework. In this game, both government and cryptocurrency investors will determine the strategies to maximize their own benefits. In order to achieve the Nash equilibrium, two-person zero sum game matrix is created. As a result, Nash Equilibrium occurs in only the one case that indicates cryptocurrency should be taxed with low tax rate and high penalty rate. Moreover to understand tax policies of cryptocurrency this chapter also investigate various countries taxation policy on cryptocurrency. It is clear that there is no consensus among countries about legal status and taxation process of cryptocurrencies.

20.1 Introduction

Cryptocurrency is a digital money and has specific characteristics include immutability, irreversibility, decentralization, persistence, and anonymity (Harwick, 2016; Puthal, Malik, Mohanty, Kougianos, & Das, 2018). It uses cryptography that involves the process of converting legible information into an almost non-breakable code by using cryptographic hash algorithms.

Cryptocurrencies have increased rapidly ever since the inception of Bitcoin and blockchain technology. Bitcoin is the first implementation of cryptocurrencies in the world. It is a kind of digital cash and online payment system which operates without going through a central bank (Hayes, 2017; Swan, 2015). Bitcoin was created in 2009 by an unknown person or group using the name Satoshi Nakamoto. The technical details were first described in the fall of 2008 as a whitepaper entitled

G. Ö. Yalaman (✉)
Department of Public Finance, Faculty of Economics and Business Administration, Eskişehir Osmangazi University, Eskişehir, Turkey

H. Yıldırım
Centre for Distance Education, Eskişehir Osmangazi University, Eskişehir, Turkey
e-mail: hayildirim@ogu.edu.tr

© Springer Nature Switzerland AG 2019

U. Hacioglu (ed.), *Blockchain Economics and Financial Market Innovation*, Contributions to Economics, https://doi.org/10.1007/978-3-030-25275-5_20

"Bitcoin: A Peer-to-Peer Electronic Cash System" and later released as open software in 2009. According to Nakamoto (2008), it is defined as "*A purely peer-to-peer version of electronic cash would allow online payments to be sent directly from one party to another without going through a financial institution*". The Bitcoin[1] was first traded in January 2009 and by June 2011, there were around 6.5 million Bitcoins in circulation among 10,000 users (Reid & Harrigan, 2013).

Cryptocurrencies are built upon the underlying substructure of the blockchain technology. A blockchain is a data structure that makes it possible to create a digital ledger of data and share it among a network of independent parties (Laurence, 2017). All confirmed transactions are involved in the blockchain, and then cryptocurrencies wallets compute their spendable balance. It is therefore approved that new transactions actually belong to the spender. Cryptography is applied for the integrity and chronological order of the blockchain and allows us to create mathematical verifications that make available a high level of security.

Some researchers claim that blockchain will be the next major disruptive technology (Laurence, 2017; Swan, 2015; Tapscott & Tapscott, 2016). There are three generations of Blockchain technology namely Blockchain 1.0 for digital currency, Blockchain 2.0 for digital finance, and Blockchain 3.0 for digital society (Zhao, Fan, & Yan, 2016). Blockchain 1.0 provides applications that allows you to make transactions of cryptocurrency with a low-cost to all over the world (e.g. Bitcoin). Blockchain 2.0 provides programmable money (e.g. Ethereum) namely smart contracts bringing an enormous innovation to Blockchain 1.0. Blockchain 3.0 provides widespread implementation for government, health, science and IoT (Casino, Dasaklis, & Patsakis, 2018).

The ongoing discussions about blockchain technology are undertaken especially on cryptocurrencies, blockchain is not only has a great potential for global financial system but also has a great potential for education (Blockcert), health care sector (MedicalChain, Nano Vision), transportation systems (IBM Blockchain, De Beers, Food Industry), and social services and entertainment (Spotify), insurance (Accenture), real estate (BitProperty) etc.[2]

Cryptocurrencies and blockchain technology attracted much attention from many academics and practitioners. There are a lot of recent work on cryptocurrencies and background technology of blockchain in various areas in the literature. For example, Casino et al. (2018) provides a systematic literature review of blockchain-based applications and they emphasize that there is a still numerous research gaps for both academics and practitioners. Hawlitschek, Notheisen, and Teubner (2018) provides

[1]*Bitcoin can be definedasin many different ways.* For example, Simser (2015) defines Bitcoinas"*crypto-currency based on open-source software and protocols that operates in peer-to-peer networks as a private irreversible payment mechanism*". Shcherbak (2014) defines Bitcoin as"*a novel decentralised payment mechanism functioning under the Bitcoin protocol which is practically impossible to amend in a way that contradicts the interests of the majority of Bitcoin stakeholders*".

[2]https://www.forbes.com/sites/bernardmarr/2018/05/14/30-real-examples-of-blockchain-technology-in-practice/#3a8e153d740d

another systematic literature on blockchain technology and trust in the sharing economy. They mainly revealed that a huge differs between the contexts of blockchain and the sharing economy, and they highlight the blockchain technology might change trust in platform providers. Yli-Huumo, Ko, Choi, Park, and Smolander (2016) create a map on Blockchain technology by reviewing 41 papers from scientific databases. The results show that more than 80% of the papers focus on Bitcoin system and less than 20% deals with other Blockchain applications like smart contracts and licensing.

Li and Wang (2017) investigate the determinants of the Bitcoin exchange rate by using co-integration and ARDL model, taking into account both technology and economic factors. The results show that there is significant relationship between Bitcoin exchange rate and economic fundamentals in the short run period. But interestingly, the results differ in a long term, now the Bitcoin exchange rate is more sensitive to economic fundamentals and less sensitive to technological factors.

Fry and Cheah (2016) recommend an econophysics model for financial crashes by using the relationship between statistical physics and mathematical finance. They show empirical application of their model by using the highest global market cap coins as Bitcoin and Ripple. They highlight Bitcoin and cryptocurrency markets are extremely volatile and there is significant speculation in the market. Scott (2016) examines the role of cryptocurrency and blockchain technology in building social and solidarity finance and they identify many different research gap for further studies.

It is clear that there is a few paper focus on taxation of cryptocurrencies or regulations. Important point here, each government has separate legal arrangements and governments have different perspectives about taxation or regulation about cryptocurrencies. For example, while some governments are treated cryptocurrency as barter transaction, some others are treated as property. It is clear that there is no consensus of taxation framework of cryptocurrency among the countries. Thus many paper in the literature investigate the taxation and regulation possibilities under governments own legislation. Moreover they also investigate how taxpayers should report cryptocurrency transactions (Akins, Chapman, & Gordon, 2014; Boehm & Pesch, 2014; Bal, 2015; Kaplanov, 2012; Lambert, 2015; Litwack, 2015; McLeod, 2013; Ram, 2018; Small, 2015; Tsukerman, 2015).

Boehm and Pesch (2014) focus on the state's legal framework about cryptocurrencies. They compare German and US-American public, criminal and civil law systems. They find that state's current legal systems are not appropriate for cryptocurrencies. In addition, they suggest that the balance between the rules of law (public, criminal and civil law).

Small (2015) clarifies that it is important to regulate the cryptocurrencies, especially after the Mt. Gox's bankruptcy. In addition, he recommends that Mt. Gox's event should change the viewpoint that sees state regulations of cryptocurrencies unnecessary.

Litwack (2015) analyzes the taxation of cryptocurrencies in US. He recommends that cryptocurrencies can be classified as both asset and currency. In the same way, Wiseman (2016) criticizes the IRS decision whichadopts bitcoin as a property. The paper states that cryptocurrencies are not only an investment tool but they can be also used as a tool to purchase daily items. Thus, paper suggests that cryptocurrencies should classify as a currency and they are subject to sales tax.

Ram (2018), using correspondence analysis, investigates the taxation problems of bitcoin. According to the findings of the paper, bitcoin is considereddifferent from the currency, since the transactions with bitcoin are treated as barter transactions. Due to Bitcoin's special features, Ram (2018) suggest that Bitcoin should be arranged in the same way as a currency.

Marian (2013), Bryans (2014) and Slattery (2014) focus on tax evasion and money laundering which is one of the major disadvantages of cryptocurrencies. Marian (2013) take attention to the tax haven characteristic of cryptocurrencies. Cryptocurrencies may allow tax evasion in many ways. First, because they do not have jurisdiction in which they operate, they are not taxed at the source. Second, cryptocurrency accounts are anonymous. Individuals perform a transaction without giving any identifying information, and there is no financial intermediaries such as banks. Additionally, Lambert (2015) sign that the value of cryptocurrencies cannot be precisely known at any specific time. There are several converters, but their reliability is limited. While valuation problem can facilitate the tax evasion, honest taxpayers have a difficulty about calculating taxable income. Due to the valuation of cryptocurrency, both tax administration and taxpayers have problems.

Cryptocurrencies provide a number of advantages to its users. One of this advantages is low transaction cost. Because of the lower transaction costs, many enterprises including overstock.com, Dell, Expedia, Etsy accept cryptocurrency payments. Some of them claim that cryptocurrency as a stable currency in weak markets. The other advantage can be considered as the blockchain technology which is forming the technology of crycptocurrencies. Also, security and anonymity can be accepted as other advantages. Although there are some advantages, cryptocurrencies have some disadvantages such as tax evasion, money laundering and financing of terrorism or illegal activities such as narcotics (e.g. The Silk Road), weapons sales (Choo, 2015; Lambert, 2015; Litwack, 2015; Tsukerman, 2015). For dealing with disadvantages of cryptocurrencies, it is recommended to legalize regulations for cryptocurrencies.

Recently, Holub and Johnson (2018) analyze many different research papers related to cryptocurrencies. Their findings show that there is a huge amount of papers about cryptocurrencies across different disciplines such as, economics, law, technical area, public policy, taxation, finance, etc. They analyze 1206 papers. Of 1206 papers, 157 are related to regulation of cryptocurrencies, 39 are related to taxation of cryptocurrencies. It can be easily seen that there are few papers about taxation of cryptocurrencies. Moreover, their findings show that most of the papers in the cryptocurrency literature are either unpublished paper or proceedings.

Thus, different from the common literature this chapter investigates whether the government should taxed cryptocurrency or not by using game theoratical framework. Moreover to understand tax policies of cryptocurrency this chapter also investigate various countries taxation policy on cryptocurrency.

The following section includes the background of cryptocurrencies. Section 20.3 shows the basic game theoretical model for investigating whether the government should tax cryptocurrency and Sect. 20.4 presents the tax policy of cryptocurrencies by comparison of various countries. Section 20.5 provides the discussion and conclusion.

20.2 The Market Capitalization of Cryptocurrencies

Cryptocurrencies have increased rapidly. According to coinmarketcap.com data in January 2019, there are more than 2100 other "altcoin" (alternative coin) cryptocurrencies, like Ethereum, Litecoin and Dogecoin and the total cryptocurrency market has approximately $112B capitalization (see the details of market capitalization of top 50 coins in Table 20.1). The global market capitalization had reached

Table 20.1 Market capitalization of top 50 coins

Name	Symbol	Marketcap	Name	Symbol	Marketcap
Bitcoin	BTC	$59,504,491,178	VeChain-Token	VET	$210,679,747
Ripple	XRP	$12,365,281,803	TrueUSD	TUSD	$208,271,352
Ethereum	ETH	$10,931,833,122	Bitcoin-Gold	BTG	$169,025,954
EOS	EOS	$2,070,450,136	Qtum	QTUM	$161,070,827
Bitcoin-Cash	BCH	$1,969,649,570	HoloToken	XHOT	$149,348,192
Litecoin	LTC	$1,874,831,146	Zilliqa	ZIL	$147,388,335
Tether	USDT	$1,736,069,725	OmiseGo	OMG	$147,174,265
Tronix	TRX	$1,704,853,358	0x	ZRX	$143,433,247
Stellar	XLM	$1,547,022,771	Basic-Attention-Token	BAT	$137,989,477
Stellar	XLM	$1,547,022,771	ChainLink	LINK	$137,051,904
Bitcoin SV	BSV	$1,110,072,347	Decred	DCR	$135,729,840
Cardano	ADA	$959,985,566	Augur	REP	$133,539,579
Binance-Coin	BNB	$815,303,691	Bitcoin-Diamond	BCD	$133,224,622
Monero	XMR	$716,925,914	Arbitrage	ARB	$127,127,712
IOTA	IOT	$696,203,404	Paxos-Standard	PAX	$127,015,073
Dash	DASH	$576,182,687	Lisk	LSK	$124,987,643
NEO	NEO	$443,146,202	Nano	NANO	$112,387,774
Ethereum-Classic	ETC	$419,140,760	Bytecoin	BCN	$105,867,877
NEM	XEM	$357,384,510	DigiByte	DGB	$102,090,135
USD-Coin	USDC	$300,161,824	BitShares	BTS	$98,743,864
Zcash	ZEC	$277,791,866	Verge	XVG	$89,662,146
Waves	WAVES	$271,576,425	Gemini-dollar	GUSD	$88,687,334
Maker	MKR	$266,261,436	Siacoin	SC	$87,424,787
Tezos	XTZ	$227,721,823	ICON	ICX	$87,374,869
DogeCoin	DOGE	$223,542,439	Aeternity	AE	$84,733,936

Source: https://www.barchart.com (01.02.2019)

the highest level approximately $850B in January 2018. Commonly discussed for being the technology behind Bitcoin, blockchain technology is well known due to the current cryptocurrency hype (Angelis & da Silva, 2018).

20.3 Should Cryptocurrency Be Taxed? The Basic Game Theoretical Framework

There is a fact that people who have cryptocurrency, rise their gross income.[3] If people has income, this income is taxable. However, it does not mean that it is actually taxed. People who have cryptocurrency income do not pay tax on that income for several reasons. First, they are not aware that such income is taxable or the government hasn't defined cryptocurrency as income categories yet; e.g. in the scheduler tax system, the government have to define income categories to collect tax. If the cryptocurrency is not define one of the income categories, there is not tax liability. Second, the taxpayer can avoid or evade (Bal, 2015). To find an answer of the following questions as "*should cryptocurrency be taxed*?" and "*how to tax*?", we use a basic game theoratical framework, because there is a mutual competition and conflict of interest between government and cryptocurrency investors.

Game theory was invented by John von Neumann and Oskar Morgenstern in 1944 and has come a long way since then and scientists have been awarded the Nobel Prize in Economic Sciences for their contributions to game theory. Game theory is applied in a number of fields, including business, finance, economics, political science and psychology. Game theory is a method of determining the most accurate strategy against conflicting interests. That is, it will be necessary to make the most correct strategic decision in this conflict. As the decision of one player depends on the decision of other player, there will be competition between them. The special mathematical techniques developed in order to make the right decision at this point (Fudenberg & Tirole, 1991).

In the game theory model, the players adopt to increase their benefits as much as possible or lose as little as possible. Thus, the game theory strategy provides a player with the opportunity to develop the best strategy against the other player. In other words, when equilibrium is achieved in the game theory strategy, the player do not want to go to another place. Because, they provide maximum benefit according to the possible strategies available at this equilibrium point.

In order to achieve the Nash equilibrium, two-person zero sum game matrix is created as Table 20.2.

[3]Kristoffer Koch invested 150 kroner ($26.60) in 5000 bitcoins in 2009; 4 years later, in 2013, it reached NOK5m ($886,000), https://www.theguardian.com/technology/2015/dec/09/bitcoin-forgotten-currency-norway-oslo-home

Table 20.2 The Nash equilibrium between government and cryptocurrency investors

	Cryptocurrency investor	
Government	Low tax rate (15%)	High tax rate (40%)
Tax applied		
High tax penalty	**Y∗tr, Y∗tr** **G = +3000[a], I = −3000[a]** **(Case-1)**	Y∗tr, Y∗tr G = +8000, I = −8000 (Case-3)
Low tax penalty	X∗tr, X∗tr G = +1500, I = −1500 (Case-2)	X∗tr, X∗tr G = +4000, M = −4000 (Case-4)
Tax not applied		
No tax penalty	Y∗tr, Y∗tr G = −3000, I = +3000 (Case-5)	Y∗tr, Y∗tr G = −8000, I = +8000 (Case-6)

[a]Nash equilibrium point (indicated in bold)

20.3.1 The Model and Hypothetical Example

The model help us to understand in which way maximizes both cryptocurrency investors and any government benefits. The model previously used Yalama and Çelikkaya (2007) for examining the relationship between optimal tax rate, tax penalty and tax audit for Turkey, i.e. this chapter has adapted Yalama and Çelikkaya (2007) model to cryptocurrency applications.

The model is not belong to any individual countries taxation system and has some restrictions.

To analyze whether the cryptocurrency should be taxed, we created a hypothetical example. Suppose cryptocurrency investors get 20,000 $ revenue from their investment and the lower and higher tax rate represents as 15 and 40%[4] respectively. If the tax penalties are not deterrent sufficiently, let's assume that the cryptocurrency investor will declare half of the revenue as 10,000$.

Case-1 If the government applies a minimum tax rate as 15%, the cryptocurrency investor will choose the full payment of tax by declaring the whole income for the case where the tax penalties are high. In this case, the government will benefit $Y*tr = 20000*0.15 = +3000$ $, while the cryptocurrency investor will lose −3000 $.

Case-2 If the government applies a minimum tax rate as 15%, the cryptocurrency investor will not choose the full payment of tax where the tax penalties are not deterrent sufficiently. In this case, let's assume that the cryptocurrency investor will declare half of the revenue as 10,000$. The government will benefit $X*tr = 10000*0.15 = +1500$ $, while the cryptocurrency investor will lose −1500 $.

[4]The upper and lower income tax rate determined as 15 and 40% taking in to account income tax tariff of some countires as Australia, US, UK, Canada, Turkey.

Case-3 If the government applies a maximum tax rate as 40%, the cryptocurrency investor will choose the full payment of tax by declaring the whole income for the case where the tax penalties are deterrent enough. In this case, the government will benefit Y∗tr = 20000∗0.40 = +8000 $, while the cryptocurrency investor will lose −8000 $.

Case-4 If the government applies a maximum tax rate as 40%, the rational cryptocurrency investor will not choose the full payment of tax where the tax penalties are not deterrent sufficiently. In this case, let's assume that the cryptocurrency investor will declare half of the revenue as 10,000$. The government will benefit X∗tr = 10000∗0.40 = +4000 $, while the cryptocurrency investor will lose −4000 $.

Case-5 If the government does not applies any tax rate and tax penalties, the government will lose Y∗tr = 20000∗0.15 = −3000 $ (for the case of low tax rate) which is non-collectable tax revenue from cryptocurrency investor, while the cryptocurrency investor will benefit +3000 $ which is unpaid tax and represents the benefit of a cryptocurrency investor.

Case-6 If the government does not applies any tax rate and tax penalties, the government will lose Y∗tr = 20000∗0.40 = −8000 $ (for the case of high tax rate) which is non-collectable tax revenue from cryptocurrency investor, while the cryptocurrency investor will benefit +8000 $ which is unpaid tax and represents the benefit of a cryptocurrency investor.

As can be seen in the Table 20.2, in order to achieve the Nash equilibrium, both government and cryptocurrency investors will determine the strategies to maximize their own benefits. In this case, when the government selects the first line of the matrix (high tax penalty), the taxpayer will decide between the payment of −3000 $ and − 8000 $ tax and choose the first column (Case-1), namely the payment of −3000 $ (low tax rate, high tax penalty). When the taxpayer chooses the first column (low tax rate = 15%), the government will decide to get payment of 3000, 1500 and −3000 tax income and prefer 3000 $ tax income (Case-1). As a result, Nash Equilibrium occurs in only the Case-1 with low tax rate and high penalty rate. In other cases; for both sides, the maximization of benefits does not be realized at the same point and Nash equilibrium does not occur.

Finally if take into account the maximization for the benefits of both cryptocurrency investors and governments, Nash Equilibrium indicates that cryptocurrency should be taxed with low tax rate and high penalty rate.

20.4 Tax Policies of Cryptocurrencies: The Case of Various Countries

Moreover to understand tax policies of cryptocurrency we also investigate various countries taxation policy on cryptocurrency.

20.4.1 The Case of Various Countries

The subject of consensus among the countries is the caution that citizens should be warned against to cryptocurrencies. For example, many central banks educate their citizens about the difference between cryptocurrencies which are not issued and guaranteed by the state and actual currencies which is. Most government emphasize the high risk of cryptocurrencies. It is seen that the system has unregulated transactions process and there is no legal recourse is available to the cryptocurrencies investor in the event of loss. The cryptocurrencies also create for illegal activities, such as money laundering, terrorism and tax evasion. But recently some of the countries like Australia, Canada, and the Isle of Man have expanded their laws on cryptocurrency markets to reduce the disadvantages of cryptocurrencies market. There is no consensus of the terms used by countries to reference cryptocurrency. For example, Argentina, Thailand, and Australia use the term of *digital currency*; Canada, China, Taiwan use the term of *virtual commodity*; Germany uses the term of *crypto-token*; Switzerland use the term of *payment token*; Italy and Lebanon use the term of *cyber currency*; Colombia and Lebanon use the term of *electronic currency* and finally Honduras and Mexico use the term of *virtual asset*. As well as there is no consensus among countries about legal status of cryptocurrencies, there is no consensus how to tax cryptocurrencies. For example some countries[5] ban absolutely cryptocurrencies, while some countries[6] ban implicitly. In addition, there are some countries that do not have entirely or partially regulatory framework such as France, Finland, Belgium, Denmark, Mozambique, Namibia, South Africa. Some countries such as Spain, Belarus, the Cayman Islands, and Luxemburg see a huge potential in the technology behind cryptocurrency system and they develop a regulatory regime to attract investment in technology companies. The issue of taxation is one of the important matter. The problem arise how to categorize cryptocurrencies for the purposes of taxation. Many countries have categorized cryptocurrencies differently for tax purposes. For example Israel, Bulgaria, Switzerland taxed cryptocurrencies as asset, financial asset, and foreign currency, respectively. Argentina, Spain and Denmark subject to cryptocurrencies as income tax. In addition to that the corporations pay corporate tax, unincorporated businesses pay income tax, individuals pay capital gains tax in United Kingdom. Gains in cryptocurrency investments are not subject to value added tax in the European Union Member States. In the most of the countries the mining of cryptocurrencies is exempt from taxation. However, the mining of cryptocurrencies is taxable in Russia which exceeds a certain energy consumption threshold (The Law Library of Congress, 2018).

The taxation status of cryptocurrencies for many different countries are discussed more details as below:

[5]Algeria, Bolivia, Egypt, Iraq, Morocco, Nepal, Pakistan, United Arab Emirates.

[6]Bahrain, Bangladesh, China, Colombia, Dominican Republic, Indonesia, Iran, Kuwait, Lesotho, Lithuania, Macau, Oman, Qatar, Saudi Arabia, Taiwan.

USA The cryptocurrencies do not have any legal tender status in USA. Moreover they are considered as a property for US federal tax purposes. Whatever tax policy is applied to property transactions, the same tax principles also apply to transactions using the cryptocurrencies. If employees are paid a wage in cryptocurrencies, these wages are subject to federal income tax withholding and payroll tax. If payment is made to independent contractors and other service providers in cryptocurrencies, these payments are subject to tax and self-employment tax rules apply. If a payment made using cryptocurrencies is subject to reporting.[7]

China Central government regulators[8] announced that the initial coin offerings is banned in China, in 2017 and the government does not recognize cryptocurrencies as legal tender.[9]

Russia The cryptocurrency transactions banned by 2015, but interestingly at the beginning of 2018 the Russia's Ministry of Finance explained that they are working on legislation to regulate cryptocurrency transactions without fully banning them and through this legislation it will be possible to tax cryptocurrency transactions to support the state budget.[10]

Australia Australian Taxation Office delivered guidance on taxation of cryptocurrencies in 2014. According to guidance cryptocurrency transactions are treated like barter transactions. If individual *sell or gift cryptocurrency, trade or exchange cryptocurrency (including the disposal of one cryptocurrency for another cryptocurrency), convert cryptocurrency to fiat currency like Australian dollars, or use cryptocurrency to obtain goods or services, capital gain tax occurs.*[11] Crycrptocurrency businesses e.g. cryptocurrency trading businesses, cryptocurrency mining businesses, cryptocurrency exchange businesses (including ATMs) are subject to goods and services tax.[12] Australian Taxation Office states that each of cryptocurrency transaction should be recorded in order to determine its status against tax.[13]

Canada Canada Revenue Agency define the digital currency as virtual money that can be used to buy and sell goods or services on the Internet, and cryptocurrencies

[7]https://www.irs.gov/newsroom/irs-virtual-currency-guidance

[8]The People's Bank of China, the China Banking Regulatory Commission, the China Securities Regulatory Commission, the China Insurance Regulatory Commission, the Cyberspace Administration of China, the Ministry of Industry and Information Technology, the State Administration for Industry and Commerce.

[9]https://www.loc.gov/law/help/cryptocurrency/china.php#_ftn1

[10]https://www.reuters.com/article/us-russia-cryptocurrencies-bill/russia-ready-to-regulate-not-ban-cryptocurrencies-idUSKBN1FE0Y0

[11]https://www.ato.gov.au/general/gen/tax-treatment-of-crypto-currencies-in-australia%2D%2D-specifically-bitcoin/?page=2#Transacting_with_cryptocurrency

[12]https://www.ato.gov.au/general/gen/tax-treatment-of-crypto-currencies-in-australia%2D%2D-specifically-bitcoin/?page=3#Cryptocurrency_used_in_business

[13]https://www.ato.gov.au/media-centre/media-releases/ato-delivers-guidance-on-bitcoin/

are accepted as digital currency. Digital currency i.e. cryptocurrency can be sold or bought like a commodity. Tax liability can arise, in this context. According to Canada's implementation cryptocurrencies are subject to the Income Tax Act. In addition, the Canadian Revenue Agency should be informed about the use of the cryptocurrency, otherwise it is not legal.[14]

Cyprus In Cyprus the term of virtual currency is used to describe the cryptocurrencies. The Central Bank of Cyprus declare that the purchase, holding or trading of virtual currencies is not legal tender. Also, there is no regulatory framework about cryptocurrencies, and public warned potential disadvantages of cryptocurrencies.[15] According to data of local lawyers the profit from trade in cryptocurrency is not taxed, because "The Cyprus Tax on profits from trading is shares or other securities, including forex or bitcoins is 0%".[16]

France In the report that published by Banque de France in 2016, cryptocurrencies define as an unregulated virtual currency with no guarantee of reimbursement. Virtual currencies do not have legal status or regulations. In addition, cryptocurrencies is criticized for helping criminal activities.[17] Another report that published by Banque de France in 2018 state that the cryptocurrencies are not accepted as currency. Therefore there is no guarantee of security, convertibility or value. However, Banque de France advises the regulatory framework for dealing with disadvanteges of cryptocurrencies.[18] One-off profits made on cryptocurrencies are regarded as capital gain and taxable. Profits from cryptocurrency speculation and mining are subject to the progressive income tax Schedule. For companies, profits from cryptocurrencies are liable to tax under the general corporation tax regime. There is no specific VAT law of cryptocurrencies and no transfer taxes are payable in France on cryptocurrencies. Moreover cryptocurrencies portfolios are not taxable assets under the new French real estate wealth tax.[19]

Czech Republic On July 31, 2017, Mojmír Hampl, the Vice-Governor of the Czech National Bank state that *there is no reason for banks to fear cryptocurrencies.*[20] On February 27, 2018, Mojmír Hampl state that as Czech National Bank, they do not want to ban cryptocurrencies and they are not hindering their development, but they are also not actively helping or promoting them and they are not protecting them or

[14]https://www.canada.ca/en/revenue-agency/programs/about-canada-revenue-agency-cra/compliance/digital-currency.html

[15]https://www.centralbank.cy/en//announcements/07022014

[16]https://lawstrust.com/en/ico/pravovoj-status-kriptovalyut

[17]https://publications.banque-france.fr/sites/default/files/medias/documents/focus-10_2013-12-05_en.pdf

[18]https://publications.banque-france.fr/sites/default/files/medias/documents/focus-16_2018_03_05_en.pdf

[19]https://www.osborneclarke.com/insights/taxation-of-cryptocurrencies-in-europe/

[20]https://www.cnb.cz/en/public/media_service/interviews/media_2017/cl_17_170731_hampl_omfif.html

the customers that use them.[21] The income tax on the sale of goods and services for cryptocurrencies in the Czech Republic is governed by the same rules as when paying with conventional money.[22]

Austria The Financial Market Authority of Austria has not issued any regulations concerning Bitcoin and other virtual currencies.[23] For individuals who make speculative transactions with cryptocurrencies are taxed up to 1 year (with a tax-exempt amount of €440 per annum applying), but their income is not taxed after 1 year period.[24]

Chile The monetary authority (Central Bank of Chile) does not recognize cryptocurrencies as legal tender and cryptocurrency transactions are not subject to the regulation.[25] According to Internal Revenue Service in Chile, the cryptocurrency earnings are going to be taxed in 2019.[26]

UK Cryptocurrencies are not classified as a legal tender in UK and has no specific cryptocurrency laws.[27] If individuals in UK are holding cryptocurrencies for investment, this is considered as an asset, and the gains to be derived is subject to capital gains taxation. Individuals who trading in cryptocurrencies are taxed as income on their profits. In terms of corporations, the gains or losses on cryptocurrencies are subject to tax as income. The UK Tax Authority has published guidance on the provisional VAT treatment of cryptocurrencies. Lastly, there is no transfer taxes are payable in UK.[28]

Argentina Bitcoins are not legal currency in Argentina, since they are not issued by the government. As said by some experts a bitcoin may be considered a good or a thing under the Civil Code, and transactions with bitcoins may be governed by the Civil Code. According to the Income Tax Law, the profit derived from the sale of digital currency will be considered income and will be taxed.[29]

Spain The cryptocurrency is not authorized by any regulatory in Spain. However, the government aims to make some arrangements for the crypto currency, which will include possible tax cuts for companies in the block chain technology sector. The profits of cryptocurrencies are taxable under the Law on Income Tax of Individuals.

[21]https://www.cnb.cz/en/public/media_service/conferences/speeches/hampl_20180227_bbva.html

[22]https://sb-sb.cz/en/news/pravovoe-regulirovanie-kriptovalyut-v-chehii

[23]https://www.newsbtc.com/2016/11/15/fma-in-austria-issues-a-warning-against-fradulent-virtual-currency-schemes/

[24]http://publications.ruchelaw.com/news/2017-12/guidance-taxation-bitcoin-cryptocurrency.pdf

[25]https://cointelegraph.com/news/chilean-government-making-progress-on-crypto-regulation-says-finance-minister

[26]https://www.ccn.com/chilean-citizens-will-begin-paying-cryptocurrency-taxes-in-2019/

[27]https://complyadvantage.com/knowledgebase/crypto-regulations/cryptocurrency-regulations-uk-united-kingdom/

[28]https://www.osborneclarke.com/insights/taxation-of-cryptocurrencies-in-europe/

[29]https://www.loc.gov/law/help/cryptocurrency/argentina.php

However, the General Directorate on Taxation has recognized that transactions with bitcoins are exempt from value added tax.[30]

Croatia Cryptocurrencies are not regulated by any law in Croatia. Croatian Tax Administration has issued few opinions about tax treatment of crypto currency with reference to the judgment of ECJ (C-264/14, on 22 October 2015). Cryptocurrencies trading in Croatia is considered a financial transaction and the income generated by the sale of crypto currencies is subject to corporate income tax.[31]

Hungary Hungarian Central Bank does not recognize cryptocurrencies as legal tender and there is no any regulations on it. Additionally the Central Bank highlight that the cryptocurrency has many risks.[32]

Turkey The monetary authority in Turkey does not recognize cryptocurrencies as legal tender and cryptocurrency transactions are not subject to the regulation. Gains to be derived from cryptocurrencies are not subject to taxation as any income categories for individuals in Turkey. However, if cryptocurrencies are defined as a revenue category in Turkey then income tax will be born. In this case, the payment of VAT may arise. In terms of corporations, the gains to be derived from cryptocurrencies are subject to corporate income tax.[33]

It is clear that there is a few countries focus on taxation of cryptocurrencies or regulations and there is no consensus among countries about legal status and taxation process of cryptocurrencies. Important point here, each countries has separate legal arrangements and governments have different perspectives about taxation or regulation about cryptocurrencies.

20.5 Discussions and Conclusion

Cryptocurrency is a digital or virtual currency that uses cryptography for security. It has specific characteristics include immutability, irreversibility, decentralization, persistence, and anonymity. Nowadays, cryptocurrency transactions are remarkable research area and there is significant amount of studies that investigate cryptocurrencies in the existing literature. This chapter examines whether the government should tax cryptocurrency by using game theoretical framework. According to model, both government and cryptocurrency investors will determine the strategies to maximize their own benefits. Two-person zero sum game matrix is created to achieve the Nash equilibrium. As a result, Nash Equilibrium occurs in only the one

[30]https://cointobuy.io/countries/spain

[31]https://www.irglobal.com/article/cripto-currency-taxation-in-croatia

[32]https://bitcoinist.com/hungary-not-consider-cryptocurrency-legal-tender/

[33]http://www.spk.gov.tr/siteapps/yayin/yayingoster/1130

case that indicates cryptocurrency should be taxed with low tax rate and high penalty rate.

Moreover, to understand tax policies of cryptocurrency, this chapter also analyze various countries taxation policy on cryptocurrency. The analyses show that there is no consensus among countries about legal status of cryptocurrencies. In addition to that there is no consensus about both regulation and taxation policy of cryptocurrencies as well. While some countries do not have regulatory framework, some countries have entirely or partially regulatory framework. Countries have different taxation systems and they treated cryptocurrency as a different manner as barter, property, asset etc. In this context, it would not be realistic to propose a single model that could be applicable to each country. Thus, it is reasonable to published practical guidance about cryptocurrency by each countries for taxpayers.

References

Akins, B. W., Chapman, J. L., & Gordon, J. M. (2014). A whole new world: Income tax considerations of the cryptocurrency economy. *Pittsburgh Tax Review, 12*, 25–56.

Angelis, J., & da Silva, E. R. (2018). Blockchain adoption: A value driver perspective. *Business Horizons*. https://doi.org/10.1016/j.bushor.2018.12.001

Bal, A. (2015). How to tax cryptocurrency? In D. L. K. Chen (Ed.), *Handbook of digital currency*. San Diego, CA: Academic Press, Elsevier.

Boehm, F., & Pesch, P. (2014). Bitcoin: A first legal analysis. In *International conference on financial cryptography and data security* (pp. 43–54). Springer, Berlin.

Bryans, D. (2014). Cryptocurrency and money laundering: Mining for an effective solution. *Indiana Law Journal, 89*(1), 441–472.

Casino, F., Dasaklis, T. K., & Patsakis, C. (2018). A systematic literature review of blockchain-based applications: Current status, classification and open issues. *Telematics and Informatics, 36*, 55–81.

Choo, K. K. R. (2015). Cryptocurrency and virtual currency: Corruption and money laundering/terrorism financing risks? In *Handbook of digital currency* (pp. 283–307). Elsevier, Amsterdam.

Fry, J., & Cheah, E. T. (2016). Negative bubbles and shocks in cryptocurrency markets. *International Review of Financial Analysis, 47*, 343–352.

Fudenberg, D., & Tirole, J. (1991). *Game theory*. Cambridge: MIT Press.

Harwick, C. (2016). Cryptocurrency and the problem of intermediation. *The Independent Review, 20*(4), 569–588.

Hawlitschek, F., Notheisen, B., & Teubner, T. (2018). The limits of trust-free systems: A literature review on blockchain technology and trust in the sharing economy. *Electronic Commerce Research and Applications, 29*, 50–63.

Hayes, A. S. (2017). Cryptocurrency value formation: An empirical study leading to a cost of production model for valuing Bitcoin. *Telematics and Informatics, 34*(7), 1308–1321.

Holub, M., & Johnson, J. (2018). Bitcoin research across disciplines. *The Information Society, 34*(2), 114–126.

Kaplanov, N. M. (2012). Nerdy money: Cryptocurrency, the private digital currency, and the case against its regulation. *Loyola Consumer Law Review, 25*(1), 111–174.

Lambert, E. E. (2015). The internal revenue service and Bitcoin: A taxing relationship. *Virginia Tax Review, 35*, 88.

Laurence, T. (2017). *Blockchain for dummies*. Hoboken, NJ: Wiley.

Li, X., & Wang, C. A. (2017). The technology and economic determinants of cryptocurrency exchange rates: The case of Bitcoin. *Decision Support Systems, 95*, 49–60.
Litwack, S. (2015). Bitcoin: Currency or fool's gold: A comparative analysis of the legal classification of Bitcoin. *Temple International and Comparative Law Journal, 29*, 309.
Marian, O. Y. (2013). Are cryptocurrencies 'super' tax havens? *Michigan Law Review First Impressions, 112*, 38–48.
McLeod, P. (2013). Taxing and regulating cryptocurrency: The government's game of catch up. *CommLaw Conspectus, 22*, 379–406.
Nakamoto, S. (2008). Bitcoin: A peer-to-peer electronic cash system. *Consulted*. https://doi.org/10.1007/s10838-008-9062-0
Puthal, D., Malik, N., Mohanty, S. P., Kougianos, E., & Das, G. (2018). Everything you wanted to know about the blockchain: Its promise, components, processes, and problems. *IEEE Consumer Electronics Magazine, 7*(4), 6–14.
Ram, A. J. (2018). Taxation of the Bitcoin: Initial insights through a correspondence analysis. *Meditari Accountancy Research, 26*(2), 214–240.
Reid, F., & Harrigan, M. (2013). An analysis of anonymity in the bitcoin system. *Security and Privacy in Social Networks*, 197–223. https://doi.org/10.1007/978-1-4614-4139-7_10
Scott, B. (2016). *How can cryptocurrency and blockchain technology play a role in building social and solidarity finance? (No. 2016-1)*. UNRISD Working Paper.
Shcherbak, S. (2014). How should Bitcoin be regulated. *European Journal of Legal Studies, 7*, 41.
Simser, J. (2015). Bitcoin and modern alchemy: In code we trust. *Journal of Financial Crime, 22* (2), 156–169.
Slattery, T. (2014). Taking a bit out of crime: Cryptocurrency and cross border tax evasion. *Brooklyn Journal of International Law, 39*(2), 829–873.
Small, S. (2015). Bitcoin: The Napster or currency. *Houston Journal of International Law, 37*, 581.
Swan, M. (2015). *Blockchain blueprint for a new economy*. Sebastopol, CA: O'Reilly Media.
Tapscott, D., & Tapscott, A. (2016). *Blockchain revolution: How the technology behind bitcoin is changing money, business, and the world*. London: Penguin Books.
The Law Library of Congress, Global Legal Research Centre. (2018). *Regulation of cryptocurrency around the world*. Research Report. https://www.loc.gov/law/help/cryptocurrency/world-survey.php
Tsukerman, M. (2015). The block is hot: A survey of the state of cryptocurrency regulation and suggestions for the future. *Berkeley Technology Law Journal, 30*(4), 1127–1169.
Wiseman, S. A. (2016). Property or currency: The tax dilemma behind Bitcoin. *Utah Law Review*, (2), 417–440.
Yalama, A., & Çelikkaya, A. (2007). Türkiye Açısından Optimum Vergi Oranı, Vergi Cezası ve Denetim İlişkisinin Oyun Teorisi Modeli ileBelirlenmesi. *ESOGÜ Sosyal Bilimler Enstitüsü Dergisi, 8*(1), 131–145.
Yli-Huumo, J., Ko, D., Choi, S., Park, S., & Smolander, K. (2016). Where is current research on blockchain technology?—A systematic review. *PLoS One, 11*(10), e0163477.
Zhao, J. L., Fan, S., & Yan, J. (2016). Overview of business innovations and research opportunities in blockchain and introduction to the special issue. *Financial Innovation, 2*, 28. https://doi.org/10.1186/s40854-016-0049-2

Gamze Oz Yalaman is an Assistant Professor of Public Finance at Eskisehir Osmangazi University Department of Public Finance, Eskisehir-Turkey. Dr. Gamze Oz Yalaman has a BS in Public Finance from Eskisehir Osmangazi University (2008), an MBA from Eskisehir Osmangazi University (2011) and a Ph.D. in Public Finance from Eskisehir Osmangazi University (2015) on the informal economy. Her research interests lie in the informal economy, tax morale, and tax evasion. She has taught public finance, tax policy, research methods, and informal economy, among others, at both graduate and undergraduate levels. She has many international publications and presented many papers at international congresses.

Hakan Yildirim is a lecturer at Eskişehir Osmangazi University, Turkey. He undertook undergraduate studies in the field of Computer Education and Instructional Technology (CEIT) between the years of 2003–2007 at Anadolu University, Turkey. He holds a Master of Education in Distance Education from Anadolu University. Also, he is currently a doctoral student in the Department of Distance Education at Anadolu University. His research interests include blockchain technology, human-computer interaction, mobile learning and learning management systems.

Chapter 21
Using Smart Contracts via Blockchain Technology for Effective Cost Management in Health Services

Nihal Kalayci Oflaz

Abstract Societies evolved towards a knowledge-based construct with technology by providing a change in the methods of traditional production and service presentation. In today's information societies, digital technologies enable peers to transfer data and knowledge without needing any central authority or intermediary. Morever, blockchain technology, which is becoming more and more widespread, has begun to be renowned as Bitcoin and crypto currencies enter the markets, and practiced as a business strategy in different sectors with the technology it relies on. There are findings that Blockchain Technology will eliminate inefficiency in many sectors and will contribute to the reduction of costs by conducting transactions among parties without a central authority through smart contracts. Recently, the use of this technology in the health sector has been keeping the agenda occupied and sector specific solutions have been becoming prevalent.

In this study, the health services where E-government applications, Tele-Medicine and Artificial Intelligence are reviewed and the effects of the sharing of the data about patients and diseases among health sector parties with the Blockchain Technology through smart contracts have been investigated. The theoretical framework of blockchain technology has been investigated within the existing framework and the applications of countries such as Estonia, Sweden, and the U.S.A, who use Blockchain Technology in the health sector have been analyzed and their effects on the costs of health services were evaluated.

21.1 Introduction

Technology is rapidly transforming everything that has ever existed in every aspect of our lives. The transformation of the Industry 4.0 breeze in the field of economics is manifested as the differentiation of traditional forms of production and consumption.

N. K. Oflaz (✉)
International Trade and Finance, Medipol Business School, Istanbul Medipol University, Istanbul, Turkey
e-mail: nkalayci@medipol.edu.tr

© Springer Nature Switzerland AG 2019

U. Hacioglu (ed.), *Blockchain Economics and Financial Market Innovation*, Contributions to Economics, https://doi.org/10.1007/978-3-030-25275-5_21

This transformation also changes our way of doing business in parallel with production. In order to develop processes in the business world and to evolve into a technology-centered structure, we are experiencing a new era of new technologies and further new business technologies based on algorithms. One of the technological structures which we often hear about is blockchain technology.

Blockchain technology is a digital structure that records data and transactions, providing security, transparency and decentralization features for these processes. This structure was revealed by Satoshi Nakamoto (2008) in his article, *Bitcoin: A Peer-to-Peer Electronic Cash System*, which states the need for an electronic payment system based on cryptographic evidence that does not require the presence of a third party. Satoshi Nakamoto not only attracted attention with Bitcoin, but he also aroused curiosity amongst those for whom his article was written. However, the most important point emphasized in his article is that a payment system in which trust is ensured with cryptographic techniques has been designed. In the background of this payment system, there is Blockchain Technology. It is defined as a distributed digital ledger that is protected by cryptographic techniques, is invariable, consensus-based, verifiable, and classified (Bell, Buchanan, Cameron, & Lo, 2018). With this digital ledger, data is collected on an electronic network and stored in blocks. The data recording format and preservation of the Blockchain technology is mainly based on cryptographic techniques, algorithms and mathematical formulations. This technology is also seen as a protocol that allows transfering money, value and information safely over an internet-based network without a third party (Swan, 2015). As can be understood from the definitions, blockchain technology can actually be described as software based on encryption, an online network or a database. With this technology, as well as money and financial assets, intellectual property rights and personal information can be stored, value transfer can be done. However, the recorded information can be stored safely. Public authorities also plan and attempt to take advantage of this technology for many other applications, such as legislative functions, taxes and financial regulations, conducting the elections through the blockchain, storing credentials on this digital platform and sharing between public institutions, and structuring smart cities. As reflected on the press recently, the statements made by the leaders of Japan and Russia on the recognition of crypto investments indicate that the digital race of nations will gain momentum. One of the countries that plans to participate in this race is China. China has announced that it will benefit from blockchain technology in public areas like taxation (Johnson, 2018: 4).

It is expected that many sectors such as logistics, transportation, trade, finance, education, and health will have a transformation with the use of blockchain technology. In particular, there can be many potential usage areas of blockchain in the health sector such as medical records, disease management, invoicing and reimbursement of health services, financing of health services, drug and medical supplies, supply chain, scientific research, etc. It is possible to achieve economic gains in these areas with smart contracts based on the blockchain. The smart contract is a digital treaty in which the rules of a transaction are determined and all parties forming the mechanism automatically adhere to these rules, and the level of confidence is high. In this study, the use of smart contracts in health services is investigated. The main aim of the study is to determine whether an economic added value can be achieved through the use of blockchain technology-based smart contracts in health services. For this reason, leading studies on the use of blockchain technology in health services and the current literature have been investigated and studies on the current application examples of the sector have been made. This paper is organized into five main sections. The first section includes the introduction. In the second section, current literature on blockchain technology and smart contracts based on blockchain technology are investigated. In the third section, the methodology of the study is presented. The fourth section contains the findings and in the final section, the findings are discussed within the framework of theoretical and practical applications and suggestions are made by taking the expectations for the future into consideration.

21.2 Literature Review of Blockchain Technology

Although not having a standard definition, blockchain technology is generally referred to in the literature as "Distributed Ledger Technology" (Berg, Davidson, & Potts, 2018; Catalini & Gans, 2018; Davidson, De Filippi, & Potts, 2016; Johnson, 2018). We have started to hear, in the financial markets with Bitcoin and other crypto currencies, that this technology is seen as a technological solution to the problem of double payment for a decentralized peer-to-peer cash payment system database for digital currencies (Christidis & Devetsikiotis, 2016; Davidson et al., 2016; Nakamoto, 2008) and it is described as a revolution that will transform all sectors (Bell et al., 2018; OECD, 2018; Roubini & Byrne, 2018; Swan, 2015).

The most important innovation provided by blockchain technology is that the registration of public transactions can be established with integrity without a central authority (Mainelli & Smith, 2015: 10). In the literature, this potential has been discussed by various authors and has been evaluated under the scope of "new corporate economy" as it allows new types of contracts and organizations (Davidson et al., 2016). In another approach that deals with the subject with the perspective of corporate economics, this new technology can be addressed to the context of social welfare (Berg et al., 2018).

The use of the blockchain system in various fields such as management, legislation, and public investments is thought to contribute to efficiency amongst governments. Moreover, as in all management processes for the achievement of the system, government endorsement and cooperation are seen as one of the primary components (Berg et al., 2018). In the private sector, the expectation of a sectoral transformation will be provided through the implementation of blockchain technology, which has been implemented in different areas, and has been discussed in the literature within the framework of Schumpeter's creative destruction approach (Davidson et al., 2016). Roubini and Byrne (2018) acknowledged the potential for creative destruction by blockchain technology in many industries, and they criticized the system by stating that recording data onto a distributed database could cause problems such as scale and energy usage, and it is not possible to eliminate intermediaries with such a system. In the studies carried out, it usually constitutes the common point that blockchain technology and smart contracts reduce transaction costs (Lamberti, Gatteschi, Demartini, Pranteda, & Santamaria, 2017; Schatsky & Muraskin, 2015). However, this is the case in the opposite direction. Initially, the cost of using technology is high due to the cost of verification and high network costs. Provision of software, exchange platforms, and reputation systems increases costs during installation (Catalini & Gans, 2018). As pointed out in Roubini and Byrne (2018), a system of scale to serve everyone on the network will be required to be configured and the energy costs necessary to ensure a healthy operation of this system will have to be met.

Blockchain technology, along with the supply chain of products, intellectual property rights, finance and commerce, software and the internet, government applications, taxation, real estate, and health sector is considered to be especially useful for medical records, prescription drugs monitoring and clinical research (Alexis & Frapper, 2017; Lamberti et al., 2017). In addition to this, with the blockchain technology, digital signatures, digital contracts, automobiles and houses, stocks, bonds, and digital currencies are considered within the scope of digital rights that can be transferred (Sundararajan, 2016). Considering the use of blockchain technology in health services, some studies suggest that engineering solutions including the functioning of the model in the sector (Peterson, Deeduvanu, Kanjamala, & Boles, 2016), some of them will contribute to the development of new business models with potential benefits and that health services are one of the most important application areas of blockchains (Dhillon, 2016; Kuo, Kim, & Ohno-Machado, 2017). Some of them think that they will have difficulties in adapting blockchain technology to the health system and therefore they tend to be more cautious (Lamberti et al., 2017; Till, Peters, Afshar, & Meara, 2017).

In this study, in the framework of literature on blockchain technology and blockchain based smart contracts, what blockchain technology is, how it works, which areas it been used in, application examples in different countries and usage areas in the health sector have been investigated.

21.2.1 Methodology

It is thought that blockchain technology will affect the functioning of institutions in business processes and even in state mechanisms. According to the literature review related to blockchain technology and smart contracts, which can also be used in the health sector, an attempt was made to explain the basic concepts and working mechanism of this technology by taking account of the figures. In this study, the use of blockchain technology in the health sector and possible areas of usage in health services have been classified as scientific research, data sharing, financing, monitoring of drugs and medical devices. While discussing the contribution of smart contracts to the sector, compliance with this classification has also been tried maintained.

In this study, cost analysis studies have been reviewed because the gains that can be achieved with the use of smart contracts based on blockchain technology in health services were important, but no quantitative data or economic analysis was found in what indicates the dimension of cost-effectiveness. There is not enough data in the scope of this subject, and it is aimed to determine the areas where cost-effectiveness can be achieved by using blockchain technology in the health sector. Therefore having examined the literature for studies conducted previously, an analysis was performed by utilizing data and reports from various institutions. While analyzing these data, a descriptive analysis method was used and also interpretations and summaries of these data references from related publications were included. Thus, besides the benefits that can be obtained from health services through blockchain technology and smart contracts, there is an attempt to put forward the possible effects on the costs of health services.

21.2.2 Empirical Data

In this section, the findings of the study in line with the conceptual framework of blockchain technology and the examples in the health sector are explained.

21.2.3 The Concept of Blockchain Technology

In the white paper "*Blockchain Beyond the Hype a Practical Framework for Business Leaders*" which the World Economic Forum published in April 2018, blockchain technology is defined as a technological structure that enables individuals to change their currencies and other assets without the need for a third party to carry out their transactions. According to Greenspan (2015), blockchain technology, which was firstly found being used in financial markets, is an innovation based on both economics and computer science, and is unthinkable apart from crypto currencies. Blockchain technology has demostrated how to conduct peer-to-peer transactions, especially with Bitcoin, without a third party (Sundararajan, 2016).

Blockchain technology has the characteristics such as that it relies on cryptography, that the data created in the system cannot be easily changed, and that it eliminates the intermediaries. In addition to these features, it has features such as data integrity, transparent and reliable execution of transactions, and distribution of responsibility, in other words decentralized management (Gainfy White Paper 2018: 13). According to Wright and De Filippi (2015), blockchain technology is expressed as follows:

> Blockchain technology enables the creation of decentralized currencies, self-executing digital contracts (smart contracts) and intelligent assets that can be controlled over the Internet (smart property). The blockchain also enables the development of new governance systems with more democratic or participatory decision-making, and decentralized (autonomous) organizations that can operate over a network of computers without any human intervention.

For this reason, blockchain technology can be used not only for routine transactions but also as a registry and inventory system in the fields of financial, economic or money related matters, related to physical or intellectual property rights, intangible assets, in public transactions such as voting and taxation and for recording, monitoring, and transferring data (Swan, 2015: 10).

There is a decentralization approach in blockchain technology, so there is no centralized authority. This technological infrastructure also serves to provide a social benefit such as the elimination of information asymmetry by the participation of all parties of cooperation without a central institution. Having the same information will help actors to make more accurate decisions by increasing confidence in the transactions (Wu, 2018).

Blockchain technology is expected to contribute to cost savings via the shared ledger. The reason for is that it eliminates intermediaries and therefore works without a third party in transactions made (Halaburda, 2018: 2). It is expected to save costs from the system, but the cost-effectiveness of such a platform, which contains significant data in production environments, has not yet been empirically proven (Angraal, Krumholz, & Schulz, 2017).

According to the report *Banking On Blockchain: A Value Analysis For Investment Banks* (2017); with blockchain technology a potential cost savings of 50% could be acheived. In addition, annual cost savings are estimated to be around $10 billion. Blockchain, as well as contributing to the reduction of transaction costs by eliminating intermediaries, also reduces the risk due to the fact that the trust is process-based in the transactions performed (Avital, Beck, King, Rossi, & Teigland, 2016: 4). Because transactions are being carried out step by step, and it is necessary to overcome a complex approval process to make transactions. However, it is also possible to recover lost incomes and create new revenues with blockchain technology (Carson, Romanelli, Walsh, & Zhumaev, 2018).

According to the findings, it can be said that the biggest expectation from this technology is to provide cost savings in economic terms. In addition, it is expected that the applied technology will create an impact that will transform all sectors. However, it should not be ignored that it is unclear what the possible advantages and disadvantages of the implementation of blockchain technology are.

21.2.4 How Blockchain Works

To understand the logic of how blockchain works, it will be useful to explain how the network works. A blockchain refers to the digital registering of a process that users perform on a network. The transactions made in this system enable people to save their transactions into a digital block by using the passwords they have.

The operation of blockchain technology is based on the following elements: providing peer-to-peer transfer over a distributed database, which is a distributed database between peers, and using computational logic and transparency (De Filippi, 2017). In this context, how data is recorded to the blocks in the recording system of blockchain technology, how to transfer digitally to the parties and how the transactions are carried out over a shared network may be explained.

21.2.5 Digital Transactions

In the blockchain, data are recorded to blocks in sequence. Blocks are like a list of actions. Each block has its own cryptographic document summary, which includes a summary of the previous block (Christidis & Devetsikiotis, 2016: 2293). This document, also called hash, encrypts data belonging to a block and marks it with a digital signature to add the block to the chain (Jones & Jones, 2017: 18).

Hash refers to the summary or fingerprint of the document. The hash value of each block also affects the hash value of the next block (Christidis & Devetsikiotis, 2016: 2293).

When a block is filled, that is, when it fills its data capacity, the records begin to be added to the next block and the blocks are connected together in a chain. The first block is described as a genesis block. From the genesis block, it is necessary to carry out a validation process consisting of several retrospective steps for each block added to the chain until the last block is processed. Validation includes a long reference number and the hash of the previous block. This verification process is also called mining and allows the chain to be processed in chronological order (Mainelli & Gunten, 2014: 10). What enables this chronological order is the non-central timestamp algorithm (Hawlitschek, Notheise, & Teubner, 2018: 51). Figure 21.1 demonstrates the operation of a simple blockchain network.

The proof of work mechanism is complementary to the blockchain. For adding a network to the blockchain, the security of the network is ensured by making it compulsory to solve algorithmic problems with the work proof mechanism (Pazaitisi, De Filippi, & Kostakis, 2017: 109). The operation of the system is

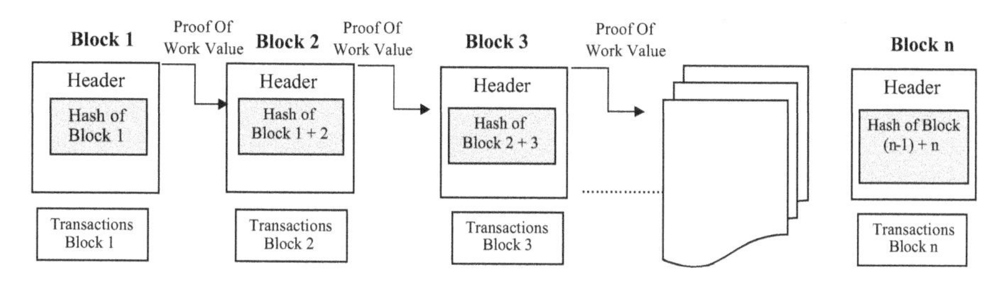

Fig. 21.1 Blockchain network. Adapted from Christidis, K. & Devetsikiotis, M. (2016: 2293)

extremely complex, and this complex structure is also one of the key elements of the safety of the system. In the event of any data breach in the system, users are faced with a long and complicated process that requires considerable processing power.

21.2.6 Distributed Ledger System and Shared Network

The blockchain exhibits a decentralized peer-to-peer (P2P) structure. This distributed logging system allows data to be stored and approves of multiple locations at the same time. Through a common network, the data is shared among all participants and the functioning of the blockchain is carried out on an ecosystem as shown in Fig. 21.2.

It is possible to express this ecosystem in six steps:

It starts with the need to perform an operation.

The transaction is published to the peer-to-peer online network called the node.

Participants in the nodes network confirm the operation and the status of the users using the data contained in the algorithm. The verified transaction may include a monetary transaction, contracts, registration of any data, and much more.

The approved process is combined with other operations to create a new data block in the digital registry.

The new block is added to the current blockchain in a way that is permanent and unchanged.

Thus, the processing is completed (blockgeeks.com 2018).

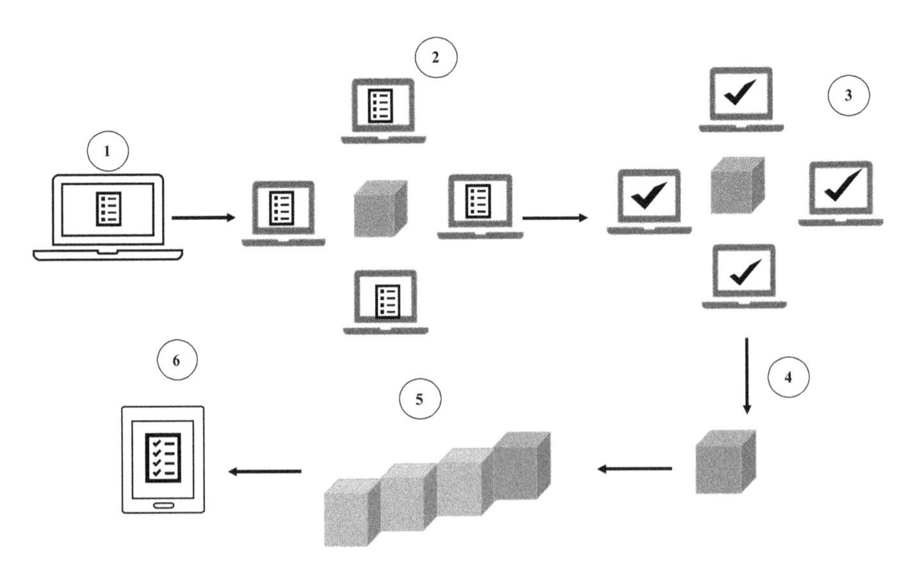

Fig. 21.2 Shared network and distributed ledger system (Source: Adapted from *What is Blockchain Technology? A Step-by-Step Guide For Beginners*, https://blockgeeks.com/guides/what-is-blockchain-technology/ (23 Nov 2018))

Depending on the structure of the system, all members in the distributed network store a digital copy of the processes and changes are reflected in all copies in the case of any changes (Linn & Koo, 2016: 2). However, each blockchain does not function in the same way. The platform's openness (whether open or private) and the level of permission to add information to blocks (allowed or unauthorized) affect the level of access to the system (OECD, 2018). In the open source blockchain, the blocks are open to everyone who is viewing them and a there is network that can be joined by anyone like Bitcoin (Swan, 2017: 10).

In the blockchain that is publicly available, all members see this change because the hash value in all blocks changes when there is any change in the data shared on the network. This is because all copies of the document are spread between users and synchronized automatically (OECD, 2018). In the specially attended blockchain, the form of participation in the system is divided as permitted and without permission. Only the people allowed by the system can add data to blocks or validate blocks with their passwords (OECD, 2018).

It can be said to be that the change made in documents on a shared network is the most important feature in ensuring the transparency of the system, which is ultimately reflected in all sides. The transparency of the system, as well as security, is one of the most important features that stand out. In the blockchain, security is provided by a structure that requires only the process to change the records added to the system, and the cryptographic encrypts that are specific to them, and that requires validation in each backward block for the actions performed.

Blockchain technology is thought to have the potential to secure data in all sectors with this non-centralized structure (Swan, 2017: 7). However, the Global Cyberattack Trends Report (2017) points out that we will witness new methods of attack on the cryptocurrencies and the blockchain. Among these methods cryptocurrency theft, digital identity theft, malicious software is seen. The report also mentions some violations. The first of these violations occured in August 2016. 120,000 Bitcoin units (worth $66 million) were stolen due to a weakness in Bitfinex's account structure. Secondly, in October 2017, came malware named CryptoShuffler, which monitors the victim's machine dashboard and replaced any cryptographic ID with the attacker's address (a theft event worth US$140,000); Third, in November 2017, a cryptocurrency issued by Tether Limited was stolen from the company's treasury wallet (worth $30 million).

Creating additional layers of security through smart contracts can be considered a solution to prevent security breaches in blockchain technology. Smart contracts can mediate the execution of all transactions between the parties in a manner based on the trust relationship.

21.2.7 Smart Contracts

A smart contract is a computer protocol that aims to digitally facilitate, validate or enforce a contract's negotiation or performance (Janin, 2018: 30). The person who defines this concept is Nick Szabo. Szabo defines smart contracts as one of the key building blocks of the free market economy and sees them as a computerized trading protocol to meet the requirements of a contract (Szabo, 1996). Smart contracts are often compared to vending machines for drinks/groceries in exchange for a coin. As when you insert the coin through the coin compartment of the machine and it approves, you are able to carry out your shopping, it contributes to the realization of many business processes within the framework of a contract that is pre-coded in contracts (Szabo, 1996).

The findings obtained from the report on the articulation of the basic features of the smart contract between Bitcoin were published by Capgemini Consulting (2017). The development process and the future of these agreements between 2012 and 2014 have been shown in Table 21.1.

As stated in the table, it is possible to express that the use of smart contracts will be widespread by 2020 and that new products and services provided by smart contracts will be introduced after this year.

Blockchain-based smart contracts represent files that hide commands in the blockchain. The processing order of smart contracts, which is an encoding that automatically executes the terms of the agreement between the parties, and that works directly with the codes, is carried out over a network. Because this agreement and the parties of the contract both agreeing on transactions, the risk of error is reduced. A number of factors come to the forefront such as speed in transactions and

Table 21.1 Development and future of smart contracts

Origin	1996	Nick Szabo the idea of smart contracts
	2009	Nakamato introduced the blockchain
	2012–2014	Smart contract features added to Bitcoin
Experimentation	2014–2015	Smart contract solutions introduced
		Banks and other companies set up labs to develop proof of concept (POCs)
	2015–2017	R3CEV initiative consortium of banks, insurers and IT service providers is formed
		Several POCs succeed, implementation gathers speed
Take-off	2018–2019	Regulations and laws to bring blockchain and smart contracts under the purview of law arrive on scene
		Expected first in production implementation of smart contracts by financial services firms
Mainstream adoption begins	2020	Mainstream adoption of smart contracts begins
		The emergence of new products and services enabled by smart contracts

Source: Based on Capgemini Consulting Report (2017: 16)

updates, the reduction of intermediaries, less manpower and less cost (Deloitte, 2016b: 2). Because the financial institution that performs the transaction for immovable acquisition/transfer, bank transactions, and many other transactions, the buyer or seller will be able to obtain the information that provides the confidence element in line with the information registered in the blockchain and transactions can be carried out without requiring intermediaries and unnecessary procedures in accordance with the process.

The phases of executing any process with a blockchain-based smart contract begin with the need to perform a transaction as stated in Fig. 21.2, which describes the way the block works. The process is published to a peer-to-peer online network called a node. The participants in the network verify the smart contracts by using the data contained in the algorithm and the transactions described in the smart contract are executed according to the order of operation.

The potential uses and contributions of a smart contract can be summarized as:

It can enable individuals to have a digital identity with data, reputation and digital assets and to control them.

It can be the execution of commercial transactions digitally and used for filing.

Digitalizing workflows related to tangible assets and possible help to reduce operational risks.

It can provide trade finance. Also it can increase the efficiency of financing for the buyer and supplier institutions.

It can ensure accurate and transparent recording of financial data. It can contribute to reducing costs by allowing the sharing of costs between institutions.

It can be used for mortgage contracts. Since the parties participate automatically, they can contribute to facilitating payment tracking.

It can reduce the risk of theft and fraud by ensuring easy monitoring of every step in the supply chain.

It can provide positive contributions such as regular processing of data sharing between institutions in healthcare services, increased inter-institutional visibility and clinical trials becoming more efficient, regular processes and increased confidence in patient confidentiality.

Data sharing with smart contracts can be made easier for cancer research (Coleman, 2016).

Considering the prevalence of various areas of use, trade, finance, accounting, banking, health, and many other areas, the implementation of smart contracts is likely to have conflicting aspects with the existing legal structure. Therefore, it will be mandatory to make new arrangements that will regulate digital transactions.

21.2.8 Use of Blockchain Technology in Health Services

According to the World Health Organization (2017), the use of the highest health standard accessible without distinction of race, religion, political belief, economic or social status is regarded as one of the fundamental human rights for everyone. Financial resources are one of the most important barriers to access to health services. For this reason, the United Nations has adopted access to Universal Health Coverage (UHC) in parallel with sustainable development objectives and aims to provide the health care that everyone needs without causing financial problems. In this respect, equality in access to health services for all, quality health care and financial risk protection is one of the main objectives of the UHC (European Public Health Alliance, 2016: 18). According to Till et al. (2017: 1), universal access to UHC and health care is possible with fair access to capital markets. If fair access to the capital market is possible by nation states, municipalities, hospitals and clinics, they will be able to make qualitative improvements such as infrastructure, supply chain and labor force. The technology that can provide this access is seen in blockchain technology. Thus, the strengthening of countries, institutions, and consumers can be ensured by providing new standards for payment and reimbursement with the multilateral financing mechanisms. Health services are a right for individuals. For the states, it represents a large-scale field of activity and therefore is a high-cost area. According to 2014 figures, health spending as the global average of total public expenditure corresponds to a cut of 11.7% (WHO, 2018: 8). The fact that public health expenditures are proportionally high makes cost control more important in the sector. Therefore, the technological structure behind Bitcoin is keeping the agenda of those who manage the health system as much as the actors on the financial market. The areas where blockchain technology can be used in health services are clinical studies, medical device monitoring, drug monitoring, health insurance (Bell et al., 2018), and patient-related health data such as clinical, biometric, nutritional, fitness-related, and psychological profiles (Swan, 2015).

The advantages of using blockchain technology in health services are:

The theft of medical data can be prevented.
By providing the coordination of medical information between health care providers, a reduction in medical errors can be achieved.
It can contribute to increased productivity and reduced costs.
Patients can be included in the control and sharing of their own health data.
Confidentiality and security of health information can be ensured (Miller, 2018).

For these reasons, the usage of the blockchain technology in the health sector and the market size are increasing. Blockchain technology in the health sector is estimated to reach 829.02 million USD in 2023 from the market size of 53.9 million USD in 2018 (MarketsandMarkets, 2018). The most important benefits of blockchain technology in healthcare are the contribution of providing a special structure for the storage and sharing of health data, as well as the creation of a low-cost public data partnership. So individuals, healthcare providers, insurance companies and other parties of the healthcare system can provide access to health records (genomics, lifestyle, medical history, etc.) with their specific passwords (Swan, 2015: 59).

Since blockchain technology in the health and life sciences industries is also an effective way of combating counterfeit drugs industry, it is envisaged to provide the protection of intellectual property rights and for medical and genomic data to be stored safely (Pilkington, 2017). According to Kuo et al. (2017: 1214) benefits derived from adaptation of biomedical and health services to blockchain technology are: (1) decentralized management, (2) invariant audit trail, (3) data source, (4) durability, (5) security/confidentiality. Considering these benefits, it can be said that the added value can be provided for all stakeholders of the health sector with blockchain technology.

The areas that can receive added value by using blockchain technology in health services can be classified as management of medical records, elderly care, and management of chronic diseases, financing, reimbursement and billing, follow-up of medicines and medical devices, and scientific research.

In terms of the management of medical records, ensuring proper management of health-related data also gives direction to the quality of healthcare and the effective implementation of health care policies in the country. According to Hillestad et al. (2005: 1103), the use of electronic medical records can save around 142–371 billion dollars. Savings for outpatient and inpatient care may be more than 77 billion USD annually. Similarly, the Mckinsey report (Manyika et al., 2011: 2) estimated that more than 300 million USD could be recovered by the more creative and effective use of health data, saving about 8% of national health spending.

Utilizing blockchain technology to provide more efficient data sharing among stakeholders of health services can contribute to cost-effective management by minimizing human errors in medical records (OECD, 2018). However, building a common health data pool with blockchain technology will also have an impact on cost effectiveness (Swan, 2015: 60).

Deloitte's (2016a: 8) report also emphasizes the cost savings. According to this report the open source technology of blockchain enables the exchange of personal health records and health information as well as eliminating intermediaries, increasing productivity and reducing transaction costs. Thus, the state and various industries are expected to save billions of dollars.

Companies that manage medical records by using blockchain technology include MedRec, MedVault, Fatcom, BitHealth, Deloitte and Accenture (Kuo et al., 2017: 1215). Another company that provides management of medical records is the Gem Health Network. This company leverages Ethereum-based blockchain technology to provide access to the same information about the health services of different parties (Mettler, 2016). The most important constraint in the management of patient data by blockchain is that health data is currently being managed by existing technological systems (Petre, 2017). Therefore, transferring to a different technological structure can initially lead to the loading of high costs. Another disadvantage to utilizing technology in health care blockchain technology also arises with data sharing. Medical information has been as one of the biggest problems due to the privacy it carries. Electronic health records can create problems in the context of joint work for doctors and nurses when they are available from different service providers (Bell et al., 2018: 3). Because health services contain personal information that is specific to the person, it is very important to follow the medical records, to manage and share the medical records, and to pay attention to security and privacy. For example, 80%

of patients in the United States have been affected by violations of health data since 2009. As a result of the stealing of health data of the patients including photographs in aesthetic operations, crimes such as blackmail, medical fraud, and counterfeit tax returns have been committed and there have been institutions exposed to digital attacks by the use of malware (Miller, 2018). Similarly, Anthem, one of the leading companies in the field of health insurance in the United States, has revealed a data breach of the personal and medical records of 78.8 million people. Such violations are also known to cause serious costs. In a study conducted in 2016, the average total cost of data violations in the field of health was found to be around $7 million (Schumacher, 2017: 6).

Using blockchain technology to store and share health data securely can help reduce or even prevent data violations in this industry. When blockchain supported applications in health services are utilized, hackers cannot easily access users' accounts. Because all users have been authenticated by adding protection to the users who take part in Peer-to-Peer transactions, the information has been also verified in the blockchain (Williamson, 2018). Even as Petre (2017) points out, additional security layers can be added to the health data with smart contracts under blockchain technology. Well then, does the system not carry any risk? Of course, even if additional layers of security are added with smart contracts, there are risks that may lead to a breach of security such as in the event that the patient has the only control of their medical records, people forgetting the password to the system, and malicious people taking possession. Blockchain technology in health services can also be used in elderly care and chronic disease management. The medical treatment process in this field is one of the predefined application areas where blockchain technology can create added value. For example, Healthcoin is one of the first blockchain-based platforms to focus on diabetes prevention. Users gain "Healthcoins" as a result of positive developments in biomarkers such as heart rate, weight, and blood glucose. This also serves to reduce the cost of insurance (Rucker, 2018). The system works like an incentive mechanism and monitors the development of people's health. When the blockchain technology is assessed in terms of financing and invoicing of healthcare services, patients, health care providers and those who finance healthcare can exchange data to verify coverage. The transactions and contracts can be stored in this system. Thus, it can increasingly contribute to the reduction of management costs (COCIR, 2017: 8). In conjunction with this, it is also possible to eliminate fraud by leveraging the blockchain system and its applications. If the institutions conducting billing and reimbursement mechanisms accept payment transactions through cryptocurrencies, payments will be tracked at all stages (Pratap, 2017).

Blockchain technology allows monitoring of medicines and medical devices, as well as the use of this system to combat drug fraud. While it is possible to monitor drugs in the whole production and supply chain process with blockchains, data sharing can be done between the producer and the supplier in all processes (Hoy, 2017: 276).

Blockchain technology can also be used in scientific research related to health services. According to Lamberti et al. (2017), blockchain can help people develop their lives in other contexts such as health and science. Thanks to this technology, it can be possible to obtain health data for wider segments. Sharing data on the health of people globally in such a system can help both with understanding diseases and to achieve positive outcomes related to the health system by mediating the development of biomedical discoveries and drugs (Linn & Koo, 2016: 8).

Although some examples are given in the explanations regarding the utilization areas of blockchain technology in health services, Table 21.2 includes some countries that use blockchain technology for health services and some of the leading companies in this field.

As indicated in Table 21.2, the USA, Estonia, and Sweden are among the leading countries that benefit from blockchain technology in health services. For example, the government of Estonia, in collaboration with Guardtime in 2007 started to use the keyless signature infrastructure of blockchain technology and took its place among the countries that store their medical records online (Nichol, 2017). Estonia confirms the identity of patients with its blockchain based system. In this system, all citizens are given a smart card that matches their electronic health records with their blockchain based identities. In case of an update in the health records of individuals, this update is signed cryptographically in a block by giving a timestamp (Angraal et al., 2017: 1). All citizens of Estonia, health care providers, and insurance companies can obtain all information about medical treatments using the Guardtime blockchain (Mettler, 2016).

In Switzerland, it is ensured that medical data collected from different sources are gathered in a safe environment, that institutional cooperation is carried out and that data is shared with these institutions. In addition to this, a digital health platform was established under the name of Healthbank in order to monitor drugs electronically (Mettler, 2016). Healthbank is a cooperative in which health data created and controlled by the public is kept in the system as approved by the users only. The system works with Etherium-based blockchain technology. By using this technology, it is expected that the cluster approach for genomic data will provide common solutions for healthcare providers, the finance sector, and patients in a geographic area (Leither, 2018).

For example, if we refer to the examples in the United States, the Youbase company uses encryption, digital signatures, digital wallets, and distributed data repositories based on peer and blockchain technology for storing and transferring health information. It provides the management of personal and medical data that focuses on the individual (Schumacher, 2017: 16). Patientory argues that the use of blockchain technology in the US healthcare system and the quantification of health data will improve the quality and efficiency of health care, while at the same time, health expenditures will decrease by billions of pounds (Schumacher, 2017: 17).

Another example is MedRec. MedRec is a blockchain based application that allows medical information to be recorded indefinitely and that patients and doctors have easy access to this information (Pilkington, 2017). MedRec enables the sharing of data between patients and service providers through smart contracts, which are organized by utilizing blockchain technology. These smart contracts refer to data ownership and display permissions. With smart contracts, members can share

Table 21.2 Countries and companies that use blockchain in health sector

Company name	Task description	Country
BitGive Foundation	Leveraging the strength of the Bitcoin Community to improve public health and the circle worldwide.	United States
Blockpharma	Blockpharma offers solutions for counterfeit drugs.	France
Blockchain Health	Blockchain Health is focused on the relationship between medical researchers, investigators, and study participants. They use blockchain to track assets and transactions during the study helping researchers maintain integrity within their research protocols and activity.	United States
Bowhead Health	The Bowhead supplement tracking appliance is connected to the internet and can monitor the health of individuals while being a resource used by healthcare professionals to provide timely advice to people in need of health attention and direction. The first medical instrument powered by blockchain.	Singapore
DeepMind Health	Helping clinicians get patients from test to treatment, faster.	UK
EyePi	Crowdsourced medicine	China
Factom	The Factom blockchain is a decentralized publication protocol for building record systems that are immutable and independently verifiable.	United States
Hashed Health	Hashed Health is leading a consortium of healthcare companies using blockchain.	United States
Healthbank	It was established in Switzerland to gather medical data collected from different sources of patients in a safe environment, to share data and to monitor the medicines electronically.	Sweden
Health Wizz	Health Wizz is a mobile application platform for individuals to aggregate, organize and share/donate/trade their medical records on a blockchain.	United States
HealthChain	Securing prescriptions on the blockchain to avoid medical prescription error and prevent fraud	United States
Healthchain LLP	Healthchain is fixing two major failings of healthcare delivery systems: personalization and peer-to-peer healthcare information management.	United Kingdom
Healthcombix	Healthcare blockchain, machine learning, smart contracts, and device integration	United States
Healthcoin	Healthcoin is a blockchain-enabled platform for diabetes prevention. Healthcoin is an incentive system that tracks a person's lifestyle choices by collecting data (like heart rate, weight, sugar level etc.) and pushing it into a database run on the blockchain.	United States
HealthNexus	Health Nexus is a blockchain-based system using data transfer, payments, and storage. It is designed for Health Services. Health Nexus is the transactional cryptocurrency of the blockchain.	Canada
Guardtime	All citizens of Estonia, health care providers, and insurance companies are able to obtain all information about medical treatments using the Guardtime Blockchain.	Estonia

(continued)

Table 21.2 (continued)

Company name	Task description	Country
IBM	Blockchain's greatest potential is enabling the interoperability of electronic medical records and allowing different electronic medical records systems to communicate in real time.	United States
PointNurse	Virtual Healthcare Delivery and Monitoring Service	United States
MedRec	Managing of medical records	United States
MediChain	The company focuses on areas such as drug and medical vehicle safety, disease and treatment heterogeneity, sensitive drug and clinical decision support, quality of care and performance measurements, public health, research applications.	Hong Kong, China
Medicalchain	Medicalchain uses blockchain technology to securely store health records. The different organizations such as doctors, hospitals, laboratories, pharmacists and health insurers can request permission to access a patient's record to serve their purpose and record transactions on the distributed ledger.	United Kingdom
Microsoft	Managing of medical records	United States
Pokitdok	A new platform for eligibility checks, claims processing, scheduling, patient access, and payment optimization	United States
Patientory	Patientory is focused on the existing problem of securing clinical data and personal health information, and/or the process that supports the status.	United States
Proof.Work	Managing of medical records	United Kingdom
Serica	Serica is a blockchain-based financial services network for healthcare.	United States
SimplyVital Health	SimplyVital Health, helps decentralize medical health data by granting access via smart contracts.	United States
YouBase	Individual-centric data exchange to enable a world where individuals have complete control over and are empowered to own their life data.	United States
QEData	Market solutions utilizing blockchain as tech for healthcare, financial services, and public sectors.	Canada

Source: Jones, G. & Jones, C. P. (2017); Bitcoin and Blockchain Industry Data, https://bitcoinmagazine.com/industry/blockchain/ (20 Nov 2018); Mettler, M. (2016); Medichain, https://medichain.online/ (10 Dec 2018); Markets and Markets - Global Forecast to 2022. https://www.marketstandmarkets.com/Market-Reports/blockchain-technology-market-90100890.html (03 Nov 2018)

medical information with service providers over a peer-to-peer network, and service providers can add new records to a patient (Ekblaw et al., 2016).

This means that the smart contracts of blockchain technology can be used to ensure that each patient has control over their personal data. In addition to medical data, the government can benefit from smart contracts for healthcare-related transactions such as between institutions/organizations, pharmaceutical manufacturers and pharmacies, medical laboratories, researchers and others that finance health care.

21.2.9 Smart Contracts and Cost Management in Health Services

By leveraging blockchain technology based smart contracts in the health sector, more efficient, faster and cost-effective outputs can be obtained. From the moment when the demand for health care is born, the smart contract on the individual/patient defined in the blockchain will act to carry out the process steps by means of pre-defined codes. The payment of the transactions by the third party (insurance company) to the health service provider who requests the payment for the applied medical treatment or examinations will be automatically transferred.

How a blockchain-based smart contract operates in a medical process is as follows in Fig. 21.3:

1. The process begins with the individual/patient's request for treatment from a health care provider.
2. Since the smart contract is precoded, the information about the person and the transactions are submitted to the parties of the contract for approval.
3. The transaction is confirmed by the parties. The verified transaction may include a monetary transmission, registration of any data, and more.
4. The approved process is combined with other operations to create a new data block in the digital registry. The new block is linked up to the existent chain.
5. After approval, the insurance company/health care institution shall pay the treatment payment to the institution providing treatment.

Thus, according to the determined rules, the smart contract ensures the completion of the transaction. Pharmacists can be involved in the provision of medication

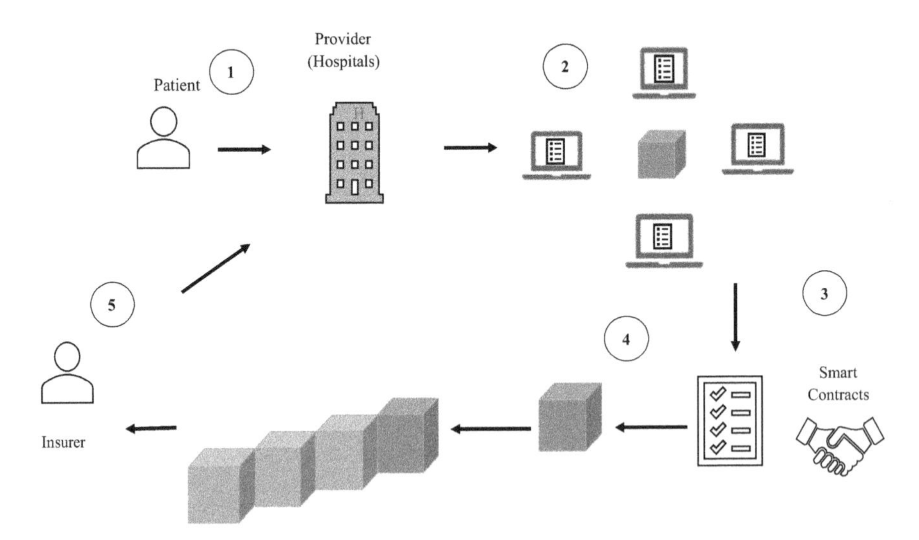

Fig. 21.3 Operation of medical treatment process with blockchain-based smart contracts

by smart contracts, and researchers by adding the clinical findings of the healthcare institution to the blockchain, the state and even the other parties in the health sector can participate in the process. By adding the patient's clinical results to the blockchain by the health service provider, pharmacies, researchers, the government and even other parties in the health sector can participate in the process through smart contracts. The parties of the health services can access data that is accessible to everyone. Because health records are encrypted in the blockchain, only the patient can allow access to information by using a personal key. Information cannot be retrieved or changed without the consent of the patient. The information cannot be taken or changed without the consent of the patient. The decentralized system of blockchain technology ensures the security of personal data in this way. Currently, there are companies that have started to design blockchain based smart contracts for the health systems, for example TIBCO Software. The company believes that it is possible to establish a more effective working system with smart contracts between service providers and those who pay for the costs of these services. With the established smart health network, the "Intelligent Healthcare Network reaches approximately 2100 government and commercial payer connections, 5500 hospitals, 900,000 physicians, and 33,000 pharmacies. The fiscal year ended on March 31, and it facilitated nearly 14 billion healthcare transactions and $2 trillion in annual healthcare expenditures" (Markets Insider, 2018).

According to Jones and Jones (2017: 23), 40% of the economic value spent on health care in existing health systems is wasted. In order to prevent this loss, a common pool can be created with the contributions collected from the members who pay for the health service and the management of the health expenditures can be provided with a blockchain. Thus, it is ensured that the third party intermediaries are eliminated. For this purpose, they recommended Universal Health Coin. Universal Health Coin is defined as a solution that uses blockchain and cryptocurrency technology to create a direct personal and financial relationship between members and healthcare providers. Thus, not only financial procedures, but also the compliance of the patients with the treatment given by the physician can be monitored. Depending on the success of the treatment, additional health coins are added to the patient's blockchain account and it is the aim to promote a healthy life with this mechanism.

In parallel with the United Nations Sustainable Development Goals, the creation of a universal currency and payment system for health care can be provided within the scope of access to Universal Health Coverage. The establishment of a system in which healthy life behaviors will be reflected as added value because the patients in the blockchain can contribute to positive outcomes of wasted economic values for health expenditures around the world. Considering that 70% of diseases today are related to individuals' lifestyles, it will be possible to manage chronic diseases, which is one of the most important problems in the world, and to improve the health literacy of individuals by using blockchain based smart contracts. Thus, additional savings can be obtained in health services.

Smart contracts are implemented independently of people's decisions, reducing both the risk of error and transaction costs (Halaburda, 2018: 5). Thus, cost-effective management can be achieved by utilizing blockchain technology and blockchain based smart contracts in health services. When the transactions amongst all parties of the health services; individuals, doctors, healthcare providers, pharmaceutical manufacturers and pharmacies, radiology and laboratory service providers, researchers are fictionalized on an open source blockchain protocol, as no intermediary is present, transactions can be performed at faster and lower rates.

Smart contracts in health services can be used for data sharing, as well as the financial transactions between health financing institutions and health care providers of medical and treatment services received by a patient, in addition to providing a mechanism for the procurement and financing of medicines prescribed by the physician regarding this treatment process. However, in addition to these basic operations, more complex processes can be achieved through smart contracts. In health services, the sides of the blockchain based smart contracts are shown in Fig. 21.4.

In this context, effective management can be provided for many factors such as reduction of clinical errors and violations, short waiting times related to operations, monitoring of drugs and medical devices, providing quality and accurate resources for medical and genomic research, combating counterfeit drugs, management of chronic diseases, management of laboratory and radiology examinations, organ

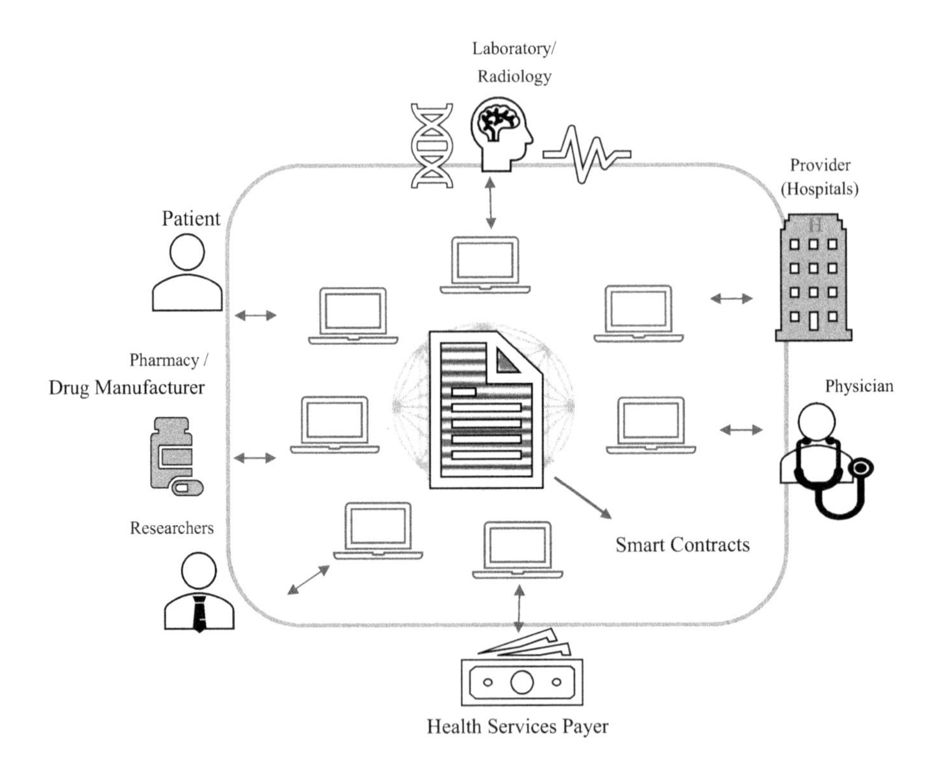

Fig. 21.4 The sides of the blockchain based smart contracts (Source: Based on Schumacher (2017: 31))

transplantation and development of health literacy. Because the information about the patients is stored digitally with the blockchain based smart contracts, it allows physicians to obtain patient information from a single source when the patient receives treatment services from different doctors. In addition, payment services will be realized automatically through smart contracts between health service providers and health care providers (state, insurance company, etc.). Smart contracts enable physicians and pharmacists to coordinate the interaction process between the parties in the pharmaceutical supply chain in health care services associated with the pharmaceutical industry and even detect counterfeit drugs, as well as the use of new drugs, the identification of side effects, and the possible effects that may occur in the case of use of other drugs. Another possible area of use for smart contracts based on blockchain technology is organ transplantation. A system in which donors and receptors can be paired by using this technology can be constructed (applicature.com, 2018). The positive and negative effects of blockchain technology and blockchain based smart contracts on the costs of health services are brought together in Table 21.3.

It is often emphasized in other parts of the study, which can be supported by findings and opinions, that it contributes to the reduction of the costs of health services by the use of blockchain technology. In addition to this, it can contribute to creating added value indirectly for sectoral resources and the economy by the benefits of blockchain technology in the areas mentioned in Table 21.3. But these technologies are fairly new to an area that needs to undergo empirical studies analyzing the economic impacts and costs.

Table 21.3 Effects of blockchain and blockchain based smart contracts on health services costs

Positive effects
The cost advantage of providing a public data partnership
The cost advantage of elimination of intermediaries
The acquisition of drug and medical equipment by the more effective implementation of follow-up
The financial advantages of medical examinations and the correct matching of the owners of these data and the reduction of medical errors
By mediating the development of health awareness, it affects the rational use of health services and decreases health expenditures.
Benefits of counterfeiting and smuggling in medicines and medical devices
Financial gains by shortening waiting times in health institutions
Providing more reliable data for medical research and mediating the production of medical data that will benefit humanity and thus mediating the reduction of health expenditures
It contributes to the reduction of security costs due to being more secure against cyber attacks.
Negative effects
Installation, networking and energy costs
Costs that have to be borne for the existing data to be transferred to electronic systems and integration into the blockchain system
Structuring of legal regulations supporting blockchain technology and smart contracts and costs to be incurred in this context

21.3 Conclusions

While health is a very specific and highly sensitive service, the health sector is a multi-voice field that is carried out with the partnership of many parties from the state and the private sector. Therefore, it has political, economic and social features within it. While the public sector aims to use resources efficiently and to ensure that health services are presented equally to everyone, the goal of the private sector is to turn to preferences where profit maximizing is achieving. The parties we describe as being in the private sector are the pharmaceutical industry, medical device suppliers, the companies producing and selling internet technologies specific to healthcare, as well as all other stakeholders of the sector concerned with health services.

Considering that the resources allocated from global output to public health expenditures are only around 10–11% every year, it is highly important to administrate the resources in a way to ensure cost-effectiveness. In this study, it has been investigated whether cost-effective management can be achieved by taking advantage of blockchain technology and blockchain based smart contracts in health services. Most academics and researchers, as well as institution reports, point to the finding that blockchain technology provides cost savings. However, there is no empirical study on how these savings will be realized. This situation causes the analysis to be insufficient at this stage. When blockchain technology is utilized to provide cost-effective management, depending on the type of service, the factors that create cost and economic gain in the sector should be determined. Table 21.3 was prepared to reveal some of these factors.

To summarize, firstly, the work and transactions carried out in health services with blockchain technology and smart contracts can be realized in a shorter time and with fewer errors and more effective decisions can be made for the sector. Secondly, due to the importance of confidentiality and security of patient data, the data can be protected by blockchain-based smart contracts. Thus, transactions that require trust and relying on information exchange between the parties involved in health services can be carried out without any problems. Thirdly, it can mediate the coordinated work of all parties in the health sector. Finally, it can contribute to the development of additional resources in the sector through the mediation of the improvement of health consciousness. For example, by enabling the management of chronic diseases with blockchain-based smart contracts, the effects can also help provide gains in areas that may arise in the long term.

As a result, the development of blockchain technology, which started with finance, has also spread to other sectors. It is possible to say that blockchain technology, which has started to find its usage area in the health sector, can be used as a new business model for many service areas in the sector. Via blockchain technology, it is possible to provide the sharing of medical data both nationally and globally. In addition, it is possible to take advantage of the management of supply chains, drug and medical supplies, and the financing of health services. With blockchain technology, many benefits are provided such as reduction of medical errors, reduction of processing time, security of data and so on. For the effective management of resources allocated to health, it is necessary to determine the economic value of these benefits.

In this study, it was investigated whether efficient cost management can be achieved via blockchain technology, in particular the effects of blockchain-based smart contracts on the economy were examined. The effects of this technology on the economy are almost negligible due to the lack of statistical data, the new technology being applied, the difficulty of identifying the cost items or the areas which are being saved and the fact that they are not fully known. For these reasons, it has an important potential for future studies.

Another aspect of blockchain technology and blockchain-based smart contracts related to the economy is emerging in relation to the market and employment. For the use of blockchain technology in the health sector, companies or public authorities need to develop a technological infrastructure. For this reason, the market size of this technology will increase with the inclusion of different companies for the establishment, installation, and integration of the infrastructure required for health services. At the same time, it may cause creative destruction in all sectors, and it will have an impact on the disappearance of some business areas as well as on new jobs and employment. As a product of this technology, new jobs such as blockchain developer and data mining are mentioned. In this framework, it is very important to determine the standards of the technological infrastructure to be used by the state and to ensure that the companies meet these standards. In addition to this, it will be necessary to revise education policies as it will require the presence of qualified personnel.

Since the digital infrastructure required by blockchain technology needs to be integrated with existing technologies already used, factors such as transfer of data stored in old systems, installation costs and the adaptation process of people to this technology can cause problems in the beginning. These problems can be prevented if confidence can be built through the determination of standards and taking measures on a legal level. If it can be proved empirically that this technology provides management at less cost and contributes to savings in many business processes, economic incentives can also be considered as a solution in order to encourage the use of blockchain technology. In line with the objective of ensuring equal access to universal health coverage in the context of financing health care, new standards integrated with blockchain technology for health care financing can be developed and a national or global health currency can be utilized. It will also be mediated to provide equal access to health services as it is possible to implement these standards automatically with smart contracts.

Health is a right that all people gain with birth. In addition to being an individual right, it is a special area that has to be presented to everyone equally, creating environmental, social, human and economic effects. Therefore, it is very important to manage the resources allocated for the sector by effective, efficient and cost-effective methods. Blockchain technology can mediate the provision of cost-effective management for health services. However, it should be kept in mind that there are uncertainties and risks for the future.

References

Accenture Consulting. (2017). *Banking on blockchain: A value analysis for investment banks.* Retrieved September 10, 2018, from https://www.accenture.com/t20171108T095421Z__w__/us-en/_acnmedia/Accenture/Conversion-Assets/DotCom/Documents/Global/PDF/Consulting/Accenture-Banking-on-Blockchain.pdf#zoom=50

Alexis, B., & Frapper, I. (2017). *10 things general counsel need to know about blockchain.* ACC Docket. Retrieved October 22, 2018, from https://www.accdocket.com/articles/10-things-general-counsel-needs-know-blockchain.cfm

Angraal, S., Krumholz, H. M., & Schulz, W. L. (2017). Blockchain technology: Applications in health care. *Circulation. Cardiovascular Quality and Outcomes, 10*(9). https://doi.org/10.1161/CIRCOUTCOMES.117.003800

applicature.com. (2018). *Implementation of blockchain healthcare smart contracts, 2018.* Retrieved December 15, 2018, from https://applicature.com/blog/blockchain-healthcare-smart-contracts-2

Avital, M., Beck, R., King, J. L., Rossi, M., & Teigland, R. (2016). Jumping on the blockchain bandwagon: Lessons of the past and outlook to the future. In *Thirty seventh international conference on information systems*, Dublin. Retrieved December 1, 2018, from https://aisel.aisnet.org/cgi/viewcontent.cgi?article=1320&context=icis2016

Bell, L., Buchanan, W. J., Cameron, J., & Lo, O. (2018). Applications of blockchain within healthcare. *Blockchain Healthcare Today, 1.* https://doi.org/10.30953/bhty.v1.8

Berg, C., Davidson, S., & Potts, J. (2018, March 2). *Some public economics of blockchain technology*. doi: https://doi.org/10.2139/ssrn.3132857

Bitcoin and Blockchain Industry Data. (2018, 20 Nov). Retrieved from https://bitcoinmagazine.com/industry/blockchain/

Capgemini Consulting. (2017). *Smart contracts in financial services: Getting from hype to reality.* Retrieved January 5, 2019, from https://www.capgemini.com/consulting-de/wp-content/uploads/sites/32/2017/08/smart_contracts_paper_long_0.pdf

Carson, B., Romanelli, G., Walsh, P., & Zhumaev, A. (2018). *Blockchain beyond the hype: What is the strategic business value?* Retrieved September 20, 2018, from https://www.mckinsey.com/business-functions/digital-mckinsey/our-insights/blockchain-beyond-the-hype-what-is-the-strategic-business-value

Catalini, C., & Gans, J. S. (2018). *Some simple economics of the blockchain*, NBER Working Paper Series 22952. Retrieved September 10, 2018, from https://www.nber.org/papers/w22952.pdf

Christidis, K., & Devetsikiotis, M. (2016). Blockchains and smart contracts for the internet of things. *IEEE Access, 4*, 2292–2303. https://doi.org/10.1109/ACCESS.2016.2566339

COCIR. (2017, December). *Beyond the hype of blockchain in healthcare. Sustainable competence in advancing healthcare.* European Coordination Committee of the Radiological, Electromedical and Healthcare IT Industry. Retrieved December 18, 2018, from https://Www.Cocir.Org/Uploads/Media/17069_Coc_Blockchain_Paper_Web.Pdf

Coleman, L. (2016). *Smart contracts: 12 use cases for business and beyond.* Retrieved December 23, 2018, from https://www.ccn.com/smart-contracts-12-use-cases-for-business-and-beyond/

Davidson, S., & De Filippi, P., & Potts, J. (2016). *Economics of blockchain.* Retrieved from https://ssrn.com/abstract=2744751 or doi:https://doi.org/10.2139/ssrn.2744751

De Filippi, P. (2017). What blockchain means for the sharing economy. *Harvard Business Review Digital Articles.* 2–5. Retrieved September 25, 2018, from https://hbr.org/2017/03/what-blockchain-means-for-the-sharing-economy

Deloitte. (2016a, August). *Blockchain: Opportunities for health care.* Retrieved November 26, 2018, from https://www.healthit.gov/sites/default/files/4-37-hhs_blockchain_challenge_deloitte_consulting_llp.pdf

Deloitte. (2016b, June). *Getting smart about smart contracts.* Retrieved November 26, 2018, from https://www2.deloitte.com/tr/en/pages/finance/articles/cfo-insights-getting-smart-contracts.html

Dhillon, V. (2016, June). Designing decentralized ledger technology for electronic health records. *Telehealth and Medicine Today, 1*(2). https://doi.org/10.30953/tmt.v1.77

Ekblaw, A. et al. (2016, August). *A case study for blockchain in healthcare: "MedRec" prototype for electronic health records and medical research data*. White Paper. Retrieved September 10, 2018, from https://www.healthit.gov/sites/default/files/5-56-onc_blockchainchallenge_mitwhitepaper.pdf

European Public Health Alliance. (2016). *Universal health coverage, sustainable development and the pillar of social rights: Implications and opportunities for the European Public Health Alliance*. Retrieved September 25, 2018, from https://epha.org/wp-content/uploads/2016/09/Universal-Health-Coverage-report_final.pdf

Gainfy White Paper v.6, Bringing trust to healthcare with blockchain AI&IOT. Retrieved September 10, 2018, from. https://theinternetofthings.report/Resources/Whitepapers/005d99fa-4385-4e42-ab7f-3ee10e1af8ec_Gainfy_WhitePaper.pdf

Global Cyberattack Trends Report. (2017). Retrieved November 5, 2018, from https://www.checkpoint.com/downloads/product-related/infographic/H2_2017_Global_Cyber_Attack_Trends_Report.pdf

Greenspan, G. (2015). *Ending the bitcoin vs blockchain debate*. Retrieved November 1, 2018, from Multichain.com, https://www.multichain.com/blog/2015/07/bitcoin-vs-blockchain-debate/

Halaburda, H. (2018). *Blockchain revolution without the blockchain*. Bank of Canada, Staff Analytical Note 2018-5, Bank of Canada, ISSN 2369-9639. Retrieved November 1, 2018, from https://www.bankofcanada.ca/wp-content/uploads/2018/03/san2018-5.pdf

Hawlitschek, F., Notheise, B., & Teubner, T. (2018). The limits of trust-free systems: A literature review on blockchain technology and trust in the sharing economy. *Electronic Commerce Research and Applications, 29*, 50–63. https://doi.org/10.1016/j.elerap.2018.03.005

Hillestad, R., Bigelow, J., Bower, A., Girosi, F., Meili, R., Scoville, R., et al. (2005). Can electronic medical record systems transform health care? Potential health benefits, savings, and costs. *Health Affairs, 24*(5), 1103–1117. https://doi.org/10.1377/hlthaff.24.5.1103

Hoy, M. B. (2017). An introduction to the blockchain and its implications for libraries and medicine. *Medical Reference Services Quarterly, 36*(3), 273–279. https://doi.org/10.1080/02763869.2017.1332261

Janin, S. (2018). Smart contracts in healthcare. *Health Management, 18*(1), 30–32, ISSN:1377-7629.

Johnson, K. D. (2018, January 22). *Blockchain technology implications for development*. Risk Innovation Lab, Arizona State University. Retrieved November 22, 2018, from https://riskinnovation.asu.edu/wp-content/uploads/2018/01/ResearchPaper-Blockchain-KevinDJohnson-Published.pdf

Jones, G., & Jones, C. P. (2017). *The Universal Health Coin White Paper V.2.3*. Retrieved December 29, 2018, from https://uhx.io/wp-content/uploads/UHCWhitePaper-V2.3.pdf

Kuo, T., Kim, H., & Ohno-Machado, L. (2017). Blockchain distributed ledger technologies for biomedical and health care applications. *Journal of the American Medical Informatics Association, 24*(6), 1211–1220. https://doi.org/10.1093/jamia/ocx068

Lamberti, F., Gatteschi, V., Demartini, C., Pranteda, C., & Santamaria, V. (2017). Blockchain or not blockchain, that is the question of the insurance and other sectors. *IT Professional*. https://doi.org/10.1109/MITP.2017.265110355

Leither, C. (2018). *Healthbank creates the first patient-centric healthcare trust ecosystem*. Retrieved December 6, 2018, from https://www.healthbank.coop/2018/10/30/healthbank-creates-the-first-patient-centric-healthcare-trust-ecosystem/

Linn, L. A., & Koo, M. B. (2016). *Blockchain for health data and its potential use in health IT and health care related research*. In ONC/NIST Use of Blockchain for Healthcare and Research Workshop. Gaithersburg, MD: ONC/NIST. Retrieved September 13, 2018, from https://www.healthit.gov/sites/default/files/11-74-ablockchainforhealthcare.pdf

Mainelli, M., & Gunten, C. (2014). *Chain of a lifetime: How blockchain technology might transform personal insurance, a long finance report prepared by Z/Yen Group*. Retrieved

November 12, 2018, from http://archive.longfinance.net/images/Chain_Of_A_Lifetime_December2014.pdf
Mainelli, M., & Smith, M. (2015). Sharing ledgers for sharing economies: An exploration of mutual distributed ledgers (Aka Blockchain Technology). *Journal of Financial Perspectives, 3*(3). Available from https://ssrn.com/abstract=3083963
Manyika, J., Chui, M., Brown, B., Bughin, J., Dobbs, R., & Roxburgh, C., & Byers, A. H. (2011, May). *Big data: The next frontier for innovation, competition, and productivity*. McKinsey & Company Report.
MarketsandMarkets – Global Forecast to 2022, (Blockchain Market by Provider, Application (Payments, Exchanges, Smart Contracts, Documentation, Digital Identity, Supply Chain Management, and GRC Management), Organization Size, Industry Vertical, and Region – Global Forecast to 2022). Retrieved November 3, 2018, from https://www.marketsandmarkets.com/Market-Reports/blockchain-technology-market-90100890.html
MarketsInsider. (2018, November 5). *Change healthcare and TIBCO to bring blockchain-powered smart contracts to healthcare*. Retrieved December 25, 2018, from https://markets.businessinsider.com/news/stocks/change-healthcare-and-tibco-to-bring-blockchain-powered-smart-contracts-to-healthcare-1027692340
Medichain. Retrieved December 10, 2018, from https://medichain.online/
Mettler, M. (2016). Blockchain technology in healthcare: The revolution starts here. In *2016 IEEE 18th international conference on e-health networking, applications and services (Healthcom)*. doi: https://doi.org/10.1109/HealthCom.2016.7749510.
Miller, K. (2018). *Will blockchain technology usher in a healthcare data revolution?* Retrieved September 23, 2018, from https://leapsmag.com/will-blockchain-technology-usher-in-a-healthcare-data-revolution/
Mulligan, C., Scott, J. Z., Warren, S., & Rangaswami, J. P. (2018, April). *Blockchain beyond the hype: a practical framework for business leaders, World Economic Forum, White Paper*. Retrieved September 8, 2018, from http://www3.weforum.org/docs/48423_Whether_Blockchain_WP.pdf
Nakamoto, S. (2008). *Bitcoin: A peer-to-peer electronic cash system [Online]*. Retrieved September 13, 2018, from https://bitcoin.org/bitcoin.pdf
Nichol, P. B. (2017). *Blockchain applications for healthcare* (November 8, 2016). Retrieved January 5, 2019, from https://medium.com/PeterBNichol/top-3-blockchain-applications-in-healthcare-d66bcfacdbd6
OECD. (2018). *The OECD blockchain primer*. Retrieved September 25, 2018, from https://www.oecd.org/finance/OECD-Blockchain-Primer.pdf
Pazaitisi, A., De Filippi, P., & Kostakis, V. (2017). Blockchain and value systems in the sharing economy in the sharing economy: The illustrative case of Backfeed. *Technological Forecasting and Social Change, 125*, 105–115. https://doi.org/10.1016/j.techfore.2017.05.025
Peterson, K., Deeduvanu, R., Kanjamala, P., & Boles, K. (2016). *A blockchain-based approach to health information exchange networks*, Mayo Clinic White Paper [online]. Retrieved September 21, 2018, from https://www.healthit.gov/sites/default/files/12-55-blockchain-based-approach-final.pdf
Petre, A. (2017, March 1). *3 Challenges for blockchain in healthcare*, Anca Petre, Blog. Retrieved December 19, 2018, from http://ancapetre.com/challenges-blockchain-healthcare
Pilkington, M. (2017, August 24). *Can blockchain improve healthcare management? Consumer medical electronics and the IoMT*. Retrieved from doi:https://doi.org/10.2139/ssrn.3025393
Pratap, M. (2017). *Blockchain in healthcare: Opportunities, challenges, and applications (August 8, 2018)*. Accessed December 19, 2018, from https://hackernoon.com/blockchain-in-healthcare-opportunities-challenges-and-applications-d6b286da6e1f
Roubini, N., & Byrne, P. (2018). *The Blockchain Pipe Dream, Project Syndicate*. Retrieved November 11, 2018, from https://www.project-syndicate.org/commentary/blockchain-technology-limited-applications-by-nouriel-roubini-and-preston-byrne-2018-03?barrier=accesspaylog

Rucker, M. (2018, February 14). *Will blockchain technology revolutionize health care?* Retrieved October 3, 2018, from https://www.verywellhealth.com/blockchain-technology-in-health-care-4158528

Schatsky, D., & Muraskin, C. (2015). *Beyond Bitcoin blockchain is coming to disrupt your industry*. Deloitte University Press. Retrieved October 20, 2018, from https://www2.deloitte.com/insights/us/en/focus/signals-for-strategists/trends-blockchain-bitcoin-security-transparency.html

Schumacher, A. (2017). *Blockchain & healthcare. 2017 Strategy guide for the pharmaceutical industry, insurers & healthcare providers*. Retrieved October 15, 2018, from https://www.researchgate.net/publication/317936859

Sundararajan, A. (2016). *The sharing economy: The end of employment and the rise of crowd based capitalism*. Cambridge, MA: The MIT Press.

Swan, M. (2015). *Blockchain blueprint for a new economy* (1st ed.). Sebastopol: O'Reilly Media.

Swan, M. (2017). Anticipating the economic benefits of blockchain. *Technology Innovation Management Review, 7*(10), 6–13. https://doi.org/10.22215/timreview/1109

Szabo, N. (1996). *Smart contracts: building blocks for digital markets*. Retrieved January 6, 2019, from http://www.fon.hum.uva.nl/rob/Courses/InformationInSpeech/CDROM/Literature/LOTwinterschool2006/szabo.best.vwh.net/smart_contracts_2.html

Till, B. M., Peters, A. W., Afshar, S., & Meara, J. G. (2017). From blockchain technology to global health equity: Can cryptocurrencies finance universal health coverage? *BMJ Global Health, 2*, e000570. https://doi.org/10.1136/bmjgh-2017-000570

What is blockchain technology? A step-by-step guide for beginners. Retrieved November 23, 2018, from https://blockgeeks.com/guides/what-is-blockchain-technology/

Williamson, S. (2018). *Blockchain solutions are changing the sharing economy*. Retrieved September 22, 2018, from https://www.nasdaq.com/article/blockchain-solutions-are-changing-the-sharing-economy-cm965635

World Health Organization. (2017, December 10). *Health is a fundamental human right*. Retrieved September 15, 2018, from http://www.who.int/mediacentre/news/statements/fundamental-human-right/en/

World Health Organization. (2018). *World health statistics 2018: Monitoring health for the SDGs, sustainable development goals*. Geneva: World Health Organization.

Wright, A., & De Filippi, P. (2015). *Decentralized blockchain technology and the rise of lex cryptographia*. Available from https://doi.org/10.2139/ssrn.2580664

Wu, K. (2018). *The promise & appeal of blockchain technology: Transforming the sharing economy as we know it*. Retrieved November 5, 2018, from https://www.katherinewu.me/writings/thepromiseofblockchain

Nihal Kalaycı Oflaz has been working as an assistant professor in the International Finance and Trade Department, School of Business, Istanbul Medipol University, Turkey. In January 2018, she earned her doctorate degree (Ph.D.) in Economics Policy, Marmara University, Turkey. Her primary research interests are in the areas of political economy, health economics, health literacy, fiscal policy, global economy, innovation, and technology. More specifically, she has been investigating health technologies that aim to minimize healthcare costs. At the same time, she has been giving various lectures about micro, macroeconomics and innovation economics.

Part V
Related Subjects, Political Agenda for Crypto Markets

Chapter 22
Cryptocurrencies in the Digital Era: The Role of Technological Trust and Its International Effects

Serif Dilek

Abstract Due to the lack of trust as a result of the 2008 global financial crisis, the leading cryptocurrency, Bitcoin, is increasingly thought of as an alternative to the prevailing financial architecture by presenting a technologically more trustworthy alternative. Because of the effect it had due to its number of users, volume, and market size, Bitcoin attracted the attention of the whole world. By using blockchain technology and mining, Bitcoin attempts to replace the services based on trust and the mediation role offered by banks in the traditional finance system.

With its characteristics such as low cost, fast transaction times, and low risks, Bitcoin first emerged in 2008 and since then, due to these characteristics, has triggered an important change and transformation. The fact that Bitcoin is used for speculative investment motives, and that the energy used for its mining is made up of conventional energy sources brought a debate about the sustainability and environmental effects of this system. This article evaluates the evolution of money in the era of digital transformation and the repositioning of cryptocurrencies with a focus on Bitcoin. The study analyzes the global effects of blockchain technology and cryptocurrencies, and the risks, opportunities and environmental effects of mining.

22.1 Introduction

The increasing scope of globalization in the 1980s brought neoliberal policies to many countries. The decreasing role of the state and the application of free-market principles during this era accelerated financial deepening on the international level. In addition to this, due to mutual interdependence, the increase of global trade and the integration of markets caused different countries to experience a series of financial crises. Due to the fact that many countries quickly entered the market economy without establishing their restrictive and regulatory organizations, the crises they experienced led to the regional crises experienced in the 1990s and later the 2008 global financial crisis (Balaam & Dillman, 2015, pp. 261–263). The

S. Dilek (✉)
Kırklareli University, Kırklareli, Turkey

© Springer Nature Switzerland AG 2019

U. Hacioglu (ed.), *Blockchain Economics and Financial Market Innovation*,
Contributions to Economics, https://doi.org/10.1007/978-3-030-25275-5_22

2008 financial crisis, which has been the biggest financial crisis of this century so far, was not only an outcome of the period experienced before it, but also led to the breakdown of pre-accepted ideals. The 2008 financial crisis led to many changes in the global finance system; with the collapse of state stock markets, credit rating agencies lost their reputation, financial institutions and companies came to the brink of bankruptcy and some big banks went bankrupt. However, above all, the perception people had against the finance sector changed and their trust in banks was seriously damaged. Defined as a cryptocurrency, Bitcoin emerged under such circumstances.

While there are many different views about the emergence of cryptocurrencies, one of the most noteworthy of these is that cryptocurrencies emerged due to the lack of trust in central banks and financial institutions after the 2008 financial crisis (Dahan & Casey, 2016; Varoufakis, 2013; Weber, 2016).

Emerging after the collapse of the financial markets, Bitcoin presented an alternative model to the failure of the financial system and the powerful reserve currencies (the U.S. dollar and the euro) (Vondráčková, 2016). In this time period, where the financial crisis had been experienced and people lost trust in the system, Bitcoin was increasingly perceived as a viable alternative, with its technological security and mathematical precision based on blockchain technology. Bitcoin and other cryptocurrencies, whose foundations lie in blockchain technology, do not have a central administrator. Based on an advanced technological infrastructure, the composition, alteration, and inspection of Bitcoin is pursued through principles of cryptography and is controlled by software algorithms. With the emergence of Bitcoin, many other cryptocurrencies, also known as altcoins, entered the market.

As will be explained below, there are differing views about this first cryptocurrency. Some research conducted in this area highlights the creativity of Bitcoin and the positive effects of the transformation it has brought on a global scale (Andreessen, 2014; Grinberg, 2011; Kostakis & Giotitsas, 2014; Woo, Gordon, & Laralov, 2013). In contrast, many others, including the famous economists *Stiglitz* and *Krugman,* compare Bitcoin to a Ponzi scheme and view cryptocurrencies as a threat (Ahmed, 2017; Costelloe, 2018; Krugman, 2013; Trugman, 2014). Other researchers argue that Bitcoin can be used as an appropriate portfolio diversification tool in some stock markets and as a hedge against the dollar (Bouri, Molnár, Azzi, Roubaud, & Hagfors, 2016; Dyhrberg, 2015). However, as can be perceived through the studies in this area, many do not see Bitcoin as a payment system but rather perceive it as a tool for investment (Glaser, Haferkorn, Weber, & Zimmermann, 2014). While this is the case, studies conducted show that when comparing the volatility of Bitcoin and other investment methods, the Bitcoin market is highly speculative and exhibits excessive fluctuations (Baek & Elbeck, 2015; Fry & Cheah, 2016). In this sense, with its unique structure, Bitcoin has been addressed as both a speculative and standard financial asset (Kristoufek, 2015).

One of the most emphatic criticisms against cryptocurrencies in general is that they are often used in illegal activities such as money laundering. Some research claims that Bitcoin is an ideal instrument to conduct financial transactions for illegal activity (Foley, Karlsen, & Putnins, 2018; Yelowitz & Wilson, 2015). It has been highlighted that there is a strong possibility for Bitcoin and other cryptocurrencies to become the next tax havens due to the opportunities they provide for tax evaders (Marian, 2013). There have been a high number of studies conducted on how Bitcoin and cryptocurrencies should not be thought of as alternative currencies but rather should be thought of as speculative investment tools (Baur, Hong, & Lee, 2018; Dorfman, 2017; Glaser et al., 2014; Luther & White, 2014). In addition to these, there are arguments calling for the IMF to take action due to the threat posed to the stability of the global economic system due to Bitcoin's decentralized and agentless structure (Plassaras, 2013).

Moreover, due to Bitcoin's architecture, the first users have formed a "Bitcoin aristocracy" and it has been highlighted that they can become a threat to the whole system due to the number of Bitcoin they own and their mining activities (Kostakis & Giotitsas, 2014, p. 437).

In recent times, Bitcoin has found itself plenty of room on the agenda both in terms of its value and the amount of energy that it uses. Just like gold mining, in order for Bitcoin to be structurally produced, it goes through a process of digital "mining", and it is for this reason that some predict that Bitcoin prices will remain relatively stable (Cheah & Fry, 2015). However, its current level of energy consumption and the fact that this consumption is going to increase brings with it many adverse effects. The fact that approximately 80% of the world's energy consumption is connected to fossil fuels, and that this situation will not change much in the future, causes serious problems for the environment. In such an environment, the amount of energy consumed by Bitcoin is seen as an effect that will accelerate the depletion of fossil reserves, which will be exhausted within a certain period of time. However, research indicates that a large amount of Bitcoin mining is pursued through renewable energy sources. By considering the historical transformation of money, this article examines the leading cryptocurrency, Bitcoin, and the blockchain technology that it relies upon. Analyzing both the risks and opportunities presented by the effects of cryptocurrencies, this article argues that in the medium and long term, cryptocurrencies have a high potential. In a context where the consequences of the financial crisis are still felt, and where the issue of whether states can use blockchain technology in order to control their own crypto currency is being discussed, cryptocurrencies and blockchain technology have the potential to bring immense change to the global currency system.

22.2 The Evolution of Money: The Digitalization and Emergence of Cryptocurrencies

The story of money is as old as the story of civilization. Throughout history, people used various objects as means of payment before using gold, silver and other precious metals to obtain the objects they needed. Many different objects, from animals to cereals, and from seashells to precious metals have been used as a means of payment in the past, and it has been stated in the literature that the history of money comes to the forefront with exchange transactions. The basic philosophy of exchange transactions shows us that on the basis of money, there is an element of trust with the debt-recipient relationship. Thus, rather than simply being a piece of metal, money is the measure of the faith and trust relationship between a lender and a borrower (Ferguson, 2008; Graeber, 2011).

Before going into the arguments about cryptocurrencies, it is important, first of all, to see whether this kind of digital money fits with the academic definition of "money". The debates in the literature argue whether Bitcoin is a commodity, currency, or a financial investment tool. It can be observed that the amount of energy consumed by the Bitcoin mining procedure has also been recently added to these debates. First, in order for Bitcoin to be classified as a currency, it has to carry some basic characteristics. Therefore, among the many functions of money, the following features are always prominent: (1) it acts as a medium of exchange, (2) is used as a unit of account, and (3) functions as a store of value (Mankiw, 2003, p. 76).

In light of this definition, it can be stated that neither the academic literature nor national or international organizations can provide a consensus as to the status of Bitcoin. Some studies claim that cryptocurrencies do not fulfill these roles (Bariviera, Basgall, Hasperué, & Naiouf, 2017; Grym, 2018; Yermack, 2013), while others argue that cryptocurrencies can be thought of as units of currency (Ali, Barrdear, Clews, & Southgate, 2014). Similarly, it can be seen, in the approaches of different countries and institutions, that there is no global consensus on the state of cryptocurrencies. The *Internal Revenue Service (IRS)*—the revenue service of the U.S. federal government—sees Bitcoin as an instrument of change; Finland sees it as a product that can be priced; Germany sees it as a private currency unit; while the European Central Bank and European Banking Authority (EBA) perceive Bitcoin as a digital currency (Kancs, Ciaian, & Miroslava, 2015).

When we look at the historical transformation of money, it can be seen that the coin has sustained its sovereignty, and that since the seventeenth century the use of paper money has increased. The banking system gradually developed within this period, and, with the aim of states being able to control money in the new economic system, "authority banks," better known as central banks, were formed. The circulation of paper money in this period was indexed against the gold reserves in central banks, but this only lasted until the outbreak of World War I. Both during and after the war, the Gold Standard was abandoned by many countries. While there were some efforts to return to gold standardization after the war, these efforts were unsuccessful. In this process, with the development of the banking sector, money gained a registered status beyond the representation determined by central banks. In 1944, with the index of the U.S. dollar to gold, state foreign exchange rates were not indexed to gold but were indexed to the U.S. dollar. As a result, the U.S. dollar became accepted as the "reserve" currency. However, what came to be known as the Bretton Woods System came to an end in 1971 when the U.S. stopped indexing the dollar to gold, and the global gold standardization system came to an end. With this, the U.S. Federal Reserve Bank was able to release as many dollars as it wanted to onto the international market.

On the other hand, technological development, international trade and the expansion of finance has brought new innovations for money. With American banks pioneering the electronic transfer of money with the Electronic Fund Transfer (EFT) system, they paved the way for the digitalization of money with the increasing use of credit cards and Automated Teller Machines (ATMs) (Yüksel, 2015, p. 176). In addition, the Society for Worldwide Interbank Financial Telecommunication (SWIFT) came into use in the 1970s, for fast and trustworthy transactions between banks, and is now used in over 200 countries and in over 10,000 institutions (Swift, 2017). The U.S., which is the dominant force in the global system, is still dominant today in terms of currency. Since 2014, 51.9% of international trade is conducted in U.S. dollars, 30.5% with Euros, and 5.4% with the British Pound Sterling (Swift, 2015). While money transfers and movements of capital around the world can be monitored by the SWIFT system, this system is not open to everyone.

After the millennium, rather than holding cash in hand, people have tended to use bank cards, and with the development of technology money has also entered an age of digitalization/virtualization.

Particularly in developed countries, it is possible to observe a decreasing number of people using cash for their day-to-day transactions and an increasing number using digital currency. In one sense, we are witnessing a period in which paper money is being replaced by digital currency. With the digitalization of money, many different electronic payment methods have also emerged. The volume of electronic payment methods (credit cards, PayPal, e-cash, e-wallets, mobile payments, PcPay, First Virtual, etc.) have increased (Yüksel, 2015, p. 176). Within this framework, money is becoming less physical and more digitalized. The process of digitalization in the transformation of money can be seen as a product of 60 years of technology. With the development of technology, the digital transformation of money brings some difficulties. For instance, the systems behind banking and credit cards are open to hacking and there is an increased likelihood of money being stolen. With the widespread use of electronic money, new problems, such as information security and high transfer costs, have emerged. From this point of view, it can be said that people need systems that can perform their transactions as quickly, safely, and as cost-effectively as possible.

Appearing as a cryptocurrency during 2008 (the year of the global financial crisis), Bitcoin has had an enormous international impact in the ten years since the crisis. The popularity of Bitcoin among its users and those that see it as an alternative investment area has increased rapidly, reaching a peak in 2017. Coming into being after the financial crisis, Bitcoin first gained prominence after an article written by Satoshi Nakamoto entitled *Bitcoin: A Peer-to-Peer Electronic Cash System.* This article defined Bitcoin as an electronic payment system based on encryption, in which the two parties are directly connected with each other. By criticizing the current banking system and the mediation role that they play, the article highlighted the growing trend of electronic trade and argued that there was no longer any need for banks to pursue transactions (Nakamoto, 2008).

The emergence of Bitcoin was generally perceived as a product of ideas that emerged from crypto anarchists; a cryptocurrency without a central authority and a digital value with an encrypted system. However, it can be observed today that even these crypto anarchists have lost control of Bitcoin (Vigna & Casey, 2017; Wirdum, 2016). Without doubt, one of the main characteristics that makes Bitcoin different is the fact that it does not need a third-party authority to mediate the transaction between the buyer and seller; the transaction takes place exclusively between two parties. This characteristic is due to the blockchain technology used. The blockchain technology that Bitcoin rests upon replaces the third-party mediator and the "trust relationship" between the two parties with a mathematically precise technology, and therefore fulfills a new "trust mechanism" (The Economist, 2015).

Digital/virtual currencies, which have been defined as cryptocurrencies, are perceived as both transfer methods and investment areas due to their cost-effectiveness and ability to conduct fast transactions. Some of the reasons why there was a need for this type of digital/virtual currency are due to the fact that people wanted to be freer, wanted to transfer their money from one place to another without limitations, and wanted to conduct their transactions in a more cost-effective and trustworthy manner (Kurtuluş, 2018). Another reason could be related to people trying to compensate for their confidence in the banking system through virtual money.

Bitcoin brings a great wind of change and if successful, "it will not only radically transform our banking and trade systems, but also has the digital technological potential to carry millions of people to the modern, integrated, digital and globalized economy" (Vigna & Casey, 2017).

Following the increasing demand for Bitcoin in financial markets, a variety of "altcoins" such as Litecoin, Namecoin, and Swiftcoin, also emerged. The most popular altcoins after Bitcoin—Ethereum, Ripple, Bitcoin Cash, and Litecoin—are also traded on the markets. Bitcoin emerged in 2008 and entered the market in 2010 at a value of $0.07. While its maximum value was $1 in 2011, the value of Bitcoin increased to $960 in 2017 growing exponentially until it reached a high of $20,000 in December 2017. The fact that the value of Bitcoin increased to such an extent, it was considered as "digital/virtual gold." The reason why the price of Bitcoin increased at such a rate was because people were increasingly opting for cryptocurrencies. As of the end of January 2019, there are currently 2120 cryptocurrencies being traded in 15,511 market. A daily total transaction volume of $20 billion is realized with a total market value of $111 billion (coinmarketcap.com, 2019). On the other hand, only Bitcoin's market share is about 53 percent with a market size of 60 billion dollars and 17.5 million Bitcoins currently circulating in the market (coinmarketcap.com, 2019). Thus, even though the value of Bitcoin increased up to $20,000, it has still not recovered from its decline.

Limited to a supply of 21 million, if the current algorithm system operates correctly Bitcoin will reset its supply in the year 2140 (Kancs et al., 2015). The reason why Bitcoin's supply is limited is that gold is determined according to the supply-demand relationship in the markets where the price is included. Due to its limited supply, the increase in demand increases the value of Bitcoin. As it can be easily bought and transferred, Bitcoin uses the principles of cryptography, known as blockchain technology. The feature that distinguishes Bitcoin from other currencies is that the algorithm is not controlled by any person, group, company, central authority, or government. Recently, Bitcoin has started to be used as a means of payment in the international arena and the increase in the number of enterprises accepting it is notable. For example, as of the end of January 2019 there are 4243 Bitcoin ATMs in 76 countries, and the fact that 95% of these ATMs are in North America and Europe attracts attention (coinatmradar.com). When looking at these percentages, it is not wrong to say that Bitcoin is spreading at a faster rate in the West and has reached more users in those regions.

22.3 Blockchain Technology and Bitcoin Mining

With the recent financial and banking crises in the global system damaging society's trust in the banking system, questions about the basis of money have begun to be asked. The feeling of trust in the banking sector is the basic building block for the system to function. Today, while it is the state authority behind paper money that makes it secure, the factor that makes electronic money trustworthy is its financial regulation. Likewise, blockchain technology is the reason that cryptocurrencies are perceived as more trustworthy.

When monetary transactions are made in the financial system, banks are mediators, and this creates a sense of trust. However, cryptocurrencies present a service based on technological trust. In other words, the trust one has in the bank in the traditional financial system is replaced with trust based on blockchain technology (Blundell-Wignall, 2014).

Blockchain technology provides the operation of cryptocurrencies and their infrastructure, does not require tools, and is transparent. However, its most distinguishing factor is that it is highly secure. Blockchain has great potential due to its many positive characteristics, including a rapid transaction process, lower costs, increased trust, and facilitating operational functioning. With the increasing demand for cryptocurrencies, blockchain technology has eliminated the need for third parties, has allowed for transfers to be anonymous and can perform at a very low cost. Therefore, it has the potential to have a reformative or disruptive effect on the financial system (Shin, 2015). Even though it is not an unknown technology, blockchain—which can be defined as an ever-growing distributed database of records in which the records are connected to each other by cryptographic elements (hash functions)—has gained a new dimension with Bitcoin (Piscini, Hyman, & Henry, 2017). In addition to its most important characteristic of not having a central authority, data is stored by users who are also integrated into the system. Blockchain uses a distributed database that provides encrypted transaction tracking and is defined as Distributed Ledger Technology (Collomb & Sok, 2016). All these processes are kept together with blockchain technology based on encrypted scattered data-based mathematical algorithms instead of being recorded in separate books (Knezevic, 2018).

Blockchain technology provides unprecedented control over digital identity to individual users. A globally open account book, blockchain is not only used in the production of cryptocurrencies but is also used in many other areas such as storage and management (Atabaş, 2018). The opportunities it provides to digital identity makes it key to the economy of confidence (Piscini et al., 2017). Blockchain can act as a digital record store and can store, hide, process, and manage information on real estate, can store evidence and record of tools and valuable assets, and can store birth and death certificates, and smart contracts. With its de-centralized structure and distributed and transparent trust architecture, blockchain is highly secure and has an irreversible infrastructure. Thus, as a result of the opportunities that it provides in related processes, blockchain brings a profound smart/digital transformation in many areas from trade to finance and from logistics to transportation (Ganne, 2018). In this respect, it can be pointed out that blockchain technology is slowly evolving and it is claimed that it is at the center of the fourth industrial revolution (Mougayar, 2016).

With blockchain, important steps are being taken to ensure that different sectors in the international arena are integrated with the technology. In this context, several examples of blockchain technology usage will be provided here. With the aim of utilizing the opportunities provided by technology, technological giant IBM tried to create a trade consortium using blockchain technology with Europe's biggest banks, namely Deutsche Bank, HSBC, KBC, Natixis, Rabobank, Société Generale and Unicredit (Kharpal, 2017).

IBM further illustrated its desire in this direction with its statement that it will seek to create a blockchain-based trade finance platform with UBS, the Bank of Montreal, CaixaBank, Erste Group, and Commerzbank (Arnold, 2017; Kharpal, 2017). Likewise, Wells Fargo and the Commonwealth Bank of Australia used blockchain technology in cotton transactions regarding the delivery of cotton from the U.S. to China. These developments illustrate that in the upcoming period, the finance sector in particular will adapt to the opportunities provided by blockchain technology and will affect the "financial technology" (fintech) trend (Kharpal, 2016; Yağcı, 2018).

The production of Bitcoin, the records on blockchain and the verification of the transaction are all done by the miners themselves. At the same time, miners ensure the security of the blockchain system and the realization of money transfers, which are registered in the distributed data book. Blockchain technology has an open account feature which gives all users the opportunity to check their transactions. Known as Bitcoin mining, the production of Bitcoin, when it first appeared on the agenda, was easy and did not take a long time due to the lack of people who knew how to mine. However, due to the increasing demand for Bitcoin, the number of people mining Bitcoin increased meaning that the price of producing Bitcoin increased and its system has become more difficult. When Bitcoin mining first emerged, it was possible to do so using a fairly modest computer and could only be done by a small number of people. However, nowadays, due to increasing demand and the increasing number of people mining Bitcoin, the process has become more difficult and less profitable, and is only possible using more powerful computers. Since Bitcoin users and institutions want to gain more profit, investments have been made in Bitcoin mining for the high-powered machines and directed to areas with low energy costs (Vigna & Casey, 2017).

22.4 The Role and Effect of Bitcoin Mining in Energy Consumption

Bitcoin mining is a complex crypto puzzle, and as this process requires a large amount of computing power, the energy used to facilitate the calculation used in mining can cost a considerable amount of money. People involved in producing Bitcoin, sometimes with computer networks that they have installed in factory-sized buildings, can benefit significantly from the incentives system. The increase in the number of peers participating in the system is an important factor in increasing the reliability of the Bitcoin system (Price, 2015). According to the Bitcoin energy consumption index, as of the end of January 2019, the average annual consumption of Bitcoin is 47.3 TWh (Digiconomist, 2018a).

When comparing the energy consumption of countries and that of Bitcoin mining, it can be seen that Bitcoin uses more energy than some countries' entire output. According to the International Digiconomist Energy Agency, Bitcoin outperforms many countries, including Portugal and Singapore, and ranks 53rd in terms of the energy consumption of countries (IEA, 2018). It is argued that the reason why Bitcoin consumes so much energy is because of "security." Due to the lack of a central authority in its transactions, Bitcoin is compelled to secure its system against any attacks or against corruption (Fig. 22.1).

The index in which Bitcoin consumption is calculated shows that the annual global Bitcoin mining revenue is $2.4 billion, and the associated mining costs are $2.31 billion. It consumes 440 KWh of electricity per Bitcoin process and constitutes 0.21% of the world's electricity consumption. Bitcoin's energy consumption is equivalent to 12.3% of France's total consumption, 17.4% of the UK's, and 43.2%

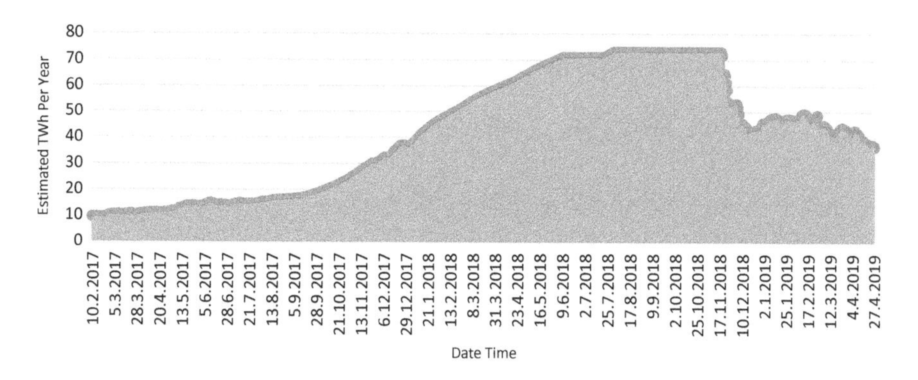

Fig. 22.1 Bitcoin Energy Consumption (TWh). (Source: "Bitcoin Energy Consumption Index", *Digiconomist* (Retrieved: April 29, 2019))

of the Netherlands' energy consumption (Digiconomist, 2018a). This massive consumption of energy for Bitcoin processes contributes to an increased carbon footprint and to increased global warming (Mora et al., 2018). The amount of energy that Bitcoin consumes is astronomical, and the fact that this consumption relies on hydrocarbon sources raises concerns (Hern, 2018). In his study, Alex de Vries (2018) uses a new methodology in order to understand Bitcoin's growing energy problem and its increasing energy consumption. His study argues that this increasing energy consumption will not help achieve climate goals, and that if Bitcoin energy consumption increases in this way, it will amount to 5% of the world's energy consumption (de Vries, 2018). Another study compares the amount of energy required to produce the same amount of gold and Bitcoin and shows that Bitcoin mining requires more energy than gold mining. However, what needs to be highlighted here is the fact that Bitcoin and gold do not serve the same purpose (Digiconomist, 2018a, 2018b).

Bitcoin's electricity consumption and its high costs have meant that miners have often relocated to countries where electricity is cheaper. The over-consumption of energy after a Bitcoin process is largely based on coal and is concentrated in countries where the cost of electricity consumption is very low (Digiconomist, 2018a, 2018b). According to the latest studies, more than 60% of miners are in China and the miners are more prevalent in places that have low electricity costs and fast internet. However, in contrast to what is known, Chinese Bitcoin miners have been shown to use mainly renewable energy (Bendiksen, Gibbons, & Lim, 2018). Countries following China's lead, including, the U.S., Canada, Sweden, Iceland, and Georgia, have also stated that they are using renewable energy sources.

Thus, in contrast to what is assumed to be known, 77.6% of Bitcoin mining relies on renewable energy sources while the rest is made up of fossil/nuclear energy (Bendiksen et al., 2018). Due to the fact that a majority of Bitcoin mining processes are pursued with renewable energy resources, it can be argued that Bitcoin will alter the negative perception that has been established in terms of its energy consumption, climate change and the environmental effects that this will have.

22.5 Cryptocurrencies in the Global Economy

22.5.1 Security Based on Technological Trust

As emphasized in the previous section, the phenomenon of trust is a determining factor in the historical transformation of money. When looked at from the trust issue, while state authority is what attributes a sense of trust for paper money, the regulation of electronic money corresponds to this trust. In times of political and economic crisis and a possible crisis of confidence against electronic money, as previous experiences have shown, people are more likely to turn to paper/metal money. With the collapse of financial institutions after the 2008 financial crisis, which has been the biggest crisis the global economy has faced in the twenty-first century so far, the failure of central banks to act on time and the decrease in confidence against banks in general have directed people toward more transparent systems. While Bitcoin and other cryptocurrencies have become widespread and are seen as an alternative in times of crisis, the crisis experienced by the banking sector in many European countries also raised the profile of cryptocurrency. For instance, during the banking crisis in Cyprus in 2013, many people used Bitcoin as an alternative currency to send money to other countries or to take their money out of the country (Cox, 2013; Vigna & Casey, 2017, pp. 163–164). Being one of the most affected countries during the euro crisis, Greece's Finance Minister threatened to move to Bitcoin during negotiations with the Troika (Papapostolou, 2015). The interest in Bitcoin increased after the crisis in Cyprus; and when looking at the case of Greece it is clear that Bitcoin became a viable option for people who live in countries, such as Italy, Spain, and Portugal, where banks could not provide for their citizens. Bitcoin therefore became more popular in these countries as people used it to withdraw and transfer money independently of the financial system (Adamopoulos, 2015; Chatfield, 2014; Kelly, 2015).

When looked at from this perspective, it can be reasonable to assume that people want to be able to pursue their transactions independently, and in a secure, low-cost, and efficient manner. The biggest benefits of Bitcoin and other cryptocurrencies is that they cut out the third party and can complete a transaction rapidly, securely, and at low cost (Doguet, 2013; Üzer, 2017, p. 67). On the other hand, with the spread of electronic money, the problem of information security and high costs in transfers has also emerged. For instance, the commission taken during international money transfers can be as high as 30% and the transfer itself can take several days, whereas with Bitcoin, one does not need a bank and can send money through their own wallet. In the current international economy, this means that transactions can be performed far more rapidly, and independently of any regulatory authority.

Questions such as whether cryptocurrencies are sustainable and safe, or whether they will disappear in the future are also discussed. At this moment in time, the areas in which Bitcoin can be used are limited, it is not controlled by any regulatory authority, and has no guarantees. The amount of Bitcoin in the market can be monitored and the number of transfers between wallets can always be seen. Since Bitcoin and other cryptocurrencies have no physical form, they only have value in the virtual environment and there is no authority to secure them. For this reason, each individual has to protect his/her own money, and this is realized through what are called *wallets,* which operate in a similar manner to bank accounts. If one forgets their password for their wallet or if Bitcoin are sent to another account by mistake, there is no way to recall the transaction and there is no legal entity that can address such problems. The biggest risks and disadvantages, therefore, include the possibility of being hacked or the forgetting of passwords. In order to address these problems, each wallet is given what is called a QR code. Thus, by using a QR reader, one can accurately and quickly send Bitcoin to any account they want.

Due to the fact that Bitcoin users do not pay taxes and pursue their transactions in a cost and time-effective manner and without any regulatory authority, this system poses a great challenge to the current banking system. The destructive effect of cryptocurrencies will undoubtedly trigger a major change in the virtualization of money in the banking sector. Therefore, it can be estimated that the banking and finance sectors will not ignore this development and will adapt to the cryptocurrency concept in the near future. Amazon's leading role in retail, Uber's taxi alternatives, and the replacement of hotels by Airbnb, shows that alternatives in today's world can have a destabilizing effect as they are more flexible, secure, faster, and cheaper (DigitalTalks, 2017).

22.5.2 Global Effects of Cryptocurrencies

Bitcoin has recently been increasingly mentioned along with trade wars. As a result of these trade wars, which began with U.S. President Donald Trump, countries have increasingly perceived Bitcoin and other cryptocurrencies as being viable alternatives. The sanctions on countries that the U.S. considers as targets make trade in cryptocurrencies more attractive. In recent years, countries such as Russia, Venezuela, and China have been searching for alternative methods in order to break the U.S. hegemony in the international financial system. Cryptocurrencies also draw attention in the search for alternatives. For instance, after Trump kick-started his trade war with China, it can be observed that Chinese investors now hold more Bitcoin than before (Bovaird, 2018). Likewise, as Russia is moving toward alternative methods such as cryptocurrencies and its own payment methods, China is re-invigorating the use of the Renminbi as an international currency. Also, in order to circumvent sanctions, Venezuela has created an oil-backed cryptocurrency (Johnson, 2018). North Korea, a country that has in recent times been perceived as a potential threat by the U.S., is considering virtual currency alternatives due to economic sanctions, and it is claimed that the country has invested in cryptocurrency mining. In addition, North Korea has also been blamed for the recent cyberattacks on South Korea's cryptocurrency exchange market (Chapman, 2017; Chen, 2017).

Countries that have been severely affected by U.S. sanctions, such as Russia, Iran, and Venezuela, are still continuing their efforts to form alternative finance systems and to resist the sanctions by using blockchain technology. With the U.S. re-initiating a two-wave sanction scheme on Iran, the country has been investing in technologies with a blockchain foundation. In an attempt to create their own cryptocurrency, the Central Bank of Iran has established its own blockchain laboratory (Fanusie, 2018; Motamedi, 2018; Meyer, 2018). Only time will show how realistic it is going to be for Iran to pursue trade in cryptocurrencies using blockchain technology with its former trade partners. More research needs to be conducted in order to determine whether Iran is ready for such a system and whether it has the necessary foundations.

The use of Bitcoin in deferred trading on large exchanges has created a positive atmosphere for the future of cryptocurrencies for international investors. After U.S. exchange operators CME and CBOE decided to allow the processing of Bitcoin in future transactions; NASDAQ also followed suit (Baker & Massa, 2017). On the day that CME began futures trading, Bitcoin broke new records. The fact that Bitcoin and other cryptocurrencies began to be traded in the world's largest stock markets is an important development—as an asset that emerged as an alternative to banks and states had entered the world financial system.

22.5.3 Risks and Threats

As the popularity of Bitcoin has increased, it is possible to see both the positive and negative news being made about it. Most of the articles about Bitcoin either make people more excited or more cautious about using Bitcoin. For instance, while some universities in the U.S. see Bitcoin as an opportunity to pay their staff and for students to pay their tuition fees (Bitcoinist, 2016; Nghiem, 2017), U.S.-based JP Morgan's CEO's definition of Bitcoin as a "fraud" and his claim that he will "sack anyone who buys Bitcoin" (Monaghan, 2017) affects the demand and volatility of the cryptocurrency. With the increasing popularity of Bitcoin, many other altcoins have emerged. However, it is argued that many of these do not have a future. Inspired by the decentralized structure of Bitcoin, it is pointed out that these altcoins have negative features, such as fueling arms races, excessive electricity consumption, hash process power, and the monopolization of the Bitcoin mining industry (Vigna & Casey, 2017, p. 221).

Although it has advantages and opportunities, the use of Bitcoin and other digital currencies carries some risks. The fact that cryptocurrencies do not have a legal foundation, they are excessively wavy and speculative, can rapidly increase/decrease in value, there is no authority to go to in case of a lost password or other queries, and they are often used in illegal transactions are all factors to take into consideration. First of all, the value increase experienced in the case of Bitcoin illustrates the risk of excessive fluctuation of cryptocurrency values. In this respect, the fact that the first thing people request is a "stable price" in the money they use as a means of payment, and the fact that this is less likely in cryptocurrencies in comparison to conventional money, is something that makes it more difficult to get their general acceptance in the short term. Since Bitcoin fluctuates wildly in value, historical examples are provided which say that it is a bubble.

Among these examples are the *Tulip Mania*, *Mississippi* and *South Sea Bubble*, the *Internet Bubble*, and the *Mortgage Crisis* (Partington, 2017). However, at the end of 2017, the high price increases in Bitcoin and other cryptocurrencies shows the high impact of international speculators.

Table 22.1 shows the difference between Bitcoin, gold and, fiat money. According to this study, cryptocurrencies could become a mainstream payment solution within the next decade (Gurguc & Knottenbelt, 2018). The excessive fluctuation of cryptocurrencies is a big handicap. For instance, in a system where salaries are paid by Bitcoin, workers will not receive a stable income due to the fact that the amount they are paid will be affected by the fluctuation of Bitcoin, which is a result of the supply-demand chain.

Table 22.1 The difference between Bitcoin, Gold and Fiat Money

	Gold	Bitcoin	Government money (Fiat money)
Production mechanism	Mineral mining using electrically-powered extraction devices. Electricity in, physical commodity out.	Cryptocurrency mining using electrically-powered extraction devices. Electricity in, digital commodity out.	Physical notes are printed but most money is created electronically. Typically issued by commercial and central banks of nation states.
Maximum supply	Finite (but unknown). Supply has consistently increased at a rate of c. 1.5% p.a. for more than 100 years.	Finite (and known). Supply currently increasing at c. 4% p.a. but rate of increase from year to year is always decreasing and will drop to 0 by 2140.	Theoretically unlimited. Supply has increased at an average rate of c. 11.5% p.a. over the last 40 years.
Concentration of resource	Varies by geography but is fixed within specific locations. Independent of global mining power deployed.	Dynamic concentration, dependent on global mining power deployed, and adjusted every 2016 blocks (±2 weeks).	Dynamic, dependent on government and central bank policies.
Storage	Expensive. Requires a secure physical location. Can be held directly or via nominees.	Inexpensive. Requires secure storage for private keys, which can be offline or online. Can be held directly or via nominees.	Usually inexpensive. Requires wallet, secure physical storage or bank account. Can be held directly or via nominee.
Unit of trade	Priced per Troy Ounce (31.1 g). Typically, available in quantities ranging from 0.5 g (±$30) to 1 kg (±$40000).	Priced per BTC. Typically, available in quantities ranging from 1 mBTC (±$7.60) to 100 BTC (±$760000).	USD, EUR, GBP, etc. which are further subdivided into 100 units (cents/pence).
Licensing requirements for production	Typically requires a mineral extraction license issued by the government.	Typically, none, although certain jurisdictions have imposed moratoriums on new commercial operations.	Production rights for physical representation are exclusive to government.
Price volatility	Moderate	Extreme	Variable depending on currency and government
Environmental impact of production process	Negative	Negative	Neutral

Source: Zeynep Gurguc and William Knottenbelt (2018). "Cryptocurrencies: Overcoming Barriers to Trust and Adoption", *Imperial College London Consultants*

It is highly unlikely that cryptocurrencies will spread and supplant the role of traditional money in an environment where inflation is controlled and monitored and the role of central banks are ignored in order to ensure price stability (Smith, 2017).

It is highly emphasized that the blockchain technology Bitcoin relies upon is extremely secure. However, the fact that cryptocurrency wallets and their stock markets can be hacked highlights a serious security weakness. One of the latest hacking incidents occurred when a Slovenian cryptocurrency stock exchange market, NiceHash, was hacked and 4700 Bitcoin worth $65 million were stolen (Gibs, 2017). Likewise, a South Korean exchange market, Youbit, was attacked for the first time in 2017. While only approximately four Bitcoin were stolen during the first attack, subsequent attacks stole 17% of Youbit assets (Reuters, 2017). These attacks on cryptocurrency exchange markets have increased security concerns in this area.

After these hacking incidents, the fact that Youbit applied for bankruptcy is reminiscent of the bankruptcy of Bitcoin's oldest and biggest stock exchange market. As can be remembered, after the attack against Tokyo-based Mt. Gox, it was announced that 850,000 Bitcoin had been lost. The company filed for bankruptcy in 2014 and the company's CEO was later arrested on claims of fraud. Likewise, 120,000 Bitcoin were stolen after an attack on Hong-Kong-based Bitfinex (Dunkley, 2017; Elgot, 2015). These kinds of digital thefts damage the security dimension of cryptocurrencies and users have increasingly appealed for greater security. In this sense, therefore, although the most attractive feature of cryptocurrencies has been the lack of authority and a P2P "peer-to-peer" digital transfer system, the extent to which they can be trusted is uncertain on an international level.

Since the security network of blockchain technology in which cryptocurrencies rest is very advanced, cybercriminals have been deploying alternative methods to crack the system. According to various studies, the cryptocurrency wallets of Bitcoin users have been infiltrated and Bitcoin worth approximately $140,000 have been stolen through this method (Kaspersky, 2017). Another handicap of Bitcoin and other cryptocurrencies is the fact that they are suitable for use in money laundering and other illegal transactions, such as the drug trade. The areas in which the use of cryptocurrencies have increased includes trade on the black market, the drugs trade, and money laundering (Smith, 2017). The most pressing issue for states is the fact that cryptocurrencies have been used for illegal activities on the "deep web" and the dark net. It was uncovered that virtual drug trafficking was carried out by a U.S. website, Silk Road, which, by using Bitcoin instead of real money, was able to carry out a drug trade worth $1.2 billion over the past two and a half years (Greenberg, 2013).

Another issue that is not yet certain, and is something that is being worked upon, is the taxation of Bitcoin. This issue of taxation initially emerged due to the fact that Bitcoin still has not been properly defined. Whether it is a commodity, currency unit, or movable property is important with regard to its taxation. Some leading countries, such as the Swedish Tax Office, have defined and taxed Bitcoin as "another entity," due to the fact that it does not consider it to be connected to the central bank (Kowalski, 2015, p. 149).

Brazil, Canada, Finland, Bulgaria, and Denmark have all made legal arrangements regarding the taxation of Bitcoin. Singapore has accepted Bitcoin as an asset and has added value-added tax from local purchases (Çarkacıoğlu, 2016, p. 67).

Since the legal status of cryptocurrencies is yet to be secured, the direction in which they will evolve in the future is debatable. There are discussions around the fact that non-state actors, including terrorist groups and illegal organizations, can use cryptocurrencies as a strategy to add to their political power, and this may give them the potential to confront the authority of the nation state (Üzer, 2017, p. 98). Especially in 2017, a year in which the value of Bitcoin massively increased, many arguments as to whether Bitcoin can replace conventional currencies such as the dollar or euro have also emerged. As the Bitcoin wave continues in the international arena, the debates on the future and legal status of cryptocurrencies also continue. Countries such as the U.S., Canada, and Japan, who are trying to place cryptocurrencies within a legal framework, have taken the first steps in this regard. While the U.S. is on the road to defining Bitcoin as a commodity, Japan released a virtual money law, which for the first time has legitimized cryptocurrencies as a payment method, and Canada is on the search for an attractive center for cryptocurrencies that will regulate and supervise their use. Europe is also paying close attention to cryptocurrencies and is formulating laws in order to prevent their use in money laundering and terror financing (Nuroğlu, 2018). On the other hand, some Asian countries have initiated prohibition practices for investors and miners in recent years due to their uncontrolled structure. While China has already banned cryptocurrency trading, there is also a possibility that South Korea will start an embargo on cryptocurrency exchanges (Meyer, 2017). Apart from these countries, many other states are still unsure about how to act on this issue. Some countries are turning to cryptocurrencies just to finish the race first and are implementing policies in this regard. Other countries that see cryptocurrencies as risks and threats are not only closely monitoring this process but have also warned about the over-valuation of cryptocurrencies as a global bubble in the face of an increasing number of users.

While discussions about whether cryptocurrencies will become an alternative currency are already in the mainstream, the question is still open as to how it will challenge one of the most important forces of the nation state, which is the monopoly of printing money. With this in mind, there have been studies on whether central banks will produce their own cryptocurrencies (Bech & Garratt, 2017, pp. 55–70; Bordo & Levin, 2017). The managing director of the IMF, Christine Lagarde, stated how, on the one hand, cryptocurrencies carry a certain number of risks, but how on the other hand they have the potential to be used by central banks in the long term. She emphasized that digital currencies should not be undermined, and that rather than trying to block them, central banks should observe how this technology can be used more effectively and how it can be made more cost-effective. According to Lagarde, states that have weak institutions and an unstable currency can utilize cryptocurrencies rather than adopt a different country's currency (for instance, the U.S. dollar). Perceiving cryptocurrencies as a viable alternative, Lagarde (2018) called this approach 'dollarization 2.0'. Many central banks across the globe have highlighted how cryptocurrencies will become more popular and that countries should not refrain from evaluating possible opportunities in this respect (Cox, 2017).

22.6 Conclusion

The development and diffusion of technology has given the digitalization of money a different dimension. As a result of this, the process and emergence of Bitcoin, which has taken its first big step in terms of both volume and market value, as well as the transformation of money, blockchain technology, and the global dimensions of cryptocurrencies were examined in this article. When we look at our daily lives and the way we do business, the effects of the digital transformation can easily be seen. From smartphones to robots, three-dimensional printers and big data, digital transformation is continuing at a rapid rate in an increasingly globalized world economy. Cryptocurrencies are an important component and consequence of this digital process. Bitcoin and other cryptocurrencies, known as altcoins, have had an immense effect on a global scale in a short period of time. Being the pioneer of cryptocurrencies in recent years, Bitcoin has become a phenomenon due to its prevalence and market value. As can be seen in the case of Bitcoin, despite risks such as the fact that cryptocurrencies are new, the fact that they do not have a legal foundation, and that they are used in illegal transactions, cryptocurrencies are also attractive as a payment method due to their speed and low cost. Major investments are being made in the production and infrastructure of cryptocurrencies, which are not subject to the control of a central authority and are realized through passwords alone. In this respect, it can be observed that states will not remain silent in the face of the opportunities and risks presented by Bitcoin.

With the increasing use of Bitcoin in the international market, payment methods have changed. Unlike fiat money such as the U.S. dollar, the euro, and the yen, if manufactured products were valued, distinct from inflation and on a fixed Bitcoin rate, this could cause the democratization of money and create a global breakthrough. The fact that cryptocurrencies function as alternatives in the global economy in which there are exchange and trade wars can turn cryptocurrencies into safe harbors. This transformation could shake state institutions and the states themselves. However, the level of demand will be decisive as to whether or not this will occur. In fact, it has been observed that many countries, inspired by Bitcoin, are trying to create their own cryptocurrency or are in the process of creating different works based on blockchain technology. In addition, the increase in energy consumption in Bitcoin mining has led to the development of alternative energy sources. While the astronomical energy consumption used in Bitcoin mining and the damage to the environment has sparked debates as to the sustainability of Bitcoin, it can be said that this problem has been resolved in recent years through the use of renewable energy.

Hence, despite the fact that Bitcoin has yet to prove itself, this article has illustrated how it has serious potential in the medium and long term. In this respect, with the increasing number of users and investors in this field, states have had to create legal apparatuses in order to adapt to this change. Related to this, the increasing expectancy of cryptocurrencies will undoubtedly challenge state power of controlling money, and as this article has underlined, central banks will not remain silent in this regard.

The most pressing concern is how the emergence of cryptocurrencies during the process of the digitalization of money, and the global effect this will have, will alter the power of central banks to print money and how it will change their monetary policies. It can be predicted that in the near future, governments will take steps to use blockchain technology to control cryptocurrencies. When looked at from this perspective, in the not-too-distant future, central banks may take the initiative to create their own digital currencies based on blockchain technology, and we may witness the consequences that this will have for the global monetary system.

References

Adamopoulos, A. (2015). *Greeks protect their money by converting them to Bitcoin*. Greece Greek Reporter. Retrieved November 12, 2018, from https://greece.greekreporter.com/2015/07/03/greeks-protect-their-money-by-converting-them-to-bitcoin/

Ahmed, K. (2017). Regulator warns Bitcoin buyers: Be ready to lose all your money. *BBC*. Retrieved January 24, 2019, from https://www.bbc.com/news/business-42360553

Ali, R., Barrdear, J., Clews, R., & Southgate, J. (2014). The economics of digital currencies. *Bank of England Quarterly Bulletin, 54*(3), 276–286.

Andreessen, M. (2014). Why Bitcoin matters. *The New York Times*. Retrieved December 12, 2018, from https://dealbook.nytimes.com/2014/01/21/why-bitcoin-matters/

Arnold, M. (2017). Banks team up with IBM in trade finance blockchain. *Financial Times*. Retrieved September 23, 2018, from https://www.ft.com/content/7dc8738c-a922-11e7-93c5-648314d2c72c

Atabaş, H. (2018). *Blokzinciri teknolojisi ve kripto paraların hayatımızdaki yeni yeri*. Istanbul: Ceres Press.

Baek, C., & Elbeck, M. (2015). Bitcoins as an investment or speculative vehicle? A first look. *Applied Economics Letters, 22*(1), 30–34.

Baker, N., & Massa, A. (2017, November 29). *Nasdaq plans to introduce Bitcoin futures*. Bloomberg. Retrieved November 12, 2018, from https://www.bloomberg.com/news/articles/2017-11-29/nasdaq-is-said-to-plan-bitcoin-futures-joining-biggest-rivals

Balaam, N. D., & Dillman, B. (2015). *Uluslararası ekonomi politiğe giriş*. Ankara: Adres Press.

Bariviera, A. F., Basgall, M. J., Hasperué, W., & Naiouf, M. (2017). Some stylized facts of the Bitcoin market. *Physica A: Statistical Mechanics and its Applications, 484*, 82–90.

Baur, D. G., Hong, K., & Lee, A. D. (2018). Bitcoin: Medium of exchange or speculative assets? *Journal of International Financial Markets, Institutions and Money, 54*, 177–189.

Bech, M. L., & Garratt, R. (2017). Central bank cryptocurrencies. *BIS Quarterly Review*, 55–70.

Bendiksen, C., Gibbons, A., & Lim, E. (2018). *The Bitcoin mining network: Trends, composition, marginal creation cost, electricity consumption & sources*. Retrieved December 24, 2018, from https://coinshares.co.uk/wp-content/uploads/2018/11/Mining-Whitepaper-Final.pdf

Bitcoinist. (2016). *Can Bitcoin be used to pay for a college education?* Retrieved November 15, 2018, from https://bitcoinist.com/can-bitcoin-be-used-to-pay-for-a-college-education/

Blundell-Wignall, A. (2014). The Bitcoin question. *OECD Working Papers on Finance, 37*, 1–21.

Bordo, M. D., & Levin, A. T. (2017). Central bank digital currency and the future of monetary policy. (No. w23711). *National Bureau of Economic Research, 237*(11), 1–30.

Bouri, E., Molnár, P., Azzi, G., Roubaud, D., & Hagfors, L. I. (2016). On the hedge and safe haven properties of Bitcoin: Is it really more than a diversifier? *Finance Research Letters, 20*, 192–198.

Bovaird, C. (2018, July 3). Could Bitcoin benefit from a trade war? *Forbes*. Accessed November 26, 2018, from https://www.forbes.com/sites/cbovaird/2018/07/31/could-bitcoin-benefit-from-a-trade-war/#6f4ba2b644cb

Çarkacıoğlu, A. (2016). Kripto-para bitcoin. *Sermaye Piyasası Kurulu Araştırma Dairesi Araştırma Raporu.*

Chapman, B. (2017). Bitcoin latest: North Korea suspected of South Korean cryptocurrency exchange hack. *Independent.* Retrieved December 23, 2018, from https://www.independent.co.uk/news/business/news/bitcoin-latest-updates-north-korea-south-youbit-exchange-hack-cryptocurrency-a8121781.html

Chatfield, T. (2014). Bitcoin and the illusion of money, *BBC.* Retrieved December 17, 2018, from http://www.bbc.com/future/story/20130412-bitcoin-and-the-illusion-of-money

Cheah, E. T., & Fry, J. (2015). Speculative bubbles in bitcoin markets? An empirical investigation into the fundamental value of bitcoin. *Economics Letters, 130*, 32–36.

Chen, O. (2017, September 13). Bitcoin mining: A new way for North Korea to generate funds for the regime. *CNBC.* Retrieved November 21, 2018, from https://www.cnbc.com/2017/09/13/bitcoin-mining-a-new-way-for-north-korea-to-generate-funds-for-the-regime.html

Coin Atm Radar. (2018). *Bitcoin ATM industry statistics/chart*. Retrieved January 3, 2018, from https://coinatmradar.com/

Coinmarket. (2019). Retrieved February 8, 2019, from www.coinmarketcap.com

Collomb, A., & Sok, K. (2016). Blockchain/distributed ledger technology (DLT): What impact on the financial sector? *Digiworld Economic Journal, 103*, 93–111.

Costelloe, K. (2018). Bitcoin ought to be outlawed' Nobel prize winner Stiglitz says. *Bloomberg.* Retrieved January 23, 2019, from https://www.bloomberg.com/news/articles/2017-11-29/bitcoin-ought-to-be-outlawed-nobel-prize-winner-stiglitz-says-jal10hxd

Cox, J. (2013). Bitcoin bonanza: Cyprus crisis boosts digital dollars. *CNBC.* Retrieved December 25, 2018, from https://www.cnbc.com/id/100597242

Cox, J. (2017). Federal reserve starting to think about its own digital currency, Dudley says. *CNBC.* Retrieved September 25, 2018, from https://www.cnbc.com/2017/11/29/federal-reserve-starting-to-think-about-its-own-digital-currency-dudley-says.html

Dahan, M., & Casey, M. (2016). Blockchain technology: Redefining trust for a global, digital economy. *World Bank.* Retrieved November 27, 2018, from http://blogs.worldbank.org/ic4d/blockchain-technology-redefining-trust-global-digital-economy

De Vries, A. (2018). Bitcoin's growing energy problem. *Joule, 2*(5), 801–805.

Digiconomist. (2018a). *Bitcoin energy consumption index*. Retrieved April 29, 2019, from https://digiconomist.net/bitcoin-energy-consumption

Digiconomist. (2018b). *Bitcoin mining is more polluting than gold mining*. Retrieved January 23, 2018, from https://digiconomist.net/bitcoin-mining-more-polluting-than-gold-mining

DigitalTalks. (2017). *Kripto para burada ve biraz korkutucu*. Retrieved November 25, 2018, from https://www.digitaltalks.org/2017/08/23/kripto-para-burada-ve-biraz-korkutucu/

Doguet, J. J. (2013). The nature of the form: Legal and regulatory issues surrounding the Bitcoin digital currency system. *Louisiana Law Review, 73*(4), 1120–1153.

Dorfman, J. (2017). Bitcoin is an asset, not a currency. *Forbes*. Retrieved November 16, 2018, from https://www.forbes.com/sites/jeffreydorfman/2017/05/17/bitcoin-is-an-asset-not-a-currency/#1259d4452e5b

Dunkley, E. (2017). Problems at two cryptocurrency exchanges raise security concerns. *Financial Times.* Retrieved November 26, 2018, from https://www.ft.com/content/aa9fdd64-e536-11e7-97e2-916d4fbac0da

Dyhrberg, A. H. (2015). Hedging capabilities of Bitcoin. Is it the virtual gold? *Finance Research Letters, 16*, 139–144.

Economist, T. (2015). The promise of the blockchain: the trust machine. *The Economist.* Retrieved November 22, 2018, from https://www.economist.com/leaders/2015/10/31/the-trust-machine

Elgot, J. (2015, August 1). Ex-boss of MtGox Bitcoin exchange arrested in japan over lost $390m. *The Guardian*. Retrieved November 25, 2018, from https://www.theguardian.com/technology/2015/aug/01/ex-boss-of-mtgox-bitcoin-exchange-arrested-in-japan-over-lost-480m

Fanusie Y. (2018, August 15). Blockchain authoritarianism: The regime in Iran goes crypto. *Forbes*. Retrieved December 12, 2018, from https://www.forbes.com/sites/yayafanusie/2018/08/15/blockchain-authoritarianism-the-regime-in-iran-goes-crypto/#5c2f87273dc6

Ferguson, N. (2008). *The ascent of money: A financial history of the world*. New York: Penguin.

Foley, S., Karlsen, J., & Putnins, T. J. (2018). Sex, drugs, and Bitcoin: How much illegal activity is financed through cryptocurrencies? *Review of Financial Studies*, 1–62.

Fry, J., & Cheah, E. T. (2016). Negative bubbles and shocks in cryptocurrency markets. *International Review of Financial Analysis, 47*, 343–352.

Ganne, E. (2018). Can blockchain revolutionize international trade? In *World Trade Organization Report*, Geneva, 1–145.

Gibs, S. (2017). Bitcoin: $64m in cryptocurrency stolen in 'sophisticated' hack, exchange says. *The Guardian*. Retrieved December 15, 2018, from https://www.theguardian.com/technology/2017/dec/07/bitcoin-64m-cryptocurrency-stolen-hack-attack-marketplace-nicehash-passwords

Glaser, F., Haferkorn, M., Weber, M., & Zimmermann, K. (2014). How to price a digital currency? Empirical insights on the influence of media coverage on the Bitcoin bubble. *MKWI 2014 (Paderborn) & Banking and Information Technology, 15*(1), 1–14.

Graeber, D. (2011). *Debt-updated and expanded: The first 5,000 years*. Brooklyn: Melville House.

Greenberg, A. (2013). End of the Silk Road: FBI says it has busted the web's biggest anonymous drug black market'. *Forbes*. Retrieved November 26, 2018, from https://www.forbes.com/sites/andygreenberg/2013/10/02/end-of-the-silk-road-fbi-busts-the-webs-biggest-anonymous-drug-black-market/#1a7e93d55b4f

Grinberg, R. (2011). Bitcoin: An innovative alternative digital currency. *Hastings Science and Technology Law Journal, 4*, 159–208.

Grym, A. (2018). The great illusion of digital currencies. *BoF Economics Review*, 1–17.

Gurguc Z. & Knottenbelt W. (2018). *Cryptocurrencies: Overcoming barriers to trust and adoption*. Imperial College London Consultants.

Hern, A. (2018). Bitcoin's energy usage is huge – we can't afford to ignore it. *The Guardian*. Retrieved November 12, 2018, from https://www.theguardian.com/technology/2018/jan/17/bitcoin-electricity-usage-huge-climate-cryptocurrency

IEA. (2018). *Statistics*. Retrieved December 12, 2018, from https://www.iea.org/

Johnson, K. (2018). The buck stops here: Europe seeks alternative to US-dominated financial system. *Foreign Policy*. Retrieved December 26, 2018, from https://foreignpolicy.com/2018/09/05/europe-seeks-alternative-to-us-financial-system-germany-france-sanctions/

Kancs, D. A., Ciaian, P., & Miroslava, R. (2015). The digital agenda of virtual currencies. Can Bitcoin become a global currency? *Information Systems and e- Business Management, 14*(4), 883–913.

Kelly, J. (2015). Fearing return to drachma, some Greeks use Bitcoin to dodge capital controls. *Reuters*. Retrieved December 21, 2018, from https://www.reuters.com/article/us-eurozone-greece-bitcoin-idUSKCN0PD1B420150703

Kharpal, A. (2016). Major banks trade cotton using blockchain in a move that could transform a major industry. *CNBC*. Retrieved November 12, 2018, from https://www.cnbc.com/2016/10/24/major-banks-blockchain-trade-cotton-in-a-move-that-could-transform-a-major-industry.html

Kharpal, A. (2017). Blockchain technology is moving into the financial mainstream with IBM and seven European banks. *CNBC*. Retrieved September 28, 2018, from https://www.cnbc.com/2017/06/26/ibm-building-blockchain-for-seven-major-banks-trade-finance.html

Knezevic, D. (2018). Impact of blockchain technology platform in changing the financial sector and other industries. *Montenegrin Journal of Economics, 14*(1), 109–120.

Kostakis, V., & Giotitsas, C. (2014). The (A) political economy of Bitcoin. Triple C: Communication, Capitalism & Critique. *Open Access Journal for a Global Sustainable Information Society, 12*(2), 431–440.

Kowalski, P. (2015). Taxing Bitcoin transactions under Polish tax law. *Comparative Economic Research, 18*(3), 139–152.

Kristoufek, L. (2015). What are the main drivers of the Bitcoin price? Evidence from wavelet coherence analysis. *PLoS One, 10*(4), 1–15.

Krugman, P. (2013). Bitcoin is evil. *The New York Times*. Retrieved December 3, 2018, from https://krugman.blogs.nytimes.com/2013/12/28/bitcoin-is-evil/

Kurtuluş, K. (2018). *Fokus programı* [TV Programme]. Istanbul: *Bloomberg*. Retrieved November 19, 2018, from https://www.youtube.com/watch?v=VuIudxQZiP8

Lagarde, C. (2018). Central banking and fintech: A brave new world. *Innovations: Technology, Governance, Globalization, 12*(1–2), 4–8.

Luther, W., & White, L. (2014). Can Bitcoin become a major currency? In *GMU Working Paper in Economics*, (14–17), 1–9.

Mankiw, N. G. (2003). *Macroeconomics*. New York: Worth Publishers.

Marian, O. Y. (2013). Are cryptocurrencies 'super' tax havens? *112 Michigan Law Review First Impressions, 38*, 1–57.

Meyer, D. (2017). South Korea follows china by banning icos. *Fortune*. Retrieved November 15, 2018, from http://fortune.com/2017/09/29/south-korea-china-bitcoin-ethereum-icos-ban/

Meyer, D. (2018). Iran is planning to launch its own cryptocurrency in order to bust U.S. sanctions. *Fortune*. Retrieved November 12, 2018, from http://fortune.com/2018/07/26/iran-sanctions-cryptocurrency/

Monaghan, A. (2017). Bitcoin is a fraud that will blow up, says JP Morgan boss. *The Guardian*. Retrieved December 26, 2018, from https://www.theguardian.com/technology/2017/sep/13/bitcoin-fraud-jp-morgan-cryptocurrency-drug-dealers

Mora, C., Rollins, R. L., Taladay, K., Kantar, M. B., Chock, M. K., Shimada, M., et al. (2018). Bitcoin emissions alone could push global warming above 2°C. *Nature Climate Change, 8*(11), 931–933.

Motamedi, M. (2018). US sanctions accelerate Iran's blockchain drive. *Al Monitor*. Retrieved December 29, 2018, from https://www.al-monitor.com/pulse/originals/2018/12/iran-cryptocurrency-blockchain-swift-russia-agreement.html

Mougayar, W. (2016). *The business blockchain: Promise, practice, and application of the next internet technology*. New York: Wiley.

Nakamoto, S. (2008). Bitcoin: A peer-to-peer electronic cash system. *Bitcoin.org*. Retrieved September 12, 2018, from https://bitcoin.org/bitcoin.pdf

Nghiem, A. (2017). Bitcoin: Would you want to get paid in cryptocurrency? *BBC*. Retrieved November 24, 2018, from https://www.bbc.com/news/business-42435838

Nuroğlu, E. (2018). Dünya Bitcoin'i tartışıyor. *Anadolu Ajansı*. Retrieved January 24, 2018, from https://www.aa.com.tr/tr/analiz-haber/dunya-bitcoini-tartisiyor/1028051

Papapostolou, A. (2015). Yanis Varoufakis: Greece will adopt the Bitcoin if Eurogroup doesn't give us a deal. *Greece GreekReporter*. Retrieved December 29, 2018, from https://greece.greekreporter.com/2015/04/01/yanis-varoufakis-greece-will-adopt-the-bitcoin-if-eurogroup-doesnt-give-us-a-deal/

Partington, R. (2017). Bitcoin bubble? The warnings from history. *The Guardian*. Retrieved November 23, 2018, from https://www.theguardian.com/business/2017/dec/02/bitcoin-bubble-the-warnings-from-history

Piscini, E., Hyman, G., & Henry, W. (2017). Blockchain: Trust economy. *Tech Trends*.

Plassaras, N. A. (2013). Regulating digital currencies: Bringing Bitcoin within the reach of IMF. *Chicago Journal of International Law, 14*, 377.

Price, R. (2015). The 21 companies that control Bitcoin. *Business insider*. Retrieved December 25, 2018, from http://uk.businessinsider.com/bitcoinpools-miners-ranked-2015-7

Reuters. (2017). *South Korean cryptocurrency exchange to file for bankruptcy after hacking.* Retrieved January 24, 2019, from https://www.reuters.com/article/us-bitcoin-exchange-southkorea/south-korean-cryptocurrency-exchange-to-file-for-bankruptcy-after-hacking-idUSKBN1ED0NJ

Shin, L. (2015). Bitcoin blockchain technology in financial services: How the disruption will play out. *Forbes*. Retrieved January 24, 2019, from https://www.forbes.com/sites/laurashin/2015/09/14/bitcoin-blockchain-technology-in-financial-services-how-the-disruption-will-play-out/

Smith, N. (2017). Start worrying when investors borrow to buy Bitcoins. *Bloomberg*. Retrieved November 12, 2018, from https://www.bloomberg.com/opinion/articles/2017-12-20/start-worrying-when-investors-borrow-to-buy-bitcoins

Swift. (2015). Worldwide currency usage and trends. *SWIFT Situation Reports*, 1–20.

Swift. (2017). *SWIFT history*. Retrieved January 24, 2019, from https://www.swift.com/about-us/history

Trugman, J. M. (2014). Welcome to 21st-century Ponzi scheme: Bitcoin. *New York Post*. Retrieved December 5, 2018, from https://nypost.com/2014/02/15/welcome-to-21st-century-ponzi-scheme-bitcoin/

USA Kaspersky. (2017). *CryptoShuffler: Trojan stole $140,000 in Bitcoin*. Retrieved November 20, 2017, from https://usa.kaspersky.com/blog/cryptoshuffler-bitcoin-stealer/13137/

Üzer, B. (2017). *Sanal para birimleri.* (Uzmanlık Yeterlik Tezi), TCMB.

Varoufakis, Y. (2013). Bitcoin and the dangerous fantasy of apolitical money. *Yanisvaroufakis.EU*. Retrieved December 28, 2018, from https://www.yanisvaroufakis.eu/2013/04/22/bitcoin-and-the-dangerous-fantasy-of-apolitical-money/

Vigna, P., & Casey, J. M. (2017). *Kriptopara çağı*. Ankara: Buzdağı Yayınevi.

Vondráčková, A. (2016). Regulation of virtual currency in the European Union. *Charles University in Prague Faculty of Law Research Paper, 3*, 1–17.

Weber, B. (2016). Bitcoin and the legitimacy crisis of money. *Cambridge Journal of Economics, 40* (1), 17–41.

Wirdum, A. V. (2016). Bitcoin returns to its cypherpunk roots: An interview with Lupták and sip of hackers' congress paralelní polis. *Bitcoin Magazine*. Retrieved November 18, 2018, from https://bitcoinmagazine.com/articles/bitcoin-returns-to-its-cypherpunk-roots-an-interview-with-lupt-k-and-sip-of-hackers-congress-paraleln-polis-1475087856/

Woo, D., Gordon, I., & Laralov, V. (2013). *Bitcoin: A first assessment, fx and rates*, Research Report from Merrill Lynch.

Yağcı. M. (2018). Yükselen finansal teknolojilerin ekonomi politiği: fintek ve bitcoin örnekleri. *İnsan ve Toplum Dergisi, 88*, 17–24.

Yelowitz, A., & Wilson, M. (2015). Characteristics of Bitcoin users: An analysis of Google search data. *Applied Economics Letters, 22*(13), 1030–1036.

Yermack, D. (2013). Is Bitcoin a real currency? An economic appraisal. *NBER Working Papers, 19747*, 1–22.

Yüksel, A. E. B. (2015). Elektronik para, sanal para, bitcoin ve linden doları'na hukuki bir bakış. *İstanbul Üniversitesi Hukuk Fakültesi Mecmuası, 73*(2), 173–220.

Serif Dilek obtained his BA in Business from Anadolu University (2009) and an MA degree in Maritime Economy from Istanbul University (2012). Dilek earned his Ph.D. in the Political Economy of the Middle East at Marmara University, Istanbul. Dilek continues his studies in Kırklareli University, International Trade and Logistics department as assistant professor. His fields of interest include interdisciplinary studies, and fields of research include International Political Economy, International Economics, Middle East Economy, and Economic Development.

Chapter 23
Existence of Speculative Bubbles for the US at Times of Two Major Financial Crises in the Recent Past: An Econometric Check of BitCoin Prices

Sovik Mukherjee

Abstract The last decade stands as a witness to a large number digital currencies coming into existence, such as—LiteCoin, BitCoin, Ripple, AuroraCoin, DogeCoin, etc. BitCoin being the most prominent among them, both in terms of its impressive price development and the price volatility it has. In this paper, the author uses a speculative bubble tracker, based on Wiener stochastic process, at times of two major financial crises, i.e. during the 2008–2009 US Subprime Mortgage Market Crisis and the Global Recession that started from 2010 onwards. The data used has been for the daily closing prices (converted into monthly after taking a geometric mean) from July 2008 to July 2010. Then from July 2010—July 2012, 2012–2014 and finally, 2014–2016. Using such data, the author traces out the price movements and points out periods of mass hysteria i.e. a 'speculative bubble' over the period concerned by comparing the results derived using a Brownian motion equation form used in physics and hence, tries to correlate the price fluctuations of BitCoins with fluctuations in the crisis index (as constructed by the author) for USA. Intriguingly, significance of speculative bubbles is prevalent at times of these financial crises.

23.1 Introduction

In 1999, Professor Milton Friedman, stated:

> I think the internet is going to be one of the major forces for reducing the role of government. The one thing that's missing but that will soon be developed, is a reliable e-cash.

Nine years down the line, witnessed the birth of BitCoins.

A group of programmers in 2008–2009 brought into existence a digital currency under the pseudonym Satoshi Nakamoto, or BitCoins as a cryptocurrency alternative to the government-backed currencies. Because of security issues, BitCoins are not

S. Mukherjee (✉)
Department of Economics, Faculty of Commerce and Management Studies, St. Xavier's University, Kolkata, India

© Springer Nature Switzerland AG 2019

U. Hacioglu (ed.), *Blockchain Economics and Financial Market Innovation*, Contributions to Economics, https://doi.org/10.1007/978-3-030-25275-5_23

issued by Central Banks and even banned in some countries.[1] The essentials of demand-supply are not always applicable for BitCoins. As argued in Buchholz, Delaney, Warren, and Parker (2012), the BitCoin prices are mainly maneuvered by the demand-supply interaction (as with other currencies) but the level of demand does not get fully determined as the expectations regarding future price component becomes crucial. As a result, short-run speculative expectations are a reflection of mass hysteria, thus, comes the question of capturing the pulse of the community and studying its impact on the evolution of BitCoin prices.

For a currency to become a currency, it is required to possess a number of functions, such as - standard of measurement, medium of exchange and most importantly, store of value (Cheah & Fry, 2015). Also, a certain level of confidence needs to be enjoyed by any currency among its user base but recent fluctuations in BitCoin are not an indication of stability in the currency itself with the speculative component under a bubble potentially undermines the 'unit of account role' (Dowd, 2014). During the last decade, loads of digital currencies, such as BitCoin and its peer group alternatives like PeerCoin, LiteCoin, Ripple and DogeCoin among others have significantly gained ground. In the competition space, Ripple and LiteCoin have noteworthy market capitalization value worth US$441 million and US$66 million, respectively, are giving BitCoins a run for its money. However, BitCoin, at present, comprises of around 83 per cent of the total estimated cryptocurrency capitalization standing at $3.9 billion.

Significantly for over a decade, there has been a wide range of fluctuations, starting from zero in 2009, when it was initiated, to around $1100 at the end of 2013 (see Fig. 23.1). Entering into 2014, its price has dropped to around $250, since then has been steadily on the rise with minor fluctuations.[2] From Fig. 23.2, it is clear that in the past 1 year has also seen price volatility in BitCoins. Contrary to the usual form of currencies, BitCoins have huge price fluctuations (Ciaian, Rajcaniova, & Kancs, 2016), suggesting the need to look into the determinants of price formation which are specific to digital currencies. The fact that BitCoin prices fluctuate in line with the "boom-bust" pattern analogous to the economic cycles we are aware of, is the motivation for asking the research questions as follows. First, is there a bubble component in the price movements happening for BitCoins following the boom–bust pattern? Second, are the results comparable with the fluctuations that we have in economic theory?

Countries across globes have had experience of financial crises. Of late, thebalance.com reports,[3]

[1]https://www.investopedia.com/articles/forex/041515/countries-where-BitCoin-legal-illegal.asp

[2]Read At: https://cointelegraph.com/news/BitCoins-falling-price-nothing-more-than-perception-or-is-there-manipulation

https://www.draglet.com/BitCoin-ethereum-volatility/

[3]Read the full story at https://www.thebalance.com/is-BitCoin-the-answer-in-a-financial-crisis-391275

Fig. 23.1 Price formation of BitCoins from its inception till 2015. Source: adapted from https://www.blockchain.com/prices

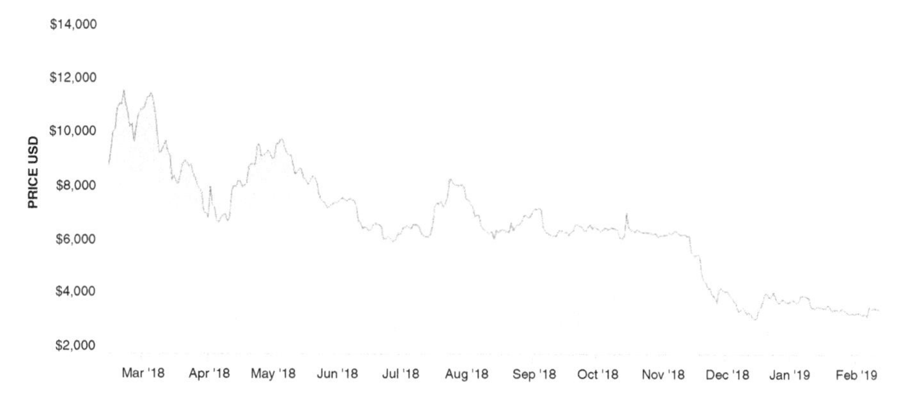

Fig. 23.2 Prices of BitCoins over a 1 year period—recent trends. Source: compiled by the author based on data accessed from https://www.coindesk.com/price/BitCoin

When Cyprus was in the thick of its banking crisis in April 2013, prices of the cryptocurrency reached record highs. BitCoin prices surged to even new heights in 2017. Other places imposing capital controls have also seen populations flee to BitCoin. Argentina is a case in point. The country's government stopped its population from buying U.S. dollars after suffering its own financial crisis. Reports suggest that Argentina has become a hotspot for BitCoin activity as banks there stagnate. Prices there are higher than in other countries.

As Mukherjee and Karmakar (2018) states, in the last decade of the last century and the first decade of the new millennium has been a witness to a record number of financial crises. These periods got characterized by both output and price volatility and given the linkages via the economic, financial and trade channels, the US Subprime crisis and the Eurozone crisis became contagious for the rest of the world, in total for some countries and partially for the others.

In this backdrop, the author traces out the price movements of BitCoins and points out periods of mass hysteria i.e. a 'speculative bubble' over the period concerned by comparing the results derived using a Brownian motion equation form used in physics and hence, tries to correlate the price fluctuations of BitCoins with fluctuations in the crisis index for USA. This is how the rest of the paper has been organized. In Sect. 23.2, the literature review discusses the drivers of BitCoin prices. In Sect. 23.3, the author introduces a customized bubble model as is in Econophysics. Section 23.4 discusses the empirical results and finally, it concludes.

23.2 A Brief Review of Select Literature

The highlight of the academic literature on issues related to BitCoins and other digital currencies has been a list of few pioneering articles and a host of unpublished works. Only recently, there have been rich contributions coming in for different aspects of BitCoins, starting from price formation movements/fluctuations (Bouoiyour & Selmi, 2016; Buchholz et al., 2012; Kristoufek, 2013; Lee, 2014), investor attractiveness (through sentiments in social media and stock markets, Kaminski, 2014; Kristoufek, 2013; Zhang, Fuehres, & Gloor, 2011), legal aspects and regulation (Grinberg, 2012; Plassaras, 2013), etc.

As is existing in the literature, there are three different schools of thought characterizing the process of price formation. First, the interaction of demand-supply market forces, second, attractiveness of BitCoin among investors, and third, the macroeconomic and financial market parameters It was pointed out by Buchholz et al. (2012) that BitCoin prices are primarily driven by its value as an exchange medium contrary to the intrinsic values that conventional commodities possess. Fundamentally, one needs to clearly understand the difference between gold and BitCoin as a medium of exchange. The demand for BitCoins is solely driven by its future value in exchange contrary to the case of gold or any other commodity currency where value determination at both the intrinsic and future exchange level. Coming to the supply side of BitCoins, supply is exogenously given, contrary to endogenous supply of gold (where it "depends on mining technology" as Ciaian et al. (2016) puts it. The dynamics take place through people's behaviour regarding expectations of future use in exchanges on the demand side (Luther & Olson, 2013).

Second, comes the issue of trust worthiness. For a currency to be trustworthy, it has to be accepted as a medium of exchange (Greco, 2001). BitCoin being a new currency, investor attractiveness becomes crucial for establishing credibility among market participants. Also, security concerns keep on lingering with BitCoins being more prone to cyber-attacks (Barber, Boyen, Shi, & Uzun, 2012; Böhme, Christin, Edelman, & Moore, 2015). In Moore and Christin (2013), they show that out of 40 BitCoin exchanges taken in their analysis, 18 closed down on account of cyber-attacks. Next, the role of social media (through Google search, Facebook, Twitter handles, etc.) in spreading BitCoin price information acts as an attractiveness criteria for investment. Also, investors' decisions may be affected based on the attention component in news media (Lee, 2014), search costs for finding out investment opportunities in stock exchanges (Barber & Odean, 2007; Gervais & Odean, 2001). Third, coming to the macroeconomic parameters, favourable stock exchange indices (Dimitrova, 2005),[4] lower oil prices and hence control of inflation parameter (Palombizio & Morris, 2012), exchange rates (through currency appreciation or depreciation) have an impact on BitCoin price formations (van Wijk, 2013).

In this paper, the author provides empirical validation for the existence of speculative bubbles in BitCoin markets and its role in the backdrop of a crises for the US economy. The contribution belongs to the recent literature on Econophysics (see Brée & Joseph, 2013) and thus far, no researcher has attempted to relate the BitCoin prices with a US crisis index and econometrically justify as to why prices of BitCoins rise at times of financial crises.

23.3 Data and Methodology

23.3.1 The Methodology

For characterizing the speculative bubble phenomenon, let, AP_t denote the price of an asset at time period t and the price formation function takes the form of a Wiener (Brownian Motion) process with a drift μ and infinitesimal variance, σ.[5]

$$dAP(t) = \mu(t)t + \sigma(t)AP(t) - \theta AP(t) \tag{23.1}$$

θ fraction of the asset value is lost or, in other words, there is a fall in the value of the asset of the magnitude, θ on account of the crash. The decomposition follows from Shreve (2004) and θ (t) denotes the jump process having the following properties,

[4] As Dimitrova (2005) puts it, "A decline in the stock prices induces foreign investors to sell the financial assets they hold. This leads to a depreciation of the underlying currency, but may stimulate BitCoin price, if investors substitute investment in stocks for investment in BitCoin. Generally, investors' return on stock exchange may capture opportunity costs of investing in BitCoin. Hence, in this case the stock exchange indices are expected to be positively related to BitCoin price."

[5] See https://galton.uchicago.edu/~lalley/Courses/313/BrownianMotionCurrent.pdf

$$\theta(t) = 0 \text{ before the crash or no crash}$$
$$= 1 \text{ after the crash.}$$

If there is a bubble then $-\theta AP(t)$ is active, otherwise, it becomes zero. Then, one can estimate using Eq. (23.1), the Brownian motion bubble component, B_t as,

$$B_t = \frac{dAP(t)}{AP(t)} \tag{23.2}$$

Now, to derive the bubble component over a given time period following Cheah and Fry (2015), we have,

$$B = \int_{t0}^{tn} \frac{dAP(t)}{AP(t)} dt \tag{23.3}$$

where,

$$t0 = \textit{starting point and,}$$
$$tn = \textit{ending point of the period}$$

Next, comes the issue of construction of a crisis index for the US economy. This paper draws inputs from the crisis index characterization in one of the earlier papers by the author, Mukherjee and Karmakar (2018) but in a different arrangement. Here, the author uses data pertaining to the US economy to characterize the transmission channel of two major crises in the recent past, the 2008 US Subprime Market crisis and the Eurozone debt crisis in 2011 (which eventually culminated into a great depression) coupled with the global recession.

1. At the outset, the paper builds up a crisis index (*CI*), at monthly frequency as, percentage change in REER devaluation (as compared to the last month) + the loss in foreign exchange reserves, in per cent (relative to the last month) + GDP lost, in per cent (relative to the last month). The choice of components in the index has been closely drawn from the ones used by Frankel and Rose (1996), Eichengreen, Rose, and Wyplosz (1994) and Frankel and Wei (2005) in the context of the US economy.
2. For characterizing a particular month as a crisis month, say, *m*, under the both the sets of crisis, the criteria is —

   (a) $CI(m) \geq 15 \textit{percent}$
   and
   (b) $CI(m) - CI(m-1) \geq 5 \textit{percent}$

Table 23.1 The summary statistics of the bubble component series

Mean	0.0681
Median	0.0523
Standard deviation	0.7111
Skewness	− 0.304
Kurtosis	12.998
Jarque–Bera (normality results)[a]	0.00

Source: Author's Computations

[a]The test statistic this paper uses is of the form, $JB = n\ [(\sqrt{b1})^2/6 + (b_2 - 3)^2/24]$ where, n is the sample size, $\sqrt{b_1}$ is the sample skewness coefficient, b_2 is the kurtosis coefficient

23.3.2 Data

The data used has been for the daily closing prices[6] from July 2008 to July 2010. Then from July 2010–July 2012, 2012–2014 and finally, 2014–2016. BitCoin Coindesk Index, Mt. Gox prices (till 2014, before its security breach and liquidation) in terms of US Dollars has been taken from www.coindesk.com/price/BitCoin for the aforementioned period to compile the data. The US monthly data for the other components has been compiled from the database of Federal Reserve of St.Louis, Center of the Eighth District of the Federal Reserve System database and the OECD database.

23.4 Results and Discussions

There has been a strand of literature looking at speculative bubbles, either belonging to the class of rational or irrational bubbles. The cause of the formation of rational bubbles is when investors anticipate the possibility of selling at an even higher price an overvalued asset while irrational bubbles are formed out of investor sentiments driven by psychological factors (see Dwyer, 2015; Weber, 2014 among others). Given in Table 23.1 is the descriptive statistics of the Bubble Component Series and shows that BitCoin prices point towards the richness in the dynamics of market volatility of BitCoin prices. In a similar way, the descriptive statistics for the crisis index (monthly data) has been reported in Table 23.2.

Next, the author carries out a BDS test for checking the linear dependence of the log returns in the bubble component and the volatility in the crisis component. In particular, the BDS test under very large samples perform very well (Broock, Scheinkman, Dechert, & LeBaron, 1996) usually needs a large sample to ensure proper performance. In the context of defining BDS test results, methodologically,

[6]For conversion with the crisis index, the bubble component has been converted into monthly data after taking a geometric mean of the daily prices. However for calculation of the significance of the bubble component, daily prices have been taken.

Table 23.2 The summary statistics of the crisis index

Mean	0.0675
Median	0.0362
Standard deviation	0.4428
Skewness	– 22.226
Kurtosis	658.341
Jarque–Bera (normality results)[a]	0.00

Source: Author's Computations

[a]The test statistic this paper uses is of the form, JB = n $[(\sqrt{b1})^2/6 + (b_2 - 3)^2/24]$ where, n is the sample size, $\sqrt{b_1}$ is the sample skewness coefficient, b_2 is the kurtosis coefficient

Table 23.3 BDS test results

BDS test for bubble component (daily prices)			BDS test for crisis index (monthly)		
ε/σ	Embedding dimension (*m*)	Bubble Component	ε/σ	Embedding dimension (*m*)	Bubble Component
2	2	14.22*	2	2	15.42*
2	3	15.61*	2	3	18.24*
2	4	19.33*	2	4	20.88*
2	5	17.21*	2	5	22.99*
1.5	2	13.96*	1.5	2	17.21*
1.5	3	18.21*	1.5	3	19.33*
1.5	4	19.99*	1.5	4	29.88*
1.5	5	20.38*	1.5	5	31.01*

Source: Author's Computations

following Lin (1997), if the ratio $\frac{N}{m}$ is greater than 200, the values of $\frac{\varepsilon}{\sigma}$ range from 0.5 to 2 and the values of m are between 2 and 5, $[C_{\varepsilon,m} - (C_{\varepsilon,1})^m]$ follows an asymptotically normal distribution with a zero mean and a variance $V_{\varepsilon,m}$ and the test statistic for the two-tailed test gets defined as:

$$BDS_{\varepsilon,m} = \frac{\sqrt{N}[C_{\varepsilon,m} - (C_{\varepsilon,1})^m]}{\sqrt{V_{\varepsilon,m}}}$$

The essence of carrying out the BDS test is to examine the bubble like price rise for the speculative component and the non-linearity in the crisis index. Results were significant as p = 0.000 under all possible combinations of dimensions, ranging from 2 to 5, are significant as in Table 23.3. This is suggestive of hidden nonlinearity or non-stationarity in the system, here for BitCoin prices and the crisis indices derived. The results point out the "chaotic behaviour" in terms of both BitCoin prices and crisis indices. The parameters of the decomposed Wiener process in Eq. (23.1) have been given below in Table 23.4.

From Table 23.4, the parametric value of μ is found to be statistically insignificant and hence, Eq. (23.1) gets validated following Lucey and O'Connor (2013). This result, in spirit, is similar to Cheah and Fry (2015) result of basic price of BitCoin

Table 23.4 Parameter estimates of the stochastic bubble decomposed model

Parameter	Estimates	t-value	p-value
$-\theta$	0.546	6.060	0.000
$\hat{\mu}$	0.163	1.212	0.883

Source: Author's Computations

being zero. It is argued that the value of the stochastic bubble component falls over time and has a zero mean to allow for the long run value to be zero taking into account dramatic fluctuations. In other words, towards the end of a bubble, the expected return is something not significantly different from zero.

The next set of results involve the comparison of the bubble component with the crisis index during probable times of crisis. In Table 23.5, the author estimates two models, first, with the time gap as specified at equal intervals of 2 years from 2008 to 2016. Second, testing for the bubble component over a moving time structure (see Filimonov, Demos, & Sornette, 2017; Geraskin & Fantazzini, 2013 among others) not at equally spaced intervals. Starting from 2008 to 2009, where the crisis was at its peak, bubble component is significant, followed by another significant bubble component towards the beginning of 2013–2014 while there is as such no bubble component in that 2010–2012 span because global recession actually started to claw into the system towards the latter half of 2012 and from 2010, the US economy showed signs of settling down following the aftermath of the crisis (Mukherjee & Karmakar, 2018). The latter case minutely does not get reflected in Model 1 on account of the assumption of equispaced intervals. However, in general, the result in Table 23.5 has put in evidences of the existence of bubbles during times of financial crises irrespective of the time gap chosen (i.e. although, not minutely observed, the correlation values between the crisis index and the bubble component are consistent across both the models).

There are two distinct periods when there exists a significant bubble component as per this model specification. There are two distinct phases,—the first one occurs because of the Subprime crisis and the other, primarily occurs on account of the confiscation from the "Silk Road website" following the US and the German declaration that BitCoins functioned like currencies. But, surprisingly, from 2015 onwards (see, the last two rows in Table 23.5), even after the explosions in BitCoin prices, the sustained bubble component seems to have fizzled out with correlational values becoming insignificant,[7] indicating the absence of any major financial breakdown.

[7]The very recent swings in BitCoin prices have not been taken into consideration and the author here does not comment on such recent features. That has been left for further research. This paper is designed to look through the crisis periods with an eye on BitCoin prices and hence the motivation.

Table 23.5 Likelihood ratio tests of the speculative bubbles and the correlation with the crisis index

Model 1 (Equispaced Time Frame)#				
Bubble component		Crisis index		Correlation (significance)
Time points	p-value	Time points	p-value	
July 2008–July 2010	0.000*	July 2008–July 2010	0.000*	0.85 (p = 0.00*)
July 2010–July 2012,	0.088	July 2010–July 2012,	0.152	0.81 (p = 0.98)
July 2012–July 2014	0.004*	July 2012–July 2014	0.005*	0.91 (p = 0.00*)
July 2014–July 2016	0.998	July 2014–July 2016	0.900	0.55 (p = 0.12)
Model 2 (moving time frame at daily prices)[a]				
Bubble component		Crisis index		
Time points	p-value	Time points	p-value	Correlation (significance)
July 2008–July 2009 (bubble)	0.000*	July 2008–July 2009 (US subprime crisis)	0.000*	0.90 (p = 0.00*)
August 2009–July 2010 (bubble)	0.000*	August 2009–July 2010 (US subprime crisis)	0.000*	0.95 (p = 0.00*)
November 2012–January 2013 (bubble fizzles out)	0.556	November 2012–January 2013 (crisis recovery phase)	0.152	0.88 (p = 0.00*)
January 2013–March 2013 (bubble)	0.000*	January 2013–March 2013 (crisis- global recession)	0.005*	0.91 (p = 0.00*)
March 2013–May 2013 (bubble)	0.000*	March 2013–May 2013 (crisis- global recession)	0.000*	0.85 (p = 0.00*)
July 2013–September 2014 (bubble)	0.000*	July 2013–September 2014 (crisis- global recession)	0.000*	0.80 (p = 0.00*)
November 2014–December 2015 (bubble fizzles out)	0.411	November 2014–December 2015 (crisis recovery phase)	0.090	0.59 (p = 0.04*)
November 2015–December 2016 (no bubble)	0.991	November 2015–December 2016 (absence of any major financial crisis)	0.883	0.87 (p = 0.00*)

Source: Author's Computations
#: at monthly values
*: denotes significance at 5 per cent level
[a]For going back to the daily data from the monthly values of crisis for comparisonal purpose, we do a simple Newtonian Interpolation

23.5 Concluding Remarks

Leo Tolstoy in his novel Anna Karenina, begins with these lines—"Every happy family is alike, but every unhappy family is unhappy in their own way." In this context, the author believes that a small twist in this quote will sum up the result as follows, it is a fact that each financial crisis is distinctively different but a crisis has striking features (i.e ... unhappy families are also similarly unhappy) when it boils down to doing some number crunching and deriving very high magnitudes of correlation values (refer to Table 23.5) of financial crises with fluctuations in prices of digital currencies like BitCoins.

The novelty of the paper lies in the contribution it makes to correlate a crisis index with the price fluctuations of BitCoins, an area which has remained unexplored in the veil of the rich literature on predicting movement in returns, price volatility and inefficiency of BitCoin into USA market.

References

Barber, B. M., & Odean, T. (2007). All that glitters: the effect of attention and news on the buying behavior of individual and institutional investors. *The Review of Financial Studies, 21*(2), 785–818.

Barber, S., Boyen, X., Shi, E., & Uzun, E. (2012). Bitter to better — how to make BitCoin a better currency. In *International conference on financial cryptography and data security* (pp. 399–414). Berlin: Springer Heidelberg.

Böhme, R., Christin, N., Edelman, B., & Moore, T. (2015). BitCoin: economics, technology, and governance. *J Econ Perspect, 29*(2), 213–238.

Bouoiyour, J., & Selmi, R. (2016). BitCoin: a beginning of a new phase. *Econ Bull, 36*(3), 1430–1440.

Brée, D. S., & Joseph, N. L. (2013). Testing for financial crashes using the log periodic power law model. *Int Rev Financ Anal, 30*(C), 287–297.

Broock, W. A., Scheinkman, J. A., Dechert, W. D., & LeBaron, B. (1996). A test for independence based on the correlation dimension. *Econ Rev, 15*(3), 197–235.

Buchholz, M., Delaney, J., Warren, J., & Parker, J. (2012). Bits and bets, information, price volatility, and demand for BitCoin. Retrieved from http://www.bitcointrading.com/pdf/bitsandbets.pdf

Cheah, E. T., & Fry, J. (2015). Speculative bubbles in BitCoin markets? An empirical investigation into the fundamental value of BitCoin. *Econ Lett, 130*(C), 32–36.

Ciaian, P., Rajcaniova, M., & Kancs, D. A. (2016). The economics of BitCoin price formation. *Appl Econ, 48*(19), 1799–1815.

Dimitrova, D. (2005). The relationship between exchange rates and stock prices: studied in a multivariate model. *Issues in Political Economy, 14*(1), 3–9.

Dowd, K. (2014). New private monies: a bit-part player?. Retrieved from https://papers.ssrn.com/sol3/papers.cfm?abstract_id=2535299

Dwyer, G. P. (2015). The economics of BitCoin and similar private digital currencies. *Journal of Financial Stability, 17*(C), 81–91.

Eichengreen, B., Rose, A. K., & Wyplosz, C. (1994). *Speculative attacks on pegged exchange rates: an empirical exploration with special reference to the European monetary system (no. w4898)*. Stanford, CA: National Bureau of economic research.

Filimonov, V., Demos, G., & Sornette, D. (2017). Modified profile likelihood inference and interval forecast of the burst of financial bubbles. *Quantitative Finance, 17*(8), 1167–1186.

Frankel, J. A., & Rose, A. K. (1996). Currency crashes in emerging markets: an empirical treatment. *J Int Econ, 41*(3), 351–366.

Frankel, J. A., & Wei, S. J. (2005). Managing macroeconomic crises: policy lessons. In J. Aizenman & B. Pinto (Eds.), *Managing economic volatility and crise: a practitioner's guide*. Cambridge: Cambridge University Press.

Geraskin, P., & Fantazzini, D. (2013). Everything you always wanted to know about log-periodic power laws for bubble modeling but were afraid to ask. *The European Journal of Finance, 19*(5), 366–391.

Gervais, S., & Odean, T. (2001). Learning to be overconfident. *The Review of Financial Studies, 14* (1), 1–27.
Greco, T. (2001). *Money: understanding and creating alternatives to legal tender*. Chelsea: Green Publishing.
Grinberg, R. (2012). BitCoin: an innovative alternative digital currency. *Hastings Sci & Tech LJ, 4*, 159.
Kaminski J (2014) Nowcasting the BitCoin market with twitter signals. Retrieved from https://arxiv.org/pdf/1406.7577.pdf
Kristoufek, L. (2013). BitCoin meets Google trends and wikipedia: quantifying the relationship between phenomena of the internet era. *Sci Rep, 3*, 3415.
Lee T (2014) These four charts suggest that BitCoin will stabilize in the future. Retrieved from www.washingtonpost.com/blogs/the-switch/wp/2014/02/03/these-fourchartssuggest-that-BitCoin-will-stabilize-in-the-future/
Lin, K. (1997). The ABC's of BDS. *Journal of Computational Intelligence in Finance, 97*(July/August), 23–26.
Lucey, B. M., & O'Connor, F. A. (2013). Do bubbles occur in the gold price? An investigation of gold lease rates and Markov switching models. *Borsa Istanbul Rev, 13*(3), 53–63.
Luther WJ, & Olson J (2013). BitCoin is memory. Retrieved from http://pricesandmarkets.org/wp-content/uploads/2015/02/Luther-Olson-4.pdf
Moore, T., & Christin, N. (2013, April). Beware the middleman: Empirical analysis of Bitcoin-exchange risk. In *International Conference on Financial Cryptography and Data Security* (pp. 25–33). Berlin, Heidelberg: Springer.
Mukherjee, S., & Karmakar, A. K. (2018). A hardheaded look: how did India feel the tremors of recent financial crises? In H. Dincer, Ü. Hacioglu, & S. Yüksel (Eds.), *Global approaches in financial economics, banking, and finance* (pp. 25–51). Cham, Switzerland: Springer.
Palombizio, E., & Morris, I. (2012). Forecasting exchange rates using leading economic indicators. *Open Access Scientific Reports, 1*(8), 1–6.
Plassaras, N. A. (2013). Regulating digital currencies: bringing BitCoin within the reach of IMF. *Chi J Int'l L, 14*, 377.
Shreve, S. E. (2004). *Stochastic calculus for finance II: continuous-time models* (Vol. 11). Berlin: Springer Science & Business Media.
van Wijk, D. (2013). What can be expected from the BitCoin. Retrieved from https://Final-version-Thesis-Dennis-van-Wijk.pdf
Weber, B. (2014). BitCoin and the legitimacy crisis of money. *Cambridge Journal of Economics, 40*(1), 17–41.
Zhang, X., Fuehres, H., & Gloor, P. A. (2011). Predicting stock market indicators through twitter "I hope it is not as bad as I fear". *Procedia Soc Behav Sci, 26*, 55–62.

Sovik Mukherjee is an Assistant Professor in Economics (at present), Department of Economics, Faculty of Commerce and Management Studies, St. Xavier's University, Kolkata, India. His educational qualifications comprises of a B.A. (Honours Distinction), an M.A. and an M.Phil in Economics from Jadavpur University, Kolkata and also UGC's National Eligibility Test (NET) for Assistant Professorship. He was awarded with the Ujjayini Memorial University Gold Medal Award by Jadavpur University in 2013 for academic excellence. The Mother Teresa Gold Medal Award 2017 was awarded to him for his contributions in the field of Economic Growth Research by the Global Economic Progress and Research Association, New Delhi. His research interests are in the areas of Application of Game Theory in IO Models, Public Economics, Applied Econometrics and Economics of Financial Crises. He is a Visiting Research Fellow at NISPAcee, Slovakia for 2019–2020. Besides, he is an Author-Affiliate at Information Resources Management Association, USA and served as a reviewer in the Journal of the International Academy for Case Studies. His membership in professional /academic bodies includes—Life Member of The Indian Econometric Society (TIES), the Indian Economic Association (IEA), the Alumni Association—(NCE Bengal & Jadavpur University) and the Bengal Economic Association (BEA).

Chapter 24
Analysis of Relationship Between International Interest Rates and Cryptocurrency Prices: Case for Bitcoin and LIBOR

Serdar Erdogan and Volkan Dayan

Abstract In this study, the change in weekly USD LIBOR Rate and USD Bitcoin Price for 2013–2018 were analyzed. LIBOR stands for London Interbank Offered Rate and it is a benchmark rate in which some of the world's leading banks charge each other for short-term loans. Bitcoin is the most traded currency in the cryptocurrency market. Vector Autoregressive Model (VAR) and Autoregressive Distributed Lag Models (ARDL) were used in this study. Impulse-response functions used and variance decomposition tests were made. Granger Causality Analysis was performed. Pairwise Granger and VAR Granger Wald tests were used for causality analysis. Additionally stationary test was also performed before the analysis. The stationary analysis of the variables will be made with ADF 79 and Perron's 89 breakpoint unit root tests. According to the results of the study; the variables are stationary at the I(0) level. VAR model was stationary and significant. According to ARDL model; short-term deviations have stabilized in the long run. The Granger causality test was one-way significant.

24.1 Introduction

In the international financial market, cryptocurrencies are gaining more importance day by day. Cryptocurrencies have been the new payment tools for both digital and other platforms. Bitcoin (BTC) is the first and the most used decentralized digital

S. Erdogan (✉)
Department of Accounting, Uzunkopru School of Applied Sciences Trakya University, Edirne, Turkey
e-mail: serdarerdogan@trakya.edu.tr

V. Dayan
Department of Banking and Insurance, Uzunkopru School of Applied Sciences Trakya University, Edirne, Turkey
e-mail: volkandayan@trakya.edu.tr

© Springer Nature Switzerland AG 2019

U. Hacioglu (ed.), *Blockchain Economics and Financial Market Innovation*, Contributions to Economics, https://doi.org/10.1007/978-3-030-25275-5_24

currency in the world. London Interbank Offered Rate (LIBOR) is the most popular benchmark rate for international banking.

This study provides important contributions to the literature about cryptocurrencies. It will give a different perspective to Bitcoin, which is mentioned frequently in the financial market. In addition, LIBOR which is one of the most important reference rates for financial institutions and investors and its relationship between cryptocurrencies and importance of this relationship will be analyzed.

In the analysis of this study, first of all, the information about the data set will be given and a graphical analysis of the series will be made. The stationary analysis of the variables will be made with ADF 79 and Perron's 89 breakpoint unit root tests. Finally, VAR, Granger Causality and ARDL models, which are the main methods for this study, will be used.

24.2 Literature

24.2.1 Related Studies

There are several studies related with Bitcoin.

Hileman and Rauchs (2017), in their studies investigated key cryptocurrency industry sectors by collecting empirical, non-public data. The study is made up of key industry sectors that have emerged and the different entities that inhabit them.

Franklin (2016), in this study; discussed the taxation issues surrounding the Bitcoin as a virtual currency.

Todorov (2017), in this study; targeted to introduce the Bitcoin's technology. It is found that despite the Bitcoin benefits over the currency of central authority, people do not believe in this cryptocurrency because of its speculative character.

Carrick (2016), in this study; analyzed the value and volatility of Bitcoin relative to emerging market currencies and explored ways in which Bitcoin can complement emerging market currencies.

Sahoo (2017), in this study; examined the comprehensive idea about the growth and future sustainability of Bitcoin as a cryptocurrency.

D'Alfonso, Langer, and Vandelis (2016), in their studies; analyzed Bitcoin and Ethereum to develop the ideal investment strategy. It is concluded that Bitcoin offered a higher expected value, but the volatility and speculative nature of cryptocurrencies indicated a need for diversification across platforms.

Shapiro (2018), in this study; considered the U.S. federal income tax treatment of loans and prepaid forward contracts denominated in cryptocurrencies and Bitcoin loans and other cryptocurrency tax problems were examined.

Davies (2014), in this study; analyzed the periods of Bitcoin market volatility. It was seeked to forecast this volatility with online searches from Google Trends and Twitter. It is concluded that Google Trends had forecasting power on the realized volatility of Bitcoin.

Bhattacharjee (2016), in this study; made a statistical analysis of Bitcoin transactions between 2012 and 2013 in terms of premier currencies.

Icellioglu and Ozturk (2018), in their studies; features and function of Bitcoin are discussed and the relationship between Bitcoin and selected exchange rates are investigated. According to this study, the existence of a long and short-run relationship between Bitcoin and other currencies was not found.

Sovbetov (2018), in this study; examined factors that influence prices of Bitcoin and other four cryptocurrencies.

Ciaian and Rajcaniova (2018), in their studies; examined interdependencies between Bitcoin and altcoin markets in the short and long-run. According to this study, Bitcoin and altcoin markets are interdependent. The Bitcoin-altcoin price relationship is significantly stronger in the short-run than in the long-run. In the long-run, macro-financial indicators determine the altcoin price formation to a slightly greater degree than Bitcoin does.

Erdogan, Dayan, and Erdogan (2018), in their studies; determined the use of cryptocurrencies in developing European countries and measured the factors affecting this usage. There is a nominally strong and accurate relationship between Bitcoin demand and Bitcoin price. However, in the case of elimination of the trend effect in Bitcoin prices, this relationship was moderate and positive.

24.2.2 Current Status of International Interest Rates (Benchmark Interest Rates)

International interest rates are used in the global financial system as a reference for short-term borrowings and some derivative transactions. The use of these benchmark interest rates in financial contracts reduces complexity in international markets.

In addition, if the financial instrument based on the benchmark interest rate is widely used, transaction costs decrease and liquidity increases. Interest rates on loans, asset-backed securities, deposits, bonds, securities, interest rate swaps and over the counter derivatives can be taken as the basis for benchmark interest rates (TCMB, 2018).

Although there are many benchmark rates in the financial market, most commonly used benchmark rates are; European Central Bank (ECB) interest rates, The Euro Interbank Offered Rate (Euribor), Euro Overnight Index Average (Eonia), and London Interbank Offered Rate (LIBOR).

The Governing Council of the European Central Bank sets European Central Bank (ECB) interest rate. ECB interest rate is used for the euro area. Inflation is taken into account when adjusting interest rates. Refinancing operations, deposit facility and marginal lending facility are main topic for this rate (Key ECB interest rates, 2018).

The Euro Interbank Offered Rate (Euribor) is the rate at which Euro interbank term deposits are offered by one prime bank to another prime bank within the Economic and Monetary Union (EMU) zone (About Euribor®, 2018).

Euro Overnight Index Average (Eonia) is the 1 day interbank interest rate for the Euro area (Eonia, 2018). Eonia can thereby be viewed as the overnight Euribor rate (Eonia interest rate, 2018).

Due to the increase in the demand of banks to wholesale and secured funds after the global crisis, the depth of interbank fund markets decreased significantly. In addition, some banks have misrepresented their balance sheets. Confidence in LIBOR decreased. Therefore, investors have turned to alternative reference rates.

A reform is needed for major interest rate benchmarks. According to Financial Stability Board, some alternative rates for major interest rates are listed below.

LIBOR has been strengthened since the Official Sector Steering Group (OSSG) was established. But authorities have warned that the publication of LIBOR could end with the withdrawal of official sector support by the end of 2021.

According to Financial Stability Board (2018) in the future, there will be new benchmark rates for the financial market. Table 24.1 shows reference interest rate recommendations by currencies.

As shown in Table 24.1 separate rates are recommended for each currency. For Australian Dollar (AUD) current benchmark rate is Bank Bill Swap Rate (BBSW), alternative rate is Reserve Bank of Australia (RBA) Cash Rate. For Brazilian Real (BRL) Overnight interbank offered rate (DI rate) is used but alternative rate will be Average Interest Rate on Overnight Repurchase Agreements (Selic). For Canadian Dollar (CAD) current benchmark rate is Canadian Dollar Offered Rate (CDOR), alternative rate is Enhanced Canadian Overnight Repo Rate Average (CORRA). For Swiss Franc (CHF) London Interbank Offered Rate (LIBOR) is used but alternative rate will be Swiss Average Rate Overnight (SARON). For Euro (EUR) current benchmark rates are London Interbank Offered Rate (LIBOR), Euro Interbank Offered Rate (EURIBOR) and Euro Overnight Index Average (EONIA); alternative rates are Euro Short-Term Rate (ESTER) or Euro Interbank Offered Rate (EURIBOR), respectively. For British Pound (GBP) London Interbank Offered Rate (LIBOR) is used but alternative rate will be Sterling Overnight Index Average (SONIA). For Hong Kong Dollar (HKD) current benchmark rate is Hong Kong Interbank Offered Rate (HIBOR) but alternative rate is not determined. For Japanese Yen (JPY) London Interbank Offered Rate (LIBOR), Tokyo Interbank Offered Rate (TIBOR), Euroyen Tokyo Interbank Offered Rate (TIBOR) are used but alternative rates will be mostly Tokyo Overnight Average Rate (TONA). For Singapore Dollar (SGD) current benchmark rates are Singapore Interbank Offered Rate (SIBOR) and Singapore Dollar Swap Offer Rate (SOR) but alternative rate is not determined. For United States Dollar (USD) London Interbank Offered Rate (LIBOR) is used but alternative rate will be Secured Overnight Financing Rate (SOFR). For South African Rand (ZAR) current benchmark rate is Johannesburg Interbank Average Rate (Jibar) but alternative rate is not determined.

Table 24.1 Reference interest rate recommendations by currencies

Currency	Current Interest rate Benchmark	Alternative reference rate
Australian Dollar (AUD)	Bank Bill Swap Rate (BBSW)	Reserve Bank of Australia (RBA) Cash Rate
Brazilian Real (BRL)	Overnight interbank offered rate (DI rate)	Average Interest Rate on Overnight Repurchase Agreements (Selic)
Canadian Dollar (CAD)	Canadian Dollar Offered Rate (CDOR)	Enhanced Canadian Overnight Repo Rate Average (CORRA)
Swiss Franc (CHF)	London Interbank Offered Rate (LIBOR)	Swiss Average Rate Overnight (SARON)
Euro (EUR)	London Interbank Offered Rate (LIBOR)	Euro Short-Term Rate (ESTER) or Euro Interbank Offered Rate (EURIBOR)
Euro (EUR)	Euro Interbank Offered Rate (EURIBOR)	Euro Short Term Rate (ESTER)
Euro (EUR)	Euro Overnight Index Average (EONIA)	Euro Short Term Rate (ESTER)
British Pound (GBP)	London Interbank Offered Rate (LIBOR)	Sterling Overnight Index Average (SONIA)
Hong Kong Dollar (HKD)	Hong Kong Interbank Offered Rate (HIBOR)	To be determined
Japanese Yen (JPY)	London Interbank Offered Rate (LIBOR)	Tokyo Overnight Average Rate (TONA) or Tokyo Interbank Offered Rate (TIBOR)
Japanese Yen (JPY)	Tokyo Interbank Offered Rate (TIBOR)	Tokyo Overnight Average Rate (TONA)
Japanese Yen (JPY)	Euroyen Tokyo Interbank Offered Rate (TIBOR)	Tokyo Overnight Average Rate (TONA)
Singapore Dollar (SGD)	Singapore Interbank Offered Rate (SIBOR)	N/A
Singapore Dollar (SGD)	Singapore Dollar Swap Offer Rate (SOR)	To be determined
United States Dollar (USD)	London Interbank Offered Rate (LIBOR)	Secured Overnight Financing Rate (SOFR)
South African Rand (ZAR)	Johannesburg Interbank Average Rate (JIBAR)	To be determined

Source: Financial Stability Board (2018)

Although it is in the transition period, the most commonly used benchmark rate in the international financial market is LIBOR. Therefore, LIBOR data was used in the study.

24.2.3 London Interbank Offered Rate (LIBOR)

In this part of the study, firstly the definition of LIBOR then LIBOR scandal lastly process of calculating LIBOR will be explained.

24.2.3.1 Definition of LIBOR

LIBOR is the reference interest rate applied by the world's most trusted banks to each other for short-term borrowing. LIBOR stands for "London Interbank Offered Rate". This rate is used for interest prices of securities like interest rate swaps, currency swaps, or mortgages.

24.2.3.2 What Is LIBOR Scandal?

In 2012, it was revealed that banks made false or misleading statements when determining the interest rates. This event is called LIBOR scandal. The scandal was the manipulation of the LIBOR and the misleading rates. The interest rate was removed from transparency according to bank's liquidity status.

After this scandal, an investigation was started on 16 banks. The investigation concluded that all banks in this scandal have paid fines. Some banks have suffered big losses.

LIBOR affects consumer loans in many countries. Changes in this ratio affect the increase and decrease of credit card interest rate, car, student loans, government and private sector debt securities, mortgage rates, swap transactions and adjustable mortgage rates.

These changes should be taken into account as it provides ease of borrowing between banks, companies and consumers.

Responsibility for the administration of LIBOR was British Bankers Association (BBA) until on 31st January 2014. On February 1, 2014, ICE Benchmark Association became the official administrator of the LIBOR (ICE, 2018).

As shown in Table 24.2 currently there are 20 panel banks. LIBOR is quoted in five currencies from panel banks. These currencies are United States Dollar (USD), Great British Pounds (GBP), Euro (EUR), Swiss franc. (CHF) and Japanese Yen (JPY).

According to Table 24.2 all currencies are contributed by Barclays Bank, Deutsche Bank (London Branch), HSBC Bank, JPMorgan Chase Bank (London Branch), Lloyds Bank, MUFG Bank, National Westminster Bank, UBS.

4 currencies are contributed by Citibank (London Branch) and Société Générale (London Branch).

3 currencies are contributed by Cooperatieve Rabobank, Credit Suisse (London Branch), Mizuho Bank, Royal Bank of Canada.

2 currencies are contributed by Crédit Agricole Corporate & Investment Bank, Santander UK, Sumitomo Mitsui Banking Corporation Europe Limited, The Norinchukin Bank.

1 currency is contributed by Bank of America (London Branch) and BNP Paribas (London Branch).

Table 24.2 Panel banks for the LIBOR benchmark

Panel Banks \ Currency	USD	GBP	EUR	CHF	JPY
Barclays Bank					
Deutsche Bank (London Branch)					
HSBC Bank					
JPMorgan Chase Bank (London Branch)					
Lloyds Bank					
MUFG Bank					
National Westminster Bank					
UBS					
Citibank (London Branch)					
Société Générale (London Branch)					
Cooperatieve Rabobank					
Credit Suisse (London Branch)					
Mizuho Bank					
Royal Bank of Canada					
Crédit Agricole Corporate & Investment Bank					
Santander UK					
Sumitomo Mitsui Banking Corporation Europe Limited					
The Norinchukin Bank					
Bank of America (London Branch)					
BNP Paribas (London Branch)					

Source: ICE (2018)

24.2.3.3 Process of Calculating LIBOR

The following table lists panel composition of banks and methodology for calculation of LIBOR. LIBOR is calculated by an arithmetic mean after a few corrections. Each panel banks determines an interest rate. All determined interest rates are ranked from high to low. The first quarter and the last quarter of the number of participating banks are removed. The remaining interest rates are collected and divided by the number of remaining banks. Table 24.3 shows methodology for calculation of LIBOR.

According to Table 24.3, if the number of banks are 16, first highest 4 and last lowest 4 rates are removed and average of 8 contributor rates are calculated. If there are 15 banks for calculation, first highest 4 and last lowest 4 rates are removed and average of 7 contributor rates are calculated. If the number of banks are 14, first highest 3 and last lowest 3 rates are removed and average of 8 contributor rates are calculated. If there are 13 banks for calculation, first highest 3 and last lowest 3 rates are removed and average of 7 contributor rates are calculated. If the number of banks are 12 banks for calculation, first highest 3 and last lowest 3 rates are removed and average of 6 contributor rates are calculated. If there are 11 banks, first highest 3 and last lowest 3 rates are removed and average of 5 contributor rates are calculated. If the number of banks are 11, 12, 13 or 14, first highest 3 and last lowest 3 rates are removed. If the number of banks are 15 and 16, first highest 4 and last lowest 4 rates are removed.

24.2.4 Bitcoin

In this part of the study, firstly the definition of Bitcoin then Bitcoin Mining Costs, lastly Bitcoin ATM installations will be explained.

24.2.4.1 Definition of Bitcoin

Bitcoin (BTC) is the first decentralized digital currency, created in 2009. It was invented by Satoshi Nakamoto based upon open source software and allows users to make peer-to-peer transactions via the Internet that are recorded in a decentralized, public ledger (Bitcoin (Cryptocurrency), 2018).

The cryptocurrency market has 2090 cryptocurrencies and 15.678 markets (All Cryptocurrencies, 2018). Figure 24.1 shows cryptocurrencies by market capitalization.

As seen in the Fig. 24.1 Bitcoin accounts for nearly 50% of the total $113,316,966,816, Ethereum accounts for close to 10% of the total and $22,615,152,829, XRP accounts for nearly 10% and $21,352,238,610, Bitcoin Cash accounts for 5% and $11,018,617,273, EOS accounts for 2% and $5,141,945,457, Stellar accounts for 2% and $4,885,598,865, Litecoin accounts for 2% and $3,244,324,937, Cardano accounts for 1% and $2,055,151,825, Others account for 1% and $36,047,142,840.

Table 24.3 Methodology for calculation of LIBOR

Number of Banks	Methodology	Number of Banks	Methodology	Number of Banks	Methodology	Number of Banks	Methodology	Number of Banks	Methodology	Number of Banks	Methodology
1	4 highest rates	1	4 highest rates	1	3 highest rates	1	3 highest rates	1	3 highest rates	1	3 highest rates
2		2		2		2		2		2	
3		3		3		3		3		3	
4		4		4	8 contributor rates averaged	4	7 contributor rates averaged	4	6 contributor rates averaged	4	5 contributor rates averaged
5	8 contributor rates averaged	5	7 contributor rates averaged	5		5		5		5	
6		6		6		6		6		6	
7		7		7		7		7		7	
8		8		8		8		8		8	
9		9		9		9		9		9	3 lowest rates
10		10		10		10		10	3 lowest rates	10	
11		11		11		11	3 lowest rates	11		11	
12		12	4 lowest rates	12	3 lowest rates	12		12			
13	4 lowest rates	13		13		13					
14		14		14							
15		15									
16											

Source: https://www.theice.com

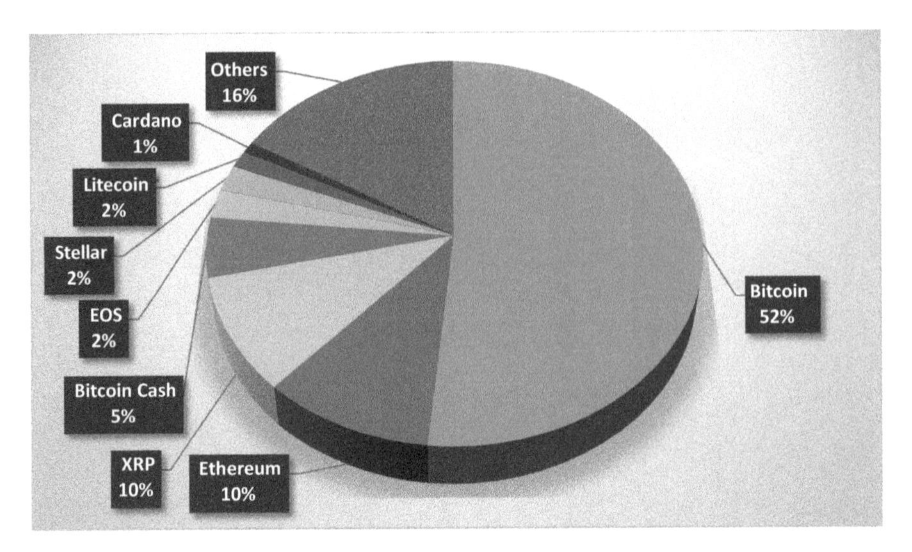

Fig. 24.1 Cryptocurrencies by market capitalization (Source: All Cryptocurrencies 2018)

24.2.4.2 Bitcoin Mining Costs

The Bitcoin network is a peer-to-peer network that monitors and manages both the generation of new Bitcoins and the consistency verification of transactions in Bitcoins. This network is composed by a high number of computers connected to each other through the Internet. They perform complex crypto graphic procedures which generate new Bitcoins (mining) and manage the Bitcoin transactions register, verifying their correctness and truthfulness (Cocco & Marchesi, 2016). Mining is essential for the circulation of Bitcoin but electricity is one of its biggest costs.

Figure 24.2 shows information about cost to mine one Bitcoin. It is based on average electricity rates of countries (Bitcoin Mining Costs Throughout the World, 2018).

According to the Fig. 24.2 highest cost of Bitcoin mining is in South Korea ($26,170), in Niue $17.57, in "Bahrain and Solomon Islands" average $16.49 in Cook Islands $15.86, in "Denmark, Germany, Marshall Islands, Tonga, Turks and Caicos Islands, Tuvalu" average $14.42 in "Belgium and Vanuatu" average $13.28, in "Kiribati and Western Samoa" average $12.83, in "Curaçao, Ireland, Spain, Sri Lanka, Tahiti" average $11.37, in "American Samoa, Guyana, Italy, Portugal" average $10.62, in "Australia, Chile, Greece, Jordan, Netherlands, Palau, Papua New Guinea" average $9.50, "Cambodia, Cyprus, Japan, Liechtenstein, Nicaragua, Rwanda, United Kingdom, Uruguay" average $8.57, in "Colombia, Finland, France, Hong Kong, Jamaica, Latvia, Luxembourg, Mexico, New Zealand, Norway, Pakistan, Philippines, Slovenia, Switzerland, Uganda" average $7.53, in "Brazil, Iraq, Israel, Malta, Montenegro, Poland" average $6.46, in "Croatia, Estonia, Fiji, Gibraltar, Hungary, Lithuania, Malaysia, Nigeria, Romania, Singapore, South Africa, Venezuela" average $5.49, "Argentina, Bosnia and Herzegovina,

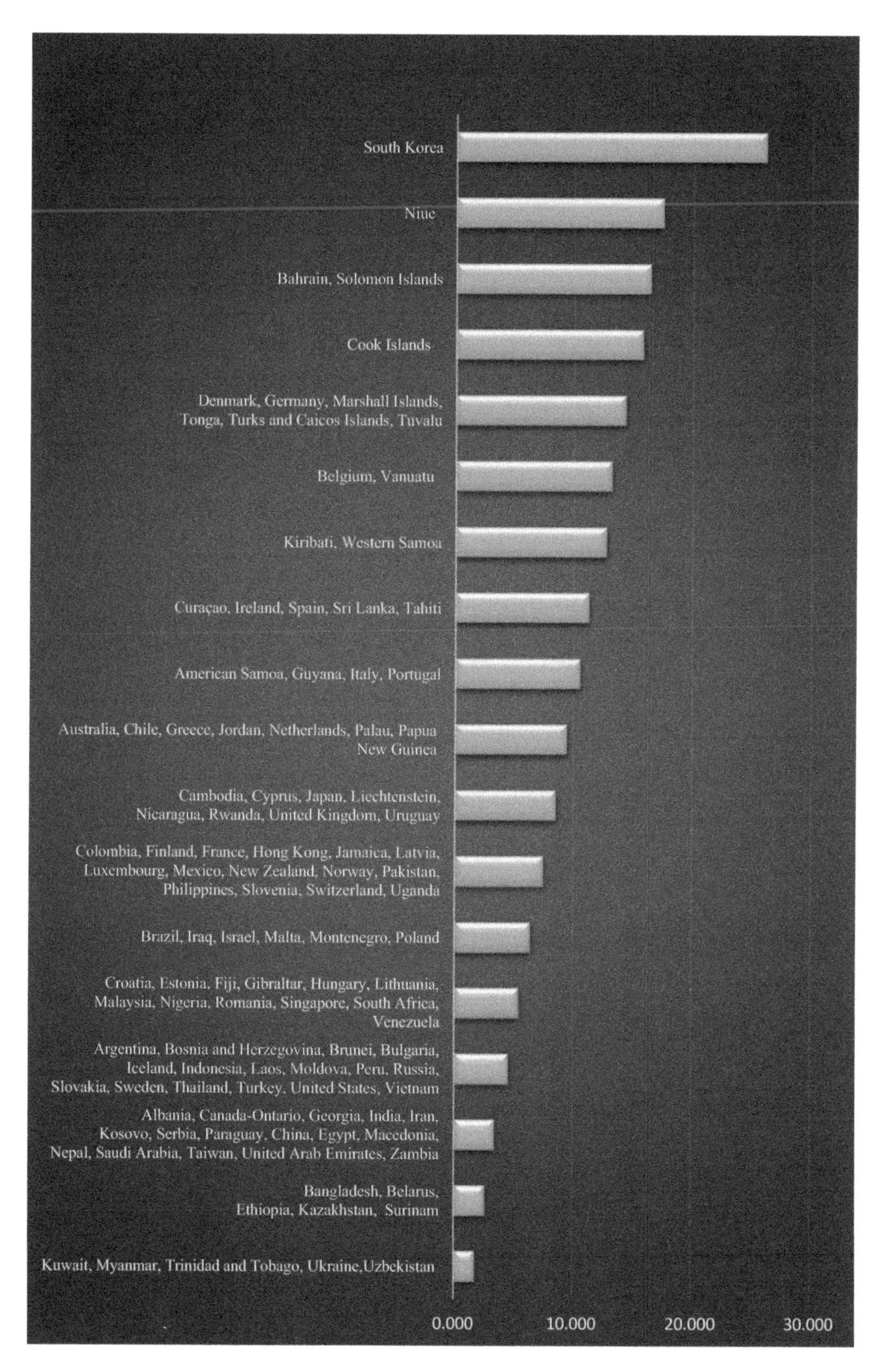

Fig. 24.2 Bitcoin mining costs by country (Source: Bitcoin Mining Costs Throughout the World 2018)

Brunei, Bulgaria, Iceland, Indonesia, Laos, Moldova, Peru, Russia, Slovakia, Sweden, Thailand, Turkey, United States, Vietnam" average $4.63, in "Albania, Canada-Ontario, Georgia, India, Iran, Kosovo, Serbia, Paraguay, China, Egypt, Macedonia, Nepal, Saudi Arabia, Taiwan, United Arab Emirates, Zambia" average $3.44, "Bangladesh, Belarus, Ethiopia, Kazakhstan, Surinam" average $2.64. Lowest cost of Bitcoin mining is in "Kuwait, Myanmar, Trinidad and Tobago, Ukraine, Uzbekistan" and average $1.76.

24.2.4.3 Bitcoin ATM Installations

Bitcoin ATM is an Internet connected kiosk that allows customers to purchase Bitcoins with deposited cash. Bitcoin ATMs are rarely operated by major financial institutions. As such, they do not connect customers to a bank account. Customers, instead, deposit cash into the Bitcoin ATM, which can then be used to purchase the cryptocurrency (Definition of Bitcoin ATM, 2018). The distribution of Bitcoin ATM installations are indicated in the Table 24.4 (Bitcoin ATMs by Country, 2018).

As shown in Table 24.4 highest number of ATMs are in USA and 2400 ATMs. There are 689 ATMs in Canada, in Austria 265 ATMs, in United Kingdom 209 ATMs, in Russian Federation 76 ATMs, in Spain 68 ATMs, in Czech Republic 67 ATMs, in Australia 54 ATMs, in Switzerland 42 ATMs, in Italy 39 ATMs, in Slovakia 36 ATMs, in Hong Kong 32 ATMs, in Poland 27 ATMs, in Colombia and Netherlands 26 ATMs, in Romania 24 ATMs, in Greece 19 ATMs, in Finland 18 ATMs, in Slovenia 13 ATMs, in "Belgium, Hungary, Mexico, Panama" 11 ATMs, in "Georgia, Japan, Singapore" 10 ATMs, in "Israel, Taiwan, Ukraine" 8 ATMs, in "Dominican Republic and Malaysia" 7 ATMs, in "Croatia, Serbia, South Africa" 6 ATMs, in Vietnam 5 ATMs, in "Argentina and Liechtenstein" 4 ATMs, in "Costa Rica, Ecuador, Kazakhstan, Kosovo, Malta, Portugal" 3 ATMs, in "Brazil, Bulgaria, Chile, China, Denmark, Estonia, France, Guam, Ireland, New Zealand, Norway, Saudi Arabia, Thailand" 2 ATMs, in "Albania, Anguilla, Armenia, Aruba, Barbados, Bosnia and Herzegovina, Cyprus, Djibouti, Germany, Guatemala, Iceland, Indonesia, Kenya, Latvia, Mongolia, Peru, Philippines, Saint Kitts and Nevis, San Marino, South Korea, Turkey, Uganda and Zimbabwe" 1 ATM.

24.3 Econometric Analysis

In this study the reasons for the selection of these analysis methods are; to compare the short and long term relationships of the variables in the model and to determine the stability and causality of the relationship between variables. Firstly, the data set will be explained in this study. Next, the correlation between the variables will be examined. Then, VAR, Granger Causality and ARDL Error Correction Model will be made. All analysis were made by using Eviews 9.0 program.

Table 24.4 The distribution of Bitcoin ATM installations

Country	Number of locations	Country	Number of locations
United States	2400 ATMs	Kazakhstan	3 ATMs
Canada	689 ATMs	Kosovo	3 ATMs
Austria	265 ATMs	Malta	3 ATMs
United Kingdom	209 ATMs	Portugal	3 ATMs
Russian Federation	76 ATMs	Brazil	2 ATMs
Spain	68 ATMs	Bulgaria	2 ATMs
Czech Republic	67 ATMs	Chile	2 ATMs
Australia	54 ATMs	China	2 ATMs
Switzerland	42 ATMs	Denmark	2 ATMs
Italy	39 ATMs	Estonia	2 ATMs
Slovakia	36 ATMs	France	2 ATMs
Hong Kong	32 ATMs	Guam	2 ATMs
Poland	27 ATMs	Ireland	2 ATMs
Colombia	26 ATMs	New Zealand	2 ATMs
Netherlands	26 ATMs	Norway	2 ATMs
Romania	24 ATMs	Saudi Arabia	2 ATMs
Greece	19 ATMs	Thailand	2 ATMs
Finland	18 ATMs	Albania	1 ATM
Slovenia	13 ATMs	Anguilla	1 ATM
Belgium	11 ATMs	Armenia	1 ATM
Hungary	11 ATMs	Aruba	1 ATM
Mexico	11 ATMs	Barbados	1 ATM
Panama	11 ATMs	Bosnia and Herzegovina	1 ATM
Georgia	10 ATMs	Cyprus	1 ATM
Japan	10 ATMs	Djibouti	1 ATM
Singapore	10 ATMs	Germany	1 ATM
Israel	8 ATMs	Guatemala	1 ATM
Taiwan	8 ATMs	Iceland	1 ATM
Ukraine	8 ATMs	Indonesia	1 ATM
Dominican Republic	7 ATMs	Kenya	1 ATM
Malaysia	7 ATMs	Latvia	1 ATM
Croatia	6 ATMs	Mongolia	1 ATM
Serbia	6 ATMs	Peru	1 ATM
South Africa	6 ATMs	Philippines	1 ATM
VietNam	5 ATMs	Saint Kitts and Nevis	1 ATM
Argentina	4 ATMs	San Marino	1 ATM
Liechtenstein	4 ATMs	South Korea	1 ATM
Costa Rica	3 ATMs	Turkey	1 ATM
Ecuador	3 ATMs	Uganda	1 ATM
		Zimbabwe	1 ATM
Total			4340 ATM

Source: Bitcoin ATMs by Country (2018)

Table 24.5 Variables definition

Variables	Explanation	Source
BCV_t	Bitcoin price variation	Bitcoin Price (BTC) (2018)
BC_t	Bitcoin price	Bitcoin Price (BTC) (2018)
LR_t	LIBOR	LIBOR Rates (2018)
TLR_t	Trend LIBOR	LIBOR Rates (2018)

24.3.1 *Data Set*

The following Table 24.5 lists definition of variables which are used in this study. Weekly data from 21 Oct 2013 to 10 Sep 2018 are used for the analysis.

According to the Table 24.5, Bitcoin price variation rate valuable was taken from Bitcoin price data of "Bitcoin Price (BTC), 2018". This variable represents cryptocurrencies. LIBOR valuable was taken from "LIBOR Rates, 2018". This variable represents international interest rates.

According to Graph 24.1, trend effect is observed for LIBOR. Therefore, this variable is extracted from trend and a new series has been obtained. Graph 24.1 shows the changes in the series between periods. In this study, the stability test was performed before the analysis. Conditions such as constant and trend effects and lack of these effects were observed from the graphs. According to results, the model of the unit root test was determined.

As seen in the Graph 24.1, Bitcoin price change rate variable was obtained from Bitcoin price variable and included in the study. There is no trend, constant and seasonal effects in Bitcoin price variation variable. On the other hand, the LR variable, which is the first form of the LIBOR variable, is clearly seen in the trend effect. By extracting the trend effect on this variable, a new LIBOR series named TLR was obtained. In the new variable TLR, it is seen that the trend effect has decreased significantly. The stationary analysis of these variables will be examined in the next section.

24.3.2 *Correlation Analysis*

Information about the size and direction of the relationship between the dependent variable and the independent variable must be explained in regression analysis for econometric modelling. For this purpose, the standard unit of measurement obtained from the covariance, which is not a standard unit of measure, is called the coefficient of correlation coefficient (Guris & Caglayan, 2000).

The following Table 24.6 shows that the variables used in the correlation matrix are shown in all cases.

As shown in the Table 24.6, it is seen that the correlation relations between Bitcoin price and the natural shape of LIBOR ratio are high and positive directional as seen in the Correlation matrix. However, the relationship between the rate of

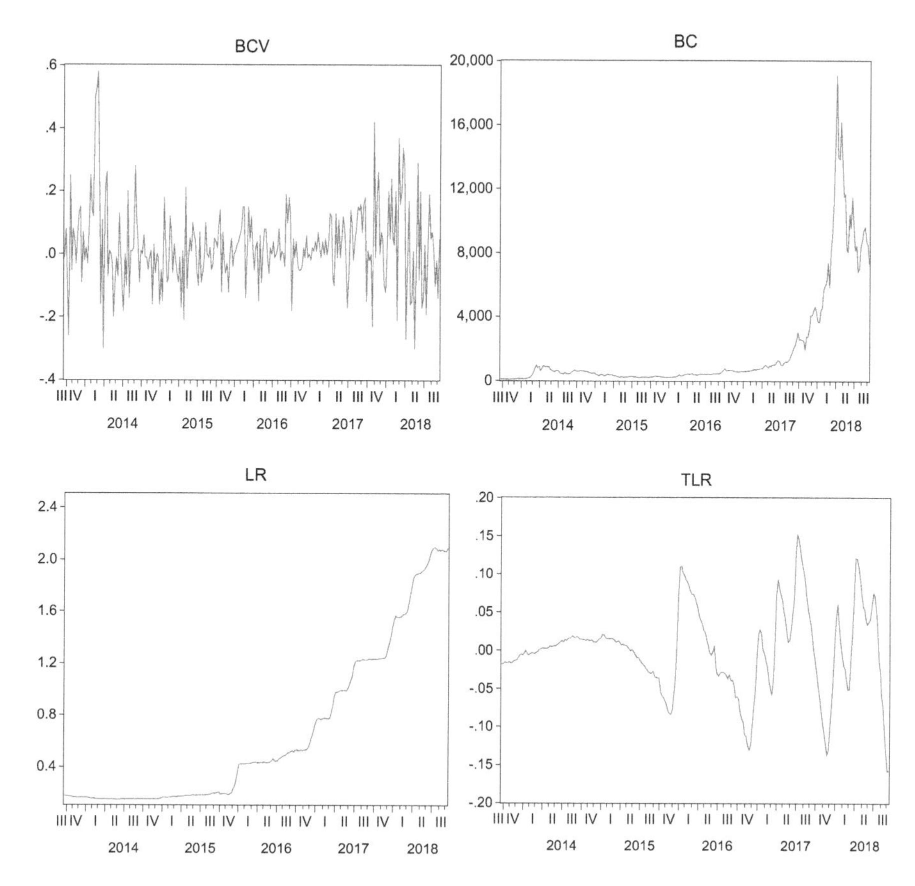

Graph 24.1 Series of views

Table 24.6 Correlation matrix

Variables	BC	LR	BCV	TLR
BC	1	0.83	−0.04	0.12
LR	0.83	1	0.04	0.09
BCV	−0.04	0.04	1	−0.01
TLR	0.12	0.09	−0.01	1

change of Bitcoin prices and the rate of LIBOR seems to be in the right direction but at a low level. Finally, the correlation between the Bitcoin price change rate and the no-trend LIBOR ratio was found to be negative and low.

Table 24.7 Stationary analysis and unit root test

	Tests								
	ADF 79				Perron 89 (Breakpoint)				
	Exogenues								
	None		Constant + Trend		Model A		Model B		
Variable	Test	Prob	Test	Prob.	Test	Prob.	Test	Prob	Break Date
LR_t			−0.83	0.96			−4.40∗	0.06	2015.4
TLR_t	−5.68∗	0.00			−6.12∗	0.01			2017.3, 4
BC_t			−2.20	0.48			−3.87	0.22	2017.4
BCV_t	−8.67∗	0.00			−15.10∗	0.01			2014.1

Critical values according to ADF 79 unit root test; in no constant + no trend model 1%: −2.573, 5%: −1.942 and 10%: −1.615; in constant + trend model 1%: −3.994, 5%: −3.427, 10%: −3.137. Critical values according to Perron 89 (Breakpoint) unit root test; in Model A, 1%: −4.949, 5%: −4.443 and 10%: −4.193 in Model B and 1%: −5.067, 5%: −4.524 ve 10%: −4.261
*indicates that the series is stationary

24.3.3 Stationary Analysis

In time series analysis, the change in economic data is dependent on stochastic process. This situation has made stability analysis a basic condition (Maddala & Lahiri, 2009). There are two methods used to determine the stability of the series. These; the graphical method is the correlogram method and the statistical method which is the unit root tests (Johnston & Dinardo, 1997).

In the unit root test estimations used in this study, if the $\delta = 0$ space hypothesis is established and the δ test statistic value is greater than the McKinnon critical value as absolute value, if H_1 is accepted, there is no unit root and the series is stationary. Otherwise, if H_0 is accepted, there is a unit root in the series and it is decided that the series is not stationary (Gujarati, 2003).

If deviations in the national economies are not taken into account in the context of regression models, some deviant results can be seen in the results of the empirical studies for the time series.

In this case, it is important to take into account the breaks that occur with the level and trend effect of the series which are normally static. According to the unit root test developed by Dickey and Fuller (1979), it can be seen that the series is stationary when it is known that the refraction time developed by Perron is still stable (Perron, 1989).

Table 24.7 reports stationary analysis of the study and the results of the unit root test.

As shown in Table 24.7, TLR_t and BCV_t variables with ∗ marked test values are stationary at I(0) level. Also break periods according to break test are also seen. In the unit root tests of the variables, Graph 24.1 were taken into consideration in determining the model type.

Table 24.8 Var analysis lags criterion indicates

VAR Lag Order Selection Criteria
Endogenous variables: BCV TLR
Exogenous variables:
Date: 11/09/18 time: 23:15
Sample: 9/23/2013–9/10/2018
Included observations: 252

Lag	LogL	LR	FPE	AIC	SC	HQ
1	856.5026	NA	3.95e−06	−6.765893	−6.709871	−6.743351
2	969.6915	222.7846	1.66e−06	−7.632472	−7.520427[a]	−7.587387[a]
3	974.4119	9.216056	1.65e−06	−7.638190	−7.470122	−7.570563
4	980.3847	11.56638[a]	1.63e−06[a]	−7.653847[a]	−7.429756	−7.563677
5	982.0979	3.290425	1.66e−06	−7.635698	−7.355584	−7.522986
6	985.8682	7.181566	1.66e−06	−7.633875	−7.297739	−7.498621
7	988.1035	4.222240	1.68e−06	−7.619869	−7.227710	−7.462073
8	989.4440	2.510779	1.72e−06	−7.598762	−7.150581	−7.418423

[a]Indicates lag order selected by the criterion
LR sequential modified LR test statistic (each test at 5% level); *FPE* final prediction error; *AIC* Akaike information criterion; *SC* Schwarz information criterion; *HQ* Hannan-Quinn information criterion

In the following section, in the analysis of VAR, Granger Causality and ARDL models, Bitcoin price change rate (BCV_t) and no trend LIBOR LIBOR (TLR_t) variables, which are the stationary states of the variables in the study, were used.

24.3.4 VAR Analysis

VAR analysis was developed by Sims (1980). It is one of the prospective time series analysis methods of the relationships between internal endogenious and external exgojen variables in which various macroeconomic values are the subject of analysis (Baltagi, 2008).

The main purpose of VAR analysis is to reveal the mutual effect between variables rather than parameter estimations in the model (Enders, 2004). The two-variable VAR model is shown with the equations below (Gujarati, 2003).

$$M_{1t} = \alpha_0 + \sum_{i=1}^{k} \beta_{1i} M_{t-i} + \sum_{i=1}^{k} \gamma_{2i} R_{t-i} + u_t \tag{24.1}$$

$$R_t = \alpha_1 + \sum_{i=1}^{k} \theta_{1i} M_{t-i} + \sum_{i=1}^{k} \gamma_{2i} R_{t-i} + u_{2t} \tag{24.2}$$

The critical values taken into account in determining the lag value of the model in the VAR analysis are shown in the Table 24.8.

Table 24.9 VAR of estimates

Vector Autoregression Estimates
Date: 11/09/18 Time: 11:42
Sample (adjusted): 10/21/2013–9/10/2018
Included observations: 256 after adjustments

	BFD	Standard errors	t-statistics	RLF	Standard errors	t-statistics
BCV (−1)	0.150795	0.06181	2.43949	0.007695	0.00506	1.52084
BCV (−2)	0.191665	0.06186	3.09849	0.005725	0.00506	1.13068
BCV (−3)	0.117571	0.06208	1.89396	−0.005464	0.00508	−1.07537
BCV (−4)	−0.137207	0.06162	−2.22658	0.001540	0.00504	0.30542
TLR (−1)	−1.928579	0.77605	−2.48511	1.774989	0.06352	27.9442
TLR (−2)	5.170312	1.58457	3.26291	−0.880699	0.12970	−6.79052
TLR (−3)	−5.129132	1.59578	−3.21419	−0.012307	0.13061	−0.09422
TLR (−4)	1.933192	0.79388	2.43512	0.062487	0.06498	0.96166
R-squared	0.088834			0.968513		
Adj. R-squared	0.063115			0.967625		
Sum sq. resids	3.739531			0.025052		
S.E. equation	0.122796			0.010051		
F-statistic	3.454094			1089.767		
Log likelihood	177.7075			818.4449		
Akaike AIC	−1.325840			−6.331601		
Schwarz SC	−1.215053			−6.220814		
Mean dependent	0.026055			0.000260		
S.D. dependent	0.126864			0.055858		
Determinant resid covariance (dof adj.)				1.52E−06		
Determinant resid covariance				1.43E−06		
Log likelihood				996.5222		
Akaike information criterion				−7.660330		
Schwarz criterion				−7.438756		

According to the Table 24.8, in general it is seen that the lag value of VAR model is usually 4. Next VAR analysis was obtained from model with lag value 4.

Table 24.9 shows the prediction results of the 4 lag VAR model.

According to the Table 24.9; in the 4 lag VAR model, the effect of the delayed values of LIBOR on Bitcoin was found to be generally significant. It is observed that

Table 24.10 Roots of characteristic polynomial

Endogenous variables: BCV TLR
Exogenous variables:
Lag specification: 1–4
Date: 11/09/18

Root	Modulus
0.744669 – 0.182844i	0.766788
0.744669 + 0.182844i	0.766788
−0.404153 – 0.459969i	0.612299
−0.404153 + 0.459969i	0.612299
0.506475 – 0.215495i	0.550414
0.506475 + 0.215495i	0.550414
0.547652	0.547652
−0.315851	0.315851

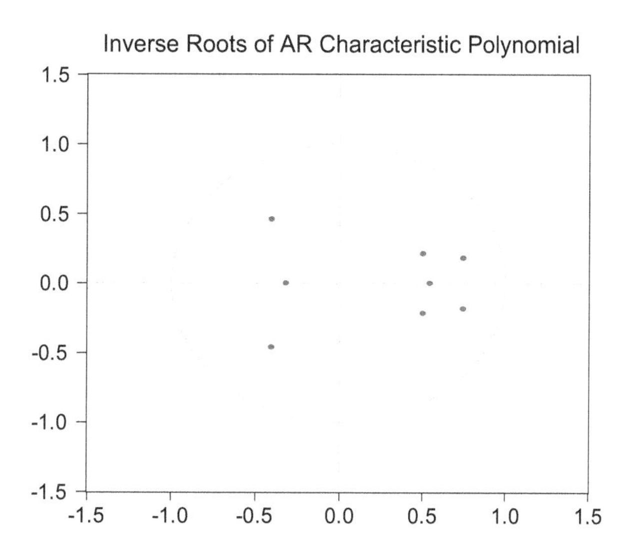

Graph 24.2 VAR Analysis Unit Circle Definition: No root lies outside the unit circle, VAR satisfies the stability condition

the coefficients showing the effect of the delayed values of Bitcoin on LIBOR are meaningless. As an indicator of the stability of the VAR model, the polynomial and VAR unit circle is also used to measure the meaning of the analysis. Table 24.10 shows, roots of characteristic polynomial and Graph 24.2 shows unit circles of VAR model.

According to the Table 24.10 and Graph 24.2, VAR model is stable and significant mean. For this reason, the VAR model has gained interpretable characteristics. Effect response functions are showed in graphic. The method used by estimators is the impulse-response function because the coefficients are difficult to interpret when making model estimation in VAR analysis (Gujarati, 2003). In the Graph 24.3, the impulse-response status of the variables in VAR model is showed.

According to Graph 24.3, LIBOR was negatively effected by Bitcoin for 2 periods. This effect turned into positive in the third period. Its effect was decreased

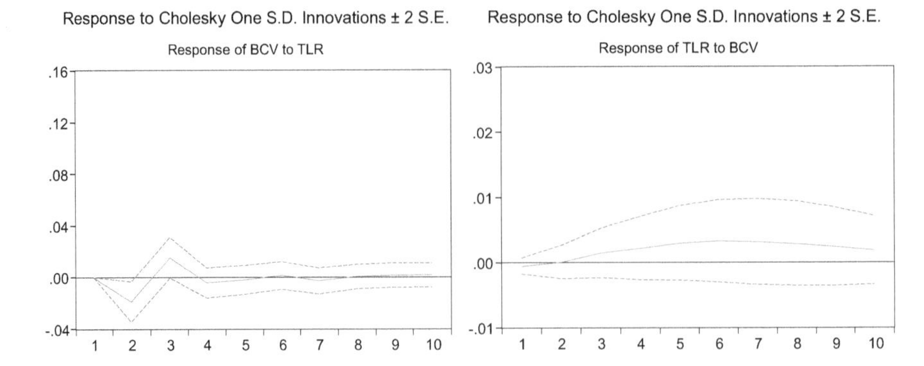

Graph 24.3 Impulse-response functions

Table 24.11 Variance decomposition

	LIBOR		BITCOIN	
Period	TLR	BCV	BCV	TLR
1	99.71154	0.288456	100.0000	0.000000
2	99.93031	0.069692	97.63425	2.365747
3	99.78598	0.214024	96.45891	3.541094
4	99.61873	0.381274	96.44310	3.556901
5	99.38208	0.617924	96.42303	3.576971
6	99.14594	0.854061	96.41170	3.588302
7	98.95385	1.046151	96.38580	3.614204
8	98.79720	1.202802	96.38229	3.617710
9	98.68471	1.315286	96.35590	3.644095
10	98.61107	1.388934	96.32609	3.673909
Average	99.26	1.23	96.88	3.11

Table 24.12 Other tests of VAR analysis

Tests name	Test	Prob.
L M test	6.12	0.18
Normality test	250.04	0.00
Heteroskedasticity tests	56.328	0.10

in later periods. Generally Bitcoin was positively effected by LIBOR. As shown in Graph 24.3, LIBOR is more efficient and prominent on Bitcoin prices.

The other one of the analysis techniques that reveal the interpretable feature of VAR model is the Variance Decomposition. This analysis tests that a variable can be effective in the changes in other variables as well as its own shocks (Sevuktekin & Cinar, 2014). Table 24.11 shows other tests of VAR analysis.

According to the Table 24.11, there are two results. First one is that Bitcoin prices are the source of changes in LIBOR and the second one is that LIBOR is the source of changes in Bitcoin prices. According to this result, LIBOR is more effective in determining Bitcoin prices. This was also demonstrated by the average values seen in the Table 24.11.

The following Table 24.12 shows other tests of VAR analysis.

According to the Table 24.12, Lagrange Multiplier (LM), normality and heteroskedasticity assumptions, VAR model was found to be statistically significant.

24.3.5 Causality Analysis

In the economics, causality tests among the variables are examined. These tests, developed by Granger, are widely used for long-term time series (Granger, 1969). In econometric literature, causality tests are divided into two as classical and Granger causality. In classical causality, variables' number of lag in the analysis may be different, Granger is considered to be the same in causality (Tari, 2005). Representation of causality analysis will be explained by the equations given below.

$$X_t = \sum_{i=1}^{n} \alpha_i X_{t-i} + \sum_{i=1}^{n} \partial_i ZX_{t-i} + e_{1t} \tag{24.3}$$

$$ZX_t = \sum_{i=1}^{n} \alpha_i ZX_{t-i} + \sum_{i=1}^{n} \partial_i X_{t-i} + e_{2t} \tag{24.4}$$

According to the these equations, only the ∂_i parameter in the equation is significant in the parameters of the equations in the determination of the one-way causality relationship. In the determination of the bi-directional causation relationship, the α_i and ∂_i parameters in the equations are all significant (Kutlar, 2009).

Table 24.13 shows granger causality tests of this study.

According to the Table 24.13, causality is significant from LIBOR to Bitcoin prices. On the other side, causality is non-significant from Bitcoin prices to LIBOR. This result shows that there is a one-way causality between Bitcoin and LIBOR.

Table 24.13 Granger causality tests

Date: 9/2372013–9/10/2018	Granger causality tests					
	Pairwise Granger causality tests			VAR Granger Wald tests		
Variables and null hypothesis	Obs	Chi-Sq	Prob	Df	Chi-Sq	Prob
TLR does not Granger cause BCV	256	2.94	0.02	4	11.85	0.01
BCV does not Granger cause TLR		1.29	0.27	4	4.93	0.29

24.3.6 ARDL Analysis

The ARDL model developed by Pesaran and Shin (1998) and Pesaran, Shin, and Smith (2001), can be used as an alternative and multifunctional according to the co-integration tests that examine other long-term relationships (Belloumi, 2014). In ARDL model, long-term prediction results are more important than other methods (Harris & Sollis, 2003). In addition, it is one of the important advantages of the fact that the number of data is low and it can give best results (Duasa, 2007). For ARDL model;

$$Y_t = \partial_0 + \partial_1 X_t + \partial_2 Z_t + e_t \tag{24.5}$$

A model equation in the form of

$$\Delta y_t = \partial_0 + \sum_{i=1}^{p} \beta_i \Delta y_{t-i} + \sum_{i=0}^{p} \delta_i \Delta x_{t-i} + \sum_{i=0}^{p} \lambda_i \Delta z_{t-i} + a_1 y_{t-1} + a_2 x_{t-1} + a_3 z_{t-1} + u_t \tag{24.6}$$

$$\Delta y_t = \partial_0 + \sum_{i=1}^{p} \beta_i \Delta y_{t-i} + \sum_{i=0}^{p} \delta_i \Delta x_{t-i} + \sum_{i=0}^{p} \lambda_i \Delta z_{t-i} + u_t \tag{24.7}$$

As seen in equations above, ϑ, β, α and λ represent the parameters, while e represents the error terms in the equation. In this method, the best model should be decided among the ARDL models firstly (Pesaran, Shin, & Smith, 2001).

The validity of some econometric tests is important prior to ARDL analysis. From these tests; LM test autocorrelation, Heteroscedasity Harvey, Ramsey Reset and specification error, CUSUM structural fracture recursive estimation and Jarque-Bera validity of normality assumptions in models is the basic conditions of model estimation in analysis (Greene, 2000).

Table 24.14 reports results of long-run estimation in ARDL model used in this study.

As shown in the Table 24.14, generally coefficients are statistically significant. Especially the effect of LIBOR variable 2, 3 and 4 delayed on Bitcoin is statistically significant. According to Lagrange Multiplier (LM), normality, heteroskedasticity assumptions, there was no problem for this model. In ARDL analysis, CUSUM test is used to determine the parameter stability and structural breaks of the models (Brown, Durbin, & Evans, 1975).

The Graph 24.4 shows a CUSUM test which is a structural break test in the ARDL model.

According to the Graph 24.4, there is no structural break in ARDL model. After making necessary transformations in ARDL model equivalents, unlimited and limited ARDL model equations, also called Bounds test, are estimated. H_0: $\alpha_1 = \alpha_2 = \alpha_3 = 0$ F test is performed for the hypothesis. Estimated F account value, Pesaran et al. (2001) the null hypothesis (H_0) is rejected and the alternative

Table 24.14 Results of long-run estimation in ARDL model

Dependent variable: BFD				
Method: ARDL				
Date: 11/09/18 time: 12:06				
Sample (adjusted): 10/21/2013–9/10/2018				
Included observations: 256 after adjustments				
Maximum dependent lags: 4 (automatic selection)				
Model selection method: Akaike info criterion (AIC)				
Dynamic regressors (4 lags, automatic): RLF				
Fixed regressors: C				
Number of models evaluated: 20				
Selected model: ARDL (4, 4)				
Variable	**Coefficient**	**Std. error**	**t-statistic**	**Prob.***
BCV (−1)	0.137621	0.061986	2.220185	0.0273*
BCV (−2)	0.176601	0.061933	2.851484	0.0047*
BCV (−3)	0.094896	0.062148	1.526925	0.1281
BCV (−4)	−0.156349	0.061628	−2.536963	0.0118*
TLR	−0.591963	0.769139	−0.769644	0.4422
TLR (−1)	−0.818049	1.566311	−0.522277	0.6019
TLR (−2)	4.573899	1.709748	2.675188	0.0080*
TLR (−3)	−5.126682	1.581134	−3.242409	0.0013*
TLR (−4)	1.966529	0.788048	2.495443	0.0132*
C	0.019482	0.008022	2.428523	0.0159*
	F tests	Prob.		
Ramsey reset	1.79	0.1311		
Heteroskedasticity	1.29	0.2392		
LM	2.40	0.0504		

*indicates that the ARDL long term coefficients are significant

hypothesis (H_1) is accepted. In this case, it is assumed that there is a long-term co-integrated relationship between the variables y, x, z, and so, it is decided that a regression model can be created with variables that are stationary at different levels (Shrestha, 2006).

The Bounds test developed by Ohtani and Kobayashi (1986) has been the solution to the long-term relationship of the variables in the model. Table 24.15 shows bound tests results in error correction model.

According to the Table 24.15, Bounds test coefficients were found to be above the upper limit of critical values. This suggests that there is a long-term mutual co-integration between variables. The following Table 24.16 shows bound test parameter estimates results.

As shown in the Table 24.16, it is observed that the effects of LIBOR's 2 and 3. delayed values on Bitcoin were statistically significant. It is observed that the changes in the explanatory variables in the ARDL model by the bound test and the R squared value explain the changes in the variable in the middle level. In addition, it was observed that F-test was significant and the parameter coefficients in the estimation results were found to be statistically significant.

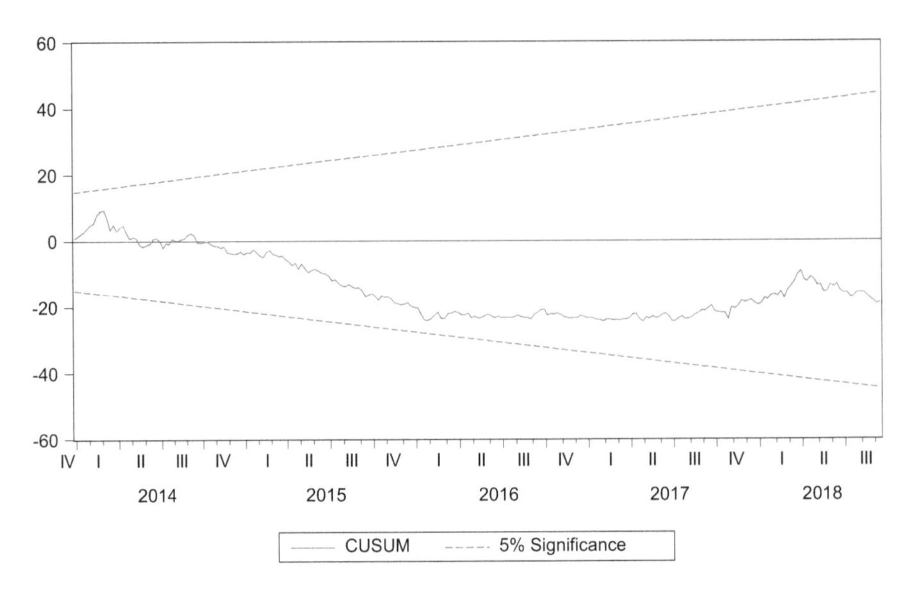

Graph 24.4 CUSUM tests

Table 24.15 Bound tests

Sample: 10/21/2013–9/10/2018		
Included observations: 256		
Null hypothesis: No long-run relationships exist		
Dependent variable: BCV		
Test statistic	**Value**	**k**
F-statistic	28.12723	1
Critical value bounds		
Significance	**I0 bound**	**I1 bound**
10%	4.04	4.78
5%	4.94	5.73
2.5%	5.77	6.68
1%	6.84	7.84

The above-mentioned variables in the table for the Bound test prediction results are given in the table below together with the error correction model (ECM) as the short term coefficients of the ARDL model. The following Table 24.17 lists short-run estimation in ARDL model.

According to the Table 24.17, error correction term in ARDL model is statistically significant. Also, it is negative and 0–1 as expected. −0.74 value found in model shows, deviations in the short run will be balanced after 1.35 periods. In addition, LIBOR ratio short-term variables are statistically significant. Furthermore, the lagged values of LIBOR were found to be effective on Bitcoin prices. In addition, the F test was significant, and the coefficients in the model were generally statistically significant. Variations in the independent variable in the R^2 value model

Table 24.16 Bound test parameter estimates results

Test Equation:				
Dependent variable: D(BFD)				
Method: Least squares				
Date: 11/09/18 time: 12:07				
Sample: 10/21/2013–9/10/2018				
Included observations: 256				
Variable	**Coefficient**	**Std. error**	**t-statistic**	**Prob.**
D(BFD (−1))	−0.115148	0.094127	−1.223326	0.2224
D(BFD (−2))	0.061453	0.082817	0.742033	0.4588
D(BFD (−3))	0.156349	0.061628	2.536963	0.0118
D(RLF)	−0.591963	0.769139	−0.769644	0.4422
D(RLF (−1))	−1.413745	0.977605	−1.446131	0.1494
D(RLF (−2))	3.160154	0.978540	3.229459	0.0014
D(RLF (−3))	−1.966529	0.788048	−2.495443	0.0132
C	0.019482	0.008022	2.428523	0.0159
RLF (−1)	0.003733	0.165254	0.022591	0.9820
BFD (−1)	−0.747231	0.099676	−7.496574	0.0000
R-squared	0.487534	Mean dependent var		0.001211
Adjusted R-squared	0.468785	S.D. dependent var		0.166930
S.E. of regression	0.121666	Akaike info criterion		−1.336795
Sum squared resid	3.641443	Schwarz criterion		−1.198312
Log likelihood	181.1098	Hannan-Quinn criter.		−1.281098
F-statistic	26.00353	Durbin-Watson stat		1.928569
Prob(F-statistic)	0.000000			

Table 24.17 Short run ARDL Model Coefficients

Variable	Coefficient	Std. Error	t-Statistic	Prob.
D(BCV (−1))	−0.115	0.094	−1.223	0.222
D(BCV (−2))	0.061	0.083	0.742	0.459
D(BCV (−3))	0.156	0.062	2.537	0.012
D(TLR)	−0.592	0.769	−0.770	0.442
D(TLR (−1))	−4.574	1.710	−2.675	0.008
D(TLR (−2))	5.127	1.581	3.242	0.001
D(TLR (−3))	−1.967	0.788	−2.495	0.013
ECM	−0.7472	0.099	−7.497	0.000
F	26.003			0.000
R^2	0.48			
D-W	1.92			

Table 24.18 Long Run Coefficients

Variable	Coefficient	Std. Error	t-Statistic	Prob.
TLR	0.004996	0.221133	0.022593	0.9820
C	0.026072	0.010189	2.558805	0.0111

indicate that the changes in the dependent variable are moderate. The Durbin Watson (D-W) test at a value of 0.92 indicates that there is no autocorrelation problem as it is around 2. The following Table 24.18 shows the long-run coefficients in the model.

According to the Table 24.18, The long-run coefficients of LIBOR ratio was statistically insignificant.

24.4 Conclusion

Bitcoin is the cryptocurrency that will be the exchange of the future. Main reasons for this situation are; number of Bitcoins is limited, on the contrary, there is no limit for its price. It can be transferred easily. It is to be regulated as a virtual asset by many countries. Bitcoin is a currency independent of macroeconomic factors. LIBOR is also most popular reference rate. The interaction of LIBOR and Bitcoin is an interesting topic. Most of studies related Bitcoin are about relationship between Bitcoin and other currencies or other cryptocurrencies. This study, which is related to analysis of relationship between international interest rates and cryptocurrencies is the first study. In this sense, it will make a significant contribution to the literature.

According to this study, LIBOR in general is effective in determining Bitcoin market value. On the other side, Bitcoin has a lower impact on the LIBOR. In the short term, a mutual interaction can be mentioned, but in the long term only LIBOR has been found to have an impact on Bitcoin. In terms of causality analysis, only one-way causality relationship from LIBOR to Bitcoin was determined. This situation supports and confirms the VAR analysis of causality. In addition, the ARDL model, which is another analysis performed in the study, showed statistically significant results. Also, Bounds coefficient and Error Mode (ECM) were found to be statistically significant. The Bounds coefficient value revealed that the long-term relationship between variables in the study was significant. It was determined that the short term deviations in the model with ECM value would be balanced after 1.35 periods.

References

About Euribor®. (2018, December 07). Retrieved from https://www.emmi-benchmarks.eu/euribor-org/about-euribor.html/

All Cryptocurrencies. (2018, November 19). Retrieved from https://coinmarketcap.com/all/views/all/

Baltagi, B. H. (2008). *Econometrics* (4th ed.). Heidelberg: Springer.

Belloumi, M. (2014). The relationship between trade, FDI and economic growth in Tunisia: An application of the autoregressive distributed lag model. *Economic Systems, 38*(2), 269–287.

Bhattacharjee, S. (2016). A statistical analysis of Bitcoin transactions during 2012 to 2013 in terms of premier currencies: Dollar, Euro and Rubles. *Vidwat, 9*(1), 8.

Bitcoin (Cryptocurrency). (2018, April 05). Retrieved from https://www.tradingview.com/ideas/Bitcoin/

Bitcoin ATMs by Country. (2018, December 01). Retrieved from https://coinatmradar.com/countries/

Bitcoin Mining Costs Throughout the World. (2018, December 01). Retrieved from https://www.elitefixtures.com/blog/post/2683/Bitcoin-mining-costs-by-country/

Bitcoin Price (BTC). (2018, September 17). Retrieved from https://www.coindesk.com/price/bitcoin/

Brown, R. L., Durbin, J., & Evans, J. M. (1975). Techniques for testing the constancy of regression relationships over time. *Journal of the Royal Statistical Society Series B (Methodological)*, 149–192.

Carrick, J. (2016). Bitcoin as a complement to emerging market currencies. *Emerging Markets Finance and Trade, 52*(10), 2321–2334.

Ciaian, P., & Rajcaniova, M. (2018). Virtual relationships: Short-and long-run evidence from BitCoin and altcoin markets. *Journal of International Financial Markets, Institutions and Money, 52*, 173–195.

Cocco, L., & Marchesi, M. (2016). Modeling and simulation of the economics of mining in the Bitcoin market. *PLoS One, 11*(10), e0164603.

D'Alfonso, A., Langer, P., & Vandelis, Z. (2016). *The future of cryptocurrency: An investor's comparison of Bitcoin and Ethereum*. Ryerson University.

Davies, D. C. (2014). *The curious case of Bitcoin: Is Bitcoin volatility driven by online search?* PhD thesis, Department of Economics, University of Victoria.

Definition of Bitcoin ATM. (2018, April 05). Retrieved from https://www.investopedia.com/terms/b/Bitcoin-atm.asp/

Dickey, D. A., & Fuller, W. A. (1979). Distribution of the estimators for autoregressive time series with a unit root. *Journal of the American Statistical Association, 74*(366a), 427–431.

Duasa, J. (2007). Determinants of Malaysian trade balance: An ARDL bound testing approach. *Global Economic Review, 36*(1), 89–102.

Enders, W. (2004). *Applied econometric time series* (2nd ed.). Hoboken, NJ: Wiley.

Eonia. (2018, December 08). Retrieved from https://www.euribor-rates.eu/eonia.asp/

Eonia interest rate. (2018, December 08). Retrieved from https://www.global-rates.com/interest-rates/eonia/eonia.aspx/

Erdogan, S., Dayan, V., Erdogan, A. (2018, 12–14 September). *The use of cryptocurrencies in emerging and developing Europe. A case study of Bitcoin*. Business and Organization Research (International Conference).

Financial Stability Board. (2018). *Reforming major interest rate benchmarks*. Basel, Switzerland.

Franklin, M. (2016). A profile of Bitcoin currency: An exploratory study. *International Journal of Business & Economics Perspectives, 11*(1), 80–92.

Granger, C. W. (1969). Investigating causal relations by econometric models and cross-spectral methods. *Econometrica: Journal of the Econometric Society, 37*, 424–438.

Greene, W. H. (2000). *Econometric analysis* (4th ed.). Upper Saddle River, NJ: Prentice-Hall.

Gujarati, D. N. (2003). *Basic econometrics* (4th ed.). New York: McGraw-Hill/Irwin Companies.

Guris, S., & Caglayan, E. (2000). *Ekonometri Temel Kavramlar* (1st ed.). Istanbul: Der Yayinlari.

Harris, R., & Sollis, R. (2003). *Applied time series modelling and forecasting* (1st ed.). Hoboken, NJ: Wiley.

Hileman, G., & Rauchs, M. (2017). Global cryptocurrency benchmarking study. *Cambridge Centre for Alternative Finance, 33*.

ICE. (2018). Retrieved November 6, 2018, from https://www.theice.com

Icellioglu, C. S., & Ozturk, M. B. E. (2018). Bitcoin ile Secili Doviz Kurlari Arasindaki Iliskinin Arastirilmasi: 2013-2017 Donemi icin Johansen Testi ve Granger Nedensellik Testi. *Maliye Finans Yazilari*, (109).

Johnston, J., & Dinardo, J. (1997). *Econometric methods* (4th ed.). New York: Mc Graw-Hill Companies.

Key ECB interest rates. (2018, December 07). Retrieved from https://www.ecb.europa.eu/stats/policy_and_exchange_rates/key_ecb_interest_rates/html/index.en.html/
Kutlar, A. (2009). *Uygulamali Ekonometri* (3rd ed.). Nobel Yayın Dagitim, Nobel Yayin No: 769, 3.
LIBOR Rates. (2018, September 17). Retrieved from https://fred.stlouisfed.org/categories/33003/
Maddala, G. S., & Lahiri, K. (2009). *Introduction to econometrics* (4th ed.). Hoboken, NJ: Wiley.
Ohtani, K., & Kobayashi, M. (1986). A bounds test for equality between sets of coefficients in two linear regression models under heteroscedasticity. *Econometric Theory, 2*(2), 220–231.
Perron, P. (1989). The great crash, the oil price shock, and the unit root hypothesis. *Econometrica: Journal of the Econometric Society, 57*, 1361–1401.
Pesaran, M. H., & Shin, Y. (1998). An autoregressive distributed-lag modelling approach to cointegration analysis. *Econometric Society Monographs, 31*, 371–413.
Pesaran, M. H., Shin, Y., & Smith, R. J. (2001). Bounds testing approaches to the analysis of level relationships. *Journal of Applied Econometrics, 16*(3), 289–326.
Sahoo, P. K. (2017). Bitcoin as digital money: Its growth and future sustainability. *Theoretical & Applied Economics, 24*(4), 53–64.
Sevuktekin, M., & Cinar, M. (2014). *Ekonometrik zaman serileri analizi: EViews uygulamali* (4th ed.). Dora Published.
Shapiro, D. C. (2018). Bitcoin loans and other cryptocurrency tax problems. *Journal of Taxation of Investments, 35*(2), 33–44.
Shrestha, M. B. (2006). *ARDL Modelling approach to cointegration test*. University of Wollongong, New South Wales, Australia and Nepal Rastra Bank (the Central bank of Nepal).
Sims, C. A. (1980). Macroeconomics and reality. *Econometrica: Journal of the Econometric Society, 48*, 1–48.
Sovbetov, Y. (2018). Factors influencing cryptocurrency prices: Evidence from Bitcoin, Ethereum, Dash, Litcoin, and Monero. *Journal of Economics and Financial Analysis, 2*(2), 1–27.
Tari, R. (2005). *Ekonometri, Gozden Gecirilmis ve Genisletilmis* (3rd ed.). Kocaeli Universitesi Yayinlari, Kocaeli, 263.
TCMB. (2018, November). *Financial Stability Report*. Central Bank of the Republic of Turkey.
Todorov, T. (2017). Bitcoin—An innovative payment method with a new type of independent currency. *Trakia Journal of Sciences, 15*(1), 163–166.

Serdar Erdogan is an Assistant Professor in the Department of Accounting, Uzunkopru School of Applied Sciences at Trakya University, Edirne (Turkey). He received a Ph.D. in Economic Theory at Marmara University, Turkey. His research interests cover topics in the field of economy and finance, such as inflation, foreign direct investment, unemployment, fiscal policy, econometric analysis, and cryptocurrencies. He took part in different international conferences.

Volkan Dayan is an Associate Professor in the Department of Banking and Finance, Uzunkopru School of Applied Sciences at Trakya University, Edirne (Turkey). He received a Ph.D. in Finance at Celal Bayar University, Turkey. His research interests cover topics in the field of economy and finance, such as Basel regulations, credit risk management, cryptocurrencies, and sustainable finance. He took part in different international conferences. He performed research projects at the national and international levels. He took part in different international conferences and in the Erasmus mobility programme at Huelva University.

Chapter 25
Cryptocurrency Derivatives: The Case of Bitcoin

Yakup Söylemez

Abstract The effects of digitalization on the business ecosystem and business models are increasing day by day. Businesses and individuals intensely benefit from the services that FinTech platforms offer. As a result, there are significant changes in the structure of financial institutions and financial instruments. Blockchain technologies play an important role in the transformation of business ecosystems. In particular, cryptocurrencies are recognized by individuals, institutions, and governments as an economic asset. However, the high price volatility of cryptocurrencies shows that they have significant risks. Cryptocurrency derivatives are used to hedge against and benefit from price movements. This study aims to provide a basic framework for cryptocurrency derivatives. In this study, the most traded cryptocurrency type, Bitcoin derivatives, are used.

25.1 Introduction

Since the last quarter of the twentieth century, business activities and business models have begun to change rapidly. The main dynamics of this transformation are technological innovations. Changes in technology necessitate differentiation in many areas, from business models to the customers of the business. These changes in technology have brought about the concept of Industry 4.0. Together with Industry 4.0, companies face disadvantages, as well as certain advantages (see, Türkmen, 2018). Firstly, companies do not operate only in the national arena, anymore. It should be acknowledged that the elimination of the barriers in capital and digitalization pave the way for many enterprises to operate in the international arena. Besides, the fact that establishing a connection with international businesses is a requirement for even companies operating in a local arena because they have to maintain their power in a competitive environment.

Y. Söylemez (✉)
Department of Accounting and Tax, Zonguldak Bülent Ecevit University, Zonguldak, Turkey
e-mail: yakup.soylemez@beun.edu.tr

© Springer Nature Switzerland AG 2019

U. Hacioglu (ed.), *Blockchain Economics and Financial Market Innovation*,
Contributions to Economics, https://doi.org/10.1007/978-3-030-25275-5_25

The globalization of the activities of the companies means that they meet with certain risks. New financial instruments are gaining importance in capital markets to protect enterprises against emerging risks. Undoubtedly, the most important ones among these financial instruments are derivative financial instruments. Derivative financial instruments are products used to reduce financial risks. The value of these financial products is depending on primary assets such as equity shares, bonds, foreign currency, interest, and properties. In this context, basic derivative instruments can be counted as futures, forward, option, and swap (Chambers, 2007).

Futures and forward contracts include the determination of the price and quantity in advance for a commodity to be delivered in the future (Johnson, 1960). Therefore, these contracts aim to reduce the risk to the parties by determining the price and amount of the goods that can be traded in a future date. The option, which is another derivative product, can be defined as a contract that gives a right to buy or sell a certain commodity at a predetermined date (Merton, 1973). Therefore, the purchaser of the option obtains the right to buy or sell the goods at a certain price and protects them against the risks that may occur in the market. Lastly, swaps can be defined as changing the cash flows of a particular financial asset resulting from two parties within a given system (Duffie & Huang, 1996). As a result of a swap transaction, the parties may transfer the risks that may occur to a financial intermediary. This is what makes swaps an indispensable product of financial markets (Chambers, 2007).

Financial risks have increased as a result of globalization which has begun to affect financial markets significantly since the 1980s. However, the digitalization emerged in this period has been affecting the financial markets at ever-mounting rate especially since the beginning of the 2000s. Today, Fintech platforms which perform activities of traditional financial institutions with new business models have emerged. Financial Stability Board (2018) defines the concept of Fintech as "technologically enabled innovation in financial services that could result in new business models, applications, processes or products with an associated material effect on financial markets and institutions and the provision of financial services."[1] The basic services offered by Fintech platforms can be listed as follows; (1) B2B solutions, (2) Peer-to-peer markets, (3) Personal financial management instruments, (4) Mobile wallets, (5) Solutions offered to the customers with low credit rating, (6) Big data analysis, (7) Digital currency and other blockchain technologies, (8) Insurance (InsurTech), and (9) Consultancy (Robo-advisor) (Anagnostopoulos, 2018).

The influence on Fintech platforms on financial instruments, institutions, and business models is increasing day by day. Technology that enables Fintech platforms to operate in a digital environment is blockchain technology. Blockchain is a database which has a dispersed structure providing data management as encrypted. This database records transactions and events between the parties. Verification of each transaction realizes in the case of consensus of the majority of the participants in the system. The information in the Blockchain system cannot be deleted.

[1]http://www.fsb.org/work-of-the-fsb/policy-development/additional-policy-areas/monitoring-of-fintech, 15/03/2019.

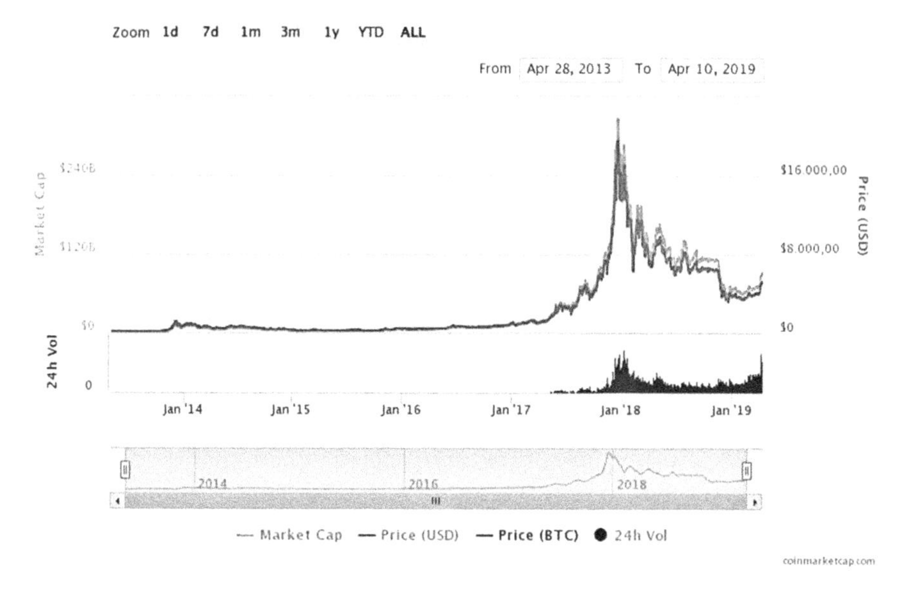

Fig. 25.1 Bitcoin price and Market Cap. (Source: https://coinmarketcap.com/currencies/bitcoin/. 10.04.2019)

Therefore, the record of each transaction done is kept by the system (Crosby, Pattanayak, Verma, & Kalyanaraman, 2016). Thus, the system is an encryption technology which can be observed by many users but cannot be interfered, which constitutes one of the most important strengths of Blockchain technology.

Blockchain technology is especially used in the field of cryptocurrency. There are many platforms performing in the field of cryptocurrency. Among the cryptocurrencies, the most traded ones in the market can be counted as Bitcoin, Ethereum, and Litecoin. The most popular one among these cryptocurrencies is Bitcoin, which has been on the market since January 3, 2008 (Top 100 Cryptocurrencies by Market Capitalization https://coinmarketcap.com/, Accessed on 03.03.2019). Bitcoin price movements and market value are also seen in Fig. 25.1.

Nakamoto, the creator of Bitcoin, defines electronic money as a means of transferring online payments from one side to another without going through any financial institutions (Nakamoto, 2008). As it can be understood from the definition, electronic money has emerged as a new technology that will deeply affect the traditional payment system. This is because the previous payment instruments are subject to specific regulations as they transfer money with reference to an account of a financial institution. However, the transfer between the peer-to-peer electronic payment system is done without any intermediary. Moreover, since it is not possible to delete data from the blockchain system, it is almost impossible to make arrangements in this field.

Bitcoin is the most widely used electronic money across the world. As of March 2019, 34 million Bitcoin users executed 397 million transactions (Statistics, https://www.blockchain.com/, 13/03/2019). Instead of lawyers, engineers determined how a great economic power will work at this rate. In this system, transactions cannot be stored only in one driver. In place of this, it is distributed to the web consisting of Bitcoin operators. As mentioned above, Bitcoin is an electronic money system that allows for irreversible transactions and money creation (Böhme, Christin, Edelman, & Moore, 2015).

Bitcoin accounts are created for free. No security procedures are applied when creating these accounts. It doesn't even matter if system users are real people. Therefore, the system is more flexible than other forms of payment. To understand the situation better, it is worth looking at the sides of the Bitcoin ecosystem. Accordingly, the important information about Bitcoin is as follows (Vora, 2015):

- Bitcoin is a virtual currency that uses cryptography.
- Transactions are executed between the parties without any intermediary.
- Transactions are made quickly, cost-effectively, and irreversibly.
- The driving force behind Bitcoin system is privacy and multilateralism.
- Network nodes (for mining) and transactions (for wallets and personal accounts) are kept up-to-date with a software by a volunteer developer group.
- A decentralized data processing network is used for encrypting (hashing) and bookkeeping (blockchain).
- The system tries not to allow manipulation for the value of the currency by creating limited money.
- Local authorities can make arrangements for Bitcoin.
- It is a virtual currency having a fluctuating exchange rate for major currencies.

Bitcoin is a digital currency that enables payment transactions between the parties without using an intermediary. Therefore, its difference from all other payment systems is mentioned (until) here. This difference uncovers its unique risks. These risks are; (1) market risk, (2) shallow market problem, (3) counterparty risk, (4) transaction risk, (5) operational risk, (6) confidentiality risk, (7) legal and regulatory risk (Böhme et al., 2015).

Market risk is the risk arising from fluctuations in exchange rates between any currency and bitcoin. However, some studies show that since Bitcoin return is not affected by changes in the stock exchange, it provides some of the market risks to be hedged (Dyhrberg, 2016). The risk emerged as a result of transactions of Bitcoin users in the market with low trading volume is called *shallow market risk* (DeVries, 2016). In Bitcoin ecosystem, there are many intermediaries which bring the purchaser and sellers together and provide them start-up Bitcoins. The reason that there are possible security gaps of the system that intermediaries created, customers are exposed to *counterparty risk* (Abramova & Böhme, 2016). The risk arising from the irreversible nature of the transactions carried out in the Bitcoin ecosystem is called *transaction risk*. In the case of an error in the transaction carried out, the transaction can be corrected only if there is an agreement between the purchaser and the seller.

There is no central regulatory structure for the errors and tricks that occur in the system (Möser, Böhme, & Breuker, 2014).

Another important risk Bitcoin users face is the operational risk. *Operational risk* is the possibility of possible losses as a result of insufficient work processes, personnel, system, and other events (Peters, Chapelle, & Panayi, 2016). Based on the Bitcoin ecosystem, operational risk can be defined as all kinds of activities that affect technical infrastructure and security system adversely. For example, the risk that all information belonging to a user are under threat due to a security gap is out of the scope of that risk (Böhme et al., 2015). *The risk of confidentiality* can be defined as the risk that can be defined as the possibility of reaching other transactions by following the trace of a transaction executed by Bitcoin user (Goldfeder, Kalodner, Reisman, & Narayanan, 2018). It is natural for a person to become his/her information apparent during the conversion of the revenue obtained in a bitcoin system into conventional financial values. On the basis of this information, the risk of following other transactions of the user in the bitcoin system is *the risk of confidentiality*. Finally, the risk that the Bitcoin user faces due to laws and regulations that are changing from country to country is called a *legal and regulatory risk*. For example, in countries such as Germany and Japan, licensing is applied for crypto money exchanges. Some countries do not license cryptocurrencies. Therefore, this difference between countries constitutes legal and regulatory risk.

The mentioned risks above in the crypto money market became more important with the rise in the last quarter of 2017. During this period, there was an extraordinary increase in prices, and the blockchain activities also expanded considerably. This situation has led to an increase in speculative movements in the crypto money markets. On the other hand, this large movement in the market has allowed more investment and more users to enter the system. As a result of this situation, the system has become more and more legitimate. All these developments have led to the emergence of new products based on cryptocurrency in the market. Among the most remarkable ones, there are cryptocurrency derivatives (Graf, 2019).

The cryptocurrency derivative, with a brief definition, is an agreement that predicts the purchase or sale of a crypto money or an asset based on it at a predetermined price in the future. The value of a cryptocurrency derivative depends on the future expected value of the cryptocurrency. Therefore, the cryptocurrency derivative has no value alone. Cryptocurrency derivatives are basically future contracts, options, and swaps. In the following parts of this study, since each derivative is examined separately, no further details will be given in this section.

There are several reasons why cryptocurrency derivatives are used in the market. These reasons are; (1) providing protection against market volatility, (2) risk management instrument against increased risks, and (3) speculation (Zanuiddin, 2019). For a better understanding of the subject, it is useful to briefly mention these reasons.

First, derivative instruments are used to protect individuals and corporations against risk and they aren't affected by fluctuations in the price of underlying assets. There is scientific evidence that Bitcoin and its derivatives can be included in the portfolio for protection against emerging market risks (Dyhrberg, 2016; Guesmi, Saadi, Abid, & Ftiti, 2018).

Second, cryptocurrency derivatives are used by investors to hedge their investment portfolios. Thus, investors have the opportunity to balance their potential losses. Investors can use derivatives against exchange rate movements (Allayannis & Ofek, 2001). For this reason, the parties may incorporate the cryptocurrency derivatives into their portfolios and reduce the losses in their investments as a result of changes in exchange rates.

Finally, cryptocurrency derivatives are used by investors to benefit from changes that may occur in the value of crypto money. Derivative instruments are widely used in the market for speculative purposes. Some studies show that derivative financial instruments are used for speculative purposes by investors rather than hedging purposes (Bohmann, Michayluk, & Patel, 2018).

Cryptocurrency derivatives are digital currency derivatives that have recently started to be traded in the market. The first futures contracts based on Bitcoin began to be traded in the Chicago Mercantile Exchange (CME) and Chicago Board Options Exchange (CBOE) in December 2017. Therefore, in the literature, there is a lack of study regarding cryptocurrency derivatives. This study aims to provide a basic framework for cryptocurrency derivatives. The scope of the study is Bitcoin derivatives which are traded mostly among the cryptocurrency derivatives. In the following parts of the study, an examination will be made on the Bitcoin future, option, and swap contracts. In the last part, the results of this study will be discussed.

25.2 Bitcoin Futures

Futures contracts, with the simplest definition, is an agreement for the future delivery of a commodity whose price is determined today (Chambers, 2007). U.S. Commodity Futures Trading Commission (2006) lists the future contract features as follows: (1) in futures contracts, the contract price is predetermined. (2) the parties to the contract are obliged to comply with the contract. (3) the contract is used to undertake or change the price risk. (4) the contract is completed with the delivery or offset of the goods (A Guide to the Language of the Futures Industry, 2019).

The Bitcoin futures contract is an agreement on the purchase and sale of Bitcoin, which is the underlying asset, on a future date, on condition that the price is determined today. There are two exchanges where Bitcoin futures contracts are traded. These exchanges are Chicago Board Options Exchange (CBOE) and Chicago Mercantile Exchange & Chicago Board of Trade (CME). Futures contracts in relevant exchanges have been traded since December 2017. The underlying asset for CBOE futures contracts is the Gemini stock exchange bid price in US dollars for Bitcoin. For CME futures contracts, the underlying asset is accepted as the Bitcoin reference ratio obtained from the large exchanges in US dollars. Contract margins

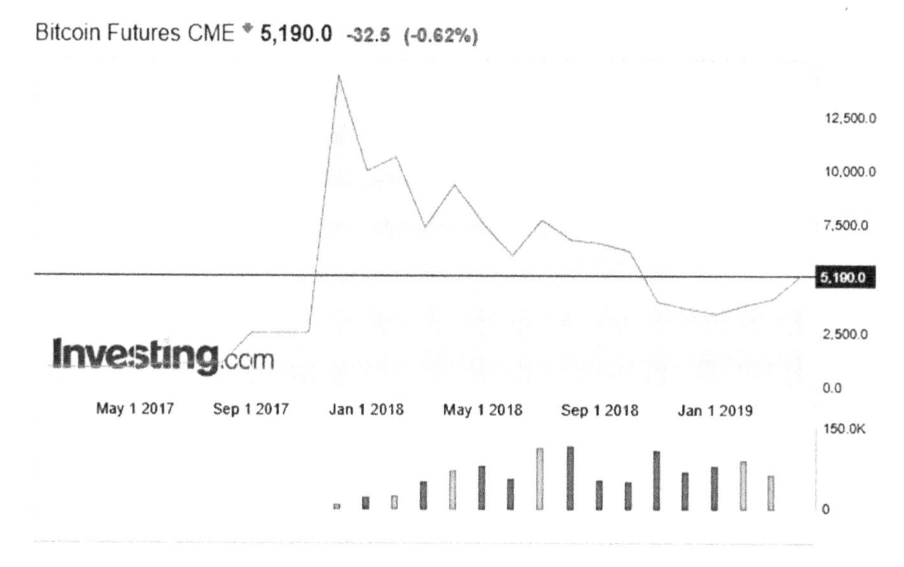

Fig. 25.2 Bitcoin futures CME. Source: https://www.investing.com/crypto/bitcoin/bitcoin-futures. 10.04.2019

are generally high and contracts are made in cash. The contract margin is 40% for BOE and 35% for CME. The main reason for this situation is that besides the high variability in Bitcoin prices, the market is not yet matured (Kapar & Olmo, 2019).

Figure 25.2 shows the prices of the Bitcoin Futures contracts traded at the Chicago Mercantile Exchange & Chicago Board of Trade (CME). When the graph is examined, it is seen that there is a positive correlation between the Bitcoin prices and the prices of Bitcoin futures contracts.

Figure 25.3 shows the Bitcoin Futures contracts traded at the Chicago Board Options Exchange (CBOE). When the graph is examined, it is seen that there is a positive correlation between the Bitcoin prices and the prices of Bitcoin futures conracts. When both graphs are evaluated together, it is seen that the prices of Bitcoin Futures contracts are very close to each other in both exchanges.

Bitcoin futures contracts can be explained with the help of an example. On 1 March, one party may agree to sell 1 Bitcoin to $700 on 1 April. This agreement is locked with BTC or bitcoin dollar exchange rate. Therefore, the investor who has a loan in terms of bitcoin and does not want to be affected by the change in bitcoin prices is protected against this risk by this agreement. However, on 1 April, even if the value of bitcoin exceeds the contract value, the parties have to fulfill their contractual obligations.

A futures contract is standardized for all conditions except the price. The parties determine the underlying asset, the amount of the asset, the place and time of delivery, and the price in the contract. CEA[2] divides the underlying assets that are

[2]The Commodity Exchange Act (CEA) organizes the commodity future transactions since 1936.

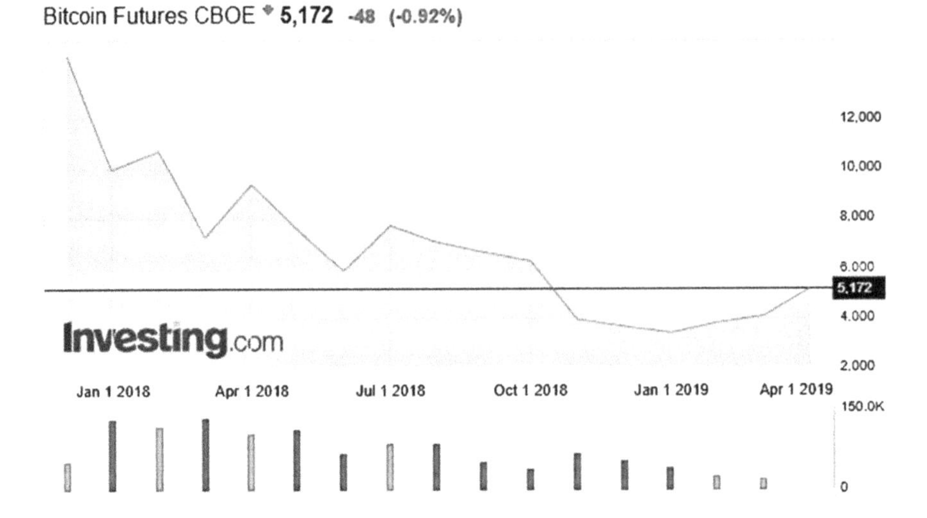

Fig. 25.3 Bitcoin futures CBOE. Source: https://www.investing.com/crypto/bitcoin/cboe-bitcoin-futures. 10.04.2019

based on futures contracts into three basic parts. Accordingly, the underlying assets on a futures contract are; (1) agricultural commodities such as wheat and soy, (2) excluded commodities such as currencies and interest rates; and (3) exempt commodities such as precious metals. Commodity Futures Trading Commission also operationalizes the underlying assets as financial and non-financial assets (Brito, Shadab, & Castillo O'Sullivan, 2014).

Another important subject for cryptocurrency futures contracts is who will be the contract parties. It is possible to examine the parties investing in Bitcoin futures contracts in three categories. Accordingly, the parties investing in contracts are as follows; (1) retail investors; (2) institutional investors such as banks, insurance companies, hedge funds, investment consulting companies, retired and aid funds, and (3) investment funds. However, due to the high volatility and the new market, cryptocurrency derivatives seem to be more suitable for institutional investors (Cindicator, 2018).

Finally, it is useful to explain how Bitcoin futures contracts are priced. Based on an active market, parties that arbitrage ensures that the price of a futures contract reflects transport costs alongside the time value of money (BSIC (Bocconi Students Investment Club), 2018). The cost of transportation results from the position that includes the interest and opportunity costs of the contracting party. If the revenue obtained in a cryptocurrency futures contract is higher than the financing cost, the transportation cost is positive. Taking into account all of these, when crypto money cannot pay dividend payment and the carrying cost is not available, the price of futures contract can be found as follows (Wilmott, 2013):

$$F(S, t) = Se^{r(T-t)} \tag{25.1}$$

Here;
F = Future price
S = Asset price (spot price)
r = Interest rate
T = Maturity date
t = Contract date

The interest rate in the formula can be found with the help of the following equation:

$$r(T-t) = \ln\left(\frac{F}{S}\right) \tag{25.2}$$

25.3 Bitcoin Options

An option is a contract that gives the holder the right to buy or sell an underlying asset at a predetermined time. In the option contract, the contract owner makes a payment called the option premium to the seller of the option (Sinclair, 2010). Options can be divided into two. These are (1) European option, and (2) American option. A European option is the contract that gives the owner the right to use the option at the expiry date. The American option can be defined as a contract that gives the buyer the right to use the option at any time up until maturity (Liu, Chen, Li, & Zhai, 2014).

According to the definition of The Financial Conduct Authority,[3] the cryptocurrency option is a contract that grants the owner the right to buy or sell cryptocurrencies (FCA, 2018). Accordingly, a Bitcoin call option grants the right to buy Bitcoin at a predetermined price until a specific date. Although the purchaser does not have a purchase obligation, the seller must sell Bitcoin when it is wanted on the due date. The put options are working on contrary to the call option. In other words, a put option allows the owner to sell at a predetermined price until a specific date. In the put option, the owner buys a right from the seller.

Within the framework of all these explanations, the Bitcoin call option may be defined as contracts that protect the buyer against the risk of overdue Bitcoin prices. Instead, Bitcoin put option protects the purchaser against the risk of a massive rise in the Bitcoin prices (Brito et al., 2014). Considering the newness of Bitcoin market and the volatility of asset prices, the option can be used as an important protection instrument in the market. It is considered that Bitcoin option contracts can play an

[3]It is the regulatory authority on financial services and institutions in the United Kingdom.

important role in the market especially when digitalization and cryptocurrency use become widespread in trade.

The popularity of cryptocurrencies is increasing day by day. This situation will increase the number of products that can be created based on these assets (Silva, 2018). The use of Bitcoin options as a financial asset is pretty new. On 02/10/2017, the Commodity Futures Trading Commission granted LedgerX the right to trade Bitcoin option contracts. Therefore, LedgerX Company executed European call and put options of the trade of Bitcoin option contracts, for the first time. Then, The US-based Coin-Desk Company and the Netherlands-based Deribit Company have also started to trade Bitcoin option contracts (Fig. 25.4).

Another issue with the crypto currency derivatives is how to price Bitcoin options. Since the trade of Bitcoin option contracts is pretty new, the number of scientific studies on this subject is relatively rare. However, Black-Scholes option model is usually used in the studies carried out (Cretarola & Figà-Talamanca, 2017; Madan, Reyners, & Schoutens, 2019). Therefore, in this study, Black-Scholes model will be mentioned as Bitcoin option valuation model.

Fig. 25.4 Bitcoin options price. Source: https://ledgerx.com/. 10/04/2019

The Black-Scholes model used in the pricing of Bitcoin options is based on some basic assumptions. According to this; (1) Bitcoin price movements show a lognormal distribution. (2) there is no transaction costs or taxes. (3) Bitcoin trading shows continuity. (4) There is no risky arbitrage opportunity in the market. (5) the risk-free interest rate is fixed, and (6) the risk-free interest rate is used among investors and this rate is the same for all parties (Chambers, 2007). According to all these assumptions, the price of Bitcoin call and put options can be calculated with the following formulas:

$$c = SN(d_1) - Xe^{-rt}N(d_2) \tag{25.3}$$

$$p = Xe^{-rt}N(-d_2) - SN(-d_1) \tag{25.4}$$

d_1 and d_2 variables in these formulas are calculated as follows:

$$d_1 = \frac{\ln(S/X) + (r + \sigma^2/2)T}{\sigma\sqrt{T}} \tag{25.5}$$

$$d_2 = \frac{\ln(S/X) + (r - \sigma^2/2)T}{\sigma\sqrt{T}} = d_1 - \sigma\sqrt{T} \tag{25.6}$$

Here;
c = The price of call option,
p = The price of put option,
S = Current price of Bitcoin,
X = Strike price of Bitcoin,
T = Option maturity,
r = Risk-free interest rate,
σ^2= Annual variation in Bitcoin prices (variance),
N(d1), N(d2) = Cumulative probability distribution function for standard normal variable (from -∞ until d1,2)
ln = logarithm and
e = Exponential represents the value of present value factor (2.7183).

25.4 Bitcoin Swaps

Swaps are financial products that have evolved especially after the 80s. With its brief definition, swap is the changing future cash flows within a certain system by two parties (Boenkost & Schmidt, 2005). With this contract, the main purpose of the parties is to change the current conditions in which they are in line with their own benefits. One of the most widely used among these products is known as foreign currency swaps. Foreign currency swaps started to be used in England in the 60s.

The interest rate swap, which is another swap type, is also widely used. The most important reason for this situation is that all parties benefit from interest swaps (Chambers, 2007).

The structures of bitcoin swaps contain great similarities with foreign currency swaps. In the foreign currency swap transaction, the parties borrow foreign currency from each other, and they guarantee to pay at a certain exchange rate in the term. The parties have two main objectives in this process. These could be categorized as (1) hedging against currency risk and (2) to speculate. In another way, bitcoin swap can be realized through cash settlement. In this method, parties do not need to trade in bitcoin or any other currency. The parties may enter into a swap contract upon changing the cash equivalent of Bitcoin and another currency at a future time. In this agreement, the party accepting Bitcoin commits to pay cash if the value of Bitcoin is decreased. In this way, it protects itself against price fluctuations. Another way to prevent the price risk in Bitcoin is to make a swap contract that takes any virtual currency index as a reference (Brito et al., 2014).

Swap transactions can also be carried out between cryptocurrencies. However, these processes do require an intermediary system. This system is the Hashed Timelock Contract which will be discussed later in this study. In addition to this, there are efforts to develop that sort of swaps. As an instance of these efforts, the swaps made between Bitcoin and ZCash cryptocurrencies can be demonstrated (Bentov, Kumaresan, & Miller, 2017). These swap transactions between cryptocurrencies have brought along a new concept. This concept is an atomic swap.

The idea of atomic swap started to be discussed in 2013. Later in 2017, the Komodo platform transformed the idea of atomic swap into a commercial product. Within this scope, swaps between some blockchain technologies have begun to be made. These blockchain technologies include Bitcoin, Litecoin and Decred. These exchange transactions started to be made with Hashed Timelock Contracts (Sopov, Purtova, & Noxon, 2019). In order to acquire a better understanding of the subject, it is worth mentioning the concepts of atomic cross chain swap and Hashed Timelock Contracts.

Atomic cross chain swap is a distributed coordination task where multiple blockchains such as Bitcoin and Ethereum enter the swap process (Herlihy, 2018). In a more explicit expression, it is a contract that allows the exchange of one cryptocurrency with the other. Atomic swap eliminates some risks compared to other swap formats. There is often a third party in swap transactions. This party is known as a financial intermediary. In the atomic cross chain swap, the swap process is performed directly between peers (Peer-to-Peer). As a result, the risks arising from the third party are being eliminated. This transfer between various cryptocurrencies is carried out through Hashed Timelock Contracts.

Hashed Timelock Contract (HTLC) is a channel which is used in order to make a secure transfer to the desired destination via multiple jump points. The channel here has a versatile but single network structure. The purpose of HTLC is to establish a multilateral and digital transaction mechanism worldwide (Poon & Dryja, 2016). In a much simpler statement, HTLC is a secure and intelligent contract that enables swap transaction between cryptocurrencies. In this system, cryptocurrency to be swapped is locked and the other currency corresponding to exchange can be taken from a specific address at the end of the swap transaction. While creating a contract, only the swapped currencies are known; however, the addresses where the exchange will be completed are not known. Figure 25.5 shows a simple HTLC system.

There are two stages in the atomic swap transaction. In the first stage, the parties agree to change cryptocurrencies. At this stage, different agreement methods can be used. These methods are (1) centralized services (2) sidechain and (3) decentralized order books. At this stage, for example, the following message is given: "I am ready to exchange 100 Etherium with 100 Bitcoin." At this point, channels continuously enter messages into the system. The recipient accepts the order simultaneously with the HTLC protocol at any time. The swap protocol starts the time the recipient signs the acceptance message (Sopov et al., 2019).

Finally, the pricing of a swap transaction will be briefly addressed. "*The pricing of a swap transaction depends on the swap rates. Swap rates are also the difference between forward rates and spot rates*[...] *By reflecting the difference between interest rates or exchange rates, swap rates determine the swap's gain or cost. It is recommended to use average spot rates in calculations* (Chambers, 2007).

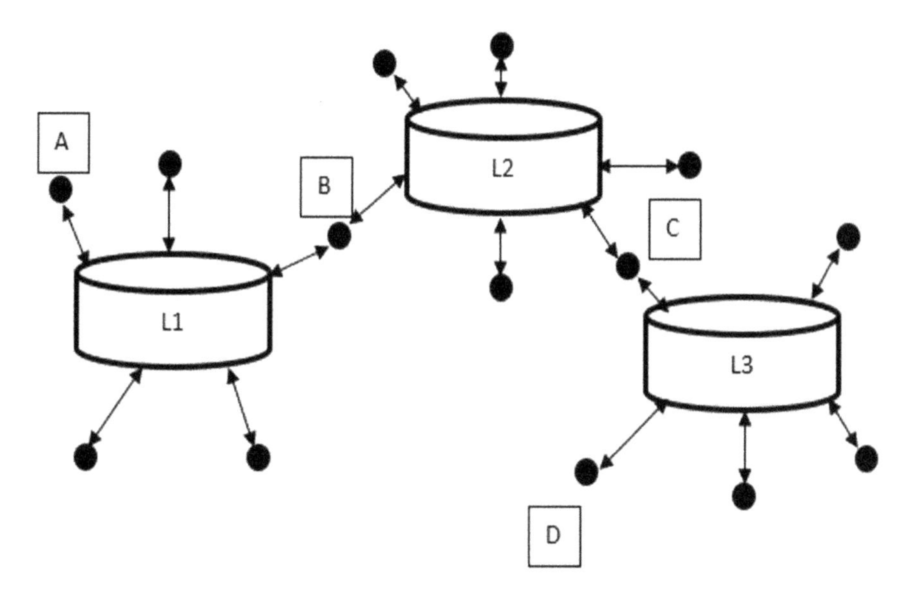

Fig. 25.5 Hashed Timelock Contract Example. Three individual ledgers L1, L2, and L3 that are connected through nodes B and C, i.e. node B participates as writer in L1 and L2 and node C participates in both L2 and L3. If each of these ledgers is a blockchain for one currency, a payment from A to D can be routed through B and C as atomic transaction, where B and C provide currency exchange. This can be achieved for example through hashed timelock contracts. Source: Wüst and Gervais (2018))

25.5 Conclusion

Since the 80s, the world has undergone a great change. One of the greatest effects of this change has been experienced in the financial markets, institutions, and instruments. The driving forces of the aforementioned change in financial markets are liberalization of capital, increased risk perception in financial markets and technological innovations. Liberalization of capital movements has resulted in companies exceeding national borders. Liberalization of capital movements enables firms to establish strong economic and technical cooperation.

In order to protect themselves against these increased risks, firms have developed new financial instruments. These tools used for hedging are called derivative instruments. Derivatives are used as an investment tool in addition to their use in reducing risk. However, in the 2008 financial crisis, derivative instruments based on mortgage loans turned into risky assets and became one of the main causes of the crisis.

Changes occurred in technology have accelerated in 2000s. Especially the widespread use of internet technologies has resulted in the increase of technological innovations in each and every field. These technological innovations have also effectively shown themselves in the field of finance.

The use of technological innovations in all areas of finance increases day by day. These technologies are commonly called as FinTech. FinTech platforms perform the work of traditional financial institutions through new and digital business models. Fields of activity of Fintech platforms; (a) back-office operations, (b) digital banking, (c) e-commerce, (d) identity management, (e) payments and (f) insurance. Within the FinTech platforms, Blockchain technologies hold the most important place.

All of these technological changes in the finance field have revealed cryptocurrencies in 2009. Cryptocurrencies have become commonly used currencies in a short period of time despite their high risk. Especially after 2017, assets developed based on crypto currencies started to be traded in financial markets.

Cryptocurrency derivatives are among the most important tools developed based on cryptocurrencies. Cryptocurrency derivatives include futures, options, and swap contracts. Since December 2017, cryptocurrency derivatives have been dealt with in stock exchanges. Cryptocurrency derivatives are used for three main purposes as follow: (1) provide protection against the risks arising from cryptocurrencies, (2) reduce risks in the market and (3) make use of these assets in a speculative manner.

Investments on cryptocurrency derivatives are increasing day by day. Today, due to the fact that the market is relatively new, and the market volatility is high, investment in cryptocurrency derivatives is relatively low. Innovations in technology, however, show that these tools will be increasingly used.

In this study, basic information about cryptocurrency derivatives is stated. The significance of the market and mechanisms of operation are presented. There is quite a low number of studies focusing on this issue in the literature. The study aims to fill the gap in the literature on cryptocurrency derivatives, which is a considerably new financial instrument. This study is based on one of the most dealt/transacted cryptocurrency, which is called Bitcoin.

References

Abramova, S., & Böhme, R. (2016). Perceived benefit and risk as multidimensional determinants of Bitcoin use: A quantitative exploratory study. In *37th ICIS*, Dublin, Ireland.

Allayannis, G., & Ofek, E. (2001). Exchange rate exposure, hedging, and the use of foreign currency derivatives. *Journal of International Money and Finance, 20*(2), 273–296.

Anagnostopoulos, I. (2018). Fintech and Regtech: Impact on regulators and banks. *Journal of Economics and Business, 100*, 7–25.

Bentov, I., Kumaresan, R., & Miller, A. (2017, December). Instantaneous decentralized poker. In *International conference on the theory and application of cryptology and information security* (pp. 410-440). Cham: Springer.

Bocconi Students Investment Club. (2018). What's wrong McFly? Chicken? Back to the Bitcoin futures. Retrieved December 20, 2018, from http://www.bsic.it/whats-wrong-mcfly-chicken-back-bitcoin-futures

Boenkost, W., & Schmidt, W. M. (2005). *Cross currency swap valuation*. Retrieved from SSRN 1375540.

Bohmann, M., Michayluk, D., & Patel, V. (2018). *Price discovery in commodity derivatives: Speculation or hedging?* Retrieved from SSRN 3055554.

Böhme, R., Christin, N., Edelman, B., & Moore, T. (2015). Bitcoin: Economics, technology, and governance. *Journal of Economic Perspectives, 29*(2), 213–238.

Brito, J., Shadab, H. B., & Castillo O'Sullivan, A. (2014). Bitcoin financial regulation: Securities, derivatives, prediction markets, and gambling. *Columbia Science and Technology Law Review, 16*, 144–221.

Chambers, N. (2007). *Türev piyasalar* (2. Baskı). İstanbul: Beta Yayınları.

Cindicator. (2018). *Bitcoin futures: Market evolution*. Retrieved March 15, 2019, from https://cindicator.com/bitcoin-futures-analysis.pdf

Commodity Futures Trading Commission. (2006). *The CFTC glossary: A guide to the language of the futures industry*. CreateSpace Independent Publishing Platform.

Cretarola, A., & Figà-Talamanca, G. (2017). *A confidence-based model for asset and derivative prices in the Bitcoin market*. arXiv:1702.00215.

Crosby, M., Pattanayak, P., Verma, S., & Kalyanaraman, V. (2016). Blockchain technology: Beyond bitcoin. *Applied Innovation, 2*(6–10), 71.

DeVries, P. D. (2016). An analysis of cryptocurrency, Bitcoin, and the future. *International Journal of Business Management and Commerce, 1*(2), 1–9.

Duffie, D., & Huang, M. (1996). Swap rates and credit quality. *The Journal of Finance, 51*(3), 921–949.

Dyhrberg, A. H. (2016). Hedging capabilities of Bitcoin. Is it the virtual gold? *Finance Research Letters, 16*, 139–144.

Financial Conduct Authority. (2018). *Cryptocurrency derivatives*. Retrieved March 13, 2019, from https://www.fca.org.uk/news/statements/cryptocurrency-derivatives

Financial Stability Board. (2018). *Monitoring of FinTech*. Retrieved March 15, 2019, from http://www.fsb.org/work-of-the-fsb/policy-development/additional-policy-areas/monitoring-of-fintech

Goldfeder, S., Kalodner, H., Reisman, D., & Narayanan, A. (2018). When the cookie meets the blockchain: Privacy risks of web payments via cryptocurrencies. *Proceedings on Privacy Enhancing Technologies, 2018*(4), 179–199.

Graf, C. (2019). *Mapping the cryptocurrency derivative landscape, is regulation needed?* Retrieved March 13, 2019, from https://www.cryptoglobe.com/latest/2019/01/mapping-the-cryptocurrency-derivative-landscape-is-regulation-needed

Guesmi, K., Saadi, S., Abid, I., & Ftiti, Z. (2018). Portfolio diversification with virtual currency: Evidence from Bitcoin. *International Review of Financial Analysis*. https://doi.org/10.1016/j.irfa.2018.03.004

Herlihy, M. (2018, July). Atomic cross-chain swaps. In *Proceedings of the 2018 ACM symposium on principles of distributed computing* (pp. 245–254). ACM.
Johnson, L. L. (1960). The theory of hedging and speculation in commodity futures. *Review of Economic Studies, 27*(3), 139–151.
Kapar, B., & Olmo, J. (2019). An analysis of price discovery between Bitcoin futures and spot markets. *Economics Letters, 174*, 62–64.
Liu, Z., Chen, L., Li, L., & Zhai, X. (2014). Risk hedging in a supply chain: Option vs. price discount. *International Journal of Production Economics, 151*, 112–120.
Madan, D. B., Reyners, S., & Schoutens, W. (2019). Advanced model calibration on Bitcoin options. *Digital Finance*, 1–21.
Merton, R. C. (1973). Theory of rational option pricing. *Theory of Valuation*, 229–288.
Möser, M., Böhme, R., & Breuker, D. (2014, March). Towards risk scoring of Bitcoin transactions. In *International conference on financial cryptography and data security* (pp. 16–32). Berlin: Springer.
Nakamoto, S. (2008). *Bitcoin: A peer-to-peer electronic cash system*
Peters, G. W., Chapelle, A., & Panayi, E. (2016). Opening discussion on banking sector risk exposures and vulnerabilities from virtual currencies: An operational risk perspective. *Journal of Banking Regulation, 17*(4), 239–272.
Poon, J., & Dryja, T. (2016). *The bitcoin lightning network: Scalable off-chain instant payments.*
Silva, F.C. (2018). *Option pricing under jump-diffusion processes: Calibration to the bitcoin options market.* Instituto Superior De Ciências Do Trabalho E Da Empresa Faculdade De Ciências Da Universidade De Lisboa. Departamento De Finanças Departamento De Matemática. Master Thesis
Sinclair, E. (2010). *Option trading: Pricing and volatility strategies and techniques* (Vol. 445). Hoboken, NJ: Wiley.
Sopov, V., Purtova D. & Noxon, A. (2019). *Swap.Online White Paper.* Retrieved March 1, 2019, from https://wiki.swap.online/en.pdf
Statistics. Retrieved March 13, 2019., from https://www.blockchain.com/
Top 100 Cryptocurrencies by Market Capitalization. Retrieved March 3, 2019, from https://coinmarketcap.com/
Türkmen, S. Y. (2018). Industry 4.0 and Turkey: A financial perspective. In *Strategic design and innovative thinking in business operations* (pp. 273–291). Cham: Springer.
Vora, G. (2015). Cryptocurrencies: Are disruptive financial innovations here? *Modern Economy, 6* (07), 816.
Wilmott, P. (2013). *Paul Wilmott introduces quantitative finance.* Hoboken, NJ: Wiley.
Wüst, K., & Gervais, A. (2018, June). Do you need a Blockchain? In *2018 Crypto Valley conference on blockchain technology (CVCBT)* (pp. 45–54). IEEE.
Zanuiddin, A. (2019). *Guide to crypto derivatives: What is cryptocurrency derivatives?* Retrieved March 10, 2019, from https://masterthecrypto.com/what-is-cryptocurrency-derivatives-guide-crypto-derivatives

Yakup Söylemez is an Assistant Professor in Finance at Zonguldak Bülent Ecevit University in Zonguldak (Turkey). Dr. Söylemez has BAs in Business Administration from Gazi University, Turkey. He received his MA in Accounting and Finance from Marmara University, Turkey. He also received a Ph.D. in Accounting and Finance at Marmara University. His research interests cover topics in the field of finance, such as corporate finance, capital market, and financial technology. He has many articles published in journals.

Chapter 26
How Is a Machine Learning Algorithm *Now-Casting* Stock Returns? A Test for ASELSAN

Engin Sorhun

Abstract This paper focus on measuring the performance of algortimic trading in now-casting of stock returns using machine learning technics. For this task, (1) nine commonly used trend indicators to capture the behavior of the stock and a binary variable to signal positive/negative returs are used as predictors and target variable, respectively; (2) the standart machine learning process (splitting data, choosing the best performing algorithm among the alternatives, and testing this algorithm for new data) is applied to ASELSAN (a Turkish defense industry company) stock traded in BIST-100. The main findings are: (1) the decission tree algoritm performs better than K-nearest Neighbours, Logistic Regression, Bernoili Naïve Bayes alternatives; (2) the now-casting model allowed to realize an 18% of yield over the test period; (3) the model's performance metrics (accuracy, precision, recall, f1 scores and the ROC-AUC curve) that are commonly used for classification models in machine learning takes values just in the acceptance boundary.

26.1 Introduction

Now-casting is defined as the prediction of the present, the very near future and the very recent past (Giannone, Reichlin, & Small, 2008). The basic principle of now-casting is the exploitation of the recent information at higher frequencies than the target variable of interest in order to obtain an early estimate before it occurs (Marta, Domenico, Modugno, & Reichlin, 2013). Growing big data treatment technology, increasing available high frequency data have resulted in a very large empirical literature in economics and finance domains especially for the last decade. The fact that many conventional predictive models in statistical environment failed to send any signal about the 2008 Global Crisis became a great impulse in this literature to update prediction techniques by incorporating the time-sensitive continuous data flow and the conventional forecasting with big data treatment. The

E. Sorhun (✉)
Department of Economics, İstanbul 29 Mayis University, Umraniye/Istanbul, Turkey
e-mail: esorhun@29mayis.edu.tr

© Springer Nature Switzerland AG 2019

U. Hacioglu (ed.), *Blockchain Economics and Financial Market Innovation*, Contributions to Economics, https://doi.org/10.1007/978-3-030-25275-5_26

technical treatment of the now-casting methods and the difference of the now-casting approach from time series econometrics are revised by Foroni and Massimiliano (2013). The incorporation of the more timely information from high frequency data flow makes the forecasts increasingly more accurate (Bragoli, 2017).

However most of the attention seems to have been given to the now-casting for GDP (see Dahlhaus, Guénette, & Vasishtha, 2017 for BRICs; Modugno, Soybilgen, & Yazgan, 2016 for Turkey; Caruso, 2018 for Mexico), CPI (see Edward & Safeed, 2017 for the US; Funke, Mehrotra, & Yu, 2015 for China), business cycles (see D'Agostino, Giannone, Lenza, & Modugno, 2016 for the US recessions periods) and assets prices (see Gilberta, Scottib, Strasserc, & Vegab, 2017 for the US Treasury bills).

One of the specific implementation areas of now-casting with high frequency data is algorithmic trading. Algorithmic trading refers to the use of sophisticated computer algorithms to automatically make certain trading decision in the trading cycle, including pre-trade analysis (data analysis), trading signal generation (buying and selling recommendations), and trade executions (order management) (Treleaven, Galas, & Lalchand, 2013). The funds that are subject to algorithmic trading reached $22 trillion in 2016 from $700 million in 2007 worldwide (Global Algorithmic Trading Market Report, 2016–2020). Although algorithmic trading gained prominence in practice in the early 1990, it has recently become more attractive in the academic literature. A literature review confirmed only 51 relevant articles from 24 journals (Hu et al., 2015) between 2000 and 2013. The same set of journals contains 418 relevant articles between 2014 and 2018.

A considerable number of empirical research articles in economics and finance literature seem to focus on developing a model to find explicit trading rules that will guide traders (Álvaro, Sebastian, & Damir, 2016; Berutich, López, Luna, & Quintana, 2016; Brogaard, Hendershott, & Riordan, 2014; Carrion, 2013; Hirschey, 2017) while a relatively small but increasing number of empirical papers investigate the impact of algorithmic trading on market efficiency (Brian, 2018; Upson & Van Ness, 2017; Yadav, 2015).

One of the advantages of algorithmic trading is the effectiveness and efficiency of machine learning techniques in financial big data analysis (Narang, 2009). Machine learning (ML) is a kind of artificial intelligence computation by means of algorithms for discovering rules (learning) from voluminous and high frequency data to make better decisions. In addition, machine learning algorithms provide powerful technical advantages over the traditional time series models based on statistics and econometrics such as ARMA, ARIMA, GARCH, etc. that suffer from limitations due to their linearity assumption (Huang et al. 2019). This technical advantage results from the fact that machine learning algorithms focus on imitating or replicating the behavior of data and obtaining the most information possible from it whereas traditional models focus on the issue such as causality relationship, significance of models and parameters, etc. that limit the predictive performance for the new cases.

Implementation of ML to stock market aims to generate trading signals based on the discovered trading rules. For this task, standard ML process is implemented: (1) splitting the data into training set and test set; (2) choosing the best performing algorithm among alternative which are engaged to learn the structural behavior on the training set; (3) developing a trading rule to the extent that this successfully replicate the behavior of the test set.

However, the data forming the explanatory variables in signal generating recommendation models stems from fundamental analysis and technical analysis. There are some empirical researches indicating lower efficiency of technical analysis compared with fundamental analysis (Lo, Mamaysky, & Wang, 2000) as there are others indicating the opposite (Park & Irwin, 2007). Although there is no consensus, technical analysis has been adopted by most of the algorithmic trading models in practical use (Boming, Yuxiang, Li, Lirong, & Zhuo, 2019). Since technical analyses have a higher frequency, they are preferably incorporated more in now-casting models than fundamental analyses.

This article presents an empirical investigation of the effectiveness of algorithmic trading in Turkish Stock Market (BİST-100) via the ASELSAN stock price behavior example during one-year period. For this purpose, (1) a set of commonly used trend indicators are selected to measure the stock return behavior on a signal generating basis; (2) a set of machine learning algorithms that are commonly used in stock price prediction modelling are run to choose the best performing one; (3) the latter is evaluated in investigating whether a machine learning algorithm "now-cast" ASELSAN daily return on the test set. To address the mentioned tasks, Python's Scikit-Learn Machine Learning package and its related libraries such as Pandas, NumPy, and Technical Analysis (TA) are utilized.

26.2 Data Definition and Preprocessing

The dataset initially covers the opening prices, closing prices, the highest prices and lowest prices of ASELSAN in the 240 trading days from 05.03.2018 to 04.03.2019 (Fig. 26.1). From the opening and closing prices, the daily return rates are calculated. During the period, the sum of positive return is 202% and the sum of negative return (loss) is 217%. The highest daily return rate is 7.58%; the highest negative return rate is 8.60%; and standard deviation is 2.3%. And, a signal variable where 1 indicates positive return and 0 indicates negative returns is generated from daily return rate variable. The distribution of return signals are more or less in equilibrium (Fig. 26.2).

Furthermore, using Python Technical Analyze (TA) library, a set of predictor variables that are related to market behavior is generated from the initial data (Table 26.1). These variables are actually the common indicators for trend following and for range trading (Achelis, 2001).

On the other hand, EMAs and MACDs do not serve as they are, since the signal comes from the price in relation to averages, or from one average in relation to the other. Three more predictors which trigger the buy/sell signal are derived (Table 26.2).

Nine explanatory variables vary over a large range but the target variable (binary variable for return) is classified as buy/sell trading signals just after opening prices.

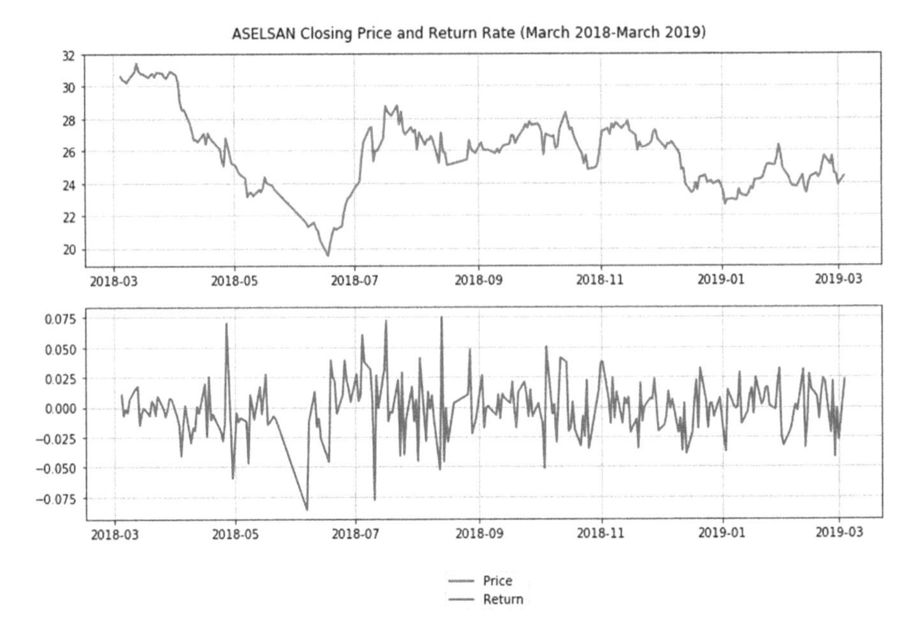

Fig. 26.1 ASELSAN price and return variation

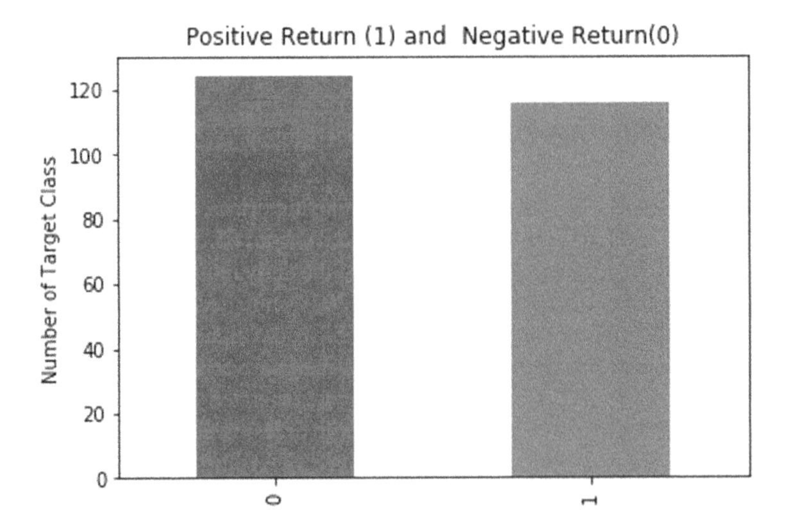

Fig. 26.2 The distribution of the target classes

Expressing in machine learning terminology, the feature variable (i.e. explanatory variable or predictors) that contains both the trend indicators and associated trend indicators is used as an integrated explanatory variable to now-caste the target variable which is return signal variable where 1 indicates the positive daily returns and 0 indicates the negative daily returns.

Table 26.1 Trend indicators as feature

Selected trend indicators	Definition	Thresholds to signal
Exponential Moving Average (EMA)	The EMA is a moving average that places a greater weight and significance on the most recent data points. Like all moving averages, this technical indicator is used to produce buy and sell signals based on crossovers and divergences from the historical average.	10 days (EMA10) and 30 days (EMA30)
Average True Range (ATR)	It provides an indication of the degree of price volatility. Strong moves, in either direction, are often accompanied by large ranges, or large True Ranges.	14 days
Average Directional Movement Index (ADX)	It measures the strength of the trend (regardless of direction) over time.	14 days
Relative Strength Index (RSI)	It compares the magnitude of recent gains and losses over a specified time period to measure speed and change of price movements of a security. It is primarily used to attempt to identify overbought or oversold conditions in the trading of an asset.	14 days
Moving Average Convergence Divergence (MACD)	Is a trend-following momentum indicator that shows the relationship between two moving averages of prices? It gives technical signals when bullish (to buy) or bearish (to sell) movement in the price is strengthening or weakening.	Calculated by subtracting the 26-period EMA from the 12-period EMA.
Moving Average Convergence Divergence (MACD Signal)	It shows EMA of MACD.	12 days for short-term, 26 days for long-term, 9 days for signal.

Table 26.2 Associated trend indicators as feature

Associated trend indicators	Signal definition	Signal
ClgtEMA10	Price > EMA10	Buy signal, otherwise sell signal
EMA10gtEMA30	EMA10 > EMA30	
MACDSIGgtMACD	MACD signal > MACD	

26.3 Distinguishing Machine Learning Approach from Conventional Econometric Approach

In this point, it is useful to distinguish machine learning approach from econometric approach based on multiple regressions in modelling. The model with multiple explanatory variables for linear regression is given by the following model:

$$y = \alpha + \beta_1 x_1 + \beta_2 x_2 + \ldots + \beta_i x_i \quad for\ i = 1, \ldots, n \tag{26.1}$$

If we actually let i = 1, . . ., n, we see that we obtain m equations:

$$\begin{bmatrix} Y_1 \\ Y_2 \\ \cdots \\ Y_n \end{bmatrix} = \begin{bmatrix} \alpha + \beta X_1 \\ \alpha + \beta X_2 \\ \cdots \\ \alpha + \beta X_n \end{bmatrix} = \begin{bmatrix} 1 & X_1 \\ 1 & X_2 \\ & \cdots \\ 1 & X_n \end{bmatrix} \times \begin{bmatrix} \alpha \\ \beta \end{bmatrix} \tag{26.2}$$

Instead of writing out the n equations, using matrix notation, the linear regression function can also be written in vector notation as:

$$Y = X\beta \tag{26.3}$$

This vector notation is equivalent to the matrix in machine learning perspective: Y is a column vector of values of the target variables and β is a column vector of the values of the model's parameters for the training examples. X is m (the number of training examples) by n (the number of features) dimensional matrix of the explanatory variables for the training examples.

Since division by a matrix is impossible, we can multiply by the inverse of X to avoid matrix division (just as dividing by an integer is equivalent to multiplying by the inverse of the same integer). We will multiply X by its transpose to yield a square matrix that can be inverted. We must find the values of β that minimize the cost function (mean squired error). We can analytically solve for β as follows:

$$\beta = \left(X^T X\right)^{-1} X^T Y \tag{26.4}$$

Our machine learning algorithm must predict the target variable (bivariate signal) from the value of ten parameters (the coefficients for the nine features (trend indicators) and the intercept term) on the train set. And with the learning indicator (β) must perform on the test set.

26.4 Train and Test Split

Data set is randomly divided to form the training and test set. The training set is composing of nine features/predictors and one target (return signal variable) over 192 trading days while the test set is composing of nine features/predictors an done target over 48 trading days. Note that 80% of 240 trading days are randomly selected as training set; and the remainder (20% of it) is randomly selected as test set. The training set is then used to train the model (i.e. learning the behavior of the training

Table 26.3 Comparative performances of the selected models

Machine Learning Algorithms	Accuracy ratios
K-Nearest Neighbors (KNN)	0.56
Logistic Regression (LR)	0.71
Bernoulli Naïve Bayes (BNB)	0.69
Decision Trees (DT)	0.73

data) and the test set is used to evaluate the performance of a model (how well the model predict the returns/losses just before it occur).

26.5 Model Selection

A typical machine learning process involves learning process through different algorithms on the training dataset and selecting the one with best performance. Since the target variable is a signal for positive or negative return, we need to choose the best performing classifier among different machine learning algorithms. For this task, K-Nearest Neighbors (KNN), Logistic Regression (LR), Bernoulli Naïve Bayes (BNB) and Decision Trees (DT) algorithms are chosen for model selection process. On the other hand, the accuracy ratio is used as performance measure to choose the best performing model. Based on the accuracy ratios, decision trees algorithm is chosen as the best performing model (Table 26.3).

26.6 Decision Tree Algorithm

In the decision tree chart, each predictor variable is used to set up a decision rule that split the data set. To optimize decision tree performance, some measures such as criterion, splitter, maximum depth are chosen among the alternatives.

Criterion: The function to measure the quality of a split. The alternative measures are the Gini impurity and the Entropy Index for the information gain.

$$Gini = 1 - \sum\nolimits_j p_j^2 \tag{26.5}$$

$$Entropy = -\sum\nolimits_j p_j \log_2 p_j \tag{26.6}$$

Splitter: The strategy used to choose the split at each node. The Alternative strategies are "best" to choose the best split (i.e. algorithm detect the best split to give the best performance score) and "random" to choose the best random split.

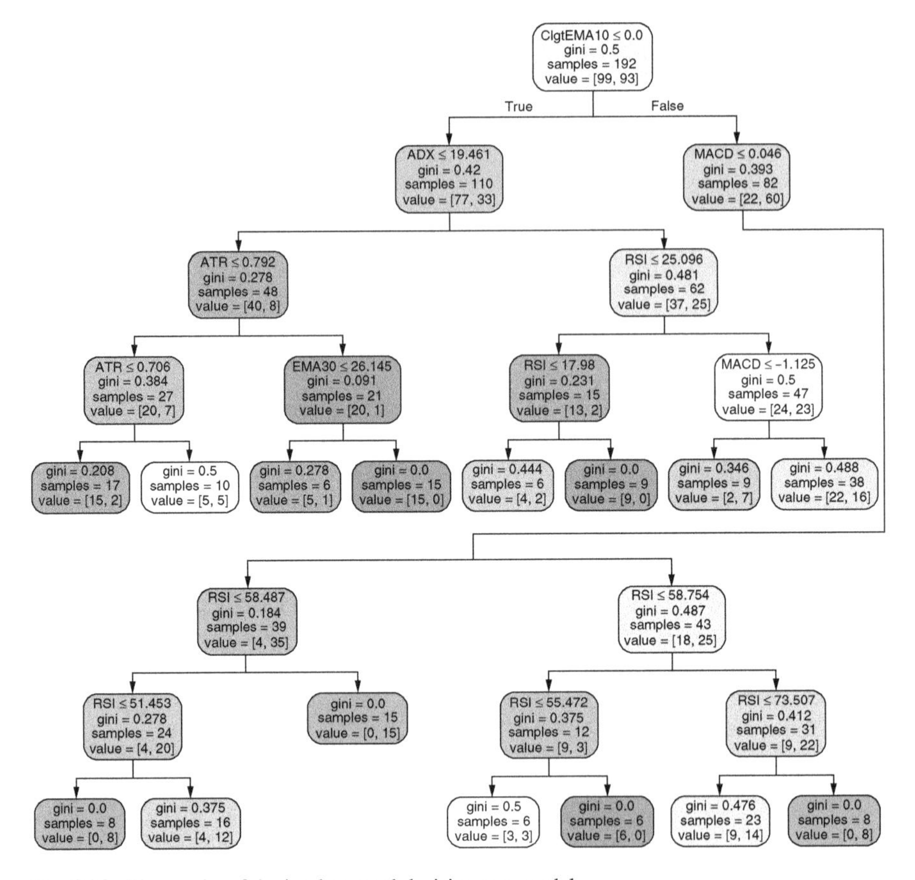

Fig. 26.3 The results of the implemented decision tree model

Maximum Depth: The maximum depth of the tree. The alternative choices are "none" and giving a number: If None, then nodes are expanded until all leaves are pure. The higher value of maximum depth causes overfitting, and a lower value causes underfitting.

Figure 26.3 shows a pair of pure nodes that allows to deduce a possible trading rules: We started with 192 samples (i.e. value shows actual signals: 99 actual buy-signal/positive return and 93 actual sell-signal/negative return) at the root and split them into two nodes with 110 samples (predicted buy-signal/positive returns) and 82 samples (predicted sell-signal/negative returns), using ClgtEMA10 cut-off ≤ 0.0. Gini referred as Gini ratio, which measures the impurity of the node. The Gini score is a metric that quantifies the purity of the node. A Gini score greater than zero implies that samples contained within that node belong to different classes. A Gini score of zero means that the node is pure, that within that node only a single class of samples exist. Notice that we have a Gini score greater than zero (0.5 in the root node); therefore, we know that the samples contained within the root node belong to different classes. The first decision is whether ClgtEMA10 indicator gives -1

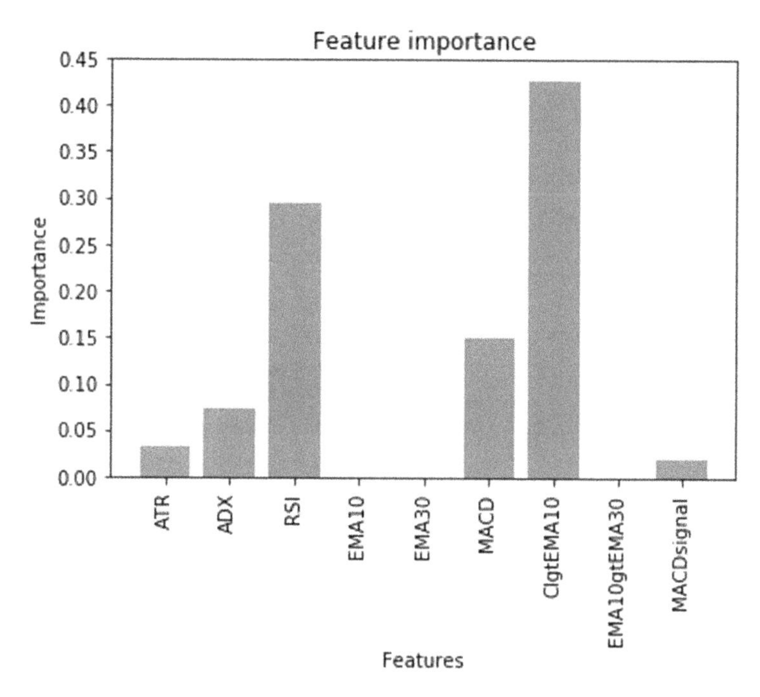

Fig. 26.4 Relative importance weights of the features

(a buy-signal) or +1 (a sell-signal). The algorithm proceeds to produce a complete tree of four levels, depth 4 (note that the top node is not included in counting the levels).

26.7 Feature Importances

As Fig. 26.4 indicates, most of the buy/sell signals in decision trees are captured by two predictors: ClgtEMA10 (43%), RSI (30%). Feature Importance plot shows the weight of each variable in model fitted on training set. It doesn't mean that the predictors/features with low weight are not important and we don't need to put it in the model. Getting out of the model one feature with low weight might reduce the weight of the others.

Table 26.4 Performance metrics definitions

Performance metrics	Definition
Accuracy score	(TP + TN)/(TP + TN + FP + FN)
Precision score	TP/(TP + FP)
Recall score	TP/(TP + TN)
F1 score	2∗[(precision∗recall)/(precision + recall)]

Table 26.5 Confusion matrix definition

Confusion Matrix		Predicted	
		Negative	Positive
Actual	Positive	True Negative (TN)	False Positive (FP)
	Negative	False Negative (FN)	True Positive (TP)

26.8 Performance Measures of the Decision Trees Model

The commonly used metrics for classification task are accuracy, precision, recall, F1 (Table 26.4) and ROC-AUC (Fig. 26.6). These metric scores are obtained from confusion matrix (Table 26.5) which shows True Negative (TN), False Negative (FN), False Positive (FP) and True Positive (TP) based on the comparison of actual target classes with predicted classes buy the model. Although metric scores varies depending on the case, as a rule of thumb, any score above 0.70–0.80 is regarded as acceptable, 0.80–0.90 as very-good.

Our decision tree classification model that has already *learned* the behavior of ASELSAN returns via the training data set (192 samples) now is used for prediction. Using the predictor's values of the test data, our model generates the target classes. We then evaluate the success of the model by comparing the predicted target classes with the actual target classes of the test data set (48 samples). 48 actual target classes seem to have an almost fair distribution: 25 sell-signals (0) and 23 buy-signals (1).

The confusion matrix of the model (Fig. 26.5) serves to evaluate the accuracy of a classification for the test data: the vertical side shows the actual classes while the horizontal side shows the predicted classes by the model: 0 indicates negative return or sell-signal; and 1 indicates positive return or buy-signal. Our model *truly* predicted 22 negative returns and 13 positive returns whereas it *falsely* predicted 10 positive return (as negative) and 3 negative return (as positive). So its accuracy score = (22 + 13)/(22 + 13 + 10 + 3) = 0.73.

All the above mentioned performance metrics are calculated from the confusion matrix. The Table 26.6 reports the scores for them both for positive and negative return predictions on the test set (i.e. "support" indicates the actual classes of the target variable).

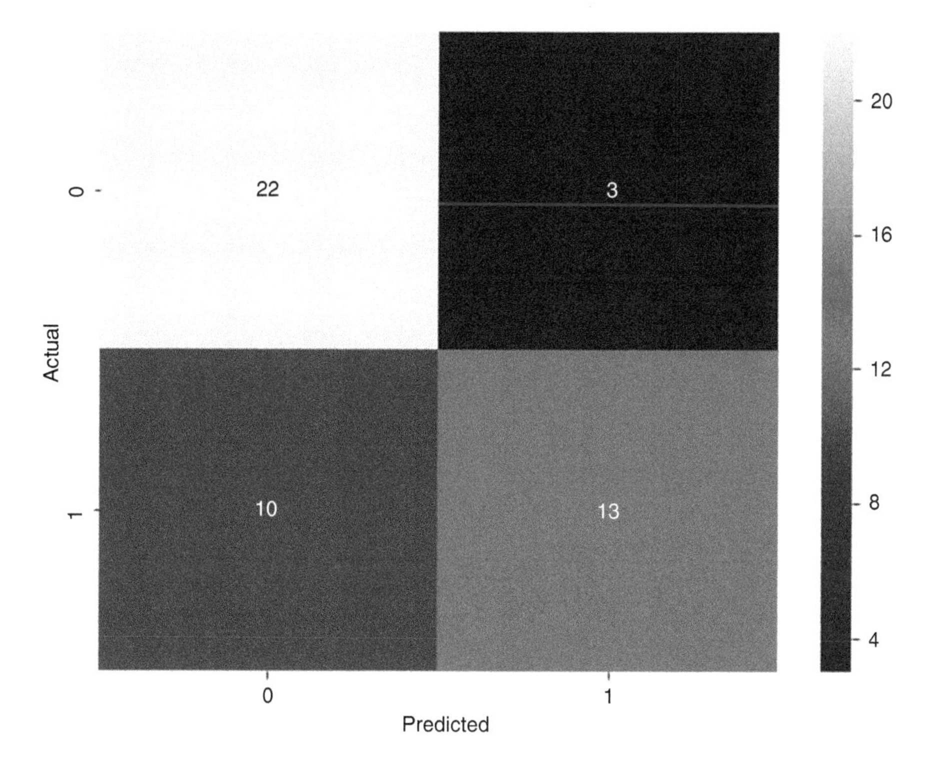

Fig. 26.5 Confusion matrix on the test set

Table 26.6 Performance scores

	Precision	Recall	f1-score	Support
0	0.69	0.88	0.77	25
1	0.81	0.57	0.67	23
micro avg	0.73	0.73	0.73	48
macro avg	0.75	0.72	0.72	48
weighted avg	0.75	0.73	0.72	48
Accuracy:0.73				

As for ROC (Receiver Operating Characteristics) and AUC (Area Under The Curve), or basically AUC-ROC curve, it tells how much model is capable of distinguishing between classes.[1] Higher the AUC, better the model is at predicting 0s as 0s and 1s as 1s. ROC is a probability curve and AUC score reduces the ROC curve to a single value that represents the expected performance of the classifier (Fig. 26.6).

[1]The ROC curve is plotted with True Positive Rates (also called recall or sensitivity) against the False Positive Rates where TPR is on y-axis and FPR (=FP/(TN + FP)) is on the x-axis. Note that the ROC does not depend on the class distribution.

Fig. 26.6 ROC-AUC curve

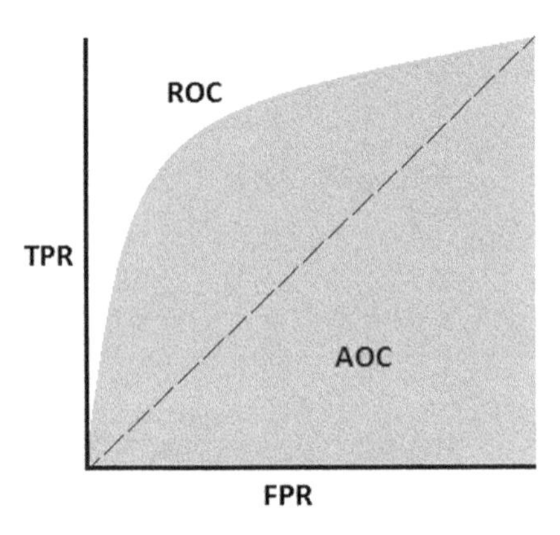

To asses ROC-AUC plot, note that a classifier that draws a curve closer to the top-left corner indicate a better performance (the closer the curve comes to the 45-degree diagonal of the ROC space, the less accurate the test) and an AUC score above 0.7 is regarded acceptable. On the other hand, the Threshold Zero represents the optimal compromise between TPR versus FPR (i.e. the most accurate in classifying the target). From ROC-AUC plot, it seems that the classifier outperforms random guessing (the dashed 45C line); most of the area of the plot lies under its curve (Fig. 26.7).

As a machine learning algorithm learns on the train set, it is expected that its performance measure improves incrementally over time. During the training of a machine learning model, the current state of the model at each step of the training algorithm can be evaluated by learning curves. This shows how well the model is "learning" on the train set and how well it uses its training experience to predict a hold-out validation dataset.

To plot learning curves, first we split (one more time) the train dataset into training set and validation dataset. Our training set that has 192 observations (or instances) is divided by a proportion of 0.8–0.2 for train set (153) and validation set (39) relatively. Second, we using each additional data from 1 to 153, the algorithm works and the accuracy scores are calculated at the each step. Evaluation on the validation dataset gives an idea of how well the model is "generalizing."

If the training and cross validation scores converge together as more data is added, then the model will probably learn from more data. If the training score is much greater than the validation score then the model probably requires more training examples in order to generalize more effectively.

Figure 26.8 shows the learning curves for our decision tree model on the training set. Our model has higher accuracy scores for train set that for validation set. It means the model needs more data to generalize an idea for the future now-casting tasks. Besides, both the training set's scores and the validation set's scores varies in a narrow range. However, the standard deviations of the training set's scores are

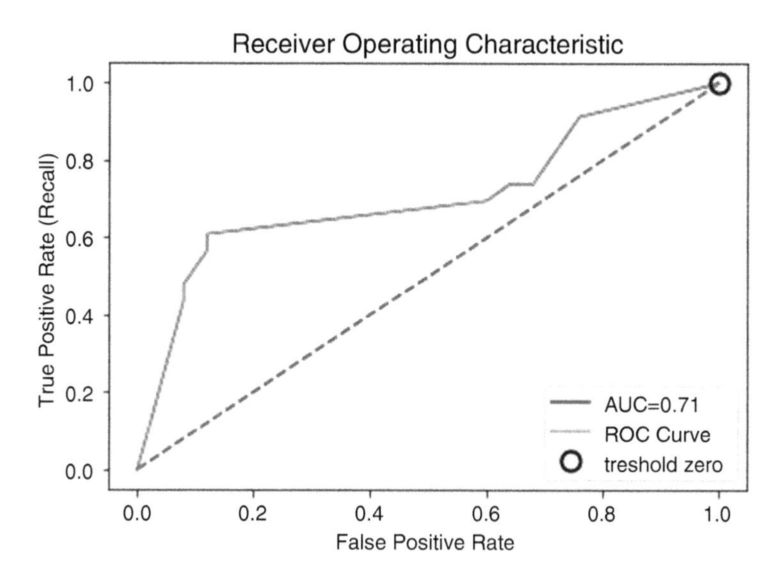

Fig. 26.7 ROC-AUC curve of the model

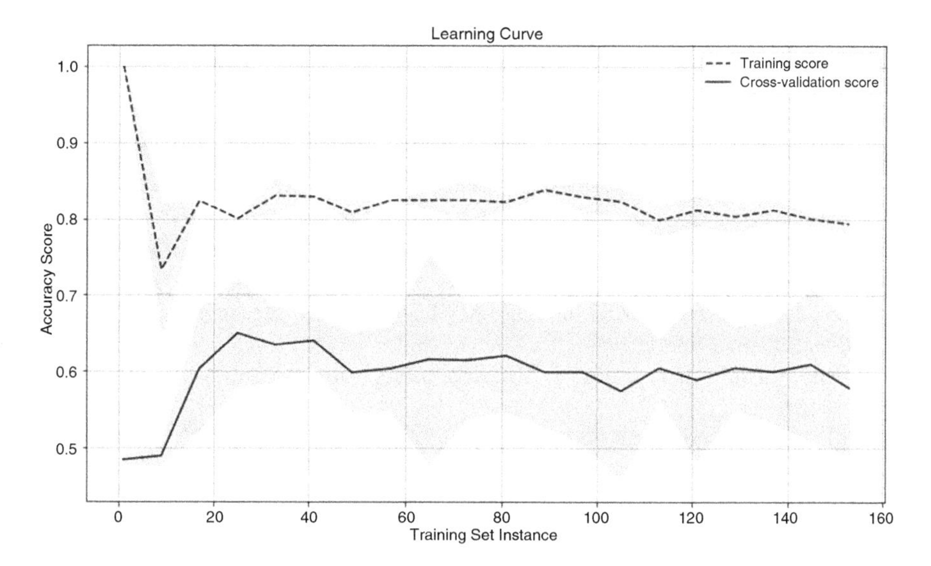

Fig. 26.8 Learning curve of the model

negligible whereas those of the validation set's scores are considerable. This is a sign for the existence of variance problem: Variance is the variability of model prediction for a given data point or a value which tells us spread of our data. Model with high variance perform very well on training data but has high error rates on test data.

26.9 Evaluating the Return Performance of Algorithmic Trading Based on the Model

Figure 26.9 allows to evaluate the model success by visualizing the comparison of the predicted return signals with the actual return classes: the upper panel shows the buy/sell signals (red triangles) predicted by the model and the actual positive/negative return classes (dashed lines) taking values 0 or 1. If they (red triangles and dashed line) overlap then the model predicts well the actual returns, otherwise it sends wrong signals.

Tables 26.7 and 26.8 lists the trading days that the train-test-split algorithm randomly assigned into the test set. And these tables also classify the return signals, captured by the decision tree model by comparing the actual target classes with the predicted ones.

At a first look, the relative weight of the sell-signals over the buy-signals. But the model performs better in capturing the buy-signals than the sell-signals.

The total return is calculated based on the scenario that a trader has bought or sold ASELSAN stock based on the model predictions. The test set contains 48 trading days that are randomly selected from 05.03.2018 to 04.03.2019. What is the total return/loss over 48 days if he/she traded ASELSAN as the model suggested?

As the confusion matrix indicates, the model has given 32 times sell-signal (0 predicted): 22 of them true and 10 of them false. Algorithmic trading based on this model saved the trader 22 times from losses. It is equivalent to 35% negative return! Nevertheless, the model caused the trader 10 times to miss positive return! It is equivalent to 16% positive return! Thus he/she missed 16% of positive return.

The model has given 16 times buy-signal (1 predicted): 13 of them true and 3 of them false. The model truly suggested the trader 13 times to buy ASELSAN shares. If he/she bought ASELSAN when the model sent 'true-buy-signal' his/her return is 21%. Nevertheless, the model falsely suggested him/her three times to buy ASELSAN stocks. If he/she bought ASELSAN when the model sent 'false buy-signal' his/her loss is 3%.

Finally, his/her net return = 21 (realized positive return) − 3 (realized loss) = 18%.

Besides, on the one hand, during the test period there has been 37% of potential total positive return. And the model helped the trader to realize nearly half of this (18/37 = 0.49). On the other, during the test period there has been 38% of potential total loss. And the model saved the trader from 92% of this (35/38 = 0.92).

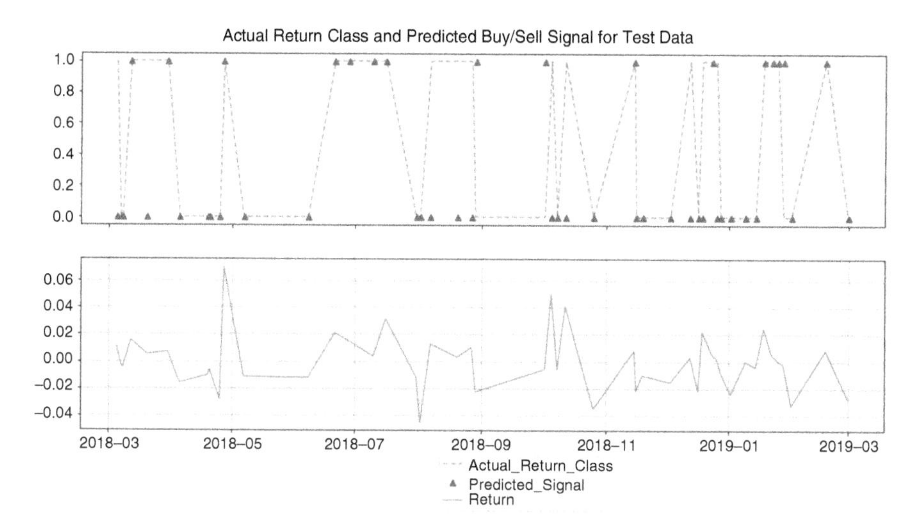

Fig. 26.9 The signal performance of the model on the test set

Table 26.7 How did the model capture the positive returns on the test set?

Date	Return rate	Actual return class	Predicted signal	Signal
12.03.2018	0.0158	1	1	True Positive
30.03.2018	0.0072	1	1	True Positive
27.04.2018	0.0703	1	1	True Positive
21.06.2018	0.0211	1	1	True Positive
28.06.2018	0.015	1	1	True Positive
10.07.2018	0.0037	1	1	True Positive
16.07.2018	0.0315	1	1	True Positive
29.08.2018	−0.0225	0	1	False Positive
02.10.2018	−0.0058	0	1	False Positive
15.11.2018	0.008	1	1	True Positive
24.12.2018	0.0049	1	1	True Positive
18.01.2019	0.0246	1	1	True Positive
22.01.2019	0.005	1	1	True Positive
25.01.2019	0.0008	1	1	True Positive
28.01.2019	−0.0024	0	1	False Positive
18.02.2019	0.0082	1	1	True Positive

Table 26.8 How did the model capture the negative returns on the test set?

Date	Return rate	Actual return class	Predicted signal	Signal
05.03.2018	0.0106	1	0	False Negative
07.03.2018	−0.002	0	0	True Negative
08.03.2018	−0.0046	0	0	True Negative
20.03.2018	0.0052	1	0	False Negative
05.04.2018	−0.0159	0	0	True Negative
19.04.2018	−0.0103	0	0	True Negative
20.04.2018	−0.006	0	0	True Negative
25.04.2018	−0.0283	0	0	True Negative
07.05.2018	−0.0114	0	0	True Negative
08.06.2018	−0.0121	0	0	True Negative
31.07.2018	−0.0117	0	0	True Negative
02.08.2018	−0.0454	0	0	True Negative
07.08.2018	0.0129	1	0	False Negative
20.08.2018	0.0032	1	0	False Negative
27.08.2018	0.0103	1	0	False Negative
05.10.2018	0.0505	1	0	False Negative
08.10.2018	−0.0059	0	0	True Negative
12.10.2018	0.0411	1	0	False Negative
26.10.2018	−0.035	0	0	True Negative
16.11.2018	−0.0216	0	0	True Negative
19.11.2018	−0.0103	0	0	True Negative
03.12.2018	−0.0154	0	0	True Negative
13.12.2018	0.0032	1	0	False Negative
17.12.2018	−0.0218	0	0	True Negative
19.12.2018	0.0222	1	0	False Negative
26.12.2018	0.0025	1	0	False Negative
28.12.2018	−0.0083	0	0	True Negative
02.01.2019	−0.0241	0	0	True Negative
09.01.2019	−0.0001	0	0	True Negative
14.01.2019	−0.0043	0	0	True Negative
01.02.2019	−0.0327	0	0	True Negative
01.03.2019	−0.0277	0	0	True Negative

26.10 Conclusion

This paper aims to test the performance of algorithmic trading in now-casting of stock returns using machine learning technics. For this task, ASELSAN (a Turkish defense industry company) traded stock in BIST-100 is subjected to the machine learning treatment.

A standard machine learning process actually involves three steps: splitting data, choosing the best performing algorithm, and testing this algorithm for new data.

First, the data set involves a target class variable that is derived from the stock's daily returns to capture the buy/sell signal (or positive and negative returns) and nine explanatory variables that are derived from commonly used trend indicators to capture the behavior of the stock (or target classes). The one-year long ASELSAN stock data containing 240 trading days is separated as training set and test set. Their contents are randomly (not consecutively) picked.

Second, several machine learning classification algorithms (K-nearest Neighbors, Logistic Regression, Bernoulli Naïve Bayes and Decision Trees) are performed on training set. It is the decision tree algorithm that sends more accurate signals relative to others. The decision tree algorithm that is trained to reveal a trading rule (specific to ASELSAN stock) is optimized by parameter setting (Gini as impurity index, "random" as split parameter and four depths as maximum depth of the tree).

Third, the decision tree classification model trained over 192 trading-days ASELSAN performance is then tested it over 48 trading-days. The actual values of ASELSAN in the BIST-100 are compared with the predicted values. The model allowed the trader to make a profit of 18% during these 48 days. Compared with the interest rate range of Turkish banking sector (8–11%) and the BIST-100 performance (7%) over the test period, an 18% of yield can be better understood.

However, though there are nine predictors, three of them (ClgtEMA10, RSI, MACD) explains 88% of the variation in the return signals. Nevertheless, the model's performance metrics (accuracy, precision, recall, f1 scores and the ROC-AUC curve) that are commonly used for classification models in machine learning takes values just in the acceptable threshold. To generalize a trading rule or a signal recommendation model for the concerned stock, higher performance scores and a stable and converging learning curve are needed.

References

Achelis, S. B. (2001). *Technical analysis from A to Z*. New York: McGraw Hill.

Álvaro, C., Sebastian, J., & Damir, K. (2016). Algorithmic trading with learning. *International Journal of Theoretical and Applied Finance, 19*(4), 1650028.

Berutich, J. M., López, F., Luna, F., & Quintana, D. (2016). Robust technical trading strategies using GP for algorithmic portfolio selection. *Expert Systems with Applications, 46*, 307–315.

Boming, H., Yuxiang, H., Li, D. H., Lirong, Z., & Zhuo, Z. (2019). Automated trading systems statistical and machine learning methods and hardware implementation: A survey. *Enterprise Information Systems, 13*(1), 132–144.

Bragoli, D. (2017). Now-casting the Japanese economy. *International Journal of Forecasting, 33*(2), 390–402.

Brian, M. W. (2018). Does algorithmic trading reduce information acquisition? *The Review of Financial Studies, 31*(6), 2184–2226.

Brogaard, J., Hendershott, T., & Riordan, R. (2014). High-frequency trading and price discovery. *Review of Financial Studies, 27*(8), 2267–2306.

Carrion, A. (2013). Very fast money: High-frequency trading on the NASDAQ. *Journal of Financial Markets, 16*, 680–711.

Caruso, A. (2018). Nowcasting with the help of foreign indicators: The case of Mexico. *Economic Modelling, 69*, 160–168.

D'Agostino, A., Giannone, D., Lenza, M., & Modugno, M. (2016). Nowcasting business cycles: A Bayesian approach to dynamic heterogeneous factor models. In E. Hillebrand & S. J. Koopman (Eds.), *Dynamic factor models (Advances in econometrics, Vol. 35)* (pp. 569–594). Bingley: Emerald Group Publishing Limited.

Dahlhaus, T., Guénette, J., & Vasishtha, G. (2017). Nowcasting BRIC+M in real time. *International Journal of Forecasting, 33*(4), 915–935.

Edward, S. K., & Safeed, Z. (2017). Nowcasting U.S. headline and core inflation. *Journal of Money, Credit and Banking, 49*(5), 931–968.

Foroni, C., & Massimiliano, M. (2013). *A survey of econometric methods for mixed frequency data.* Norges Bank Working Paper No. 2013-06.

Funke, M., Mehrotra, A., & Yu, H. (2015, June). Tracking Chinese CPI inflation in real time. *Empirical Economics, 48*(4), 1619–1641.

Giannone, D., Reichlin, L., & Small, D. (2008). Nowcasting: The real-time informational content of macroeconomic data. *Journal of Monetary Economics, 55*(4), 665–676.

Gilberta, T., Scottib, C., Strasserc, G., & Vegab, C. (2017). Is the intrinsic value of a macroeconomic news announcement related to its asset price impact? *Journal of Monetary Economics, 92*, 78–95.

Global Algorithmic Trading Market Report 2016–2020. Retrieved from https://www.technavio.com/report/global-algorithmic-trading-market-analysis-share-2018

Hirschey, N. (2017). *Do high-frequency traders anticipate buying and selling pressure?* FCA Occasional Paper No:16-2017.

Hu, Y., Liu, K., Zhang, X., Sub, L., Ngai, E. W. T., & Liu, M. (2015). Application of evolutionary computation for rule discovery in stock algorithmic trading: A literature review. *Applied Soft Computing, 36*(2015), 534–551.

Huang, B., Huan, Y., Xu, L. D., Zheng, L., & Zou, Z. (2019). Automated trading systems statistical and machine learning methods and hardware implementation: A survey. *Enterprise Information Systems, 13*(1), 132–144.

Lo, A. W., Mamaysky, H., & Wang, J. (2000). Foundations of technical analysis: Computational algorithms, statistical inference, and empirical implementation. *The Journal of Finance, 55*(4), 1705–1765.

Marta, B., Domenico G., Modugno, M., & Reichlin L. (2013). *Now-casting and the real-time data flow*. Working Paper Series NO 1564/july 2013. Retrieved from https://www.econstor.eu/bitstream/10419/153997/1/ecbwp1564.pdf

Modugno, M., Soybilgen, B., & Yazgan, E. (2016). Nowcasting Turkish GDP and news decomposition. *International Journal of Forecasting, 32*(4), 1369–1384.

Narang, K. (2009). *Inside the black box: The simple truth about quantitative trading* (1st ed.). Hoboken, NJ: Wiley.

Park, C. H., & Irwin, S. H. (2007). What do we know about the profitability of technical analysis? *Journal of Economic Surveys, 21*(4), 786–826.

Treleaven, P., Galas, M., & Lalchand, V. (2013). Algorithmic trading review. *CACM, 56*(11), 76–85.

Upson, J., & Van Ness, R. A. (2017). Multiple markets, algorithmic trading, and market liquidity. *Journal of Financial Markets, 32*, 49–68.

Yadav, Y. (2015). How algorithmic trading undermines efficiency in capital markets. *Vanderbilt Law Review, 68*(6), 1607–1671.

Chapter 27
A Comprehensive Framework for Accounting 4.0: Implications of Industry 4.0 in Digital Era

Banu Esra Aslanertik and Bengü Yardımcı

Abstract The fourth industrial revolution, namely "Industry 4.0" refers to a new digital industrial technology, digital transformation and the fourth phase of technological advancement promoting the industrial production of the future. Cyber-Physical Systems (CPS) and the Internet of things (IoT) are the main forces of industry 4.0 which are improving the manufacturing systems and business processes by leading-edge innovations. Industry 4.0 with integrated production and logistics processes, growing interaction between robots and human and data flows within global value chains will have a significant impact on all business processes. In this context, accounting systems, which have a very important function for businesses, need to adapt to industry 4.0 by redefining the whole accounting system, as well as redesigned strategies. Industry 4.0 offers a new potential for the transformation of the accounting process through digitalization and application of new tools of industry 4.0 such as big data analytics, networking, system integration. The main objective of this chapter is to offer a conceptual framework for a newly designed accounting process in terms of procedures, technology and accounting professionals.

27.1 Introduction

Technological advances have been inevitably triggering the business environment for a change since the dawn of the Industrial Revolution. The use of water and steam power with the first industrial revolution followed the realization of the second industrial revolution with the emergence of electric power. Then the emergence of the electronic and information technology, known as the third industrial revolution

B. E. Aslanertik (✉)
Department of Business Administration, Faculty of Business, Dokuz Eylul University, Izmir, Turkey
e-mail: esra.aslanertik@deu.edu.tr

B. Yardımcı
Department of Finance, Banking and Insurance, Vocational School, Yaşar University, Izmir, Turkey
e-mail: bengu.yardimci@yasar.edu.tr

© Springer Nature Switzerland AG 2019

U. Hacioglu (ed.), *Blockchain Economics and Financial Market Innovation*, Contributions to Economics, https://doi.org/10.1007/978-3-030-25275-5_27

was followed by rapid developments in digital technology and led to the rise of the fourth industrial revolution.

The fourth industrial revolution, commonly known as "Industry 4.0" refers to a new digital industrial technology, digital transformation and the fourth phase of technological advancement which creates a digital business combining advanced manufacturing and operating techniques with smart digital technologies. As a global concept, this term differs terminologically around the world. For instance, in Europe where it originated, named as Industry 4.0, on the other side, this phenomenon is known as digital supply network in the United States. Although the terminology is different, the concept encompasses the same technologies and applications (Schwartz, Stockton, & Monahan, 2017). The reason behind Industry 4.0's importance is its holistic approach integrating the digital and physical worlds. The marriage of digital and physical technologies would affect not only supply chain or manufacturing, but also its operations, revenue growth and value creation (Schwartz et al., 2017).

The concept of Industry 4.0 includes many applications. Tools for industry 4.0 adaptation are the Internet of Things (IoT&IoS), Cyber-Physical Systems (CPS), Big Data, Robotics, Simulation, Cloud, Sensors and other tools such as RFID, GPS, SMART-ID. Cyber-Physical Systems (CPS) and the Internet of things (IoT) are the main technological constituents of Industry 4.0. Cyber-physical systems (CPS) are integrated communication networks wired and wirelessly by computer-based algorithms which have an interface between the digital and physical World (Sauter, Bode, & Kittelberger, 2015; Deloitte, 2015). CPS integrates networks using multiple sensors, actuators, control processing units and communication devices (Hofmann & Rüsch, 2017). The Internet of Things is an intelligent network infrastructure that offers connectivity among devices, systems, and humans (Chen, Barbarossa, Wang, Giannakis, & Zhang, 2019; Burritt & Christ, 2016). IoT collects and shares information through the value chain, and further enables real-time decision making and Internet of Services facilitates companies a platform to offer the services to their various partners and increases the collaboration between them (Dai & Vasarhelyi, 2016). Big data refers to analytics based on large differently structured data sets often characterized using four Vs: volume (large volume of data), veracity (data from different sources), velocity (analysis of streaming data) and variety (analysis of different types of data structures) that traditional tools are inadequate to process (Sandengen, Estensen, Rødseth, & Schjølberg, 2016; Sledgianowski, Gomaa, & Tan, 2017). New technology allows robots to tackle complex assignments, interact with each other, work safely with people side by side in the future production lines. These robots will be cheaper and easier to program and more flexible in the manufacturing process (Sandengen et al., 2016; Rüßmann et al., 2015). Sensors have emerged as low-cost, low-power and multifunctional tools in digital electronics at the beginning of the twenty-first century (Dai & Vasarhelyi, 2016). Data collection, processing and communication functions of sensors have important roles in Industry 4.0. Therefore, instead of people, it is assumed that the sensors will undertake the data collection function. GPS as locations identifiers, RFID tags individual identification devices and pacemakers are the examples for sensors (Dai

& Vasarhelyi, 2016; O'Leary, 2013). RFID can identify an object in the virtual world and determines the status of a product, while GPS is used to monitor products. Additionally, smart ID cards may be employed to monitor the locations of workers in a factory (Dai & Vasarhelyi, 2016). Simulations use real-time data to project the physical world in a virtual model that will include machines, products, and people. By this means, operators can test machine settings before they change physically, allowing them to provide optimization, enabling efficient time management and quality improvement in machine settings (Rüßmann et al., 2015; Sandengen et al., 2016). While cloud technology companies are using cloud-based software for enterprise and analytics applications today, companies with industry 4.0 will have to share more data. Data collected from smart factories with higher-performance cloud technology will provide data-based services (Ernst and Young, 2017; Rüßmann et al., 2015). Therefore, cloud technology enables organizations to possess a connected, flexible system instead of outdated, fragmented and inflexible legacy systems.

Industry 4.0, which is an integrated production and logistics process, is expected to influence data flows within global value chains and business functions such as production, logistics, marketing, accounting, human resources, legislation along with growing robot and human interaction (Dai & Vasarhelyi, 2016). In this respect, accounting systems with their important roles in businesses require the adaptation of industry 4.0 by redefining the entire accounting system and redesigning corporate strategies. Industry 4.0 presents the new potential for the transformation of the accounting process through digitalization and applies new tools such as big data analytics, networking and system integration. The aim of this chapter is to propose a conceptual framework for a newly designed accounting process on the basis of procedures, technology and accounting professionals.

The chapter proceeds as follows. Section 27.2 reviews the literature on industry 4.0 and accounting related studies by laying out the theoretical dimensions of the research. Section 27.3 provides a comprehensive conceptual framework of the accounting system through digitalization. Finally, Sect. 27.4 summarizes the chapter and make significant evaluations.

27.2 Literature Review

In recent years there has been growing interest in the fourth industrial revolution, namely industry 4.0. The concept of Industry 4.0 emerged in Germany (Davies, 2015; Burritt & Christ, 2016). Industry 4.0 as a blanket term describes a group of related technological advancements that constitute the basis for the increasing digitization of the business environment (Burritt & Christ, 2016). In the literature, there are many studies regarding the concept of industry 4.0 related technological advances and how they affect the accounting processes.

Many attempts have been made in order to explain the big data in accounting. For instance, Vasarhelyi, Kogan, and Tuttle (2015) in their study provide a definition for the big data, explain its importance and present a framework by which it will cause changes in the fields of accounting and auditing (Moffitt & Vasarhelyi, 2013). Likewise, Moffitt and Vasarhelyi (2013) have prepared a framework for potential research topics related to integration of big data in accounting, auditing and standards. In their commentary, Huerta and Jensen (2017) discuss data analytics and Big Data from an accounting information systems perspective by describing risks, opportunities, and challenges for the accounting profession in all areas. Richins, Stapleton, Stratopoulos, and Wong (2017) argue in their conceptual framework based on structured/unstructured data and problem-driven/exploratory analysis that big data analytics offer accountants to complete their knowledge and skills. They conclude that to accomplish this task accounting educators, standard setters and professional bodies need to make arrangements in their curricula, standards, and frameworks. Furthermore, Warren, Moffitt, and Byrnes (2015) discuss that big data may have significant effects on managerial accounting, financial accounting and reporting. Griffin and Wright (2015) present commentaries on accounting and auditing profession and explain that educators need to modify their accounting and auditing curricula for big data. Finally, Dai and Vasarhelyi (2016) aim to imagine how industry 4.0 technologies affect audit process before it is widely applied in business.

From a different point of view, Green, McKinney, Heppard, and Garcia (2018) investigate the impact of big data on accounting by explaining the consumer's demand for accounting data, and how its effects in decision making. On the other side, Burritt & Christ (2016) explore the processes of corporate sustainability and Industry 4.0 integration through environmental accounting.

Schoenthaler, Augenstein, and Karle (2015) focus on the design and management of innovative business processes with cyber-physical systems and the Internet of things that has been further enhanced by groundbreaking innovations. From a different perspective, Erol, Schumacher, and Sihn (2016) propose a three-stage process model which shed light on their industry 4.0 vision and aid organizations to formulate their strategies.

To sum up, as explained in the literature review, previous studies have mainly focused on specific tools or processes individually. However, the present chapter offers a comprehensive conceptual framework which aims to combine the aspects of different studies, articles or applications of industry 4.0 on the accounting system and processes. This newly constructed framework supports synergy by offering the most appropriate tools for each accounting process and their sub-processes to achieve the target outcomes qualified in the framework.

27.3 Conceptual Framework

This section of the paper will provide a comprehensive conceptual framework of the accounting system through digitalization. This framework is based on an extensive literature review and analysis of accounting processes in terms of industry 4.0 tools. In order to map out the interactions and the relationships, the following Fig. 27.1 is produced.

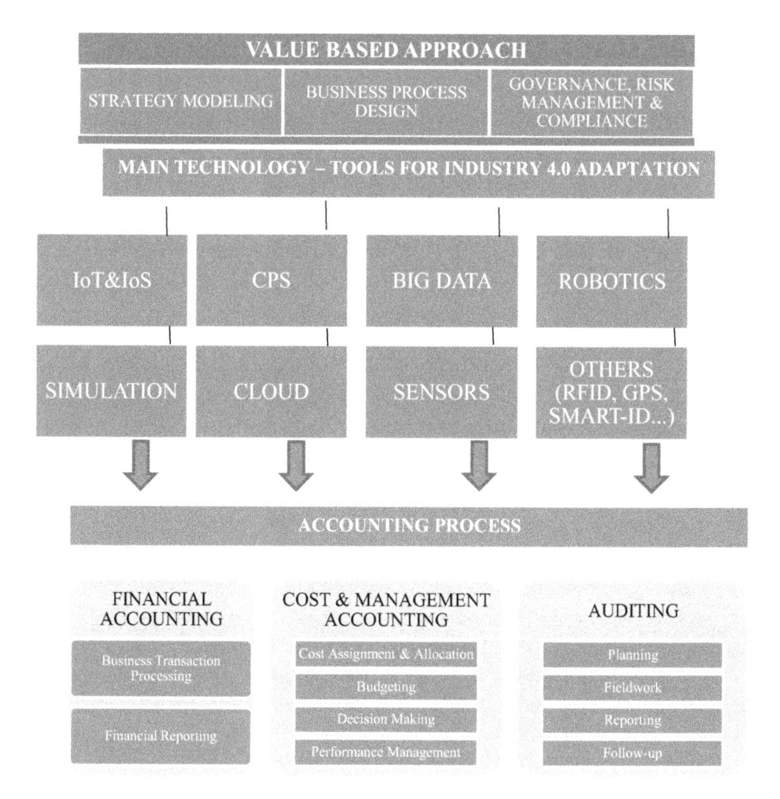

Fig. 27.1 Framework for accounting 4.0

The conceptual framework aims to combine the aspects of different studies, articles or applications in order to achieve a broader view of the implications of industry 4.0 on the accounting system and processes. The contributions of this comprehensive and conceptual framework can be stated as;

- Helps to describe the relationship between the main concepts and tools.
- It will be arranged in a logical structure in order to provide a visual display of how components relate to each other.
- Offers a broader view that shows how the processes will be held and which steps will be taken by critically examining all the aspects of various researches.
- Helps to address the knowledge gap on different areas of accounting 4.0
- By integrating the characteristics, tools and actions it will demonstrate how industry 4.0 will be effective on the components of the accounting system.

This framework offers the opportunity to explore both strategic and operational aspects of digitalization and its impact on value creation. Additionally, the framework aims to offer managers a road map to design the change in their accounting processes and to perform integrated decision making. Firstly, the components of the framework will be explained step by step.

27.3.1 Value Based Approach

The main vision of Industry 4.0 is to integrate related companies by intelligent digital communication along the value chain and also supports value creation within the functions of a company based on Industry 4.0 tools. The main driving forces for Industry 4.0 applications are; cost reduction, flexibility, stability/quality assurance and increased turnover (Sauter et al., 2015). The fundamental characteristics of the new value chain approach which will take place as a result of industry 4.0 transformation can be stated as follows (Schoenthaler et al., 2015):

- Self control: Things will operate and interact autonomously such as CPS.
- Self organization: Agents will negotiate with each other on the global IoT which will lead to decentralization of decisions.
- Less complex decentralized algorithms: Complex algorithms for centralized supply chain planning have to be replaced by less complex decentralized algorithms.
- Tight Integration: customers, suppliers and all other business partners are tightly integrated along the value chain. This means more communication and networking across organizational boundaries which results with virtual complex organizations specialized by a global network.
- Responsiveness: Transparent decisions in decentral control cycles enable fast reactions to changes and disruptions.

At the same time these characteristics are the initial inputs of the conceptual framework. Value Based Approach should be considered as a management approach which aims to create long-term value for all business partners through tight integration. Another perspective for value based approach focuses on process view and can be seen as a managerial approach focusing on value maximization through the guiding of its systems, strategies, processes, governance, performance measurements and culture. The maximization of value directs company strategy, structure and processes. At that point, analyzing the processes, determining KPIs and achieving value through elimination of non-value-added activities becomes vital (Aslanertik, 2007). Within Value Based Approach, value lies in the integration of strategic modelling, business process design, governance, risk management and compliance. Value Based Approach also offers a control mechanism that includes the understanding of structural relationships between resources and processes towards the achievement of company objectives. For the long term success of Industry 4.0 transformation, the usage of appropriate industry 4.0 tools and supporting analytical techniques is very important to achieve company's overall objectives, and management processes should be aligned to help the company maximize its value by focusing on the analysis of key drivers of management decision making.

When supported with Industry 4.0 tools, Value Based Approach can easily deal with increased complexity, greater uncertainty and risk. Measuring for value and performance is a very powerful management approach but it requires too much data or complex analysis. Extreme caution must be made when dealing with uncertainty. Advanced analytics offered by Industry 4.0 can help to prevent value destruction.

In order to maximize value, achieve competitiveness, responsiveness, higher performance and to become as efficient as possible, the companies may use different industry 4.0 tools and apply different analytical techniques or use various management approaches. But it should be noted that each of these tools have different benefits or may have some shortcomings. The most appropriate tools should be used for the related processes. Also, it can be inadequate to continue with one single tool, technique or approach that will satisfy the need to all relevant processes, problems, issues or cases. So, there will be a need for various tools in order to be successful. Considering the cost-benefit approach, the main strategy should be performing an integrated approach and simultaneous usage of different tools and techniques offered by industry 4.0 to create synergy. This framework directly supports this synergy by offering the most appropriate tools for each accounting process and their sub-processes to achieve the target outcomes specified in the framework.

27.3.2 Strategic Modeling

The rapid change for digitalisation enables a new organizational structure. Companies are facing managerial challenges because of the effects of digitalisation. These challenges demonstrate a strong need for a different vision and strategy for companies. Firstly, the strategies developed should be interlaced with performance management and this connection can be powered by industry 4.0 tools that are capable of communicating. As Industry 4.0 is based on the concept of cyber physical systems (CPS), which is mainly a technological approach, aspects such as the modification of organizational structures and processes, the adaption of existing business models, or the development of necessary employee-skills and qualifications are neglected (Erol et al., 2016). Industry 4.0 transformation requires a major change in organizational strategies. Companies need to create a strategy model that offers a roadmap to develop specific Industry 4.0 objectives along with a set of measures to reach them (Erol et al., 2016). Digitalisation and Industry 4.0 transformation can only be succeed if it was performed through a structured strategy model. Erol et al. (2016) offers a three-stage model for Industry 4.0 transformation:

Stage 1: Envision

- Common understanding of Industry 4.0
- Company-specific Industry 4.0 vision

Stage 2: Enable

- Roadmapping of Industry 4.0 strategy
- Identification of internal and external success factors

Stage 3: Enact

- Preparation of transformation
- Proposal of Industry 4.0 projects

Additionally, a value chain oriented strategy model should incorporate all business partners as early as possible into the strategy development stages.

27.3.3 Business Process Design

Business process design is a systematic tool to visualise the collaboration of all business partners within the value chain and to map communication relationships for a more structured and efficient business processes. Industry 4.0 will have a strong impact on business processes through smart factory, smart machinery and smart systems. It offers a new kind of design for business processes such as (Hitpass & Astudillo, 2019);

- Decentralized processes with greater decision-making autonomy.
- Real-time control of the automated organizational processes.
- Improved performance and quality of environment-integrated organizational processes.

In the conceptual framework, business process design is a very important stage because Industry 4.0 requires improved business processes in terms of technology and performance which is very different than the conventional business processes. This process design stage should offer a roadmap which includes;

- Collaboration of the business partners in all the stages of the design,
- Improved cyber security supported by technology based control mechanisms such as alarms, alerts, sensors or intelligent systems,
- Processes should be able to respond on time to all the changes such as capacity increases or quality specification changes, or in other words should be flexible enough,
- Processes should be able to meet the required quantity and quality within the given budget and time frame across organizational and corporate boundaries (Schoenthaler et al., 2015),
- Processes should be intelligent enough to support sustainability issues and this requires the interconnection of sustainability parameters with the process capabilities,
- Simulation tools supported by big data analytics can be used to design processes for future.

In addition, it should be mentioned that advanced analytics of various parameters is very important and the tools of industry 4.0 should be used efficiently in the business process design stages.

27.3.4 Governance, Risk Management and Compliance

The main characteristic of industry 4.0 is the interconnection between business partners, processes and information systems. The most important requirement for this kind of interconnection is the cyber security. This requirement forces corporations to (www.elevenpath.com, 2019):

- Describe the assets of the Industrial Automation and Control Systems and their interaction with Information Systems using an Enterprise Architecture model,
- Identify threats to which they can be exposed,
- Assess the risks from the business perspective and assign necessary treatments to the risk scenarios by using the most appropriate industry 4.0 tool,
- Periodically assess the degree of compliance with regulatory frameworks applicable to the sector,
- Determine corrective measures and monitor their compliance within an action plan.

Governance in digital era can be interpreted as a managerial behaviour supported by the IT aligned objectives and involves risk management and compliance mainly concentrates on cyber security.

Risk management activities in digital era should be designed in a way that supports the corporations' IT processes and all technology related functions through early warning systems.

Compliance in digital era should be characterised by data security and conformity with rules, regulations, laws, policies and/or standards within IT applications. Monitoring and control are two main functions of the compliance system. Risk oriented enterprises show the highest effort for preventing violations. From the aspect of accounting, conforming with accounting standards, auditing standards, principles, laws and regulations are more vital than for all other business functions due to high level of financial issues.

The corporations that see Governance, Risk Management and Compliance as an opportunity to improve business processes, achieve significant cost savings and competitive advantage (Schoenthaler et al., 2015). In digital era, corporations become fully digitalized and this involves mobilizing a technology portfolio that digitalizes and optimizes all risk and compliance related activities, embeds them into the organization and end-to-end processes, and engages all stakeholders based on their individual needs (consulting.ey.com, 2017).

Managers operating in a highly digitised environment engaged in Big Data analysis tend to engage in collaborative working approaches rather than command and control work styles (Bhimani & Willcocks, 2014). This kind of managerial approach really enhances the success of digitalization, motivates the partners of the system and empower the governance mechanism.

27.3.5 Accounting 4.0

This part of the conceptualframeworkaimstoofferaroadmapforindustry4.0adaptationof theaccountingprocess.Thecomponentsand/orsub-processesoftheaccountingwhichis assumedtobeaffectedatmostwillbedefinedindetail.Also,thissectiondemonstratethe expected results of adapting industry 4.0 and the appropriate usage of its tools by each component/sub-processseparately.Theoutcomesforeachcomponent/sub-processand which tools are most appropriate for which component/process are derived from the synthesis of the literature, from most common applications in practice and from the reports published by different consultancy firms.

27.3.5.1 Financial Accounting

Within the financial accounting process two different subprocesses can be considered: Financial transaction processing and financial reporting. Through the increased use of industry 4.0 tools there will be certain changes in these subprocesses.

Financial Transaction Processing (FTP) is the core process for financial accounting. It involves the recording, initial recognition and measurement of the financial transactions. Within digitalization and automation most of the data needed to record transactions can be obtained from sensors or intelligent processes directly and this enables the actual identification of items. Interconnection with the financial institutions and markets provides real-time market prices so real-time measurement can be achieved. The automatic and real-time flow of data leads automatic storage and real-time recording which directly affects relevancy. Also, the quantitative data can be enriched by the integration of qualitative data and this increases the accuracy of the recording process. For example, a phone call attached to a customer agreement about sales returns may support the decision of how to recognize an actual return. Big Data is the most important tool for FTP because it allows financial transactions to be traced, measured, recognized earlier and in detail. Various and huge amounts of data is captured and analyzed by the help of advanced analytics. This analysis highlights the structural relationships between accounts and transactions (Vasarhelyi et al., 2015). These relationships can be communicated to other accounting processes to support other operations, reporting issues or decision making. The tools that enable such communications and interconnections are IoT&IoS.

The real-time measurement offers financial reports which are more in line with fair value, current value-historical value comparisons across time can be done, valuation can be supported by many different kind of data from different kind of resources. Also, conformity to the accounting principles and standards can be done by extracting relevant data from an enormous data set. Adjustment templates, that are communicating with the related measurement sources, can be prepared. These templates that are communicating and interconnected with the source of the standards increases the financial reporting conformity and compliance. The important point is to create the best model that best reflects the real financial position of the company and results with fair presentation. For each different measurement a different model can be needed so the choice of the appropriate analytical technique becomes vital. Big data offers various data samples to give the appropriate decision for the model. For storage purposes the cloud can be used as a tool. Bhimani and Willcocks (2014) emphasized that corporations should make decisions based on verifiable financial transactions. The outcomes of integrating the tools into FTP and financial reporting process within industry 4.0 adaptation are given in Fig. 27.1.

27.3.5.2 Cost and Management Accounting

Within the cost and management accounting process four different subprocesses can be considered: cost assignment and allocation, budgeting, decision making, performance management. Especially, managerial accounting is an area that requires various kinds of quantitative and qualitative data for complex managerial decisions.

The first and the main impact of industry 4.0 on cost accounting is the change in cost structures. The digital environment and automized processes increases traceability of cost drivers so some of the indirect costs may become direct costs. Big Data analysis supported by financial intelligence enable corporations to capture the changing dynamics of cost and revenue sources (Bhimani & Willcocks, 2014). When the cost sources and revenue sources are correlated a strong relationship between the cost and the cost object can be performed. Additionally, many different parameters can be determined and measured to choose which source is the best cost driver that best represent the consumption of the related resource.

Budgeting is a process that highly requires both quantitative and qualitative data. But data will not be enough because in order to make accurate forecasting trends or patterns need to be detected. To detect trends and patterns big data analytics is required. Budgets have been criticized for being too intra-organizational due to usage of data (Green et al., 2018). Many companies prefer new budgeting techniques that reconfigures enterprise resource planning data enhanced by Big Data referred as beyond budgeting (Bourmistrov & Kaarboe, 2013).

The most important step in decision making is to gather information. Decision making also uses quantitative and qualitative data simultaneously. Before giving the decisions, the activities or processes related with the decision and their value creation potential should be evaluated. Decision making is a very complex process that involves several steps so only gathering and evaluating information will not be enough so decision support systems should be used to evaluate all alternatives with their costs and benefits. Tools such as Artificial Intelligence and Cyber Physical Systems may enhance decision making process. The CPS have the ability to communicate with both machines and people and to make autonomous decisions (Schoenthaler et al., 2015).

One significant topic for managerial accounting is performance management. A successful performance management system should involve the determination of value related key performance indicators (KPI). Balanced Scorecard is a performance management tool that evaluates both financial and nonfinancial measures within four areas: Financial, customer, internal business processes, learning and growth (Kaplan & Norton, 1996). Within each area Big Data can identify new behaviours that influence respective goal outcomes and Big Data analyses can easily facilitate the discovery of value related measures to be incorporated in management control systems (Warren et al., 2015).

The outcomes of integrating the tools into cost and management accounting process within industry 4.0 adaptation are given in Fig. 27.1.

27.3.5.3 Auditing

When all the accounting processes change due to industry 4.0 adaptation, auditing should also adapt these changes. With the use of industry 4.0 tools (Dai & Vasarhelyi, 2016);

- Auditors can leverage new technologies to collect a large range of real-time, audit-related data, automate repetitive processes involving few or simple judgements, and eventually achieve comprehensive, timely and accurate assurance,
- With the increase of the digitalization of business processes across the entire enterprise, auditors can continuously monitor business operations and abnormal behaviours in real time,
- The auditing profession significantly changes due to automation of current procedures, enlarging their scope, shorter working hours and as a result improve the overall assurance quality.

Sensors, CPS, IoT&IoS, intelligent systems, smart factories, Big Data analytics are the main tools that supports the efficiency of the audit process. When more evidence is collected it is more easy to perform an audit opinion. Analytical techniques like clustering, data mining and alarms & alerts can serve on the many stages in the auditing process (Moffitt & Vasarhelyi, 2013).

The outcomes of integrating the tools into accounting process within industry 4.0 adaptation are given as follows:

Outcomes due to usage of Industry 4.0 tools			
	Financial transaction processing	*Cost assignment and allocation*	*Planning*
Outcomes	Better processing power, ability to store more data, actual identification, ability to reach real-time market prices, real-time measurement, analysis of structural accounting relationships among accounts and transactions	Changing cost structures, more traceable data for cost drivers, more accurate product/service costing due to more stronger cause and effect relationship	Easier internal control risk measurement due to automatic capture of data streams by effective communication with the customer's information system, less labor involvement in monitoring, ability to analyse within a comprehensive population level (less effort to plan statistical sampling process), a model supported by algorithms needed to be developed for data collection, evaluation and analysis and this model should be validated on a regular basis

(continued)

Outcomes due to usage of Industry 4.0 tools			
	Financial reporting	*Budgeting*	*Fieldwork*
Outcomes	More accurate and fair presentation due to relevancy of data, more in line with fair value due to real-time market prices, standardization, textual data analytics or XBRL or data linkages used for disclosure	Predictive forecasting, efficient models for flexible budget applications based on variance calculations	Less fieldwork hours due to connection and communication between business-to-auditor databases, easier reconciliation due to business-to-customer integration, new forms of audit evidence such as text mining, alerts from machines
		Decision making	*Reporting*
Outcomes		Efficient decision making through decision support tools, integration of qualitative information into decision process, more detailed and visualized analysis through advanced analytics that supports decision making	More audit evidence for audit opinion due to big data analytics, easier detection of fraud due to continuous monitoring, improved audit efficiency through multi-model evidence, easy to incorporate legacy into reporting process
		Performance management	*Follow-up*
Outcomes		Communication between smart plants and KPI units, intelligent performance management models, integrated and correlated data for more faster and efficient performance management, integration of financial and non-financial performance measures, proactive/forecast-based approaches	Easy to determine the size and the nature of the risk through advanced analytics, easier tracking of high risky areas by setting various alarms due to alternate sources of data

27.4 Conclusion

In this chapter, a conceptual framework is suggested that aims to combine the aspects of different studies, articles or applications in order to achieve a broader view of the implications of industry 4.0 on the accounting system and processes. The first part of the framework includes a detailed explanation of value based approach and its enablers referred as strategy modeling, business process design, governance, risk management and compliance. Then, the impact of industry 4.0 on three main accounting process and their subprocesses are demonstrated with the outcomes determined separately for each subprocess. Also, which industry 4.0 tool/tools are appropriate for which subprocess is clarified supported by examples.

Industry 4.0 and its impact on accounting is a very complex topic which involves many different aspects. Within this chapter, a brief evaluation is made but more research is needed to better analyze the implications, advantages and disadvantages of the usage of tools. More examples or real-life cases may enhance the understanding of the main implications on accounting.

References

Aslanertik, B. E. (2007). Enabling integration to create value through process-based management accounting systems. *International Journal of Value Chain Management, 1*(3), 223–238.

Bhimani, A., & Willcocks, L. (2014). Digitisation, 'big data' and the transformation of accounting information. *Accounting and Business Research, 44*(4), 469–490.

Bourmistrov, A., & Kaarboe, K. (2013). From comfort to stretch zones: A field study of two multinational companies applying beyond budgeting ideas. *Management Accounting Research, 24*(3), 196–211.

Burritt, R., & Christ, K. (2016). Industry 4.0 and environmental accounting: a new revolution? *Journal of Sustainability and Social Responsibility, 1*, 23–38.

Chen, T., Barbarossa, S., Wang, X., Giannakis, G. B., & Zhang, Z. L. (2019). Learning and management for internet of things: Accounting for adaptivity and scalability. *Proceedings of the IEEE, 107*(4), 778–796.

Dai, J., & Vasarhelyi, M. A. (2016). Imagineering Audit 4.0. *Journal of Emerging Technologies in Accounting, 13*(1), 1–15.

Davies, R. (2015). *Industry 4.0 digitalisation for productivity and growth*. European Parliamentary Research Service, 10.

Deloitte. (2015). *Industry 4* (p. 0). Challenges and solutions for the digital transformation and use of exponential technologies. Deloitte AG: Zurich. https://www2.deloitte.com/content/dam/Deloitte/ch/Documents/manufacturing/ch-en-manufacturing-industry-4-0-24102014.pdf

Ernst and Young. (2017). *The role of the CFO and finance function in a 4.0 world*. UK: E&Y. https://www.ey.com/uk/en/services/assurance/ey-the-role-of-the-cfoand-finance-function-in-a-4-world-form

Erol, S., Schumacher, A., & Sihn, W. (2016, January). *Strategic guidance towards Industry 4.0–a three-stage process model*. In International Conference on Competitive Manufacturing (Vol. 9, No. 1, pp. 495–501).

Green, S., McKinney Jr., E., Heppard, K., & Garcia, L. (2018). Big data, digital demand and decision-making. *International Journal of Accounting & Information Management, 26*(4), 541–555.

Griffin, P. A., & Wright, A. M. (2015). Commentaries on big data's importance for accounting and auditing. *Accounting Horizons, 29*(2), 377–379.

Hitpass, B., & Astudillo, H. (2019). Industry 4.0 challenges for business process management and electronic-commerce. *Journal of Theoretical and Applied Electronic Commerce Research, 14* (1), I–III.

Hofmann, E., & Rüsch, M. (2017). Industry 4.0 and the current status as well as future prospects on logistics. *Computers in Industry, 89*, 23–34.

https://consulting.ey.com/agile-grc-a-new-approach-to-governance-trust-and-risk-in-the-digital-age/. 31 August 2017.

https://www.elevenpaths.com/solutions/industria-4-0/index.html. 25 April 2019.

Huerta, E., & Jensen, S. (2017). An accounting information systems perspective on data analytics and big data. *Journal of Information Systems, 31*(3), 101–114.

Kaplan, R. S., & Norton, D. P. (1996). Linking the balanced scorecard to strategy. *California Management Review, 39*(1), 53–79.
Moffitt, K. C., & Vasarhelyi, M. A. (2013). AIS in an age of big data. *Journal of Information Systems, 27*(2), 1–19.
O'Leary, D. E. (2013). Big data', The 'internet of things' and the 'internet of signs'. *Intelligent Systems in Accounting, Finance and Management, 20*(1), 53–65.
Richins, G., Stapleton, A., Stratopoulos, T. C., & Wong, C. (2017). Big data analytics: Opportunity or threat for the accounting profession? *Journal of Information Systems, 31*(3), 63–79.
Rüßmann, M., Lorenz M., Gerbert, P., Waldner, M., Justus, J., Engel, P. & Harnisch, M. (2015). *Industry 4.0 the future of productivity and growth in manufacturing industries*. Retrieved from https://www.bcg.com/publications/2015/engineered_products_project_business_industry_4_future_productivity_growth_manufacturing_industries.aspx
Sandengen, O. C., Estensen, L. A., Rødseth, H., & Schjølberg, P. (2016). High performance manufacturing—an innovative contribution towards industry 4.0. In *6th international workshop of advanced manufacturing and automation*. Atlantis Press.
Sauter, R., Bode, M., & Kittelberger, D. (2015). *How Industry 4.0 is changing how we manage value creation*. Retrieved from https://www.horvath-partners.com/en/publications/featured-articles-interviews/detail/how-industry-40-is-changing-how-we-manage-value-creation
Schoenthaler, F., Augenstein, D., & Karle, T. (2015, July). Design and governance of collaborative business processes in Industry 4.0. In *Proceedings of the workshop on cross-organizational and cross-company BPM (XOC-BPM) co-located with the 17th IEEE Conference on Business Informatics (CBI 2015)*, Lisbon, Portugal.
Schwartz, J., Stockton, H., & Monahan, K. (2017). *Forces of change: The future of work*. Retrieved May 6, 2019, from https://www2.deloitte.com/content/dam/insights/us/articles/4322_Forces-of-change_FoW/DI_Forces-of-change_FoW.pdf
Sledgianowski, D., Gomaa, M., & Tan, C. (2017). Toward integration of big data, technology and information systems competencies into the accounting curriculum. *J Account Educ, 38*, 81–93.
Vasarhelyi, M. A., Kogan, A., & Tuttle, B. M. (2015). Big data in accounting: An overview. *Accounting Horizons, 29*(2), 381–396.
Warren Jr., J. D., Moffitt, K. C., & Byrnes, P. (2015). How big data will change accounting. *Accounting Horizons, 29*(2), 397–407.

Banu Esra Aslanertik is a Professor in Accounting and Finance at Dokuz Eylul University Faculty of Business Department of Business Administration, Izmir-Turkey, and she holds a Master Degree in International Business, a Ph.D. Degree in Business Administration from Dokuz Eylul University. Her Research interests focus on Cost and Management Accounting, International Financial Reporting and Corporate Social Responsibility. She is currently giving undergraduate, master, Ph.D. courses mainly on Accounting for Decision Making, Contemporary Issues on Cost and Managerial Accounting and International Financial Reporting Standards.

Bengü Yardımcı is a lecturer (Ph.D.) at the Department of Finance, Banking and Insurance in Vocational School at Yaşar University, Izmir—Turkey. She holds her Doctoral Degree in Business Administration from Yaşar University and her MBA degree in Finance from Dokuz Eylul University. She is currently giving courses mainly on General Accounting and Financial Statement Analysis. Her research interests focus on International Financial Reporting, Corporate Governance, Corporate Social Responsibility and Sustainability Reporting.

Lightning Source UK Ltd.
Milton Keynes UK
UKHW051410030520
362622UK00001B/26